# SCALING AND DISORDERED SYSTEMS

International workshop

and collection of articles honoring

Professor Antonio Coniglio

on the occasion of his $60^{th}$ Birthday

# SCALING AND DISORDERED SYSTEMS



*Editors*

**Fereydoon Family**
*Emory University, USA*

**Mohamed Daoud**
*Centre d'Etude Nucleaire, France*

**Hans J. Herrmann**
*ESPCI, France & ICAI, Germany*

**H. Eugene Stanley**
*Boston University, USA*

World Scientific
*New Jersey • London • Singapore • Hong Kong*

*Published by*

World Scientific Publishing Co. Pte. Ltd.
P O Box 128, Farrer Road, Singapore 912805
*USA office:* Suite 1B, 1060 Main Street, River Edge, NJ 07661
*UK office:* 57 Shelton Street, Covent Garden, London WC2H 9HE

**British Library Cataloguing-in-Publication Data**
A catalogue record for this book is available from the British Library.

**SCALING AND DISORDERED SYSTEMS**



ISBN 981-02-4838-5

Printed in Singapore by Uto-Print

# Preface

It is a great pleasure to dedicate this book of articles in honor of Professor Antonio Coniglio on the occasion of his 60th Birthday. Many of the articles are based on research inspired by his ideas. Antonio is a great scientist in the tradition of the truly great scientists. His ability to select just the right questions to pose has led to the remarkably high impact of his work on many diverse branches of statistical physics. His creative approaches to solving complex problems combines deep physical intuition with mathematical prowess of the first rank.

Antonio began his scientific contributions in the field of many-body theory, concentrating on low temperature properties of superfluid helium. In the early seventies, he began his studies of critical phenomena and the renormalization group, and developed a generalized form of scaling[11,12] that anticipated multifractality and multiscaling — concepts which played a significant role in a wide variety of fields many years later.[75]

His contributions to scaling in phase transitions and critical phenomena naturally led to his classic papers on correlated percolation. He was among the first to introduce the concept of correlated percolation,[14] which he later used to develop models for the sol-gel transition.[24,26,39,176]

Antonio has played a central role in the development of the theories of random and correlated percolation by formulating a general approach for the study of continuum and correlated percolation based on Meyer cluster expansion, enabling the extension of many results from theory of fluids to percolation.[16,18] In related works, he has proven rigorous inequalities between thermodynamic quantities and percolation quantities in the Ising model. This has led to the proof that Ising clusters percolate at the Ising critical point in two dimension but not necessarily in higher dimensions.[13,15,20]

One of Antonio's most significant contributions which has led to important advances in the theory of percolation is the proof of a relation between pair connectedness and singly connected bonds (also called red bonds). In particular, he has proved that one of the consequences of this relation is that the fractal dimension of the singly connected bonds in the incipient infinite cluster in random percolation is given by the thermal scaling exponent. Based on these exact results he has made a major contribution to the characterization of the fractal structure of the percolation cluster by putting on a firm foundation the "nodes links and blobs model" for the incipient infinite cluster, which is now accepted as the standard model for percolation.[33,35,44]

As a corollary to the above results, Antonio was able to prove that the crossover critical exponents of the dilute Ising model and Heisenberg model are related, respectively, to the fractal dimension of the singly connected bonds and the resistivity exponent. This work clarified and explained the experimental results in dilute ferromagnets and gave a geometrical interpretation of why the two crossover exponents were different.[33,35,50]

In a seminal paper,[25] Antonio and Bill Klein proposed a geometrical characterization of the Ising critical point in terms of new clusters (now known as the CK droplets) which have the properties of percolating at the Ising critical point with the Ising exponents.[25,26,34,38,104] This classic work is the foundation of the widely used Swendsen-Wang cluster dynamics algorithm. More recently, this cluster definition has received attention in QCD and also in explaining cluster fragmentation in nuclear matter.

Another important work of Antonio was the analysis of the fractal structure of the CK droplets in the q-state Potts model. In this context, he provided exact values in two dimensions for the fractal dimension of the red bonds for any $q$. In particular for $q = 0$ he obtained the fractal dimension of the red bonds in the spanning tree.[95] He also related the hyperscaling breakdown, with the presence of infinitely many clusters above the uppercritical dimensionality.[69,173]

Antonio has also made significant contributions to many aspects of the theory of diffusion-limited aggregation and the theory of multifractality — research areas which have been of central importance in understanding aggregation phenomena, disordered systems and fractals. For example, in collaboration with his colleagues, he was among the first to have contributed to the development and application of the concepts of multifractality and multiscaling in diffusion limited aggregation[73,79,106,123] and in percolation.[74,83]

Antonio has also made important contributions to the theory of phase separation. With Zannetti, he provided for the first time an analytical solution for the time-dependent Ginzburg Landau model, in the limit in which the number of components of the order parameter goes to infinity.[103]

Antonio and collaborators developed a microscopic "spin-glass" type of model that elucidates the phase diagram of a class of materials displaying high-temperature superconductivity.[85] This contribution has received great attention in the field of high-temperature superconductivity and has been widely cited in the literature.

More recently, Antonio has made important contributions to a number of topics related to glasses and spin glasses.[162,163,175] He has also applied these models to granular materials,[137,141,147] by developing with collaborators the frustrated percolation model which maps exactly the spin glass model into a geometrical model.[114] These models are playing an important role in describing granular materials and glasses, which are systems of considerable complexity and great current interest.

Antonio is the voice of insight, reason, humor and honesty. Even as impressive as his contributions to physics are, what usually springs to one's mind is Antonio's great personality and friendship. The joy of a meeting is always enhanced by his presence and thoughtful discussions and comments. The fact that all of scientists who came to his Paris meeting, came from the far corners of the globe without any financial support from the organizers attests to the honor in which Antonio is held by his colleagues. He is one of the truly highly respected figures in statistical physics.

**Fereydoon Family**
*Department of Physics, Emory University, Atlanta*

**Mohamed Daoud**
*Centre d'Etude Nucleaire, Saclay, France*

**Hans J. Herrmann**
*ESPCI, Paris and ICA1, Stuttgart, Germany*

**H. Eugene Stanley**
*Center for Polymer Physics, Boston University, Boston*

Professor Antonio Coniglio

# ANTONIO CONIGLIO: Curriculum Vitae

**Name:** Antonio Coniglio

**Born:** 5 April 1940

**Education:**

— Laurea in Physics, cum laude (110/110) Naples University, November 1962
— Diploma Scuola di Perfesionamento, Naples University, November 1965

**Employment:**

(1) Chaired Professorship in Statistical Mechanics, 1980 - Present
— Research Professor of Physics, Center for Polymer Studies, Boston University, 1981 - Present
— [Visiting Professor, Ecole Superieure de Physique et Chimie Industrielles, Paris, September–October 1992, September 1996, October 1997, January 1998]
— [Visiting Professor, Service de Physique Theorique, C.E.N. Saclay, September 1989]
— [Visiting Professor, Laboratoire Leon Brillouin C.E.N. Saclay, May 1980, April–July 1988]
— [Visiting Professor, Centro Brasilero de Pesquisas Fisicas, January–February 1985, January 1987]
— [Visiting Professor, St. Francis Xavier University, Antigonish, Nova Scotia, Canada, March 1985, James Chair Professor, March–April 1987]
— Professore Incaricato Stabilizzato, 1974–1980, Naples University
— [Visiting Assistant Professor of Physics, Center for Polymer Studies, Boston University, sabbatical leave, 1977–1979]
— Professore Incaricato, 1971–1974, Naples University
— [Visiting Research Professor, King's College, London, November 1973–January 1974]
— [Awarded the title "Libera Docenza," 1972]
(2) Researcher at King's College, London, 1970–1971
(3) Professore Incaricato, 1966–1968, 1969–1970
(4) Research Assistant, 1965–1966

**Publications**

1. A. Coniglio, M. Marinaro and M. Maturi, "Stability Conditions for a Boson System Interacting with a Partly Repulsive and Partly Attractive Potential," *Nuovo Cimento* **40**, 184 (1965).
2. A. Coniglio and M. Marinaro, "On Condensation for an Interacting Boson System," *Nuovo Cimento* **48**, 249 (1967).
3. A. Coniglio and M. Marinaro, "Breakdown of Symmetry and Gapless Spectrum in Many-Boson Systems," *Nuovo Cimento* **48**, 262 (1967).
4. A. Coniglio, Marinaro and B. Preziosi, "Phase Transition Between Partially Condensed Homogeneous and Periodic Systems," *Nuovo Cimento* **61**, 25 (1969).
5. A. Coniglio, M. Marinaro and B. Preziosi, "On the Coexistence of Single and Two Particle Condensation in an Interacting Boson Gas," *Nuovo Cimento* **63**, 227 (1969).
6. A. Coniglio and R. Vasudevan, "Generalized Condensation of an Interacting Bose Gas with Pair Hamiltonian," *Nuovo Cimento* **70B**, 39 (1970).

7. A. Coniglio and M. Marinaro, "A Variant of the Scaling Hypothesis Exhibiting Asymmetry," *Lettere al Nuovo Cimento, Serie I* **4**, 391 (1970).
8. A. Coniglio and M. Marinaro, "A Natural Way of Introducing Asymmetry Through a Generalization of the Scaling Law," *Physica* **54**, 261 (1971).
9. A. Coniglio, "Scaling Parameter and Universality," *Physica* **58**, 489 (1972).
10. A. Conglio, "Universality for Three-Dimensional Ising System with Long Range Potential," *Phys. Lett.* **38A**, 105 (1972).
11. A. Coniglio and M. Marinaro, "Non-Mixing of Coupling Parameters in Cell-Site Transformation and Scaling Law," *Physica* **65**, 611 (1973).
12. A. Coniglio and M. Marinaro, "Weak and Strong Scaling as a Generalization of Kadanoff's Picture," *Physica* **66**, 385 (1973).
13. A. Coniglio, "Percolation Problems and Phase Transitions," *J. Phys.* **A8**, 1773 (1975).
14. A. Coniglio, "Some Cluster Size and Percolation Problems for Interacting Spins," *Phys. Rev.* **B13**, 2194 (1976).
15. A. Coniglio, C. Nappi, L. Russo and F. Peruggi, "Percolation in the Ising Model," *Comm. Math. Phys.* **51**, 315 (1976).
16. A. Coniglio, U. DeAngelis, T. Forlani and G. Lauro, *J. Phys.* **A10**, 219 (1977).
17. A. Coniglio, C. Nappi, L. Russo and F. Peruggi, "Percolation Points and Critical Point in the Ising Model," *J. Phys.* **A10**, 205 (1977).
18. A. Coniglio, U. DeAngelis and T. Forlani, "Pair Connectedness and Cluster Size," *J. Phys.* **A19**, 1123 (1977).
19. A. Coniglio and J. W. Essam, "Percolation Theory in Gas," *J. Phys.* **A10**, 1917 (1977).
20. A. Coniglio and L. Russo, "Cluster Size and Shape in Random and Correlated Percolation," *J. Phys.* **A12**, 545 (1979).
21. W. Klein, H. E. Stanley, P. J. Reynolds and A. Coniglio, "Renormalization Group Approach to the Percolation Properties of the Triangular Ising Model," *Phys. Rev. Lett.* **42**, 1145 (1978).
22. A. Coniglio, H. E. Stanley and W. Klein, "A Statistical Mechanical Theory of Polymer Gelation," *Phys. Rev. Lett.* **42**, 518 (1979).
23. A. Coniglio, H. E. Stanley and D. Stauffer, "Fluctuations in the Numbers of Percolation Clusters," *J. Phys.* **A12**, L345 (1979).
24. A. Coniglio and M. Daoud, "Polymer Chains and Vulcanization," *J. Phys.* **A12**, L259 (1979).
25. A. Coniglio and W. Klein, "Clusters and Ising Critical Droplets: A Renormalization Group Approach," *J. Phys.* **A13**, 2775 (1980).
26. A. Coniglio and T. Lubensky, "$\epsilon$-expansion for Correlated Percolation: Application to Gels," *J. Phys.* **A13**, 7183 (1980).
27. D. Stauffer and A. Coniglio, "Speculation on Crossover from Mean-Field to Critical Behavior for Long-Range Site Percolation," *Z. Physik* **B38**, 267 (1980).
28. A. Coniglio and D. Stauffer, "Fluctuations of the Infinite Network in Percolation Theory," *Lett. Nuovo Cimento* **28**, 33 (1980).
29. H. E. Stanley, A. Coniglio, W. Klein, H. Nakanishi, S. Redner, P. J. Reynolds and G. Shlifer, "Critical Phenomena: Past, Present and 'Future'," *Dynamics of Synergetic Systems*, ed. H. Hanken (Springer-Verlag, 1980).
31. S. Redner and A. Coniglio, "On the Crossover Exponent for Anisotropic Bond Percolation," *Phys. Lett.* **A79**, 111 (1980).

32. W. Klein and A. Coniglio, "Thermal Phase Transitions at the Percolation Threshold," *Phys. Lett.* **84**, 83 (1981).
33. A. Coniglio, "Thermal Phase Transitions of the Dilute s-State Potts and n-Vector Model at the Percolation Threshold," *Phys. Rev. Lett.* **46**, 250 (1981).
34. A. Coniglio, F. di Liberto and G. Monroy, "Site Bond Correlated Percolation in Ferromagnetic and Antiferromagnetic Ising Models: A Renormalization Group Approach," *J. Phys.* **A14**, 3017 (1981).
35. A. Coniglio, "Geometrical Structure and Thermal Phase Transitions of the Dilute s-State Potts Model at the Percolation Threshold," in *Spring Lecture Notes Proceedings Conference on Disordered Systems and Localization, Rome, 1981* eds. C. Castelleni, C. D. Castro and L. Peliti.
36. M. Daoud and A. Coniglio, "Singular Behavior of the Free Energy in the SOL-GEL Transition," *J. Phys.* **A14**, L108 (1981).
37. H. E. Stanley, A. Coniglio, W. Klein and J. Teixeira, "Connectivity and Theoretical Physics: Some Applications to Chemistry," *Proceedings of the VI Brazilian Symposium on Theoretical Physics* (*Rio De Janeiro*) (Springer-Verlag, Heidelberg and New York, 1981).
38. A. Coniglio and F. Peruggi, "Cluster and Droplets in the q-State Potts Model," *J. Phys.* **A15**, 1873 (1982).
39. A. Coniglio, H. E. Stanley and W. Klein, "Solvent Effects on Polymer Gels," *Phys. Rev.* **B25**, 6805 (1982).
40. A. Coniglio, F. Di Liberto, G. Monroy and F. Peruggi, "Clusters and Ising Droplets in the Antiferromagnetic Lattice Gas," *Phys. Lett.* **A87**, 189 (1982).
41. D. Stauffer, A. Coniglio and M. Adam, "Gelation and Critical Phenomena," *Adv. Pol. Sci.* (special volume Polymer Networks, ed. K. Dusek) **44**, 103 (1982).
42. S. Redner and A. Coniglio, "Flory Theory for Directed Lattice Animals and Directed Percolation," *J. Phys.* **A15**, L273 (1982).
43. A. Coniglio and R. V. Zia, "Analysis of the Migdal-Kadanoff Renormalization Group Approach to the Dilute s-State Potts Model: An Alternative Scheme for the Percolation ($s \to 1$) limit," *J. Phys.* **A15**, L399 (1982).
44. A. Coniglio, "Cluster Structure Near the Percolation Threshold," *J. Phys.* **A15**, 3829 (1982).
45. N. Jan, A. Coniglio and D. Stauffer, "Study of Droplets for Correlated Site-Bond Percolation in Two Dimensions," *J. Phys.* **A15**, L699–L704 (1982).
46. J. Roussenq, A. Coniglio and D. Stauffer, "Study of Droplets for Correlated Site-Bond Percolation in Three Dimensions," *J. Phys.* (*Paris*) **43**, L703–L709 (1982).
47. D. Stauffer, A. Coniglio and D. W. Heermann, "Monte Carlo Experiment for Nucleation Rate in Three-Dimensional Ising Model," *Phys. Rev. Lett.* **49**, 1299–1302 (1982).
48. A. Coniglio, "Potts Model Formulation of Branched Polymers in a Solvent" *J. Phys. Lett.* **A16**, L187–L191 (1983).
49. A. Coniglio, "Sol-Gel Transition," *Helvetica Physica Acta* [*based on invited talk at the 1983 Meeting of the European Physical Society, Lausanne, March 1983*].
50. A. Coniglio, "Percolation Effects and Disorder," in *Proceedings of Erice School on Ferromagnitic Transitions* (Springer-Verlag, 1983).
51. A. Coniglio, "Droplet Theory of Phase Transition and Metastability," *Proceedings of the Varenna School, Highlights on Condensed Matter Physics* (July 1983).

52. A. Coniglio and R. Figari, "Droplet Structure in Ising and Potts Model," *J. Phys. Lett.* **A16**, L535 (1983).
53. H. E. Stanley and A. Coniglio, "Flow in Porous Media: The Backbone Fractal at the Percolation Threshold," *Phys. Rev.* **B24**, 522 (1984).
54. D. W. Heermann, A. Coniglio, W. Klein and D. Stauffer, "Monte Carlo Simulation of Metastable States in 3D Ising Models," *J. Stat. Phys.* **36**, 447 (1984).
55. J. Kertesz, D. Stauffer and A. Coniglio, "Clusters for Random and Interacting Percolation," *Ann. Israel Phys. Soc.*, eds. Adler, Deutscher and Zallen (1983) pp. 121–148.
56. H. E. Stanley and A. Coniglio, "Fractal Structure of the Incipient Infinite Cluster in Percolation," in *Percolation Structures and Processes*, eds. G. Deutscher, R. Zallen and J. Adler, (1983) pp. 101–120.
57. C. Tsallis, A. Coniglio and S. Redner, "Break-Collapse Method for Resistor Networks–Renormalization Group Applications," *J. Phys.* **C1766**, 4339 (1983).
58. A. Coniglio and H. E. Stanley, "Screening of Deeply Invaginated Clusters and the Critical Behavior of the Random Superconducting Network" *Phys. Rev. Lett.* **52**, 1068–1072 (1984).
59. F. Family and A. Coniglio, "Geometrical Arguments Against the Alexander–Orbach Conjecture for Lattice Animals and Diffusion Limited Aggregates" *J. Phys.* **A17**, L285–L287 (1984).
60. I. Majid, N. Jan, A. Coniglio and H. E. Stanley, "The Kinetic Growth Walk: A New Model for Linear Polymers," *Phys. Rev. Lett.* **52**, 1257–1260 (1984).
61. J. Hellman, A. Coniglio and C. Tsallis, "Fracton and Fractal Structure of Proteins," *Phys. Rev. Lett.* **53**, 1195 (1985).
62. L. J. De Jongh, G. Mennenga and A. Coniglio, "Experimental Evidence for Fractal Properties of the Infinite Percolation Cluster in Randomly Dilute Magnets," *Physica* **B132**, 100 (1985).
63. L. de Arcangelis, S. Redner and A. Coniglio, "Anomalous Voltage Distribution of Random Resistor Networks and a New Model for the Backbone at the Percolation Threshold," *Phys. Rev.* **B31**, 4725–4727 (1985).
64. A. Bunde, A. Coniglio, D. C. Hong and H. E. Stanley, "Transport in a Two-Component Randomly-Composite Material: Scaling Theory and Computer Simulations of Termite Diffusion Near the Superconducting Limit," *J. Phys. Lett.* **18**, L137–L144 (1985).
65. F. Family and A. Coniglio, "Flory Theory for Conductivity of Random Resistor Networks," *J. de Physique Lett.* **46**, L9 (1985).
66. D. C. Hong, H. E. Stanley, A. Coniglio and A. Bunde, "Physics of Two-Component Randomly Composite Material: ant Limit, Intermediate Zone, and Termite Limit," *Phys. Rev.* **B33**, 4564 (1985).
67. L. de Arcangelis, A. Coniglio and S. Redner, "A Connection Between Linear and Nonlinear Resistor Networks," *J. Phys.* **A18**, L805–L808 (1985).
68. C. Tsallis, A. Coniglio and G. Schwacheim, "Simple Renormalization Group Method for Calculating Geometrical Equation of State," *Phys. Rev.* **B32**, 3322 (1985).
69. A. Coniglio, "Shapes, Surfaces and Interfaces in Percolation Clusters," in *Finely Divided Matter* [*Proc. les Houches Winter Conference*], eds. N. Boccara and M. Daoud (Springer-Verlag, New York, 1985).

70. P. Meakin, H. E. Stanley, A. Coniglio and T. A. Witten, "Surfaces, Interfaces and Screening of Fractal Structures," *Phys. Rev.* **A32**, 2364 (1985).
71. A. Coniglio, "Scaling Properties of the Probability Distribution for Growth Sites," in *On Growth and Form: Fractal and Nonfractal Patterns in Physics Proc. 1985 Cargese NATO ASI Institute*, eds. H. E. Stanley and N. Ostrowsky (Martinus Nijhoff Pub, Dordrecht, 1985), p. 101.
72. A. Coniglio, "An Infinite Hierarchy of Exponents to Describe Growth Phenomena," in *Fractals in Physics*, (Trieste 1985), eds. L. Pietronero and E. Tosatti (North-Holland, Amsterdam, 1986).
73. C. Amitrano, A. Coniglio and F. di Liberto, "Growth Probability Distribution in Kinetic Aggregation Processes," *Phys. Rev. Lett.* **57**, 1016 (1986).
74. L. de Arcangelis, S. Redner and A. Coniglio, "Multiscaling Approach in Random Resistor and Random Superconducting Networks," *Phys. Rev.* **B34**, 4656 (1986).
75. A. Coniglio, "Multifractal Structure of Clusters and Growing Aggregates," *Physica* **140A**, 51–61 (1986).
76. D. C. Hong, H. E. Stanley, A. Coniglio and A. Bunde, "Random-Walk Approach to the Two-Component Random-Resistor Mixture: Perturbing Away from the Perfect Random Resistor Network and Random Superconducting-Network Limits," *Phys. Rev.* **B33**, 4564 (1986).
77. N. Jan, A. Coniglio, H. J. Herrmann, D. P. Landau, F. Leyvraz and H. E. Stanley, "On the Relation of Kinetic Gelation and Percolation," *J. Phys. Lett.* **A19**, L399–L404 (1986).
78. F. Leyvraz, J. Adler, A. Aharony, A. Bunde, A. Coniglio, D. C. Hong, H. E. Stanley and D. Stauffer, "The Random Normal Superconductor Mixture in One Dimension," *J. Phys.* **A19**, 3683–3692 (1986).
79. P. Meakin, A. Coniglio, H. E. Stanley and T. A. Witten, "Scaling Properties for the Surfaces of Fractal and Non-Fractal Objects: An Infinite Hierarchy of Critical Exponents," *Phys. Rev.* **A34**, 3325 (1986).
80. A. Coniglio, N. Jan, I. Majid and H. E. Stanley, "Conformation of a Polymer Chain at the $\theta'$ Point: Connection to the External Perimeter of a Percolation Cluster," *Phys. Rev.* **B35**, 3617–3620 (1987).
81. D. Stauffer and A. Coniglio, "Are There Infinitely Many Independent Exponents in the Percolation Cluster Numbers?" *Physica* **143A**, 326–330 (1987).
82. A. Aharony, R. J. Birgeneau, A. Coniglio, M. A. Kastner and H. E. Stanley, "Magnetic Phases and Possible Magnetic Pairing in Doped Lanthanum Cuprate," *Physica* **C153–C155**, 1211 (1988).
83. L. de Arcangelis, A. Coniglio and S. Redner, "Multifractal Structure of the Incipient Finite Cluster," *Phys. Rev.* **B36**, 5631 (1987).
84. C. Amitrano, A. Coniglio and F. di Liberto, "Static and Dynamic Properties for Growth Models," *J. Phys.* **A21**, L201 (1988).
85. A. Aharony, R. J. Birgeneau, A. Coniglio, M. A. Kastner and H. E. Stanley, "Magnetic Phase Diagram and Magnetic Pairing in Doped $La_2CuO_4$," *Phys. Rev. Lett.* **60**, 1330–1333 (1988).
86. A. Coniglio, "Scaling Approach to Multifractality," *Phil. Mag.* **B56**, 785–790 (1987).
87. L. de Arcangelis and A. Coniglio, "Infinite Hierarchy of Exponents in a Two-Component Random Resistor Network," *J. Stat. Phys.* **48**, 935 (1987).

88. A. Coniglio, D. Stauffer and N. Jan, "Search for Multifractality in Damage Spreading for Kauffmann Cellular Automata," *J. Phys.* **A20**, L1103 (1987).
89. P. Poole, A. Coniglio, N. Jan and H. E. Stanley, "Universality Classes for the $\theta$ and $\theta'$ Points," *Phys. Rev. Lett.* **60**, 1203 (1988).
90. C. Amitrano, L. de Arcangelis, A. Coniglio and J. Kertész, "Regular versus Irregular Laplacian Growth: Multifractal Spectroscopy," *J. Phys.* **A21**, L15 (1988).
91. L. de Arcangelis and A. Coniglio, "Critical Temperature in Kauffmann Cellular Automata," *Europhys. Lett.* **7**, 113 (1988).
92. L. de Arcangelis, A. Coniglio and G. Paladin, "Information Dimension in Random Walk Processes," *Phys. Rev. Lett.* **61**, 2156 (1988).
93. A. Coniglio, "Scaling and Universality in Multifractal Growth Processes," in *Universalities in Condensed Matter*, eds. R. Julien, L. Peliti, R. Rammal and N. Boccara, Springer Proceedings in Physics 32 (Springer, Berlin, Heidelberg, 1988).
94. P. Poole, A. Coniglio, N. Jan and H. E. Stanley, "What is the Shape of a Polymer Chain near the Theta Point?" *Phys. Rev.* **B39**, 495 (1989).
95. A. Coniglio, "Fractal Structure of Ising and Potts Clusters: Exact Results," *Phys. Rev. Lett.* **62**, 3054, (1989).
96. A. Coniglio, L. deArcangelis, H. Herrmann and N. Jan, "Exact Relations between Damage Spreading and Thermodynamic Properties in the Ising Model," *Europhys. Lett.* **8**, 315 (1989).
97. A. Coniglio, M. Daoud and H. Herrmann "Conductivity and Diffusion Near the Percolation Threshold," *J. Phys.* **A22**, 4189 (1989).
98. A. Coniglio, L. de Arcangelis and H. J. Herrmann, "Fractals and Multifractals: Applications in Physics," in *ETOPIM2*, *Physica* **A157**, 21 (1989).
99. L. de Arcangelis, A. Coniglio and H. J. Herrmann, "Damage Spreading in Spin Glasses," *Europhys. Lett.* **9**, 794 (1989).
100. L. de Arcangelis, H. J. Herrmann and A. Coniglio, "Dynamical Phase Transition of Spin Glasses in a Magnetic Field," *J. Phys.* **A22**, 4659 (1989).
101. A. Coniglio, "Fractal Structure of Ising and Potts Clusters: Static and Dynamic Approach," In *Fractals: Physical Origin and Properties*, ed. L. Pietronero, Ettore Majorana International Science Series (Plenum Press, 1989).
102. A. Coniglio and H. E. Stanley, "Dilute Annealed Magnetism and High Temperature Superconductivity," *Physica* **C161**, 88 (1989).
103. A. Coniglio and M. Zannetti, "Multiscaling in Growth Kinetics," *Europhys. Lett.* **10**, 575 (1989).
104. A. Coniglio, F. di Liberto, G. Monroy and F. Peruggi, "Exact Relations Between Clusters and Thermal Quantities," *J. Phys.* **A22**, L837 (1989).
105. A. Coniglio and M. Zannetti, "Multiscaling and Multifractality," in Proceedings International Conference honoring B. B. Mandelbrot on his 65th birthday, *Physica* **D38**, 37 (1989).
106. A. Coniglio and M. Zannetti, "Novel Dynamical Scaling in Growth Kinetics," *Physica* **A163**, 325 (1990).
107. J. Lee, A. Coniglio and H. E. Stanley, "Fractal-to-Nonfractal Crossover for Viscous Fingers," *Phys. Rev.* **A41**, 4589 (1990).
108. A. Coniglio and M. Zannetti, "Renormalization Group for Growth Kinetics in the large-N limit," *Phys. Rev.* **B42**, 6873 (1990)

109. L. de Arcangelis, H. J. Herrmann and A. Coniglio, "Scaling Properties of the Damage Cloud in the 3D Ising Model," *J. Phys.* **A23**, L265 (1990).

110. A. Coniglio "Fractal Characterization of Flow in Random Porous Media," in *Hydrodynamics of Dispersed Media*, eds. J. P. Hulin, A. M. Cazabat, E. Guyon and F. Carmona (North-Holland, 1990).

111. A. Coniglio, "Correlations in Thermal and Geometrical Systems," in *Correlations and Connectivity — Geometric Aspects of Physics, Chemistry and Biology*, eds. H. E. Stanley and N. Ostrowsky, NATO ASI series Vol. 188 (1990).

112. C. Amitrano, A. Coniglio, P. Meakin and M. Zannetti, "Multiscaling in Diffusion Limited Aggregation," *Phys. Rev.* **B44**, 4974 (1991).

113. de Arcangelis, A. Coniglio and F. Peruggi "Percolation Transition in Spin Glasses," *Europhys. Lett.* **14**, 515 (1991).

114. A. Coniglio, F. di Liberto, G. Monroy and F. Peruggi, "Cluster Approach to Spin Glasses and the Frustrated Percolation Problem," *Phys. Rev.* **B44**, 12605 (1991).

115. A. Coniglio and M. Zannetti "Analytical Study of Scaling Behavior in Growth Kinetics," in *From Phase Transitions to Chaos*, eds. G. Gyorgy, I. Kondor and L. Sasvari (World Scientific, 1992).

116. A. Coniglio, Y. Oono, A. Shinozaki and M. Zannetti, "On the Nature of Dynamic Scaling in Spinodal Decomposition," *Europhys. Lett.* **18**, 59 (1992).

117. A. Coniglio and M. Zannetti, "Multiscaling in Spinodal Decomposition," in *Probabilistic Methods in Mathematical Physics*, eds. F. Guerra, M. I. Loffredo and C. Marchioro (World Scientific, 1992).

118. V. Cataudella, A. Coniglio, L. de Arcangelis and F. di Liberto, "Cluster Formulation for Frustrated Spin Models," *Physica* **A192**, 167 (1993).

119. A. Coniglio "Connectivity Properties in Complex systems," *J. de Physique IV supplement au J. de Physique II* **3**, 1 (1993).

120. J. Lee, S. Schwarzer, A. Coniglio and H. E. Stanley, "Localization of Growth Sites in DLA Clusters, Multifractality and Multiscaling," *Phys. Rev.* **E48** 1305, (1993).

121. S. C. Glotzer, M. F. Gyure, F. Sciortino, A. Coniglio and H. E. Stanley, "Kinetics of Phase Separation in the Presence of Two Disparate Energy Scales," *Phys. Rev. Lett.* **70**, 3275 (1993).

122. H. E. Stanley, A. Coniglio, S. Havlin, J. Lee and S. Schwarzer, "Disorderly Cluster Growth," in *On Clusters and Clustering*, ed. P. J. Reynolds, as part of series: Random Materials and Processes, eds. H. E. Stanley and E. Guyon (North-Holland, Amsterdam, 1993), pp. 345–356.

123. C. Amitrano, A. Coniglio, P. Meakin and M. Zannetti, "Evidence for Multiscaling in Large DLA Clusters," *Fractals* **1**, 840 (1993).

124. A. Coniglio, "Is Diffusion Limited Aggregation Scale Invariant?" *Physica* **A200**, 165 (1993).

125. S. C. Glotzer, M. F. Gyure, F. Sciortino, A. Coniglio and H. E. Stanley, "Pinning in Phase Separating Systems," *Phys. Rev.* **E49**, 247 (1994).

126. A. Coniglio, P. Ruggiero and M. Zannetti, "Scaling and Crossover in the Large N Model for Growth Kinetics," *Phys. Rev.* **E50**, 1046 (1994).

127. H. E. Stanley, A. Coniglio, S. Havlin, J. Lee, S. Schwarzer and M. Wolf, "Diffusion Limited Aggregation: A Paradigm of Disorderly Cluster Growth," *Physica* **A205**, 254 (1994).

128. V. Cataudella, G. Franzese, M. Nicodemi, A. Scala and A. Coniglio, "Critical Clusters and Efficient Dynamics for Frustrated Spin Models," *Phys. Rev. Lett.* **72**, 1541 (1994).
129. S. Prakash, A. Coniglio and H. E. Stanley "Series Expansion Method Based on the Droplet Description of Ferromagnetic and Fully Frustrated q-State Potts Models," *Phys. Rev.* **E49,** 2742 (1994).
130. S. Roux and A. Coniglio, "Interference of Directed Paths," *J. Phys. A: Math. Gen.* **27**, 5467 (1994).
131. S. C. Glotzer and A. Coniglio "Self Consistent Solution of Phase Separation with Competing Interactions," *Phys. Rev.* **E49**, 247 (1994).
132. A. Coniglio "Frustrated Percolation, Spin Glasses and Glasses," *Il Nuovo Cimento* **16D**, 1027 (1994).
133. F. Corberi, A. Coniglio and M. Zannetti, "Early Stage Scaling in Phase Ordering Kinetics," *Phys. Rev.* **E51**, 5469 (1995).
134. S. G. Glotzer and A. Coniglio, "Frustration Connectivity and the Glass Transition" *J. Comp. Mat. Sci.* **4**, 324 (1995).
135. N. Jan, S. C. Glotzer, P. H. Poole and A. Coniglio, "Clusters and Fractals in the Ising Spin Glass," *Fractals* **3**, 465 (1995).
136. A. Coniglio, in *Non-Equilibrium Phenomena in Supercooled Fluids, Glasses and Amorphous Materials*, eds. M. P. Tosi, M. Giordano and D. Leporini (World Scientific, 1996).
137. A. Coniglio and H. J. Herrmann "Phase Transitions in Granular Packings," *Physica* **A225**, 1 (1996)
138. V. Cataudella, G. Franzese, M. Nicodemi, A. Scala and A. Coniglio, "Percolation and Cluster Monte Carlo Dynamics for Spin Models," *Phys. Rev.* **E54**, 175 (1996).
139. A. Coniglio, "Fractals in the Glass Transition," *Fractals* **4**, 349 (1996).
140. P. H. Poole, S. C. Glotzer, N. Jan and A. Coniglio, "Emergence of Fast Relaxations Upon Cooling in the Ising Spin Glass Above $T_{sg}$," *Phys. Rev. Lett.* **78**, 3394 (1997).
141. M. Nicodemi, A. Coniglio and H. J. Herrmann, "The Compaction in Granular Media and Frustrated Ising Magnets," *J. Phys. Lett.* **30**, 379 (1997).
142. U. Pezzella and A. Coniglio, "Spin Glasses and Frustrated Percolation: A Renormalization Group Approach," *Physica* **A237**, 353 (1997).
143. M. Nicodemi and A. Coniglio, "The Glassy Transition of the Frustrated Ising Lattice Gas," *J. Phys. Lett.* **A30**, L187 (1997).
144. M. Nicodemi, A. Coniglio and H. J. Herrmann, "Compaction and Force Propagation in Granular Packings", *Physica* **A240**, 405 (1997).
145. S. Scarpetta, A. de Candia and A. Coniglio "Glassy Behaviour of the Site Frustrated Percolation Model," *Phys. Rev.* **E55**, 4943 (1997).
146. M. Nicodemi, A. Coniglio and H. J. Herrmann, "Frustration and Slow Dynamics of Granular Packings," *Phys. Rev.* **E55**, 3962 (1997).
147. A. Coniglio, "Frustration and Connectivity in Glass Forming Systems and Granular Materials," in Proceedings International School of Physics "Enrico Fermi" CXXXIV course, *The Physics of Complex Systems*, eds. F. Mallamace and H. E. Stanley (Varenna, 1996), p. 491.
148. S. C. Glotzer, P. H. Poole, A. Coniglio and N. Jan, "Temperature Dependence of Spatial and Dynamic Heterogeneities Above the Ising Spin Glass Transition," *Procs. MRS Fall '96* **455**, 223 (1997)

149. A. Coniglio, "Clusters and Frustration in Glass Forming Systems and Granular Materials," *Prog. Theor. Phys. Suppl.* **126**, 281 (1997).
150. S. C. Glotzer, P. H. Poole, N. Jan and A. Coniglio, "Non-monotonic Temperature Dependence of Local Dynamics and Local Energy Upon Cooling Toward the Ising Spin Glass Transition," *Prog. Theor. Phys. Suppl.* **126**, 393 (1997).
151. A. Fierro, A. de Candia and A. Coniglio, "Nonexponential Relaxation in Fully Frustrated Models," *Phys. Rev.* **E56**, 4990 (1997).
152. A. Coniglio, M. Nicodemi, H. J. Herrmann, E. Caglioti and V. Loreto, "Frustrated Models for Compact Packings," in Proceedings of the NATO ASI on "Physics of Dry Granular Media" Cargese (France) 1997, eds. H. J. Herrmann, J. P. Hovi and S. Luding.
153. E. Del Gado, L. de Arcangelis and A. Coniglio, "A percolation Dynamic Approach to the Sol-Gel Transition," *J. Phys.* **A31**, 1901 (1998)
154. A. Coniglio "Spin Glasses Glasses and Granular Materials," *Philos. Mag.* **B77**, 213 (1998).
155. E. Caglioti, A. Coniglio, H. J. Herrmann, V. Loreto and M. Nicodemi, "Segregation of Granular Mixtures in Presence of Compaction," *Europhys. Lett.* **43** 591 (1998).
156. G. Franzese, V. Cataudella and A. Coniglio, "Invaded Cluster Dynamics for Frustrated Models," *Phys. Rev.* **E57**, 88 (1998).
157. G. Franzese and A. Coniglio, "Phase Transitions in the Potts Spin Glass," *Phys. Rev.* **E57**, 2753 (1998).
158. G. Franzese, A. Fierro, A. de Candia and A. Coniglio, "Autocorrelation Functions in 3D Fully Frustrated Systems," *Physica* **A257**, 376 (1998).
159. E. Caglioti, A. Coniglio, H. Herrmann, V. Loreto and M. Nicodemi, "Geometrical Frustration: A Dynamical Motor for Dry Granular Media," *Physica* **A257**, 419 (1998).
160. A. Coniglio, A. De Candia and M. Nicodemi, "Frustration in Glass Forming Liquids and the Breakdown of the Stokes-Einstein Relation," *Il Nuovo Cimento* **D20** (December 1998).
161. M. Nicodemi and A. Coniglio, "Macroscopic Glassy Relaxations and Microscopic Motions in a Frustrated Lattice Gas," *Phys. Rev. Rapid. Comm.* **E57**, R39 (1998).
162. A. Fierro, G. Franzese, A. de Candia and A. Coniglio "Percolation Transition and the Onset of Nonexponential Relaxation in Fully Frustrated Models," *Phys. Rev.* **E59**, 60 (1999).
163. A. Coniglio, A. de Candia A. Fierro and M. Nicodemi "Universality in Glassy Systems," *J. Phys. Cond. Matt.* **11**, A167 (1999).
164. M. Nicodemi and A. Coniglio, "Aging in Out of Equilibrium Dynamics of Models of Granular Media," *Phys. Rev. Lett.* **82**, 916 (1999).
165. M. Nicodemi, A. Coniglio and H. Herrmann "Density Fluctuations in a Model for Vibrated Granular Media," *Phys. Rev.* **E59** 6830, (1999).
166. L. Cannavacciuolo, A. de Candia and A. Coniglio, "Crossover from Random Percolation to Frustrated Percolation," *Int. J. Mod. Phys.* **C10**, 555 (1999).
167. E. Caglioti, A. Coniglio, H. Herrmann, V. Loreto and M. Nicodemi, "Cooperative Length Approach for Granular Media," *Physica* **A265**, 311 (1999).
168. A. Coniglio and M. Nicodemi, "Scaling Properties in Off-Equilibrium Dynamical Processes," *Phys. Rev.* **E59**, 2812 (1999).

169. A. de Candia, V. Cataudella and A. Coniglio, "Cluster Formulation of Spin Glasses and the Frustrated Percolation Model: Statics and Dynamics," *J. Phys.* **A32**, 32 (1999)
170. G. Franzese and A. Coniglio, The Potts Frustrated Model: Relations with Glasses," *Philos. Mag.* **B79**, 1807 (1999).
171. F. Corberi, M. Nicodemi, M. Piccioni and A. Coniglio, "Off Equilibrium Dynamics of a Diffusion Equation with a Singular Mobility," *Phys. Rev. Lett.* **53**, 5054 (1999).
172. A. Beaton, C. De Coste, D. Hunter and A. Coniglio, "Equilibrium and Quenched Energy Distribution for Spin Glasses," *Int. J. Mod. Physics* **C11**, 41 (2000).
173. A. Coniglio "Geometrical Approach to Phase Transitions in Frustrated and Unfrustrated Systems," *Physica* **A281**, 129 (2000).
174. A. Coniglio and M. Nicodemi, "The Jamming Transition of Granular Media," *J. Phys.: Cond. Matt.* **12**, 6601 (2000).
175. A. Fierro, A. de Candia and A. Coniglio, "Glass Transition in the Quenched and Annealed Version of the Frustrated Lattice Gas Model," *Phys. Rev.* **E62**, 7715 (2000).
176. E. Del Gado, L. de Arcangelis and A. Coniglio, *Europhys. Lett.* **46**, 288 (1999).
177. E. Del Gado, L. de Arcangelis and A. Coniglio, "Viscosity Critical Behaviour at the Gel Point in a 3D Lattice Model," *Eur. Phys. J.* **2**, 359 (2000).
178. A. Coniglio, "Percolation Approach to Phase Transitions," *Nuclear Phys.* **A681**, 451c (2001).
179. A. Coniglio and M. Nicodemi, "A Statistical Mechanics Approach to the Inherent States of Granular Media," *Physica* **A451**, 296 (2001).
180. A. De Candia and A. Coniglio, *Phys. Rev. Lett.* **86**, 4716 (2001).

*1st row*: G. Franzese, B. Klein, M. Daoud, M. Shlesinger, F. Mallamace, ?. E. Guyon, H. Herrmann, A. Coniglio, F. Family, C. Tsallis, G. Parisi, L. Pietronero, C. de Dominicis, C. Castellano, M. Sellitto, E. Del Gado, A. Barrat, P. Grassberger, N. Jan,

*On the steps*: H. E. Stanley, B. Spagnuolo, B. Sapoval, S. Myazima, L. Cannavacciuolo, B. Derrida, A. Erzan, P. Bak, D. Leporini, J. P. Bouchaud, D. Campbell, ?. A. Gabrielli, Roux, ?,

*Last row*: ?. M. Nicodemi, I. Giardina, Tarjus, R. Swendsen, ?. L. de Arcangelis, ?, D. Stauffer, A. Gabrielli, P. Ruggiero., V. Cataudella, E. Marigliano, F. Di Liberto, A. De Candia, S. Scarpetta, Hilfer, A. Aharony, ?, F. Corberi, G. Giberti, G. Gonnella, A. Petri, M. Zannetti, A. Fierro, D. Sanchez, ?, ?, ?

# International Workshop on Scaling and Disordered Systems

*École Superiéure de Physique et Chimie Industrielles, Paris*
*13 – 14 April 2000*

## Organizing committee

F. Family, *Emory University*
M. Daoud, *CEA, Saclay*
H. Herrmann, *ESPCI, Paris and ICA1, University of Stuttgart*

## Program

Thursday morning, 13.04. Session: Percolation and fractals

09.00 Fereydoon Family, Introduction
09.10 Gene Stanley, Application of percolation theory to practical problems
09.30 Amnon Aharony, Fractal geometry of critical Potts clusters
09.50 Bertrand Duplantier, The exact distribution of potential near a 2D percolation cluster
10.10 Luciano Pietronero, Complexity in cosmology
10.30 Break
10.50 Shlomo Havlin, Scaling of polymers and optimization in disordered systems
11.10 Bernard Sapoval, Fractal growth towards equilibrium
11.30 Ayse Erzan, Novel scaling behaviour in a rough interface undergoing a delocalisation transition
11.40 Lucilla de Arcangelis, Complex viscoelastic behaviour at the sol-gel transition
11.50 Marc Barthelemy, Scaling for the critical percolation backbone
12.00 End of morning sessions

Thursday afternoon, 13.04. Session: Potts, fluids, fracture

15.00 Bernard Derrida, Ground state energy of interacting quantum particles, when is not an integer
15.20 Francesco Mallamace, Percolation effects on the viscoelasticity of a critical micellar-water mixture
15.40 Bill Klein, Clusters and the structure of fluctuations at mean-field critical points and spinodals
16.00 Yuri Feldman, Dielectric relaxation in complex liquids at mesoscale
16.10 Giuseppe Gonella, Phase separation of binary mixtures in shear flow
16.20 Manuel I. Marques, Scaling in thermally diluted Ising systems
16.30 Break
16.50 Dino Leporini, Viscous flow and jump dynamics in molecular supercooled liquids
17.00 Mario Nicodemi, Off equilibrium magnetic properties in a model of repulsive particles for vortices in superconductors
17.10 Pierre Devillard, Dimensional crossover in a pseudogap phase for high superconductors
17.20 Marco Zannetti, Multiscaling in the preasymptotics of phase ordering kinetics
17.30 Stefano Zapperi, Planar cracks in the random fuse model
17.40 Alberto Petri, Criticality in fracture of disordered media
17.50 Rudolf Hilfer, Fractional calculus and evolution equations

18.00 End of afternoon session
19.30 Banquet at restaurant Vallauris on rue Mouffetard

Friday morning, 14.04. Session: Glasses and granular media

09.00 Jean-Philippe Bouchaud, Aging in glassy systems: temperature as a microscope
09.20 Giorgio Parisi, A genetic cluster method for finding the ground state of spin glasses: methodology and results
09.40 Dietrich Stauffer, Get rich fast through percolation theory
10.00 Michael Shlesinger, Pressure effects near the glass transition: a new Vogel-like law for polymeric ionic conductivity
10.20 Hans Herrmann, Dilatancy and shear bands
10.30 Break
10.50 Alain Barrat, Response properties in a model for granular matter
11.00 Federico Corberi, Slow evolution and aging in a constrained diffusion model for glassy kinetics
11.10 Antonio de Candia, Decoupling between relaxation and diffusion in a model glass
11.20 Silvio Franz, Onsager regression principle in glassy relaxation
11.30 Giancarlo Franzese, Saddle points and relaxation times in a spin model with glassy behavior
11.40 Alexander Puzenko, Joint presentation with Yuri Feldmann. No additional presentation.
11.50 Yaroslav Ryabov, The relationship between the scaling parameter and relaxation time for non-exponential relaxation processes in disordered systems
12.00 Nicolas Vandewalle, Stripe ordering in self-stratified granular mixtures
12.10 End of morning session

Friday afternoon, 14.04. Session: Interdisciplinary science

15.00 Per Bak, Forest fires, measles, and the structure of the universe
15.20 Peter Grassberger, Go-with-the-winners: doing statistical physics with opportunistic and biased simulations
15.40 Constantino Tsallis, Microscopic mixing and statistical mechanics
16.00 Sasuke Miyazima, Dynamics in politics
16.20 Yi-Cheng Zhang, Minority game and market mechanism
16.30 Break
16.50 Alessandro Vespignani, SOC as an absorbing-state phase transition
17.00 Serge Galam, How to become a dictator: an application from statistical physics
17.10 Adam Gadomski, A discrete model of interline evolution by random rules
17.20 Enrico Scalas, Fractional calculus and continuous-time finance: scaling of the waiting time distribution in tick-by-tick financial data
17.30 Wolfgang Kinzel, Neural networks competing for the minority
17.40 Silvia Scarpetta, Hebbian learning in oscillatory associative memories neural networks
17.50 Bernardo Spagnolo, Nonlinear relaxation of population dynamics
18.00 END

# Contents

# Part I

# Critical Phenomena

# FRACTAL DIMENSIONS AND CORRECTIONS TO SCALING FOR CRITICAL POTTS CLUSTERS

AMNON AHARONY
*School of Physics and Astronomy,*
*Raymond and Beverly Sackler Faculty of Exact Sciences,*
*Tel Aviv University, Tel Aviv 69978, Israel*

JOONAS ASIKAINEN
*Laboratory of Physics, Helsinki University of Technology,*
*P. O. Box 1100, FIN-02150 HUT, Espoo, Finland*

**Abstract**

Renormalization group and Coulomb gas mappings are used to derive theoretical predictions for the corrections to the exactly known asymptotic fractal masses of the hull, external perimeter, singly connected bonds and total mass of the Fortuin-Kasteleyn clusters for two-dimensional $q$-state Potts models at criticality. For $q = 4$ these include exact logarithmic (as well as log-log) corrections.

## 1. INTRODUCTION

$q$-state Potts models, with interaction $-J\delta_{\sigma_i\sigma_j}$ ($\sigma_{i,j} = 1, 2, \ldots, q$) for the nearest neighbor (nn) sites $i$, $j$, have played an important role in condensed matter physics.[1] Here we study geometrical aspects of the critical Potts clusters, in two dimensions. For an arbitrary configuration of Potts states, one creates bonds between neighboring sites which have the same state, $\sigma_i = \sigma_j$, with a probability $p = 1 - \exp(-J/kT)$. No bonds are created between sites with $\sigma_i \neq \sigma_j$. Here we study the fractal geometry at $T_c$ of the clusters, made of sites connected by bonds.[2] Specifically, we measure the fractal dimensions $D_M$, $D_H$, $D_{EP}$, and $D_{SC}$ describing the scaling of the cluster's mass, hull, external accessible perimeter[3] and singly connected bonds,[4] respectively, with its radius of gyration $R$. As emphasized by

**Table 1 Exact theoretical predictions.**

| | $D_S$ | $q=1$ | $q=2$ | $q=3$ | $q=4$ | $c_S/a$ |
|---|---|---|---|---|---|---|
| $g$ | | $\frac{8}{3}$ | 3 | $\frac{10}{3}$ | 4 | |
| $M$ | $(g+2)(g+6)/(8g)$ | $\frac{91}{48}$ | $\frac{15}{8}$ | $\frac{28}{15}$ | $\frac{15}{8}$ | $\frac{1}{16}$ |
| $H$ | $1+2/g$ | $\frac{7}{4}$ | $\frac{5}{3}$ | $\frac{8}{5}$ | $\frac{3}{2}$ | $-\frac{1}{4}$ |
| $EP$ | $1+g/8$ | $\frac{4}{3}$ | $\frac{11}{8}$ | $\frac{17}{12}$ | $\frac{3}{2}$ | $\frac{1}{4}$ |
| $SC$ | $(3g+4)(4-g)/(8g)$ | $\frac{3}{4}$ | $\frac{13}{24}$ | $\frac{7}{20}$ | 0 | $-1$ |
| $\theta$ | $4(4-g)/g$ | 2 | $\frac{4}{3}$ | $\frac{4}{5}$ | 0(log) | |
| $\theta'$ | $4/g$ | $\frac{3}{2}$ | $\frac{4}{3}$ | $\frac{6}{5}$ | 1 | |
| $\theta''$ | $2/g$ | $\frac{3}{4}$ | $\frac{2}{3}$ | $\frac{3}{5}$ | $\frac{1}{2}$ | |

Coniglio,[5] many of these fractal dimensions have been derived analytically.[6] Some others have been found more recently.[7,8] These theoretical values are summarized in Table 1, in terms of the Coulomb gas coupling constant $g = \frac{4}{\pi}\arccos(-\frac{\sqrt{q}}{2})$.

We are currently studying the geometry of such Potts critical clusters, using numerical Monte Carlo simulations.[9] These simulations show that the asymptotic power law dependence of the various masses on $R$ is approached relatively slowly, and therefore the analysis of the data must include *correction terms*, particularly as $q$ approaches 4. The present paper contains a brief summary of analytic results for several of these corrections. Our predictions for the correction exponents (apart from the analytic ones) are also listed in Table 1.

## 2. RENORMALIZATION GROUP

The first correction relates to the *dilution field* $\psi$, which is generated under renormalization even when one starts with the non-dilute case.[10] For our non-dilute case, one expects $\psi(\ell)$ to increase under the renormalization group recursion relations (RGRR's) from $-\infty$ towards its critical value $\psi^*$ ($\leq 0$ for $q \leq 4$), while the cluster linear size, like all other lengths, rescales as $R(\ell) = Re^{-\ell}$. Following Cardy *et al.*,[10] we assume that $\ell_0$ prefacing iterations bring $\psi$ from $-\infty$ up to $\psi(\ell_0) = \psi_0$, with $|\psi_0| \ll 1$. We then expand the RGRR for $\psi$ in powers of $\psi$ and $\varepsilon = q - 4$,

$$\frac{d\psi}{d\ell} = a(\varepsilon + \psi^2 + b\psi^3 + r\varepsilon\psi + \cdots)\,. \tag{1}$$

At $q = 4$, this yields

$$a(\ell - \ell_0) \approx \psi_0^{-1} - \psi(\ell)^{-1} + b\log\frac{b + \psi(\ell)^{-1}}{b + \psi_0^{-1}}\,. \tag{2}$$

Iterating up to $R(\ell^\times) = 1$, we can express $\psi(\ell^\times)$ in terms of $\log R = \ell^\times$. For large $R$, $\psi_0/\psi(\ell^\times) \approx A(\log R + B\log(\log R) + E) + O(\log\log R/\log R)$, where $A = -a\psi_0$, $B = -b/a$ and $E$ also depends on $a$, $b$, $\ell_0$ and $\psi_0$.

For $q < 4$, an expansion to second order in $\varepsilon' = \sqrt{-\varepsilon}$ yields $\psi(\ell^\times) \approx \psi^* + \tilde{B}R^{-\theta}$, with $\psi^* = -\varepsilon'(1-(r-b)\varepsilon'/2) + \cdots$, $\theta = 2a\varepsilon'(1-b\varepsilon') + \cdots$ and $\tilde{B} \propto (\psi_0 - \psi^*)$. To order-$\varepsilon'$, one obtains a full solution,

$$\psi(\ell) = -\varepsilon' \frac{1+\hat{B}e^{-\theta\ell}}{1-\hat{B}e^{-\theta\ell}}, \tag{3}$$

where $\hat{B} = (\psi_0 + \varepsilon')/(\psi_0 - \varepsilon')$. Indeed, $\psi$ approaches $\psi^*$ for large $\ell$.

To obtain the scaling of $M_S(R)$, we write the RGRR for the field $h_S$ conjugate to the density $\rho_S \equiv M_S/R^d$ as

$$\frac{dh_S}{d\ell} = (y_S + c_S\psi(1 + e_S\psi + f_S\psi^2 + \cdots))h_S, \tag{4}$$

where the coefficients may depend on $\varepsilon$. $\rho_S$ is then found as a derivative of the free energy with respect to $h_S$. For $q = 4$, its singular part becomes

$$\begin{aligned}\rho_S(\ell) &\propto e^{-d\ell}h_S(\ell)/h_0 \\ &= \exp[(y_S - d)\ell + \int_{\ell_0}^{\ell}(c_S\psi(1+e_S\psi+f_S\psi^2+\cdots))d\ell] \\ &\propto e^{(y_S-d)\ell}[\psi(\ell)/\psi_0]^{c_S/a}(1+O(\psi(\ell)))\,.\end{aligned} \tag{5}$$

For large $\log R = \ell^\times$, this becomes

$$M_S \propto R^{D_S}(\log R + B\log(\log R) + E)^{-c_S/a}(1 + O(\log\log R/\log R))\,, \tag{6}$$

with $D_S = y_S(q = 4)$, and $c_S/a$ is to be taken from Table 1 (see below). Note that $B = -b/a$ is universal (i.e. independent of $\psi_0$), and the non-universal constant $E$ is the same for all $S$. Equation (6) generalizes the logarithmic corrections of Cardy *et al.*[10]

In practice, the numerical results are always analyzed by looking at the local logarithmic slope,

$$\begin{aligned} D_S^{\text{eff}} &= d\log M_S/d\log R = d\log h_S/d\ell|_{\ell=\ell^\times} \\ &= y_S + c_S\psi(\ell^\times)(1 + e_S\psi(\ell^\times) + f_S\psi(\ell^\times)^2 + \cdots)\,. \end{aligned} \tag{7}$$

In some cases, this expression (in which $\psi(\ell^\times)$ is related to $\log R = \ell^\times$ via Eq. (2)) gave a better fit than the derivative of the approximate expression in Eq. (6).

For $q < 4$, to leading order in $\varepsilon'$, the same procedure turns Eq. (3) into

$$M_S \propto R^{D_S}(1 - \hat{B}R^{-\theta})^{-c_S/a} \approx R^{D_S}(1 + f_S R^{-\theta})\,, \tag{8}$$

where $D_S \approx y_S - c_S\varepsilon'$ and $\theta \approx 2a\varepsilon'$. The RHS of this equation remains correct also for higher orders in $\varepsilon'$. Note that to the lowest order in $\varepsilon'$, the ratios $f_S/f_{S'}$ are universal, being equal to $c_S/c_{S'}$. This is similar to analogous ratios for thermodynamic properties in the usual $\varepsilon$-expansion.[11] Expanding the exact $D_S$ (Table 1) in $\varepsilon'$ yields $c_S$. Using also $a = 1/\pi$ (see below) yields our predictions for $c_S/a$ (given in Table 1), to be used in fitting Eq. (6). The form on the RHS of Eq. (8) is already implied by den Nijs,[12] who found that the pair correlation functions $G_H(r)$ can be expanded as a sum over $r^{-2x_n}$, implying a leading correction exponent $\theta = 2(x_{n+1} - x_n) = 4(4-g)/g$. Expanding this expression in powers of $\varepsilon'$ yields the coefficients $a = 1/\pi$ and $b = -1/2\pi$, which we use in our fits to Eq. (2). The value $a = 1/\pi$ also agrees with Cardy *et al.*[10] This expression for $\theta$ also reproduces known results for $q = 2$, 3, as listed in Table 1.

## 3. COULOMB GAS

The second source of corrections involves new contributions to the relevant pair correlation functions in the Coulomb gas representations.[12] In some of the exact derivations, the $q$-state Potts model renormalizes onto the vacuum phase of the Coulomb gas, involving 'particles' with electric and magnetic 'charges' ($e$, $m$). At criticality, the corresponding Coulomb gas has a basic 'charge' $\phi = |2-g/2| \bmod 4$. Various Potts model two-point correlation functions $G_S^P(\vec{r})$ are then mapped onto Coulomb gas analogs, which give the probability of finding two charged particles at a distance $r$ apart. Asymptotically, these are given by

$$G^{CG}_{[(e_1,m_1),(e_2,m_2)]}(\vec{r}) \propto r^{-2x^{CG}_{[(e_1,m_1),(e_2,m_2)]}}, \tag{9}$$

where

$$x^{CG}_{[(e_1,m_1),(e_2,m_2)]} = -\frac{e_1 e_2}{2g} - \frac{g m_1 m_2}{2}. \tag{10}$$

Hence one identifies $D_S = d - x^{CG}_{[(e_1,m_1),(e_2,m_2)]}$, with $d = 2$. The results in Table 1 for $S = M$ were obtained by den Nijs,[12] who noted that the spin-spin correlation function of the Potts model maps onto a Coulomb gas total electric charge $Q = -2\phi$, which splits into the two charges $e_{1,2} = \pm 1 - \phi$ (and $m_{1,2} = 0$). Continuing along similar routes, Saleur and Duplantier[6] used a mapping onto the body-centered solid-on-solid model, requiring a vortex-antivortex pair with $e_{1,2} = -\phi$ and $m_{1,2} = \pm 1/2$ or $\pm 1$ for the fractal dimensions of $S = H$ or $S = SC$. The Table also contains Duplantier's recent result[8] for $D_{EP} = 2 - x^P_{EP}$, which has not been expressed in terms of Coulomb charges. The results for $x^P_H$ and $x^P_{SC}$ are special cases of the expression $x_\ell = g\ell^2/32 - (4-g)^2/(2g)$, with $\ell = 2$ and 4 respectively.[6] For percolation ($q = 1$ and $g = 8/3$), this expression also yields $x^P_{EP} = x_3 = 2/3$ for the external perimeter and $x^P_G = x_6 = 35/12$ for the gates to fjords.[7]

We now turn to corrections to the leading behavior. den Nijs[12] derived such corrections for the order parameter correlation function. In that case, he noted that the charge $Q = -2\phi$ could also split into the pair $e_{1,2} = \pm 3 - \phi$, yielding a contribution to $G^P_M$ of the form $r^{-2x^P_{M,2}}$, with $x^P_{M,2} = x^{CG}_{[(3-\phi,0),(-3-\phi,0)]} = x^P_M + 4/g$. Since $D = d - x$ usually represents a fractal dimension, we relate each of these correction terms to some subset of the cluster, with dimension $D_{M,2} = 2 - x^P_{M,2} = D_M - 4/g$. Writing $M_M$ as a sum of powers $R^{D_i}$,[13] we have $M_M \propto R^{D_M}(1 + f'R^{-\theta'})$, with $\theta' = 4/g$.

As far as we know, there has been no discussion of the analogous corrections to the other subsets discussed here. In the spirit of den Nijs,[12] we note that the correlation function for both $H$ and $SC$ could also result from electrical charges $e_{1,2} = \pm 2 - \phi$, instead of $-\phi$. For both of these cases this would give $x\prime = x + 2/g$, hence a correction exponent $\theta\prime\prime = 2/g$. At the moment, there exists no theory for corrections to $M_{EP}$. However, in the spirit of the renormalization group it is also reasonable to interpret $\theta\prime$ and $\theta\prime\prime$ as the scaling exponent of some irrelevant perturbation (yet to be identified). If that were true then we might expect the same perturbation also to affect other quantities, like $M_{EP}$. This conjecture is supported by the 'superuniversal' relation, $(D_H - 1)(D_{EP} - 1) = 1/4$, found by Duplantier.[8]

If this relation also holds for the effective dimensions (as happens e.g. in the $\varepsilon$-expansion,[11]) then $H$ and $EP$ should have the same correction exponents.

## 4. ANALYTIC CORRECTIONS: SUMMARY

The last source of corrections involves 'analytic' terms, coming e.g. from linear cuts with dimensions $(D_S - 1)$,[13] or from replacing $R$ by $(R + a)$, since there are many possible candidates for the correct linear measure of the cluster. These would imply corrections of relative size $1/R$.

Combining all of these sources, we end up with the prediction (for $q < 4$)

$$D_S^{\text{eff}} = D_S + \sum_i f_i R^{-\theta_i}, \tag{11}$$

with $\theta_i = \theta$, $\theta'$ (or $\theta\prime\prime$) and 1. Indeed, our numerical simulations[9] basically confirm these expressions.

In summary, we have presented several general expressions for the $q$-dependent corrections to the asymptotic $R$-dependence of the mass, hull, external perimeter and singly connected bonds. Such corrections are crucial for fitting numerical data. It would be nice to have a unifying theory, which would confirm these expressions in a rigorous way. It would also be nice to obtain similar corrections for other geometrical and physical quantities, e.g. the number of gates to fjords.[7]

## ACKNOWLEDGMENTS

This paper is dedicated to Antonio Coniglio, on the occasion of his 60th birthday. Coniglio contributed a lot to our understanding of many issues discussed in this paper, both through his publications and through very stimulating personal discussions during the last quarter of a century. This project is part of a larger collaboration with Benoit Mandelbrot, who initiated it and continuously contributed via many discussions, and also with Juha Pekka Hovi and E. Rausch. We also acknowledge support from the German-Israeli Foundation.

## REFERENCES

1. F. Y. Wu, *Rev. Mod. Phys.* **54**, 235 (1982).
2. C. M. Fortuin and P. W. Kasteleyn, *Physica* (*Utrecht*) **57**, 536 (1972).
3. T. Grossman and A. Aharony, *J. Phys.* **A19**, L745 (1986); **20**, L1193 (1987).
4. A. Coniglio, *J. Phys.* **A15**, 3829 (1982).
5. A. Coniglio, *Phys. Rev. Lett.* **62**, 3054 (1989).
6. H. Saleur and B. Duplantier, *Phys. Rev. Lett.* **58**, 2325 (1987).
7. M. Aizenman, B. Duplantier and A. Aharony, *Phys. Rev. Lett.* **83**, 1359 (1999).
8. B. Duplantier, *Phys. Rev. Lett.* **84**, 1363 (2000).
9. J. Asikainen, A. Aharony, B. B. Mandelbrot, E. Rausch and J. P. Hovi, to be published.
10. J. L. Cardy, M. Nauenberg and D. J. Scalapino, *Phys. Rev.* **B22**, 2560 (1980).
11. A. Aharony and G. Ahlers, *Phys. Rev. Lett.* **44**, 782 (1980).
12. M. P. M. den Nijs, *Phys. Rev.* **B27** 1674 (1983).
13. Y. Gefen, B. B. Mandelbrot, A. Aharony and A. Kapitulnik, *J. Stat. Phys.* **36**, 827 (1984); B. B. Mandelbrot, Y. Gefen, A. Aharony and J. Peyriere, *J. Phys.* **A18**, 335 (1985).

# COMPLEX VISCOELASTIC BEHAVIOUR AT THE SOL-GEL TRANSITION

LUCILLA DE ARCANGELIS
*Dipartimento di Ingegneria dell'Informazione, Seconda Università di Napoli, INFM Gruppo Collegato SUN, UdR Napoli, via Roma 29, 81031 Aversa (Caserta), Italy*

EMANUELA DEL GADO
*Dipartimento di Scienze Fisiche, Università di Napoli "Federico II", INFM Gruppo Collegato SUN, UdR Napoli Complesso Universitario Monte Sant'Angelo, 80125 Napoli, Italy*

ANTONIO CONIGLIO
*Dipartimento di Scienze Fisiche, Università di Napoli "Federico II", INFM UdR Napoli, Complesso Universitario Monte Sant'Angelo, 80125 Napoli, Italy*

**Abstract**

We present a percolation dynamic model for the study of dynamics at the sol-gel transition. Percolation and bond-fluctuation dynamics result to be suited to study the critical behaviour of the viscoelastic properties and show a complex relaxation behaviour. The results obtained via numerical simulation on the cubic lattice are in good agreement with some theoretical predictions and experimental results.

## 1. INTRODUCTION

In gelling polymeric systems the growth of connectivity produces a dramatic change in the viscoelastic properties, with a diverging viscosity and the appearance of an elastic response.[1,2] Due to the constitution of a macroscopic polymeric network, a transition between two different viscoelastic regimes takes place, characterized by the complex relaxation behaviour: a widening of the relaxation times distribution function at the transition threshold and typical non exponential decays are observed.[3,5]

Some interesting analogies with the glassy dynamics can be seen in this phenomenon, which is actually controlled by the growth of the connectivity.The connectivity transition is naturally described in terms of a percolation transition,[6] giving the static critical exponents, on the contrary the dynamic properties at the gelation transition are intensively investigated and debated.

In the experiments both the viscosity coefficient and the elastic modulus dependence on the polymerization extent are well fitted by a power law but the experimental determination of these critical exponents is quite controversial: the results are in fact rather scattered and a better comprehension of the relevant mechanisms in the transition is requested. Recent experimental measurements of the viscosity critical exponent $k$ give values ranging from 0.7 in diisocyanate/triol to 1.5 in epoxy resins. For the elastic modulus critical exponent the values are even more scattered, ranging from 1.9 in diiscyanate/triol to 3. in polyesters.[7,13]

Within the theoretical description of the phenomenon the main results are based on the percolation model following the Flory's idea, but the viscoelastic dynamic behaviour is not simply obtained in terms of the connectivity transition. The problem is in the determination of the viscosity of a very complex medium, which is the sol at the transition threshold, a highly polydisperse polymeric solution at high concentration. The polymer dynamics is characterized by the relaxation processes over many different time scales becoming more and more complex with the increasing connectivity to produce the observed viscoelastic behaviour.

Within the Flory classical theory of gelation the viscosity remains finite or at most diverges logarithmically. Using instead the Rouse model for the polymer dynamics,[14] which neglects the entanglement effects and the hydrodynamics interactions, the viscosity in a solution of polymeric clusters, expressed in terms of the macroscopic relaxation time, grows like $\langle R^2 \rangle$ as the cluster radius $R$ grows in the gelation process. The contribution of the $n_s$ molecules of size $s$ and gyration radius $R_s$ to the average $\langle\ \rangle$ is of the order of $sn_sR_s^2$ leading to the critical exponent $k = 2\nu - \beta$,[15] where $\nu$ is the critical exponent of the correlation length diverging at the gel point and $\beta$ is the order parameter critical exponent. With the random percolation exponents the value $k \sim 1.35$ in $3d$ is found, that agrees quite well with some experimental measurements.[10,11,13] Actually this Rouse exponent could be considered as an upper limit due to the complete screening of the hydrodynamic interactions and the entanglement effects. Another approach has been proposed by de Gennes[2,15] using an analogy between the viscosity at the gelation threshold and the diverging conductivity in the *random superconducting network* model, giving an exponent $k \sim 0.7$ in $3d$. This result is in good agreement with the values experimentally obtained in gelling solutions of Refs. 8 and 9.

Our approach consists in directly investigating the viscoelastic properties at the sol-gel transition, introducing within the random percolation model a suited dynamics. Our model associates the bond-fluctuation ($BF$) dynamics[16,17] to the percolation clusters, taking in this way into account the polymer conformation changes.

## 2. THE PERCOLATION DYNAMIC MODEL

We consider a sol of tetrafunctional monomers, characterized by the the monomer density $p$ and the probability of bond formation $p_b$. In terms of these two parameters the static properties present a percolation transition and a phase diagram can be determined. We consider the case of strong gelation, i.e. of bonds quenched in time. The dynamics is given

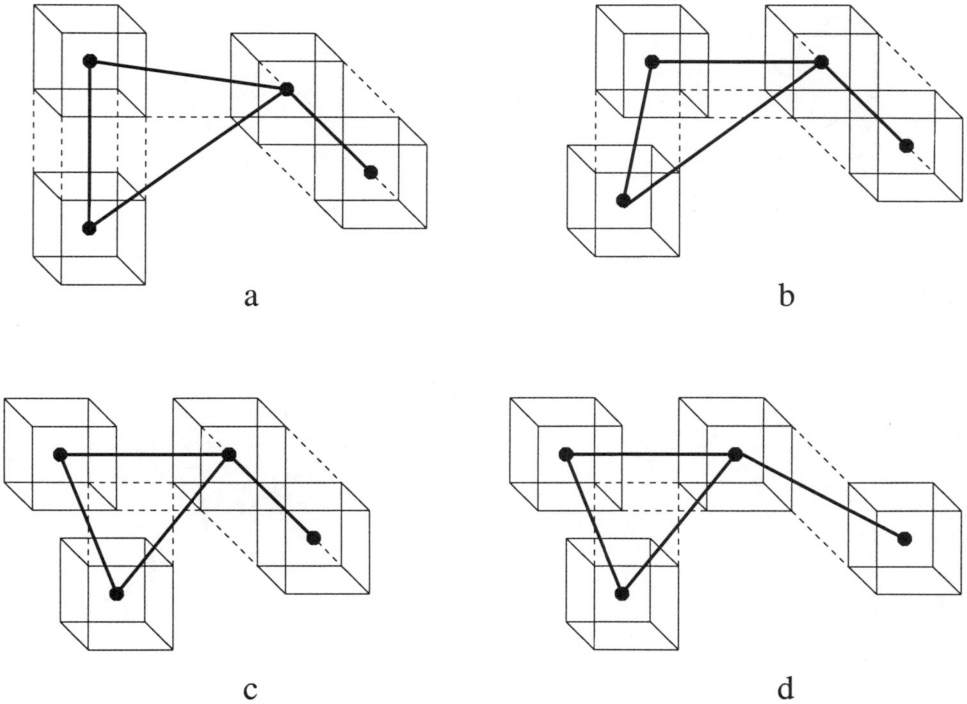


**Fig. 1** Four different possible configurations for a cluster formed by four monomers in the time evolution according to the bond fluctuation dynamics. We consider the subsequent monomer movements: in $a$ the bond lengths are (starting from the upper center bond and clockwise) $l = \sqrt{5}$, 3, 3, 2; moving forward the upper left monomer we have $b$ with $l = 2, 3, 3, \sqrt{5}$; from this configuration moving right the other left monomer we have $c$ with $l = 2, 3, \sqrt{6}, \sqrt{6}$; then moving right the front monomer on the right leads to $d$ and $l = 2, \sqrt{10}, \sqrt{6}, \sqrt{6}$.

by the $BF$ model: the monomers move via random local movements constrained by the excluded volume interaction and the $SAW$ condition. As a consequence there is a maximum allowed bond length during the dynamic evolution of the system.

We have studied this model via numerical simulation on the cubic lattice.[20] The $BF$ model can be easily expressed in terms of a lattice algorithm: a monomer occupies a lattice elementary cell, two occupied cells cannot have common sites and the $SAW$ condition produces a finite set of allowed bond lengths corresponding to a high number of different bond vectors. In Fig. 1 an example of the possible dynamic evolution of a cluster of 4 monomers is shown.

The numerical simulation have been performed on the CRAY-T3E system of CINECA, using $\sim$ 30000 hours of $CPU$ time.

We have considered lattices of size $L$ ranging between 16 and 40 and the data have been averaged over a sample of $\sim$ 30 systems with different initial site and bond configurations.

## 3. DIFFUSION PROPERTIES

We have analyzed the diffusion properties of the percolation clusters of different cluster size at different transition stages. We have studied the mean squared displacement of the center of mass and from the long time diffusive behaviour we have calculated the diffusion coefficient. In fact, as the gel point is approached, the sol is a complex viscoelastic medium made of the solvent and the different clusters, with a diverging mean cluster size. The gelation threshold is characterized by a self similar structure of the incipient percolating

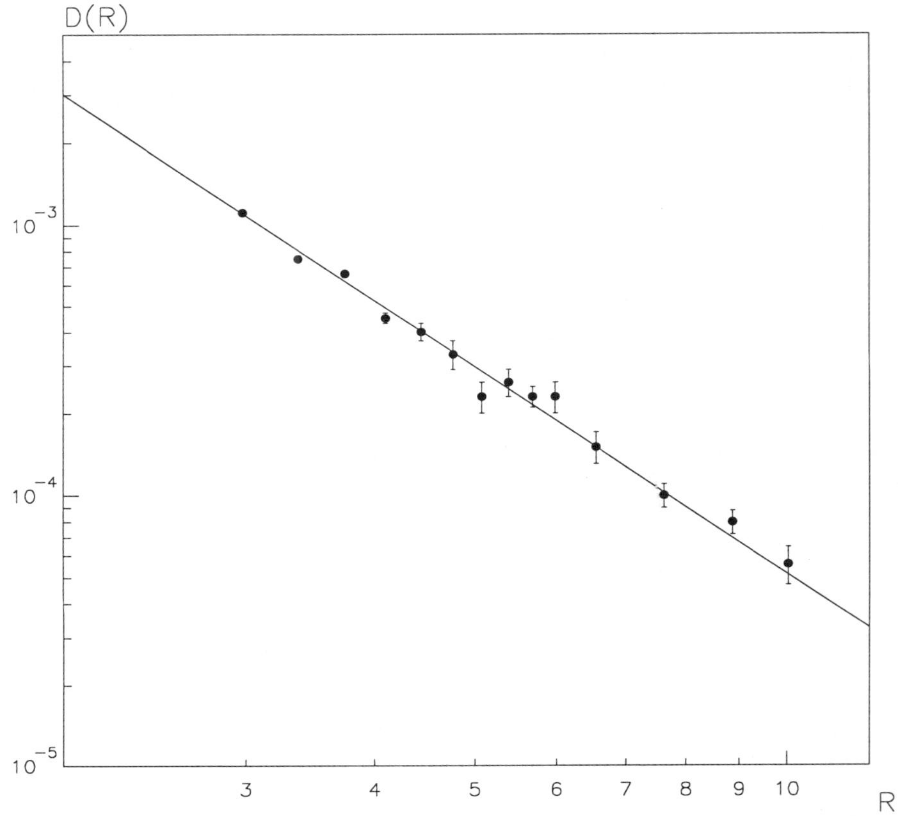


**Fig. 2** The diffusion coefficient at $p_c$ averaged over 32 different configurations for different cluster size as a function of the cluster radius of gyration $R$: supposing the scaling behaviour $D(R) \sim 1/R^{1+k/\nu}$, gives $k \sim 1.3 \pm 0.02$.

cluster and by the diverging viscosity coefficient $\eta$. If we consider a probe of size $R$ diffusing in such a medium, as long as $R \gg \xi$ (where $\xi$ is the percolation correlation length in the system), its diffusion coefficient $D(R)$ is expressed by the Stokes-Einstein relation $D(R) \sim 1/R^{d-2}\eta$. In general for a probe of size $R$ it is then reasonable to assume that it will actually diffuse in the medium made of the solvent and all the clusters of size $r \leq R$, due to the much longer relaxation times of the larger clusters. This again can be expressed following a Stokes-Einstein relation $D(R) \sim 1/R^{d-2}\eta(R)$, where now $\eta(R)$ takes into account the probe size effect.[21] At the gelation threshold $p_c$, $\eta \sim (p - p_c)^{-k}$ and the scaling behaviour $\eta(R) \sim R^{k/\nu}$ is expected. Then at the gel point the diffusion coefficient $D(R)$ of a probe of size $R$ is related to the viscosity critical behaviour

$$D(R) \sim \frac{1}{R^{d-2+k/\nu}} \,. \tag{1}$$

Within this scaling picture we have calculated the diffusion coefficients of clusters of different radius $R$ and from the scaling behaviour in terms of $R$ obtained an estimate for the viscosity critical exponent $k \sim 1.3$ (Fig. 2) in agreement with the Rouse model prediction.

We briefly mention here that on the other side the diffusion coefficients of very small clusters are not expected to be linked to the macroscopic viscosity, and in fact the diffusion coefficient of monomers does not go to zero at $p_c$, but has a definitively non-zero value for $p > p_c$ (Fig. 3). On the contrary in the $2d$ case also the very small clusters diffusion was actually blocked at the percolation threshold.[18] It is also interesting to notice that the $3d$ data seems to agree with a dependence $e^{-1/1-p}$, suggesting some cooperative mechanism in the diffusion process.

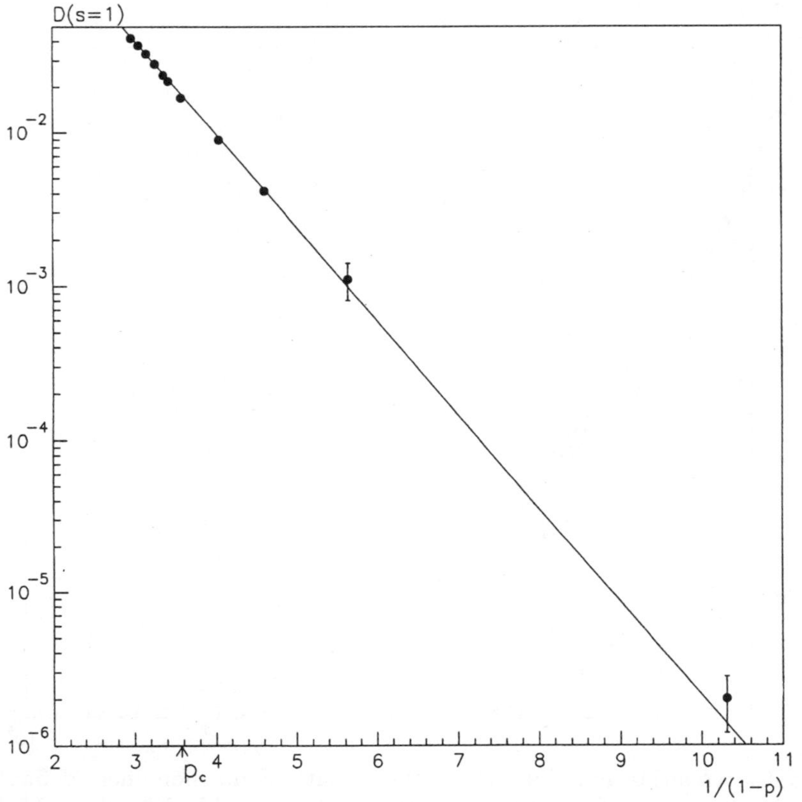


**Fig. 3** The monomer self-diffusion coefficient as function of $p$. The data show how the self-diffusion coefficient of monomers is not linked to the macroscopic viscosity diverging at the gel point. $D(s = 1)$ becomes numerically undistinguishable from zero only at $p > p_c$ and the data are well fitted with the dependence $\sim e^{-1/1-p}$ (the continuous line).

## 4. CORRELATION FUNCTIONS

Then we have independently studied the viscosity in the system via the macroscopic relaxation time. We have calculated the time autocorrelation functions $g(t)$ of the number $\varepsilon(t)$ of the pairs of nearest neighbours monomers during the dynamic evolution

$$g(t) = \left\langle \frac{\overline{\varepsilon(t')\varepsilon(t'+t)} - \overline{\varepsilon(t')}^2}{\overline{\varepsilon(t')^2} - \overline{\varepsilon(t')}^2} \right\rangle . \tag{2}$$

At different $p$ values in the critical region after a fast transient $g(t)$ decays to zero (Fig. 4) and cannot be fitted by a simple time exponential behaviour but with a stretched exponential law at long times (Fig. 5). This is a sign of the existence of a distribution of relaxation times which cannot be related to a single time and this behaviour had already been observed in the $d = 2$ study of the model.[19] It is a typical feature of polymeric systems where the relaxation process always implies the rearrangement of the system over many different length scales.[22,23] This behaviour of the relaxation functions is usually interpreted in terms of a very broad distribution of relaxation times and it is in fact experimentally observed in a sol in the gelation critical regime.[3,5]

For $p \sim p_c$ we have then fitted the $g(t)$ with a stretched exponential behaviour $e^{(-t/\tau_0)^\beta}$ (Fig. 5), where $\beta \sim 0.3$ apparently not depending on $p$. This value is quite lower than the one experimentally obtained in Ref. 3 for a gelling solution: this discrepancy could be due to

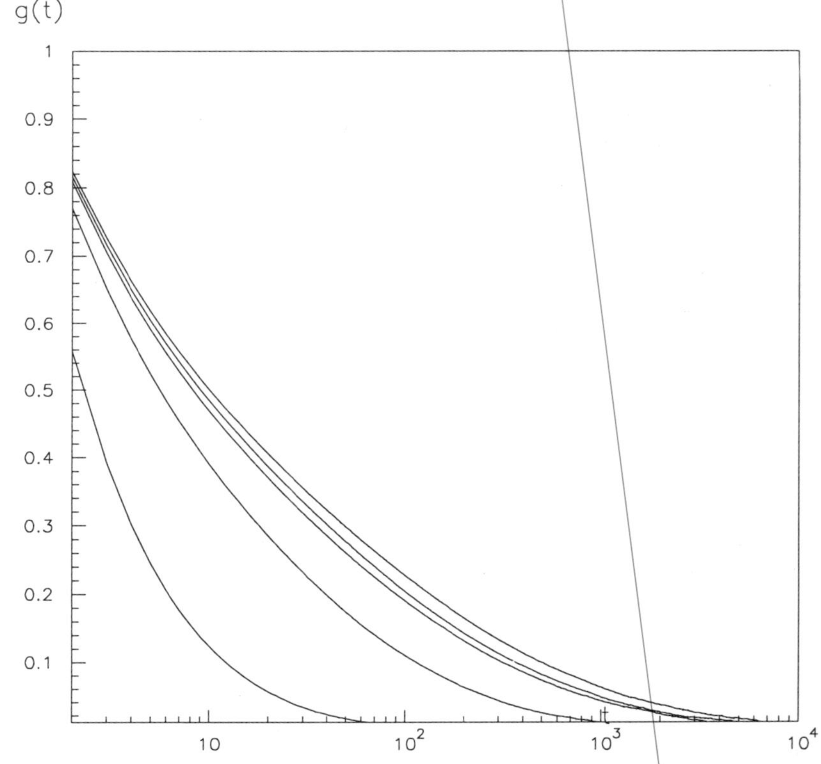


**Fig. 4** Time autocorrelation function $g(t)$ of the density of $nn$ monomers defined in Eq. (2) as function of time for various $p$. From top to bottom $p = 0.705$, 0.7, 0.69, 0.66, 0.5 (data averaged over $\sim 30$ different configurations).

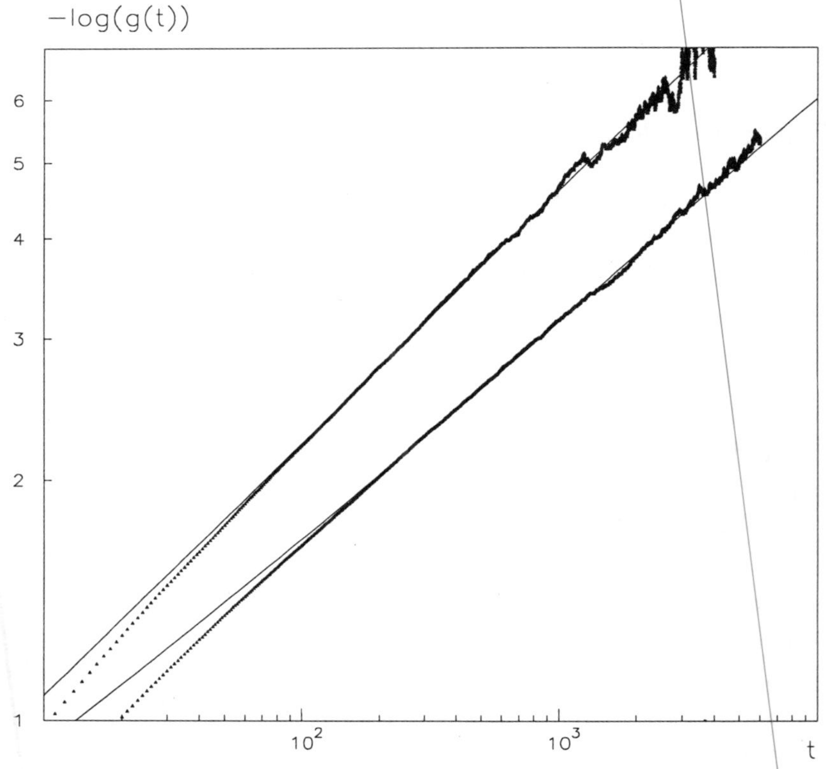


**Fig. 5** Stretched exponential behaviour $e^{-(t/\tau_0)^\beta}$ of the long time tail of the time autocorrelation function $g(t)$ for $p = 0.66$ (bottom) and $p = 0.69$ (top).

the fact that our data refer to a quite narrow region near the gelation threshold where $\beta$ is expected to assume the lowest value.

One way to characterize the distribution of the relaxation times in the system is the average characteristic time defined as

$$\tau(p) = \frac{\int_0^t t' g(t') dt'}{\int_0^t g(t') dt'} \tag{3}$$

which is a typical macroscopic relaxation time and can be directly linked to the viscosity coefficient following the Maxwell relation $\tau \sim G/\eta$.

Numerically in the Eq. (3) $t$ has been chosen by the condition $g(t') \leq 0.001$ for $t' \geq t$. This characteristic time grows with $p$ and diverges at the percolation threshold according to the critical behaviour

$$\tau \propto (p_c - p)^{-k} \tag{4}$$

with an exponent $k \simeq 1.3 \pm 0.03$ (Fig. 6), which gives the critical exponent for the viscosity at the sol-gel transition.

The critical exponent obtained for the viscosity critical behaviour is then $k \sim 1.3$ and it well agrees with the value of $k$ previously given. This value is quite close to the one experimentally measured in[11,21] and to the value obtained by recent accurate measurements in PDMS.[7]

It does not agree with the *random superconducting network* exponent predicted by de Gennes, whereas it is quite close to the Rouse exponent discussed above. We have already

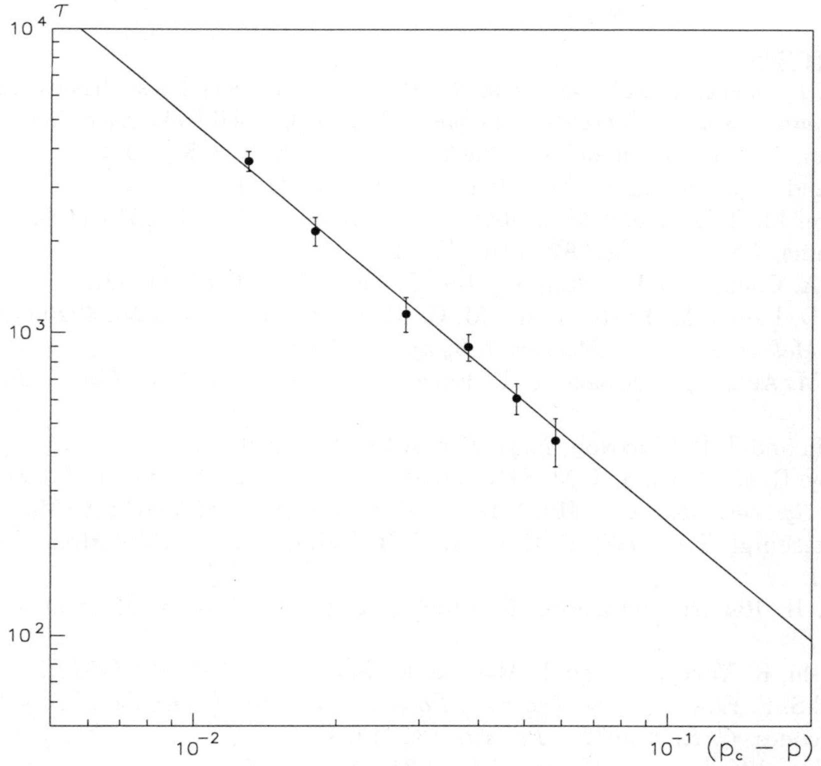


**Fig. 6** The characteristic integral time $\tau$ calculated according to Eq. (3) as a function of $(p_c - p)$. The data are well fitted by a power law with a critical exponent $k \sim 1.31 \pm 0.05$.

mentioned that a Rouse-like description of a polymer solution corresponds to a complete screening of the entanglement effects and the hydrodynamic interactions and is usually considered not realistic enough. Actually the entanglement effects could be not so important in the relaxation mechanism in the sol on the macroscopic relaxation time scale: due to the fractal structure of the gel phase the system is in fact quite fluid, probably there is no blocking entanglement yet and such temporary entanglements relaxe on a smaller time scale, not really affecting the macroscopic relaxation time.[25] Furthermore, the screening effect of the hydrodynamic interactions in a polymeric solution at high concentrations can be quite strong, drastically reducing the range of the interactions so that the Rouse model results to be in fact very satisfactory.[14] This could reasonably be the case of the sol at the gelation threshold too, and the deviation of the real critical exponent from the Rouse value would turn out to be actually very small.

## 5. CONCLUSIONS

The model we have introduced for the study of the dynamics at the sol-gel transition is essentially based on a dynamic extension of percolation with excluded volume interactions. With these ingredients it reproduces many relevant aspects of the transition. It is suited to study the viscoelastic properties critical behaviour, and the resulting critical exponent for the viscosity agrees with some experimental results, giving an interesting contribution to the discussion on the theoretical models. Our model also shows a complex dynamic behaviour, characterized by a non-exponential relaxation process close to the transition threshold with the presence of a stretched exponential decay, as it is actually observed in many gelling systems. It also suited to be extended to study other aspects of gelation phenomena, as weak gelation.

## REFERENCES


1. P. J. Flory, *Principles of Polymer Chemistry* (Cornell University Press, Ithaca, 1953).
2. P. G. De Gennes, *Scaling Concepts in Polymer Physics* (Cornell University Press, Ithaca, 1980).
3. J. E. Martin, J. P. Wilcoxon and J. Odinek, *Phys. Rev.* **A43**, 858 (1991).
4. S. Z. Ren and C. M. Sorensen, *Phys. Rev. Lett* **70**, 1727 (1991).
5. T. Norisuye, M. Takeda and M. Shibayama, *Macromolecules* **31**, 5316 (1998); F. Ikkay and M. Shibayama, *Phys. Rev. Lett* **82**, 4946 (1999).
6. M. Adam, A. Coniglio and D. Stauffer, *Adv. Polymer Sci.* **44**, 103 (1982).
7. M. Adam, D. Lairez, M. Karpasas and M. Gottlieb, *Macromolecules* **30**, 5920 (1997).
8. M. Adam, *Makromol. Chem. Macromol. Symp.* **45** (1991).
9. D. Lairez, M. Adam, E. Raspaud, J. R. Emery and D. Durand, *Prog. Colloid Polymer Sci.* **90** (1992).
10. J. E. Martin and J. P. Wilcoxon, *Phys. Rev.* **A39**, 252 (1989).
11. R. H. Colby, B. K. Coltrain, J. M. Salva and S. M. Melpolder, in *Fractal Aspects of Materials: Disordered Systems*, eds. A. J. Hurd, D. A. Weitz and B. B. Mandelbrot (Materials Research Society, Pittsburgh, PA, 1987); R. H. Colby, J. R. Gillmor and M. Rubinstein, *Phys. Rev.* **E48**, 3712 (1993).
12. T. Nicolai, H. Randrianantoandro, F. Prochazka and D. Durand, *Macromolecules* **30**, 5897 (1997).
13. M. Takahashi, K. Yokoyama and T. Masuda *J. Chem. Phys.* **101**, 798 (1994).
14. M. Doi and S. F. Edwards, *The Theory of Polymer Dynamics* (Clarendon Press, Oxford, 1986).
15. P. G. de Gennes, *C. R. Acad. Sc. Paris B*, 131 (1978).
16. I. Carmesin and K. Kremer, *Macromolecules* **21**, 2819 (1988).
17. I. Carmesin and K. Kremer, *J. Phys. (Paris)* **51**, 950 (1990).
18. E. Del Gado, L. de Arcangelis and A. Coniglio, *J. Phys.* **A31**, 1901 (1998).

19. E. Del Gado, L. de Arcangelis and A. Coniglio, *Europhys. Lett.* **46**, 288 (1999).
20. E. Del Gado, L. de Arcangelis and A. Coniglio, *Eur. Phys. J.* **E2**, 352 (2000).
21. J. E. Martin, A. Douglas and J. P. Wilcoxon, *Phys. Rev. Lett.* **61**, 2620 (1988).
22. J. Ferry, *Viscoelastic Properties of Polymers* (Wiley, New York, 1980).
23. M. Daoud, *J. Phys.* **A21**, L973 (1988).
24. P. O. Mari, I. A. Campbell, A. Alegría and J. Colmenero, *Physica* **A21**, 257 (1998).
25. K. Broderix, H. Lowe, P. Muller and A. Zippelius, *cond-mat/9811388* (1998).

# SCALING AND FINITE-SIZE EFFECTS FOR THE CRITICAL BACKBONE

M. BARTHELEMY,* S. V. BULDYREV,* S. HAVLIN,† and H. E. STANLEY*
*Center for Polymer Studies, Boston University,
Boston, 02215 MA, USA
†Minerva Center and Department of Physics, Bar-Ilan University,
Ramat-Gan 52900, Israel

### Abstract

In a first part, we study the backbone connecting two given sites of a two-dimensional lattice separated by an arbitrary distance $r$ in a system of size $L$. We find a scaling form for the average backbone mass and we also propose a scaling form for the probability distribution $P(M_B)$ of backbone mass for a given $r$. For $r \approx L$, $P(M_B)$ is peaked around $L^{d_B}$, whereas for $r \ll L$, $P(M_B)$ decreases as a power law, $M_B^{-\tau_B}$, with $\tau_B \simeq 1.20 \pm 0.03$. The exponents $\psi$ and $\tau_B$ satisfy the relation $\psi = d_B(\tau_B - 1)$, and $\psi$ is the codimension of the backbone, $\psi = d - d_B$. In a second part, we study the multifractal spectrum of the current in the two-dimensional random resistor network at the percolation threshold. Our numerical results suggest that in the infinite system limit, the probability distribution behaves for small $i$ as $P(i) \sim 1/i$ where $i$ is the current. As a consequence, the moments of $i$ of order $q \leq q_c = 0$ diverge with system size, and all sets of bonds with current values below the most probable one have the fractal dimension of the backbone. Hence we hypothesize that the backbone can be described in terms of only (i) blobs of fractal dimension $d_B$ and (ii) high current carrying bonds of fractal dimension going from $d_{\rm red}$ to $d_B$, where $d_{\rm red}$ is the fractal dimension of the red bonds carrying the maximal current.

## 1. SCALING FOR THE CRITICAL BACKBONE

The percolation problem is a classical model of phase transitions, as well as a useful model for describing connectivity phenomena, and in particular for describing porous media.[1–3]

At the percolation threshold $p_c$, the mass of the largest cluster scales with the system size $L$ as $M \sim L^{d_f}$. The fractal dimension $d_f$ is related to the space dimension $d$ and to the order parameter and correlation length exponents $\beta$ and $\nu$ by $d_f = d - \beta/\nu$.[1-3] In two dimensions, $d_f = 91/48$ is known exactly.

An interesting subset of the percolation cluster is the backbone which is obtained by removing the non-current carrying bonds from the percolation cluster.[4] The structure of the backbone consists of blobs and links.[1,5-7] The backbone can in fact be further partitioned into subsets according to the magnitude of the electric current carried.[8,9] The backbone is relevant to transport properties[1-3] and fracture.[10] The fractal dimension $d_B$ of the backbone can be defined via its typical mass $M_B$, which scales with the system size $L$ as $M_B \sim L^{d_B}$. The backbone dimension is an independent exponent and its exact value is not known. A current numerical estimate[11] is $d_B = 1.6432 \pm 0.0008$.

The operational definition of the backbone has an interesting history.[1-3] Customarily, one defines the backbone using parallel bars, and looks for the percolation cluster (and the backbone) which connects the two sides of the system.[4] A different situation arises in oil field applications,[12] where one studies the backbone connecting two wells separated by an *arbitrary* distance $r$. This situation is important for transport properties, since in oil recovery one injects water at one point and recovers oil at another point.[12] From a fundamental point of view, it is important to understand how the percolation properties depend on different boundary conditions.

We study in this first part the backbone connecting two points separated by an arbitrary distance $r$ in a two-dimensional system of linear size $L$. One goal[13] is to understand the distribution of the backbone mass $M_B(r, L)$, and how its average value scales with $r$ and $L$ in the entire range $0 < r < L$.

We choose two sites $A$ and $B$ belonging to the infinite percolating cluster on a two-dimensional square lattice (the fraction of bonds is $p = p_c = 1/2$). $A$ and $B$ are separated by a distance $r$ and symetrically located between the boundaries.[14] Using the burning algorithm, we determine the backbone connecting these two points for values of $L$ ranging from 100 to 1000. For each value of $L$, we consider a sequence of values of $r$ with $2 \le r \le L-2$. In order to test the universality of the exponents, we perform our study on three lattices: square, honeycomb and triangular lattice. For simplicity, we restrict our discussion here to the square lattice, as we find similar results for the other two lattices.

We begin by studying the backbone mass probability distribution $P(M_B)$. We show that $P(M_B)$ obeys a simple scaling form in the entire range of $r/L$,

$$P(M_B) \sim \frac{1}{r^{d_B}} F\left(\frac{M_B}{r^{d_B}}\right), \tag{1}$$

where $F(x)$ is a scaling function, whose shape depends on the ratio $r/L$.

For $r \approx L$, it seems reasonable to assume that $P(M_B)$ will be peaked around its average value $\langle M_B \rangle \sim L^{d_B}$. The data collapse predicted by Eq. (1) is represented in Fig. 1. In this case, the scaling function $F$ is peaked at approximately $L^{d_B}$. However, the case $r \ll L$ is far less clear. In fact, we expect for $r \ll L$ that the backbone mass fluctuates greatly from one realization to another, since its minimum value can be $r$ and its maximum can be of order $L^{d_f}$. $P(M_B)$ has a lower cut-off of order $r$ (since the backbone must connect points $A$ and $B$) and a upper cut-off of order $L^{d_B}$. We find good data collapse (Fig. 2), which indicates that the scaling function $F$ is a power law in the range from $r^{d_B}$ to $L^{d_B}$, with exponent approximately $\tau_B \simeq 1.20 \pm 0.03$ (there is a cut-off at $M_B \sim L^{d_B}$ not shown here). The exponent $\tau_B$ is connected to the blob size distribution[5] since typically, the two sites

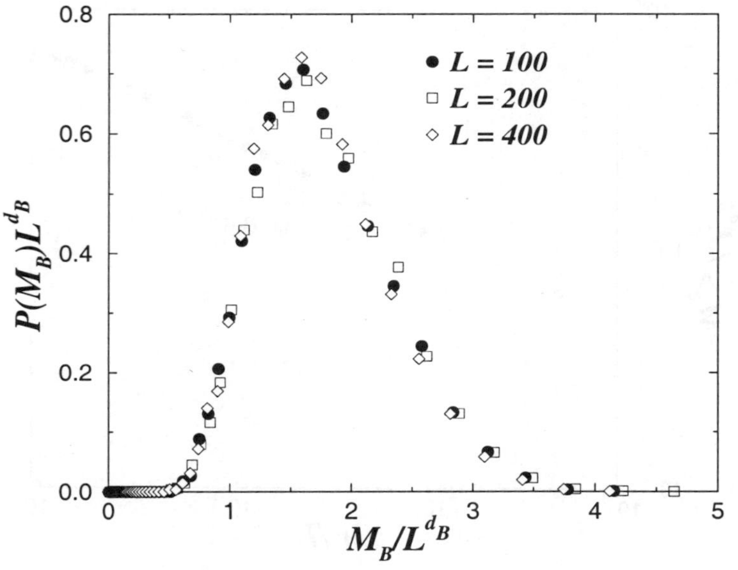


**Fig. 1** Rescaled backbone mass distribution in the case $r \ll L$.

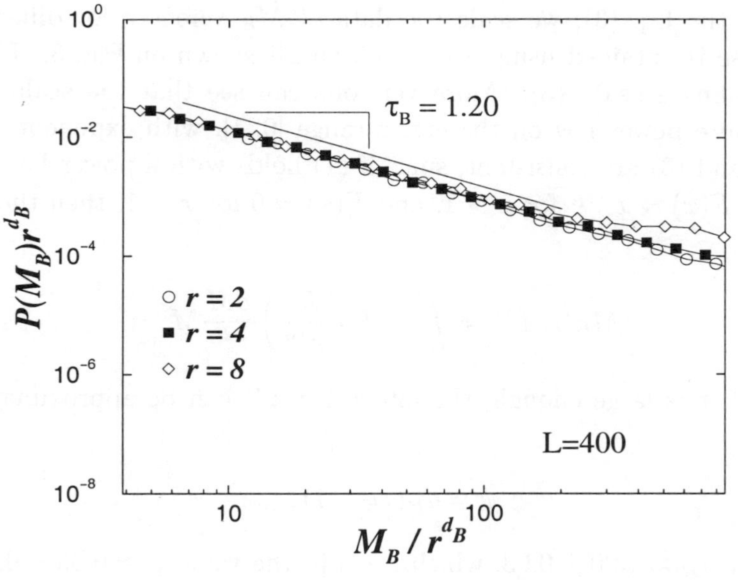


**Fig. 2** Rescaled backbone mass distribution in the case $r \simeq L$.

belong to the same blob, and the sampling of backbones is equivalent to sampling of the blobs. From,[5]

$$\frac{d}{d_B} = \tau_B \,. \tag{2}$$

This relation gives the estimate $\tau_B \simeq 1.22$ in good agreement with our numerical simulation.

We now study the average backbone mass $\langle M_B \rangle$. From dimensional considerations, the $r$ dependence can only be a function of $r/L$. We thus propose the following *Ansatz*:

$$\langle M_B(r,\, L)\rangle = L^{d_B} G\left(\frac{r}{L}\right). \tag{3}$$

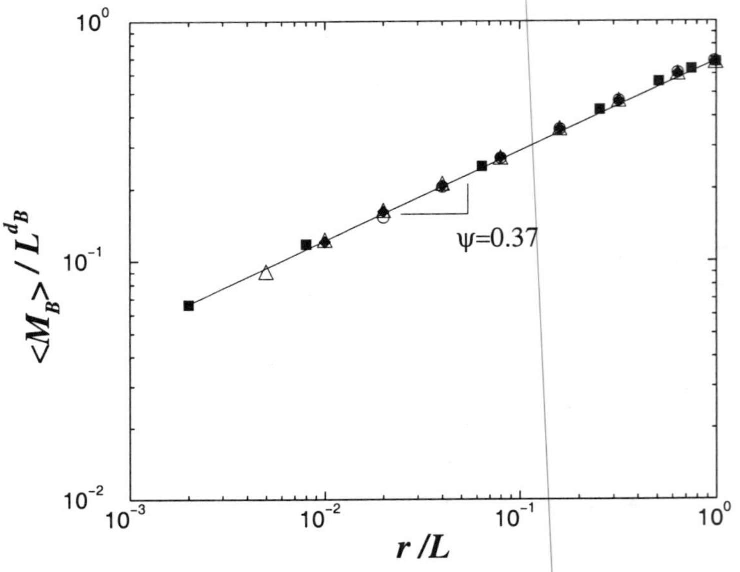


**Fig. 3** Rescaled average backbone mass versus versus $r/L$.

In order to test the Eq. (3), we scale the data of $M_B$ versus $r$ for different values of $L$. The data collapse is obtained using $d_B = 1.65$ and is shown on Fig. 3. This (log-log) plot supports the scaling Ansatz (3). Moreover, one can see that the scaling function $G$ is, surprisingly, a pure power law on the entire range [0, 1], with exponent $\psi = 0.37 \pm 0.02$. The results (1) and (3) are consistent, since if (1) holds with a power law behavior for the scaling function $F(x) \sim x^{-\tau_B}$ for $x > 1$, and $F(x) = 0$ for $x < 1$, then the average mass is given by

$$\langle M_B(r,\, L)\rangle = \int_r^{L^{d_B}} F\left(\frac{M}{r^{d_B}}\right) \frac{dM}{r^{d_B}} M\,. \tag{4}$$

Assuming that $L/r$ is large enough, the integral in (4) can be approximated as $L^{d_B-\psi} r^{\psi}$, where

$$\psi = d_B(\tau_B - 1)\,. \tag{5}$$

In our simulation $\tau_B \approx 1.20 \pm 0.03$, which leads to the value $\psi \approx 0.33 \pm 0.05$ in reasonable agreement with the value measured directly on the average mass.

Moreover, using Eq. (2) together with Eq. (5), we obtain

$$\psi = d - d_B \tag{6}$$

which means that $\psi$ is the codimension of the fractal backbone.

To summarize, we find that for any value of $r/L$, the scaling form, Eq. (1), for the probability distribution is valid. The shape of the scaling function $F$ depends on $r/L$, being a peaked distribution for $r \approx L$, and a power law for $r \ll L$. The average backbone mass varies with $r$ and $L$ according to Eq. (4). For fixed system size, it varies as $\langle M_B\rangle \simeq r^{\psi}$ (for $0 < r < L$). The value of $\psi$ is small ($\psi \approx 0.37$) indicating that the backbone mass does not change drastically as $r$ changes. On the other hand, the exponent governing the variation of $\langle M_B\rangle$ with $L$ for fixed $r$ is expected to be larger, with $\langle M_B\rangle \sim r^{d_B-\psi}$. This exponent $d_B - \psi$ is not equal to the fractal dimension $d_B$ of the backbone, but is smaller by an amount equal to $\psi$.

## 2. MULTIFRACTAL SPECTRUM AND FINITE-SIZE EFFECTS

The transport properties of the percolating cluster have been the subject of numerous studies during the last twenty years.[2,15] A particularly interesting system is the random resistor network (RRN), where the bonds have a random conductance. The random resistor network serves as a paradigm for many transport properties in heterogeneous systems as well as being a simplified model for fracture.[10]

The first studies of the RRN were devoted to effective properties of the network (conductivity, permittivity, etc.),[16,17] but for many practical applications — such as fracture, and dielectric breakdown[10] — the central quantity is the probability distribution $P(i)$ of currents $i$. For instance, in the random fuse network, it is the maximum current corresponding the hottest or "red" bonds which will determine the macroscopic failure of the system.[10]

The probability distribution $P(i)$ has many interesting features, one of which is multifractality:[9,18–20] in order to describe $P(i)$, an infinite set of exponents is needed. This idea of multifractality was initially proposed to treat turbulence[21] and later applied successfully in many different fields, ranging from model systems such as DLA[22] to physiological data such as heartbeat.[23]

It was first believed[9,20] that the low current part of $P(i)$ and of the multifractal spectrum follow a log-normal law as it is the case on hierarchical lattices. It is now clear,[25] that for small currents, the current probability distribution follows a power law $P(i) \sim i^{b-1}$ where $b \geq 0$. For large currents, the distribution quickly converges to an infinite system limit with no dependence on the system size $L$. For small currents, governed by very long paths, the distribution converges more slowly. It was suggested[24,25] that the exponent $b$ of the low-current part has a $1/\log L$ dependence. The asymptotic value $b_\infty$ of the exponent $b$ for the infinite system is of crucial importance. If $b_\infty$ is finite and positive, then a subset of bonds with low current has a fractal dimension depending on its value. On the other hand, if $b_\infty$ is zero, then the low current part of the multifractal spectrum is flat, and the subset of bonds with any low value of current has the same fractal dimension as the entire backbone. It is thus important to understand if the apparent subset structure with different fractal dimensions arises primarily from finite-size effects.

Previous estimates of $b_\infty$ include $b_\infty = 0$[24] and $b_\infty \gtrsim 0.25$,[25] but the maximum value of $L$ used was 128.[24] In this part, we study a sequence of sizes from $L = 50$ to $L = 1000$, and hypothesize[26] that $b_\infty = 0$.

We first recall the basis of multifractality applied to the percolating two-dimensional resistor network of linear size $L$. Let $n(i, L)$ be the number of bonds carrying current $i$. For large $L$, $n(i, L)$ scales as[9,20]

$$n(i, L) \sim L^{f(\alpha, L)} \tag{7}$$

where $\alpha \equiv -\log i/\log L$. The multifractal spectrum $f(\alpha, L) \equiv \log n/\log L$ can thus be interpreted as the fractal dimension of the subset of bonds carrying the current $i$. The $q$-th moment of the current is defined as $M_q \equiv \langle \sum i^q \rangle$, where the sum is over all bonds carrying a non-zero current and $\langle \cdots \rangle$ denotes an average over different disorder configurations. These moments exists for $q > q_c$, and it can be easily shown[25] that the "threshold" is $q_c = -b$. The asymptotic slope thus give the asymptotic value of the threshold $q_c$.

For the fixed current ensemble, one finds $M_q \sim L^{\tau_q}$ for large $L$ and for $q > q_c$ and where $\tau_q$ is an exponent that depends on $q$.[9,20] In particular, $\tau_0 = d_B$, $\tau_2 = t/\nu$, and $\tau_\infty = 1/\nu$[28] where $d_B$ is the fractal dimension of the backbone, $t$ the conductivity exponent, and $\nu$ is the correlation length exponent. If the behavior is monofractal, then $\tau_q$ is a linear function

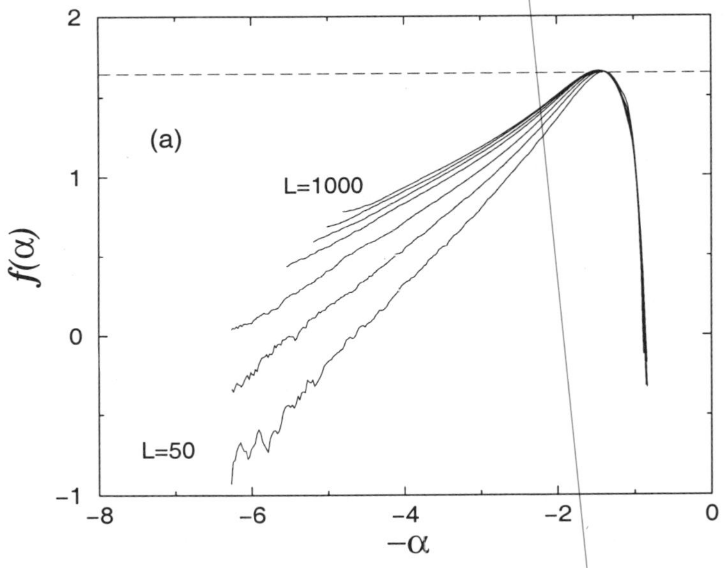


**Fig. 4** Multifractal spectrum for $r \simeq L$ going from 50 to 1000.

of $q$ with the intercept equal to the fractal dimension of the monofractal, while in the multifractal case, the exponents are not described by a simple linear function of $q$. In the $L \to \infty$ limit, knowing $f(\alpha)$ is equivalent to knowing the infinite set of exponents $\tau_q$, as $f(\alpha) = \tau_q - q d\tau/dq$ is the Legendre transform of $\tau_q$.[10]

The low current part of $f(\alpha, L)$ was found numerically to be a power law of slope $b = b(L)$. It was suggested that

$$b(L) = b_\infty + \frac{a}{\log L} + \dots \tag{8}$$

which is a strong finite-size effect since $\log L$ grows very slowly, and two possibilities for $b_\infty$ were proposed, $b_\infty = 0$[24] or $b_\infty = 1/4$.[25]

We consider the two-dimensional random resistor network at criticality, i.e. the fraction of conducting bonds $p$ is equal to its critical value $p = p_c = 1/2$. We first apply a voltage difference between two parallel bars. We compute $f(\alpha, L)$, for a fixed voltage difference, for $L = 50, \dots, 1000$, and average over $10^4$ configurations for each $L$. We show our results in Fig. 4. The slope $b$ is clearly decreasing with $L$, confirming the strong finite size effects already observed.[24,25] Next, we consider a second type of configuration, which we call the "two injection points" case, in contrast with the usual "parallel bars" case. We impose a voltage difference between two points $A$ and $B$ separated by a distance $r$, and we look for the backbone connecting these two points. This situation was studied in,[13,27] but here we keep only the backbones of linear size $L$. In this way, we have large backbones connecting the two points $A$ and $B$, and for $r \ll L$ we expect to have a large number of small currents on bonds belonging to long loops. The multifractal spectrum is then defined in the same way as for the parallel bars and we calculate the slope of the small current part of the multifractal spectrum for different values of $L$. The variation of the slope for the two-injection-points case is shown in Fig. 5. We observe that there is a large amount of small currents, and that the asymptotic limit is reached faster in this case. We expect that the low current distribution will be asymptotically the same as in the parallel bar case, so the consistency between the two configurations supports our results. For large currents there are some distinct differences in the multifractal spectrum.[29]

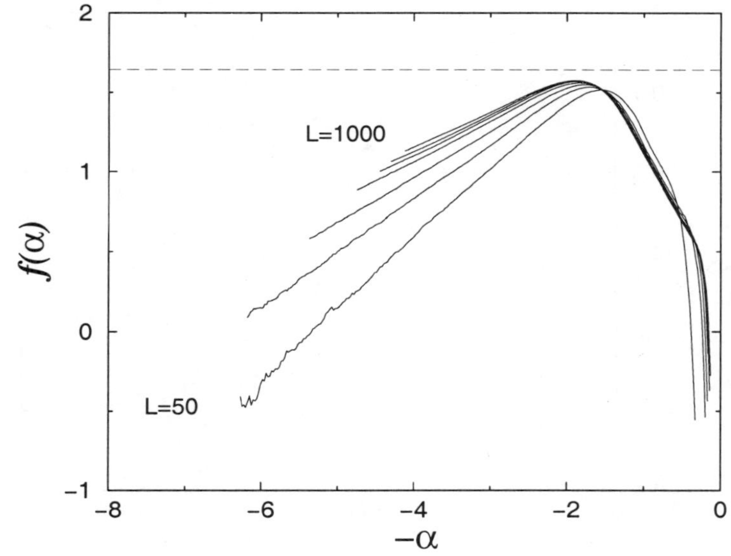


**Fig. 5** Multifractal spectrum for $r = 4 \ll L$ and $L$ is going from 50 to 1000.

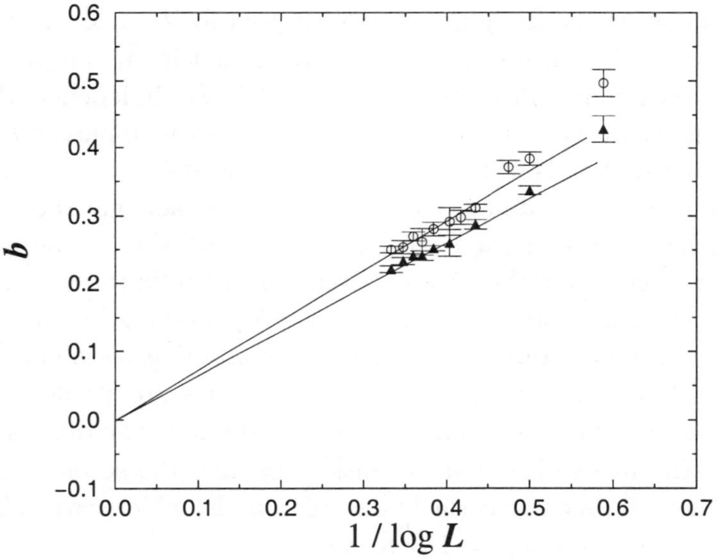


**Fig. 6** Slope $b$ versus $1/\log L$. The circles correspond to the parallel bars and the triangles to the 'two-injection' case.

Figure 6 shows the slope $b$ versus $1/\log L$ according to Eq. (8) for both multifractal spectra. The extrapolation to $L = \infty$ is consistent with $b_\infty = 0$ in both cases. This result is consistent with the behavior of the successive intercepts (Fig. 7). Another functional form of $b$ versus $L$ could lead to another value of $b_\infty$. If we replace the abscissa of Figs. 2(a) and (b) by $1/(\log L)^\alpha$, then we find that the extrapolated value for $b_\infty$ depends on $\alpha$, ranging from $b_\infty \simeq 0.10$ for $\alpha = 2$ to $b_\infty < 0$ (which is impossible) for $\alpha = 0.5$. It is numerically difficult to distinguish between a $1/\log L$ and a $1/(\log L)^2$ behavior, but the $1/\log L$ is the most commonly used.[24,25] If we accept this functional form, then our results are consistent with $b_\infty = 0$. Finally, we note that the sequence of maximum values of $f(\alpha, L)$ for the two injection points case plausibly extrapolates in the variable $1/\log L$ as $L \to \infty$ to a value of $d_B$ close to the known value 1.64.

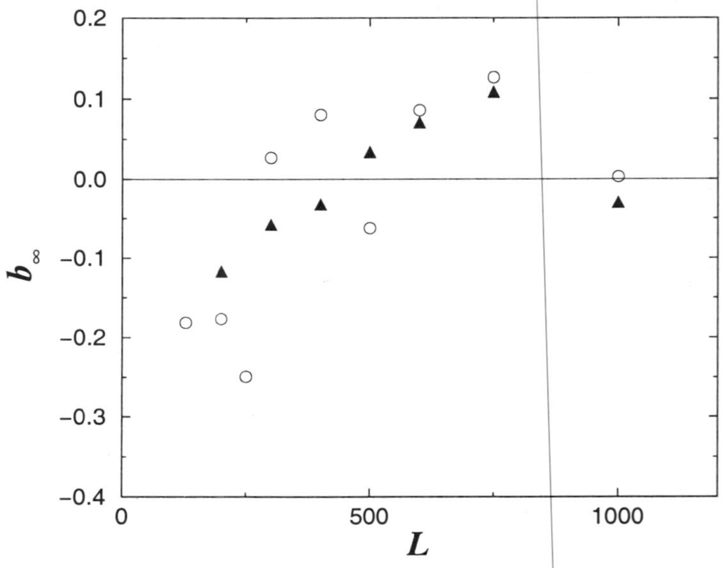


**Fig. 7** Successive intercepts computed from the datas of Fig. 6.

Thus our results suggest the intriguing possibility that for $L \to \infty$, the small current part of $f(\alpha, L)$ is a horizontal line at the value $d_B$, implying that in an infinite system the fractal dimension of the subset contributing to small current is $d_B$, independently of the value of $i$. In this sense, the small current probability distribution is apparently not multifractal. The "perfectly balanced" bonds which carry zero current have a fractal dimension equal to $d_B$.[24] Since these bonds contribute to $f(\alpha)$ for $\alpha \to \infty$, the fact that their fractal dimension is $d_B$ supports our hypothesis that $b_\infty = 0$. A related conclusion is that $q_c = 0$ or, in other words, the negative moments of the current diverge in the infinite-size limit. In particular, it shows that the first-passage time for a tracer particle travelling in a flow field in a porous medium modelled by a percolation cluster diverges in an infinite system.

For large values of the current, the multifractal features are stable against $L$ increase. This suggests that in the infinite-size limit, there are essentially two different type of subsets; the first comprises the blobs of fractal dimension $d_B$, and the second set comprises links carrying larger values of the current (including red bonds, which carry all the current), of fractal dimension ranging from $d_{\text{red}} = 1/\nu$ to $d_B$.

## ACKNOWLEDGMENTS

We thank L. A. N. Amaral for valuable help and A. Chessa, A. Coniglio, N. V. Dokholyan, P. Gopikrishnan, P. R. King, G. Paul, A. Scala, and F. W. Starr for useful discussions. We thank especially J. S. Andrade for providing the software for finding currents, and DGA and NSF for financial support.

## REFERENCES

1. A. Bunde and S. Havlin (eds.), *Fractals and Disordered Systems*, 2nd ed. (Springer, Berlin, 1996), and refs. therein.
2. D. Stauffer and A. Aharony, *Introduction to Percolation Theory* (Taylor and Francis, London, 1992).
3. M. Sahimi, *Applications of Percolation Theory* (Taylor and Francis, London, 1992).

4. S. Kirkpatrick, *AIP Conf. Proc.* **40**, 99 (1978); G. Shlifer, W. Klein, P. J. Reynolds and H. E. Stanley, *J. Phys.* **A12**, L169 (1979).
5. H. J. Herrmann and H. E. Stanley, *Phys. Rev. Lett.* **53**, 1121 (1984).
6. H. E. Stanley, *J. Phys.* **A10**, L211 (1977).
7. A. Coniglio, *Phys. Rev. Lett.* **46**, 250 (1981).
8. R. Rammal, C. Tannous, P. Breton and A. M. S. Tremblay, *Phys. Rev. Lett* **54**, 1718 (1985).
9. L. de Arcangelis, A. Coniglio and S. Redner, *Phys. Rev.* **B36**, 5631 (1987).
10. H. Herrmann and S. Roux (eds.), *Statistical Models for the Fracture of Disordered Media* (North-Holland, 1990).
11. P. Grassberger, *Physica* **A262**, 251 (1999).
12. Y. Lee, J. S. Andrade Jr, S. V. Buldyrev, S. Havlin, P. R. King, G. Paul and H. E. Stanley, *preprint Cond-Mat 9903066.*
13. M. Barthélémy, S. V. Buldyrev, S. Havlin and H. E. Stanley, *Phys. Rev.* **E60**, R1123 (1999).
14. H. J. Herrmann, D. Hong and H. E. Stanley, *J. Phys.* **A17**, L261 (1984).
15. A. Coniglio and M. Zannetti, *Physica* **D38**, 37 (1989).
16. R. Landauer, in *Proceedings of the 1st Conference on Electrical and Optical Properties of Inhomogeneous Media* (Ohio State University, 1977), *AIP Conf. Proc. No. 40*, eds. J. C. Garland and D. B. Tanner (AIP, New York, 1978).
17. D. J. Bergman and D. Stroud, *Solid State Physics* **46**, 147 (1992).
18. R. Rammal, C. Tannous and A.-M. S. Tremblay, *Phys. Rev.* **A31**, 2662 (1985).
19. R. Rammal, C. Tannous, P. Breton and A.-M. S. Tremblay, *Phys. Rev. Lett.* **54**, 1718 (1985).
20. L. de Arcangelis, S. Redner and A. Coniglio, *Phys. Rev.* **B34**, 4656 (1986).
21. B. B. Mandelbrot, *J. Fluid Mech.* **62**, 331 (1974).
22. See, e.g., J. Lee et al., *Phys. Rev.* **A39**, 6545 (1989).
23. P. Ch. Ivanov et al., *Nature* **399**, 461 (1999).
24. G. Batrouni, A. Hansen and S. Roux, *Phys. Rev.* **A38**, 3820 (1988).
25. A. Aharony, R. Blumenfeld and A. B. Harris, *Phys. Rev.* **B47**, 5756 (1993).
26. M. Barthélémy, S. V. Buldyrev, S. Havlin and H. E. Stanley, *Phys. Rev.* **E61**, R3283 (2000).
27. Y. Lee et al., *Phys. Rev.* **E60**, 3425 (1999).
28. A. Coniglio, *J. Phys.* **A15**, 3829 (1982).
29. In the "two injection points" configuration, the large currents are controlled by the small loops between $A$ and $B$. The fluctuation of the length of the shortest path between $A$ and $B$ gives thus rise to the region of power law behavior for large currents we observe in Fig. 5.
30. G. Batrouni, A. Hansen and B. Larson, *Phys. Rev.* **B53**, 2292 (1998).
31. H. E. Daniels, *Ann. Math. Stat.* **25**, 631 (1954).
32. B. B. Mandelbrot and C. J. G. Evertsz, *Physica* **A168**, 95 (1990).
33. B. B. Mandelbrot, C. J. G. Evertsz and Y. Hayakawa, *Phys. Rev.* **B42**, 4528 (1990).

# ROUGHENING TRANSITION IN BRANCHING POLYMERS

LIACIR S. LUCENA
*International Center for Complex Systems and*
*Departamento de Física, Universidade Federal do Rio Grande do Norte*
*Natal, RN 59078-970, Brazil*

LUCIANO R. DA SILVA
*Center for Polymer Studies, Boston University, USA*
*Departamento de Física, Universidade Federal do Rio Grande do Norte*
*Natal, RN 59078-970, Brazil*

STÉPHANE ROUX
*Surface du Verre et Interfaces*
*Unité Mixte de Recherche CNRS/Saint-Gobain*
*39 Quai Lucien Lefranc, 93303 Aubervilliers cedex, France*

**Abstract**

A growth model, introduced to model the development of branched polymers in an heterogeneous environment, gives rise to clusters whose boundary is either faceted or rough. We study the transition between these two morphologies as a function of the parameters of the model, impurity concentration and branching ratio. The phase diagram is first obtained by direct numerical simulations, using an original algorithm, based on a self-regulated search of a critical point. Then an analytic computation of the phase boundary is proposed based on a simple approximation. The obtained phase boundary is in good agreement with the numerical results. The nature of the transition is discussed.

## 1. INTRODUCTION

Statistical growth models have been much studied in the past ten years, and in particular the statistical properties of interfaces have been clarified in a number of models, now used as paradigma. Both their steady state properties and their temporal evolution have been characterized in great details. Reviews on this field can be found for instance in Refs. 1, 2, 3, 4. In this context a few models shows a transition from faceted growth to rough interfaces.

A model for polymer growth and branching has been introduced some time ago by Lucena *et al*[6] as a generalization of a kinetic growth model for linear polymers proposed by Coniglio and collaborators.[7] Two parameters are present in the model, one describing the branching probability $b$ of the polymer growing tip, and the second is an impurity concentration $c$ which may prevent the growth. In the parameter space $(b, c)$, a phase diagram has been established for the existence of an infinite branched polymer. A further study by Bunde *et al*[8] suggested that this transition might be in the universality class of standard percolation.

Inside the latter phase, it has been observed that the morphology of the growing polymer on a regular lattice could exhibit a faceted shape, i.e. a domain where the growth is at the maximum possible rate of the model (one lattice unit per unit time step).

In the present article we study this facetting regime, and its phase boundary in the $(b, c)$ parameter space. We first report on a direct numerical study of this problem, using an original algorithm. We then propose an approximate model which provides a good approximation to the phase boundary, and which shed some lights on the nature of the transition.

## 2. DEFINITION OF THE MODEL

We consider the growth of the polymer on a regular lattice. In the present study, we restrict ourselves to a square lattice in two dimensions.

The polymer itself grows on a lattice starting from a set of initial sites, the "seed". In the original version of this model the seed was chosen to be a point. In this case, for a square lattice, the cluster adopts (for some parameter values to be specified below) a square shape with rounded corners. The edges of the square, what we will call the facets, are along the diagonals of the lattice. In order to model these facets, it is thus more convenient to choose as a seed a line paralel to one diagonal of the square lattice.

First, the sites of the lattice are randomly occupied by "impurities" with a concentration $c$. The role of the latters is simply to prevent the polymer from growing there. The branched polymer is modeled as a tree like structure, whose endpoints called "active" sites in the following, may either grow by one unit, branch and thus produce two new active sites, or die if no more space is available for it to grow. The initial seed is a set of initially active sites. All active sites are explored sequentially, always in the same order (from top to bottom, and from left to right, as a typewriter). At each active site, one considers the free nearest neighbor sites (neither occupied by the polymer nor by an impurity). Among the latters either two (if branching is attempted with probability $b$) or one (if no branching) sites become active. If only one site is free, then branching does not take place. If no site is available then the active site dies. In all cases, once a site has grown, it is no longer active. This defines the model.

Depending on the values of the two parameters of the model, $b$ and $c$, either the polymer can grow to infinity or die after a finite time. The phase diagram of this problem has been obtained from numerical simulations. Inside the domain of infinite growth there exists a region where growth takes place at the highest permitted velocity of the model, i.e. one lattice spacing per unit time step. If a finite proportion of sites have this property, then the active boundary of the polymer exhibits a facet parallel to the seed line. A preliminary qualitative description of this phenomenon was given by us in another article.[9] The purpose of the present study is to examine the conditions of existence of this faceted phase.

## 3. NUMERICAL STUDY

We first report on a numerical study of this phase boundary obtained using an original technique, i.e. an algorithm based on the concept of self-organized criticality,[10] where a feed-back loop is introduced to regulate one control parameter so that the system approaches and remains at the critical point.

### 3.1 Algorithm

In order to study the phase boundary various strategies could be chosen. We opt here for an original method which has first been proposed in connection with the infinite growth of this branched polymer problem[11] and further extended to percolation.[12] The idea of this method is not to impose fixed values of the control parameters as usual, but rather to adjust the latter in such a way that one remains in the immediate vicinity of the critical regime.

We worked here at constant branching probability $b$. In order to maintain the system at criticality, we constantly adjust $c$. This is done by trying to maintain the proportion of active sites, $a$, close to a chosen density $a_0$. For this purpose we introduce a simple proportional feed-back control

$$c(t+1) = c(t) + \kappa \frac{(a(t) - a_0)}{a_0} . \tag{1}$$

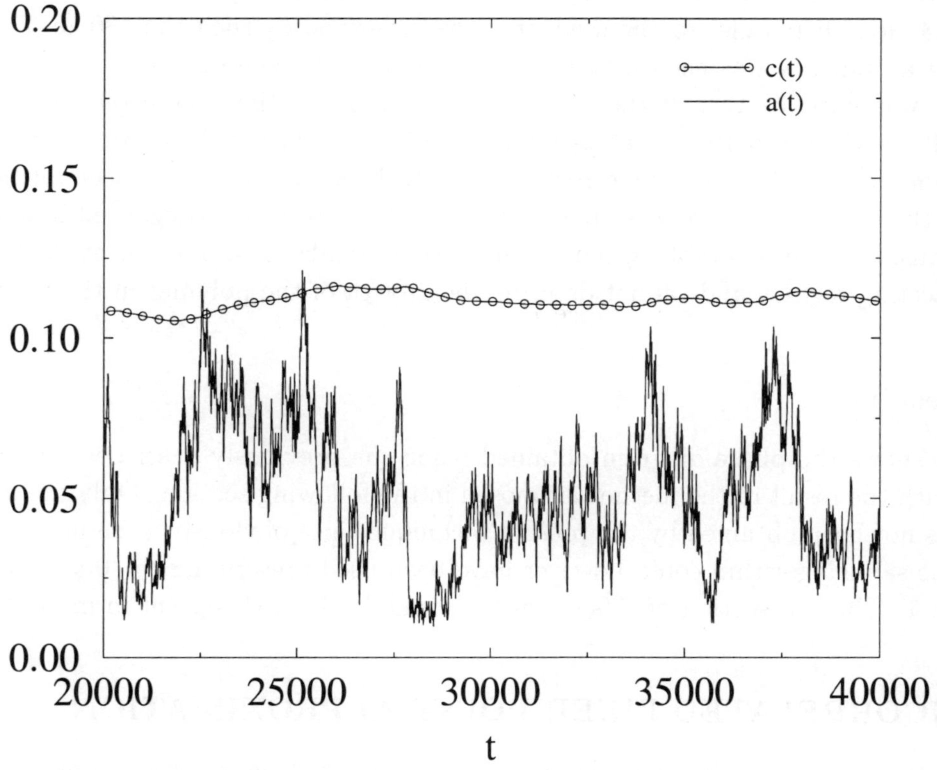


**Fig. 1** Example of the time evolution of $a(t)$ and $c(t)$ for $b = 0.5$.

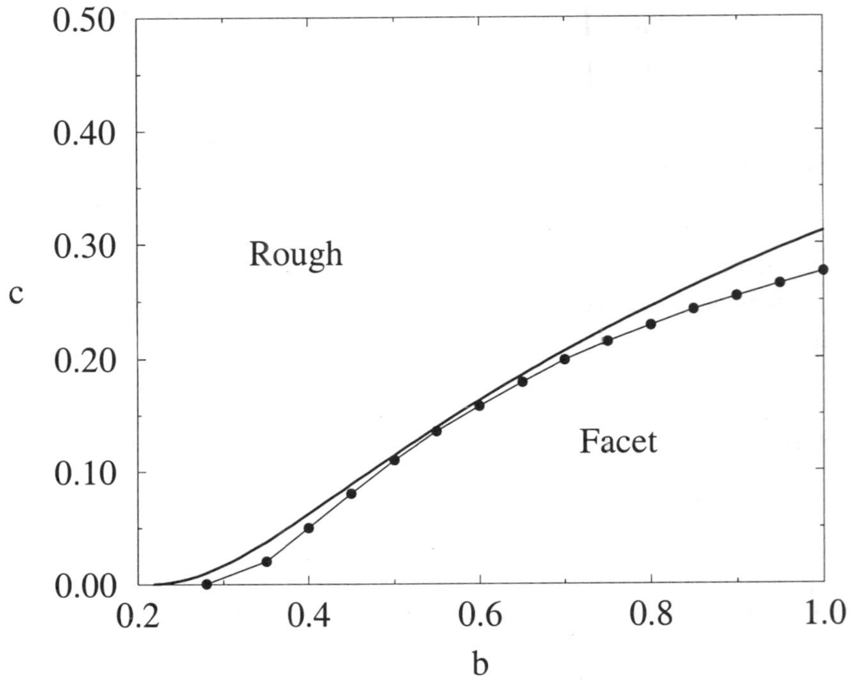


**Fig. 2** Phase diagram obtained from direct numerical simulation using the self-organization algorithm (symbol •). The fixed point method provides an approximation for the phase boundary shown as a bold dotted line.

Thus if the number of sites is larger than required, the proportion of impurities is increased, and vice-versa. The parameter $\kappa$ sets the characteristic time of the regulation. The objective value $a_0$ is chosen to be small enough for the system to be at criticality, but large enough so that the probability of extinction is small. In practice for systems of size $L = 1024$ (length parallel to the seed line), we prescribe $a_0$ such that 30 to 50 sites are active on average. $\kappa$ was chosen to be $5\ 10^{-5}$ or $5\ 10^{-6}$. The distance along the growth direction was fixed to $10^5$. Figure 1 shows an example of the time evolution of $a$ and $c$ for $b = 0.5$ with $\kappa = 5\ 10^{-6}$. Let us finally note that from the definition of the model it is sufficient to consider only three rows of sites to have access to the facet regime. Thus the growth stages which may take place below the facet layer is disregarded in our present simulations. This considerably speeds up numerical simulations, but it is evidently limited to the faceting regime, and cannot describe the growth of the polymer in the rough phase.

### 3.2 Results

Figure 2 shows the phase diagram obtained using the previously described algorithm, together with the result of the method proposed in the following section. Only the final point $c = 0$ has not been obtained by the present technique since obviously, $c$ cannot be adjusted here. The same algorithm could however have been used prescribing $c = 0$ and adjusting $b$ as in Eq. 1. The last section of this paper will come back on this point in more details.

## 4. UNCORRELATED FIXED POINT APPROXIMATION

In order to account for the phase boundary of the faceted phase, we propose now an approximate method which gives a reasonable approximation for the entire boundary.

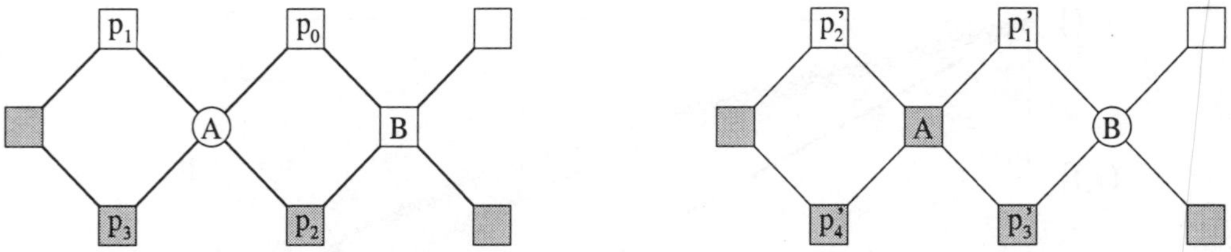


**Fig. 3** Geometry of the lattice before (left) and after (right) growing site A. The grey sites have grown. From site A four sites labeled $p_i$ can be invaded. $p_i$ denotes the probabilities that these sites are occupied by the polymer (with $p_0 = 0$). After the growth, the new probabilities are indicated with a prime. The method presented in this paragraph consists in identifying the fixed point $p'_i = p_i$.

Let us consider an elementary growth process in the geometry shown in Fig. 3. The active site which is growing is labeled $A$. The four nearest neighbor can be occupied by an impurity with a probability $c$. They can also be active with a probability $p_i$ as shown on the figure. Assuming that we sweep through the sites from left to right, $p_0$ is identically zero prior to the growth step. After the growth these probability have changed to values $p'_i$. Then the next site on the right, $B$, is considered. We can see on the figure that the corresponding probabilities can be identified in the steady state, $p'_1 = p_1$, and $p'_3 = p_3$. For $p_2$, we note that after one row has been added, $p_2$ is the probability that the growing site is active. Moreover, the same probability holds for yet another row, and thus $p'_2 = p_2$. The spirit of the method is thus to look for steady state values $p'_i = p_i$.

This is done by considering all possible growth configurations of occupied and/or active sites with their corresponding probabilities. At this stage we make the approximation that the activities of the sites can be treated as independent, an approximation which is rather severe. A cumbersome but straightforward computation leads to the following estimates:

$$\left\{\begin{aligned} p'_1 = {} & \frac{p_2}{12}(1-c)\{12r_1r_2r_3 + (1+b)[6(s_1r_2r_3 + r_1s_2r_3 + r_1r_2s_3) \\ & + 4(s_1s_2r_3 + s_1r_2s_3 + r_1s_2s_3) + 3s_1s_2s_3]\} \\ p'_2 = {} & p_1 + \frac{p_2}{12}[6s_1r_2r_3(1+b+c-bc) + s_1s_2s_3(3+3b+c+bc) \\ & + 2(s_1s_2r_3 + s_1r_2s_3)(2+c+2b-2bc)] \\ p'_3 = {} & \frac{p_2}{12}[12 + 6r_1s_2r_3(1+b+c-bc) + s_1s_2s_3(3+3b+c+bc) \\ & + 2(s_1s_2r_3 + r_1s_2s_3)(2+2b+c+bc)] \end{aligned}\right. \tag{2}$$

where we have introduced the notations $r_i = p_i + c$ for the probability that site $i$ is occupied, and $s_i = 1 - r_i$ the probability that the site is free. We have also assumed in the above computation that the facet sites are updated first, so that information flows from the surface towards the inside. In the rough domain this may not be true, and thus the present method is adequate for studying the facets, but not the growth of the polymer outside the faceted regime.

There are at most two stable fixed points: one is $p_i = 0$ (always stable), (if no site is active, then there is no way to regenerate a facet). A second stable fixed point may exist depending on the values of the parameters $b$ and $c$. This second stable fixed point corresponds to a situation where there is a non-zero probability for the growing sites to lie on a facet. Thus the point where the basin of attraction of this point vanishes is considered to be the critical point. This is the way we extracted the phase boundary $c_c(b)$ plotted on Fig. 2.

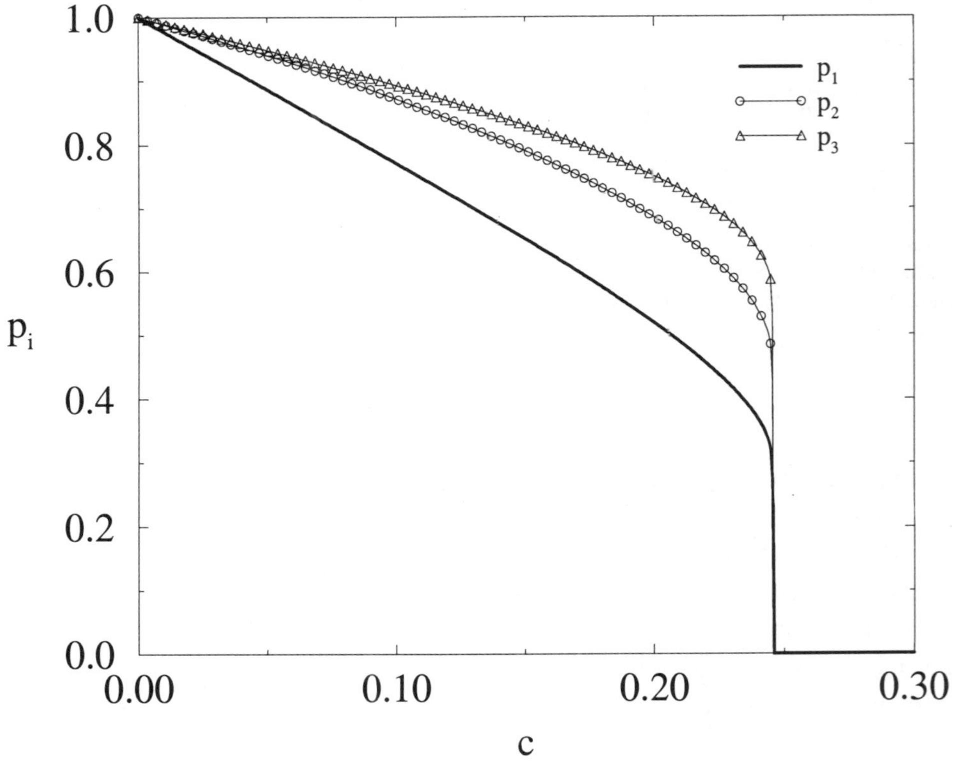


**Fig. 4** Values of the probabilities $p_i$ *versus* $c$ at the fixed point for $b = 0.8$. Note that the probabilities $p_i$ display a discontinuity at $c_c(b)$, suggesting a first-order phase transition.

Figure 4 shows the values of $p_i$ as a function of $c$ for $b = 0.8$ at the non-trivial fixed point. Above a critical value of $c$, this fixed point disappears and the system converges towards the trivial stable fixed point $p_i = 0$, i.e. the facts disappear. The most salient feature of this graph is that the $p_i$ undergo a jump at the critical value of $c_c$. This may suggest that the transition is first order. We will discuss this point in the following section.

It is also to be noticed that as $b$ decreases, the transition takes place for smaller and smaller values of $c$, and simultaneously, the probabilities $p_i$ becomes confined to $p = 1$. Indeed for $c = 0$, $p_i = 1$ is a fixed point, which is stable for $b$ larger than the particular value $b^*$ such that $c_c(b^*) = 0$.

## 5. NATURE OF THE TRANSITION

Albeit the phase boundary of the faceted domain is reasonably well reproduced by the semi-analytic treatment proposed above, the nature of the transition seems to be quite different. Indeed, the direct numerical simulation rather suggests a second order phase transition. To illustrate this point, we show on Fig. 5 the space-time diagram of the facets.

One can notice on this figure that as $b$ decreases, the facets have a tendency to develop preferentially to the right hand side, i.e. along the direction of updating. This is in agreement with the previous approach which indeed revealed such a bias through the significant difference between $p_1$ and $p_2$.

More importantly, we also note that the way the transition takes place shed some lights on the previous puzzle concerning the first order phase transition. Indeed it seems that the two stable phases predicted by the fixed point approximation do exist. However, the way the facets disappear as $c$ is increased is not along the scenario of the above model,

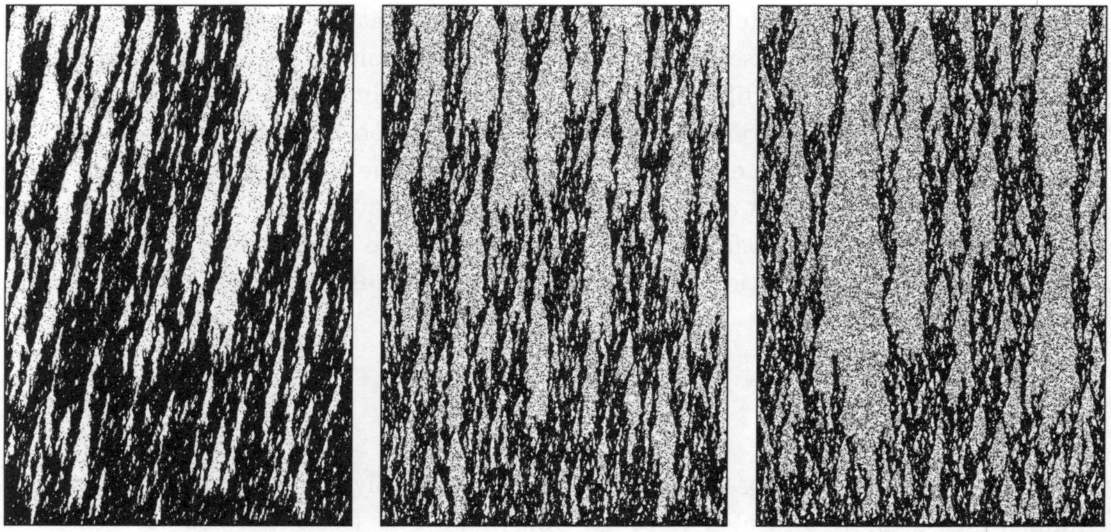

**Fig. 5** Space (horizontal axis) *versus* time (vertical axis) diagrams of the facets, for three different values of $b$, (0.4, 0.7 and 1.0 from left to right respectively), and the corresponding critical value of $c = c_c(b)$. The black points indicate the sites occupied by the polymer, and the grey dots are the impurities.

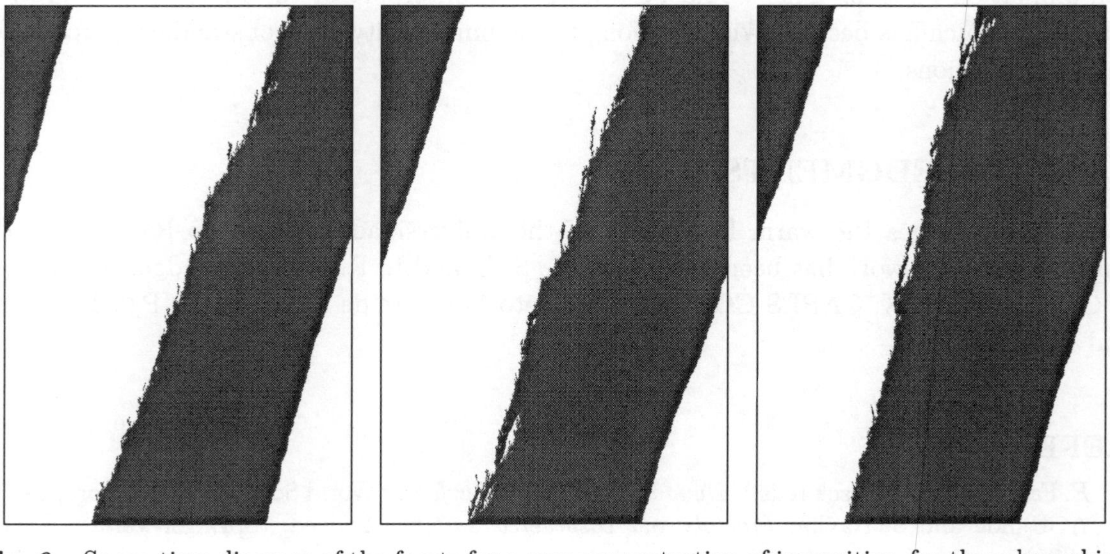

**Fig. 6** Space-time diagram of the facets for a zero concentration of impurities, for three branching probabilities, ($b$ =0.26, 0.28 and 0.3, from left to right). The initial condition was such that only the middle part of the seed line was initially occupied. One notes that the facet size decreases slowly in time on the left, remains stable in the center and expands slowly on the right figure. A quantitative analysis based on this observation allows to estimate the critical value of $b^*$ such that $c_c(b^*) = 0$.

but rather through a competition between these phases taking place along their interface. The dynamics of the latter may be ruled by a directed percolation process as the picture suggests. However, this has not been checked more thoroughly. One although understand that this is out of reach of the fixed point approximation since the latter ignores correlations, and thus cannot be used faithfully in this context.

Pushing the above reasonning to its extreme, we are lead to study the case free of impurities $c = 0$, but choosing a particular initial condition where the active sites occupy initially only the center half of the base line. We show in Fig. 6, the space time diagram of the facets for $c = 0$ and three values of $b$. One sees indeed that the compact phase would be stable for ever in the three cases, if there was no competition with the empty region. This competition is seen to affect differently the left and the right border of the compact facet, and this results from the conventional updating order chosen here. But more importantly, depending on $b$, the facet zone either expands or recedes. The critical value $b^*$, is the one for which it remains stationary. A quantitative analysis of the latter problem leads to $b^* \approx 0.28$ (central part of Fig. 6). Thus finally, albeit the order of the transition is incorrectly given by the fixed point approximation, it appears that a large number of features are correctly reproduced.

## 6. CONCLUSION

We have introduced a model for the growth of a branching polymer model in an heterogenous medium, which displays a faceted to rough transition. This transition has been characterized using an original algorithm based on the notion of self-regulated search of a critical point. Finally, we have presentedan approximate analytical treatment, which accounts for a number of features, in spite of its simplicity.

It would be of interest to study further the transition, and check in particular if it belongs to the directed percolation universality class as it has been found for a number of other contamination models. This would be a natural counterpart of the infinite/finite growth transition which has been shown[8] to belong to the universality class of standard percolation in two dimensions.[13]

## ACKNOWLEDGMENTS

S. R. acknowledges the warm hospitality of the Universidade Federal do Rio Grande do Norte, where this work has been performed. L. S. L. and L. R. S. acknowledge the support of CNPq, PRONEX, CAPES-COFECUB, Projeto Nordeste de Pesquisa, ANP/CTPETRO and FINEP.

## REFERENCES

1. F. Family and T. Vicsek (eds.), *Dynamics of Fractal Surfaces* (World Scientific, Singapore, 1991).
2. A. Bunde and S. Havlin, *Fractals and Disordered Systems*, 2nd ed. (Springer-Verlag, Berlin, 1995).
3. A. L. Barabasi and H. E. Stanley, *Scaling Concepts in Surface Growth* (Cambridge University Press, Cambridge, 1995).
4. P. Meakin, *Scaling and Growth Far From Equilibrium* (Cambridge University Press, Cambridge, 1998).
5. P. G. de Gennes, *Soft Interfaces* (Cambridge University Press, Cambridge, 1994).
6. L. S. Lucena, J. M. Araújo, D. M. Tavares, L. R. da Silva and C. Tsallis, *Phys. Rev. Lett.* **72**, 230 (1994).
7. I. Majid, N. Jan, A. Coniglio and H. E. Stanley, *Phys. Rev. Lett.* **52**, 1257 (1984).
8. A. Bunde, S. Havlin and M. Porto, *Phys. Rev. Lett.* **74**, 2714 (1995).
9. L. S. Lucena, L. R. da Silva and S. Roux, *Physica* **A266**, 86 (1999).
10. P. Bak, *How Nature Works* (Springer-Verlag, New-York, 1996).
11. J. S. Andrade, L. S. Lucena, A. M. Alencar and J. E. Freitas, *Physica* **A238**, 163 (1997).
12. A. M. Alencar, J. S. Andrade and L. S. Lucena, *Phys. Rev.* **E56**, R2379 (1997).
13. H. H. Aragao-Rego, J. Elias de Freitas, Liacir S. Lucena and G. M. Viswanathan, preprint (2000).

# PERCOLATION AND CRITICAL PHENOMENA OF AN ATTRACTIVE MICELLAR SYSTEM

F. MALLAMACE and S. H. CHEN
*Department of Nuclear Engineering, Massachusetts Institute of Technology, Cambridge MA 02139-4307, USA*

P. GAMBADAURO, D, LOMBARDO and A. FARAONE
*Dipartimento di Fisica and INFM, Universita' di Messina, Vill. S. Agata C.P. 55, 98166, Messina, Italy*

P. TARTAGLIA
*Dipartimento di Fisica and INFM, Università di Roma, La Sapienza, P.le A. Moro 5, I-00185 Rome, Italy*

## Abstract

In this work we study an attractive micellar system for which the percolation curve terminates near the critical point. We have studied such an intriguing situation by means of scattering (elastic and dynamical) and viscoelasticity experiments. Obtained data are accounted by considering in a proper way the fractal clustering processes typical of percolating systems and the related scaling concepts. We observe that the main role in the system structure and dynamics it is played by the cluster's partial screening of hydrodynamic interaction. This behaves on approaching the percolation threshold dramatic effects on the system rheological properties and on the density decay relaxations. The measured correlation functions assume a stretched exponential form and the system becomes strongly viscoelastic. The overall behavior of the measured dynamical and structural parameters indicates, that in the present micellar system, the clustering process originates dilute, polydisperse and swelling structures. Finally, this originates an interesting situation observed in the present experiment. As it has been previously, proposed by A. Coniglio et al., percolation clusters can be considered to be "Ising clusters" with the same properties as the Fisher's critical droplets. Therefore at the critical point the percolation connectedness length ($\xi_p$) can be assumed as the diverging correlation length ($\xi_p \equiv \xi$) and the mean cluster size diverges as the susceptibility.

## 1. INTRODUCTION

Complex fluids, such as polymers, colloidal and biological macromolecules, are of central interest in different fields of science and technology, in particular in areas that are at the borderline between physics, chemistry and biology.[1,2] They are characterized by a structural order, determined by the spontaneous or induced buildup of large supramolecular assemblies. Depending on the system thermodynamics (concentration, temperature and interparticle interactions) colloidal and polymer solutions originate a very rich phase behavior characterized by diverse structural orders (e.g. liquid, glass, liquid crystal and solids with different structures, e.g. cubic or hexagonal crystals). In many situations these structures have a remarkable degree of symmetry (fractals) described by different aggregation kinetics ranging from the diffusion limited to the percolation processes.[3] Understanding the role played by these structures and the related clustering phenomenon, in determining the physical behavior of complex fluids, is the essential first step for understanding more complex situations of basic and technological relevance.

Clustering effects on the properties of complex fluids are, therefore, a stimulating subject in the recent theoretical and experimental research. In particular, many studies have been devoted to colloids or colloidal like systems (micelles, microemulsions and globular proteins). A special care was applied, on this respect, to adhesive hard sphere (AHS) colloids in which clustering can originate a percolation phenomenon[4] and a special kinetic glass transition,[5] both processes in which there is a transition from ergodic to non-ergodic phases. A short range attractive tail of the interparticle potential, besides the hard core repulsion, well accounts for the very slow and complex dynamics characterizing these processes by using properly clustering concepts. Such an attractive contribution is usually represented by an exponential or a square well interaction; typical examples are the Yukawa potential or the sticky spheres potential, introduced by Baxter[6] (defined by means of the stickiness parameter $1/\tau$). For these attractive systems, theory[7] proposes a phase diagram (in the $T$ (or $\tau$), $\phi$ plane) characterized by the existence of a percolation locus (temperature concentration dependent) and a spinodal line, a gas-liquid phase transition with a critical point. The percolation loci have been derived analytically by using clustering concepts and the idea of a pair-connectedness function $P(r)$, both introduced by Coniglio et al. in 1977, in connection with the development of a continuous percolation theory.[8] Given a particle at the origin, $4\pi r^2 rP(r)$ is the number of particles in the spherical shell $(r,\ r+dr)$ which are connected to this central particle and belong to the same cluster. Coniglio et al. showed also that $P(r)$ satisfies an Ornestein-Zernike type equation with a modified direct correlation function. By invoking the short range nature of this direct correlation function, it has been shown that the onset of percolation can be identified as the point where the average cluster size diverges. The validity of such clustering approach in the determination of the percolation loci has been successfully proved for attractive microemulsions and micellar systems.[9] At the same time the experimental coexistence curve, in the colloidal like systems, has been fitted by means of a relationship between the stickiness parameter $(1/\tau)$ and the temperature.

The experimental observation in these attractive systems gives indication that the critical point is located in the gel phase so that at the critical concentration the percolation threshold temperature $T_p$ and the critical one $T_c$ differ only of few degrees.[9,10] A similar situation has been proposed some time ago from Tanaka[11] for smart cross-linked gelatin gels and discussed in terms of a mean field approach. In such a gel system, as far as for attractive colloids, the spinodal line and the gel transition line (percolation) co-exist, and by changing the gel composition (or the solvent properties, or polymer interaction), the consolute critical point can be located in the sol or in the gel phase, or just on the gelation curve, where

$T_c = T_p$ ($T_p$ is the threshold percolation temperature). Both these different systems, colloid and gels, undergo the phase separation and the liquid-gel transition reversibly. When the critical point is located very near (or just on) the percolation line an intriguing situation arises. The system properties (structure and dynamics) are characterized by two diverging lengths: one is the usual concentration correlation length $\xi$ which diverges at the consolute point, and the other, linked to the clustering process is the characteristic linear dimension of the finite cluster $\xi_p$, which diverges at $T_p$. As a result, the formation of percolating transient polydisperse fractal clusters should have profoundly affected the critical process (or the viceversa). Some years ago such a problem has been studied, in a model developed for polymer gelation, by Coniglio et al. in the 1979[12] with the use of a mean field approach based on bond correlated percolation. Besides the fair agreement with the experimental phase diagrams of the Tanaka gels[11] the main result of such a theoretical investigation has been the following: when the gelation curve terminates at the critical point ($T_c = T_p$) both bond connectivity and concentration fluctuations are critical, just as a random magnet at the percolation threshold. Such a situation has been properly formalized by using a renormalization group approach for site-bond percolation;[13] such an approach gives the definitive indication that the Ising critical point can be considered as a high order critical point for percolation. This latter result is obtained by using the concept of clustering. It has been shown that exists a thermodynamical situation for which percolation clusters are polydisperse and not compact (diluted clusters) and behave like critical droplets.[14] In such a situation percolation clusters (or Ising clusters) have the same properties as the droplets studied in the Fisher's droplet model,[14] i.e. at the critical point their connectedness length can be assumed as the diverging correlation length ($\xi_p \equiv \xi$) and the mean cluster size diverges as the susceptibility.

However, despite of their relevant interest in statistical physics, these theoretical predictions for the critical behavior of complex fluids characterized by the co-existence in a narrow temperature range of both the percolation and the critical phenomena have never been experimentally investigated. The main aim of this work is just the study of structure and dynamics in an attractive micellar system in which, along the critical isochore, the onset of the clustering related with percolation precedes of few degrees the critical phase separation. Hoping in such a way to give a definitive insight on the reported theoretical suggestions, based on clustering concepts, we will use elastic (light and neutrons), quasielastic (photon correlation spectroscopy) scattering techniques and viscoelasticity.

## 2. EXPERIMENTS

The system we study is a non-ionic triblock copolymer L64 belonging to Pluronic (BASF trademark) family.[10] Pluronic is made of polyethylene oxide (PEO) and polypropylene oxide (PPO), with the two PEO chains arranged symmetrically on each ends of the PPO chain. The chemical structure of L64 is [$PEO_m$-$PPO_n$-$PEO_m$], with a molecular weight of 2900 *Da* and is comprised of 60wt% of PPO. The importance of these polymers derives from their properties as surface active agents in aqueous solutions. As a consequence, they find widespread industrial and biomedical applications. PPO tends to become less hydrophilic than PEO at higher temperatures. Thus the copolymer acquires a surfactant property and self-assembles spontaneously forming spherical micelles in aqueous solutions at sufficiently high temperatures and above certain concentrations. The micelles exist in a wide range of copolymer concentrations from few $wt\%$ up to more that 50 $wt\%$ for $T > 30°C$.[15]

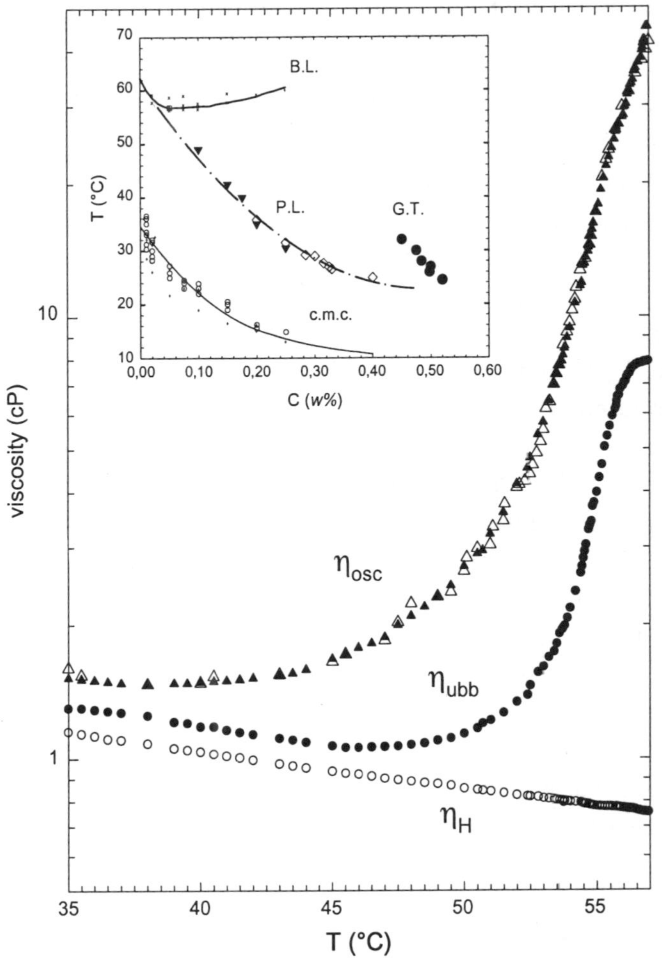


**Fig. 1** The measured viscosities, of the studied $L64/D_2O$ system along the critical isochor as a function of the temperature. In particular, are reported the viscosities obtained by using: the oscillating rheometer at a frequency of $\omega = 1.36\ rad/\sec(\eta_{osc})$, the Ubbelhode viscometer ($\eta_{ubb}$) and the hydrodynamic viscosity $\eta_H$. In the inset is reported the $L64 - D_2O$ phase diagram.

By using viscoelasticity, small angle neutron (SANS) and light scattering[10] we have recently investigated the phase behavior, the micelles microstructure and their mutual interactions in such a system obtaining a phase diagram, shown in the inset Fig. 1, that is the same of that proposed by the Baxter theory.[10] I.e. a phase diagram with an inverted binodal curve (with a critical point at $c = 5\ wt\%$ and $T_c = 330.32\ K$) and a temperature-concentration dependent percolation line cutting across the phase diagram near the critical point. More precisely, from the SANS intensity distribution fitting performed by using the sticky adhesive hard sphere model for the micellar structure factor, the existence of the percolation loci and of the critical demixing point is inferred from a precise temperature dependence of the stickiness parameter $1/\tau$. Along the isochor line $1/\tau$ increases with the temperature and approaches the critical value 10.2 at $T = 330\ K$, whereas by crossing the percolation line at different $c$ the sticking parameter presents an abrupt increase. The percolation process is confirmed by an observed increase of some order of magnitude of $G'$ and $G''$ (at certain temperature and composition) and by a well defined frequency scaling of these quantities. Scaling exponents, determined for frequency-dependent complex moduli satisfy the scaling relations predicted by the scalar elasticity percolation theory.[16] In addition, the indication that attraction dominates the properties of such a system has been recently confirmed by the existence, at high concentration of a kinetic glass transition (reported in the inset Fig. 1 as fully dots) along a temperature and concentration dependent line.[17] The glass transition and the percolation lines are located in different regions of the phase diagram. The structural arrest typical of supercooled liquids at the glass transition as far as the percolation are related to clustering processes due to the short range intermicellar attraction. In agreement with proper theoretical models, for glass transition in attractive systems, the intermediate scattering functions show a logarithmic time dependence attributed to a high order glass transition (cusp or $A3$) singularity predicted by mode coupling theory.

$L64$ copolymer solutions were prepared by using a standard procedure.[15] We worked at the critical concentration ($c = 5\ wt\%$) previously determined by using SANS and elastic light scattering (ELS).[10] Viscosity experiments were performed with an Ubbelhode viscometer and a strain controlled oscillating rheometer; in this latter case we used a double wall Couette geometry by working in the frequency range $0.0924 < \omega < 60$ $(rad/\text{sec})$. To ensure a linear response in the rheometer measurements a low strain deformation, $\gamma = 0.05$, was maintained. In such a condition viscosities were measured in the shear rate range of $0, 3 < \dot{\gamma} < 10$ $\text{sec}^{-1}$; this corresponds to Peclet numbers $Pe < 1$, where the Peclet number is defined as $Pe = \dot{\gamma}\xi^2/2D_0$, which characterizes the amount of distortion of structures with linear dimension of $\xi$ ($D_0$ is the particle short-time diffusion coefficient at low concentration where hydrodynamic effects dominate). These distortions are responsible for shear thinning in the system structures which is expected to set at $Pe \approx 1$. In other words such a condition constitutes an important experimental constrain in the study of the elasticity of systems characterized by clustering processes like polymer solutions and attractive colloids. This is also very important in viscosity measurements in phenomena dominated by long-range critical correlations. Just in clustering processes or approaching the critical point shear thinning can originates a "levelling off" of the measured viscosities. The shear thinning cannot be studied with usual viscometers like Ubbelohde, vibrating wire, rolling ball, or rotating viscometers. With these instruments a whole range of shear rates is applied. In the contrary of atomic or molecular liquids, complex fluids are characterized by diverging long ranged structures able to support energy storage and hence an increase in the elastic component; this dramatically affects their viscoelasticity. For example, as temperature (and concentration) approaches the percolation transition, the structure of the suspension results in a strongly frequency (or concentration) dependence to both the storage modulus, $G'(\omega)$, and the loss modulus, $G''(\omega)$. On the basis of our previous discussion, such a phenomenon can interfere with the critical transition especially in systems where percolation and critical transitions fall in the same area of the phase diagram. Also to account for this we performed viscosity measurements by using the Ubbelhode viscometer and the oscillating rheometer. Typical results of the measured viscosities, of the studied $L64/D_2O$ system along the critical isochor, are shown in Fig. 1 for the different studied temperatures; the viscosities obtained with the oscillating rheometer are measured at a frequency of $\omega = 1.36$ $rad/\text{sec}$. In the figure is reported for comparison the hydrodynamic contribution to the viscosity ($\eta_H = \eta_0(1 + 2.5\phi + 6.2\phi^2)$), where $\eta_0$ is the $D_2O$ viscosity.

ELS measurements were made at the scattering angle $\theta = 90°$ by using a 20 $mW$ $HeNe$ laser ($\lambda$ = 6328 Å) and an optical scattering cell of a diameter of 1 cm in a refractive index matching bath. The intensity data were also corrected for turbidity and multiple scattering effects. At the same scattering angle we have performed the dynamical scattering experiments by using the photon correlation spectroscopy (PCS). The PCS data have been taken using a Brookhaven digital correlator with a logarithmic sampling time scale. The latter feature allows us to describe accurately both the short time region and the very long time one, up to times, if necessary, of the order of seconds. We explored a temperature range from 30°C to $T_c$ with samples thermostated, by means of a circulating water bath, within $\pm 10mK$.

## 3. RESULTS AND DISCUSSION

### 3.1 Viscosity

We will discuss our results starting with the measured viscosities. Figure 1 gives evidence, from both the Ubbelhode viscosity ($\eta_{ubb}$) and the oscillatory viscosity at $\omega = 1.36\ rad/\sec$ ($\eta_{osc}$) of a dramatic increase increasing the temperature, certainly not related with hydrodynamic effects; in addition their comparison evidences the presence of shear thinning effects. On considering this latter process we will use only the viscosity data coming out from the oscillating rheometer. We relate the $\eta_{osc}$ behavior with the percolation process (the flex point in the Fig. 1 data suggests that $T_p$ is located at about 54°C).

Besides these technical problems about the measure of the viscosity in complex fluids, there are serious difficulties to treat in a complete way viscoelasticity in percolating system. These difficulties arise from the large polydispersity in their structures and a very complex dynamics characterized by the relaxation processes over many different time scales. In addition, there is a competition between relaxation processes with the increasing connectivity to produce the observed viscoelasticity. The simplest approach consists in considering the sol as a polydisperse suspension of solid spheres neglecting the cluster-cluster interactions and generalizing the Einstein formula for the viscosity of a monodisperse suspension of solid spheres,[18] which corresponds to a highly diluted regime. Within the Flory classical theory of gelation the viscosity remains finite[19] or diverges mostly logarithmically. Using instead the Rouse model for the polymer dynamics,[20] which neglects the entanglement effects and the hydrodynamic interactions, the viscosity in a solution of polymeric clusters, expressed in terms of the macroscopic relaxation time, grows like $\langle R^2 \rangle$ as the cluster radius $R$ grows in the gelation process. The contribution of the $n_s$ molecules of size $s$ and gyration radius $R_s$ to the average $\langle\ \rangle$ is of the order of $sn_sR_s^2$ leading to the critical exponent $k = 2\nu - \beta$,[21] where $\upsilon$ is the critical exponent of the correlation length, $\xi_p$, diverging at the gel point and $\beta$ is the critical exponent describing the growth of the gel phase. Let $p$ be the control parameter of the percolation transition (it may be the concentration, the temperature, etc.) and $p_c$ its value at the threshold, for $\varepsilon = |(p - p_c)/p_c| < 1$, $\xi_p$ scales as $\xi_p \sim \varepsilon^{-\nu}$ ($\nu = 0.88$). With the random percolation exponents the value $k \sim 1.35$ is found.[22]

Actually this Rouse exponent could be considered as an upper limit due to the complete screening of the hydrodynamic interactions and the entanglement effects.[23] Another approach has been proposed by de Gennes[16] using an analogy between the viscosity at the gelation threshold and the diverging conductivity in the random superconducting network model, giving an exponent $k \sim 0.7$. The Zimm approach,[24] where the monomer correlation due to the hydrodynamic interactions are not completely screened, would give a smaller exponent.

The effects of the screening of hydrodynamic interactions on the viscoelasticity at the gelation must be carefully considered to describe experimental results. These processes have been investigated on considering the diffusion properties of the percolating systems by using of scaling arguments.[22] On these bases an interesting and clear picture emerges that describes both dense and dilute gels and furnishes precise suggestions on the behavior of relaxations, the diffusion coefficients and on the viscoelasticity before, near and above the percolation threshold. Essentially the situation can be described in such a way: the sol at the sol-gel transition is a heterogeneous medium formed by solvent and all other clusters of different sizes, with the mean cluster size rapidly growing near the percolation threshold. A fractal cluster with a gyration radius $R$ can be seen as a probe diffusing in such a medium. As long as its radius is much greater than the value of the percolation correlation length in the sol the Stokes-Einstein relation is expected to be valid, and the probe diffusion coefficient, $D(R)$, will decrease proportionally to the inverse of the medium viscosity, $D(R) \sim 1/R^{d-2}\eta$. Then the generic probe of size $R$ diffuses in a medium with a viscosity coefficient depending on $R$, $\eta(R)$, and it should be expect to hold a *generalized* Stokes-Einstein, $D(R) \sim 1/R^{d-2}\eta(R)$ (where $d$ is the dimension of space). As the percolation transition is approached the probe diffuses in a medium where a spanning cluster appears with a self-similar structure. At the percolation threshold the sol viscosity will be $\eta \approx \varepsilon^{-k}$ and for the viscosity coefficient depending on $R$ holds the scaling behavior $\eta(R) \sim R^{k/\nu}$. When $R$ is of the order of the correlation length, then $\eta(R) \sim \eta$. Therefore at the threshold will be $D(R) \sim 1/R^{d-2+k/v}$.

The radius-dependent (fractal) viscosity is physically due to the screening of hydrodynamic interactions. If hydrodynamic interactions between monomers on a cluster are completely screened by smaller clusters the Rouse diffusion coefficient $D(R) \sim 1/R^{d_f}$ should apply ($d_f$ is the fractal dimensionality of the percolating cluster). The use of the Rouse model would consist in taking $D(s) \sim 1/s$ for a cluster of size $s$. Then for large enough cluster sizes $s$ in percolation is $s \sim R^{d_f}$. Taking $D(R) \sim 1/R^{d_f}$ leads to $k = \nu(d_f + 2 - d)$, which again gives $k = 2\nu - \beta$, i.e. the Rouse exponent. On the other hand, if hydrodynamic interactions are unscreened, the Zimm diffusion coefficient $D(R) \sim 1/R^{d-224}$ leads to $k = 0$. In general, a partial screening of hydrodynamic interactions gives the inequality $0 \leq k \leq \nu(d_f+2-d) ldots$ Being $d_f = (d+2)/2$ for $d = 6$ (the mean field limit) $k$ must vanish and $\eta$ must diverge logarithmically[18] like $\eta \sim \ln(1/\xi)$. Therefore, experiments should give, on approaching the percolation threshold, a viscosity exponent in the range $0 \leq k \leq 1.3$.

Together with the dynamic viscosity $\eta$ the complex shear modulus $G^* = G' + iG''$ constitutes the rheological property characterizing viscoelastic systems. Percolation theory describes the measured viscoelasticity singularities near $p_c$ by using other exponents; in addition to $k$, is considered the exponent $t$ that accounts for the divergence of the static elastic modulus $G_0 \sim \varepsilon^{-t}$ (more precisely $G_0$ is the elastic modulus of monomers at some microscopic time scale $\tau = \omega_0^{-1}$). Whereas a general scaling form, as a function of the frequency $\omega$, has been postulated for both the storage modulus, $G'(\omega)$, and the loss modulus, $G''(\omega)$[16,25] in the limit when $\omega$, $\varepsilon \to 0$:

$$G' \sim G'' \sim \omega^{\Delta} \tag{1}$$

with $\Delta = t/(k+t)$. The latter power law, that furnishes an additional procedure to evaluate viscoelasticity at percolation, holds in the frequency range $\omega_0\varepsilon^{k+t} \ll \omega \ll \omega_0$; in addition, it has a remarkable consequence: $G''/G'$ assumes an universal critical value; considering that $\tan\delta = G''/G'$, it will be $\delta_c = \pi\Delta/2 = \pi t/2(t+k)$. Very below the percolation threshold the correct frequency dependences, for $\omega \to 0$, of the elastic and viscous parts of $G^*$ are $G' \sim \omega^2$ and $G'' \sim \omega$ respectively, whereas very above the threshold it is $G' \sim const$ and $G'' \sim \omega$. However, whereas $\Delta$ seems to represent an universal exponent for the percolating viscoelasticity, $k$ and $t$ are non universal depending on the system properties.

On the bases of these discussions the unperturbed viscosity grows with the control parameter and diverges on approaching percolation threshold from below according to: $\eta \approx |p_c - p|^{-k} \approx |T_p - T|^{-k}$. Figure 2(a) where we plot, in a log-log scale, the measured $\eta_{osc}$

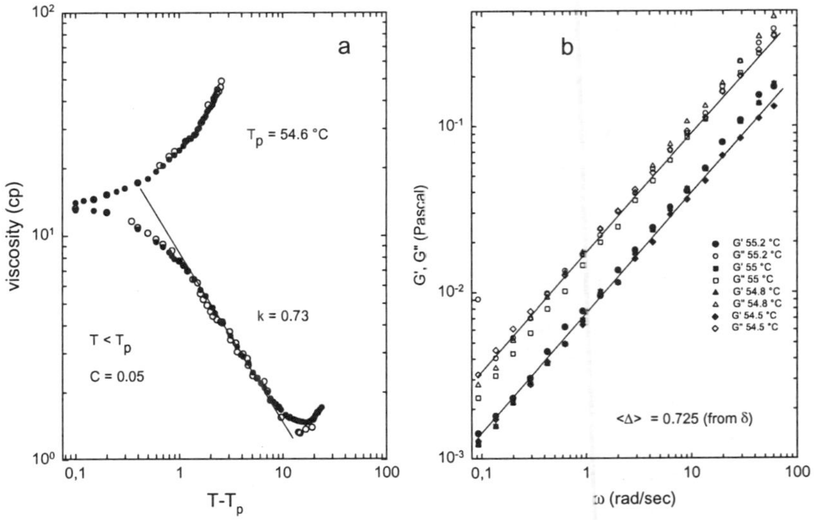


**Fig. 2** **(a)** A log-log scale of $\eta_{osc}$ vs. $|T_p - T|$ for $T_p = 54.6°\text{C}$ the straight line has a slope $k = 0.73 \pm 0.04$. **(b)** The $\omega$ dependence of $G'$ and $G''$ in the whole studied frequency domain for four different temperatures just near $T_p$. The two continuous lines, that represent the moduli power law, have a slope $\Delta = 0.72 \pm 0.03$; the exponent $\Delta$ measured separately by the corresponding loss angles ($\delta_c = \pi\Delta/2$) is also reported.

vs. $|T_p - T|$ well reproduces such a situation; as it can be observed we measure $k = 0.73 \pm 0.04$ and $T_p = 54.6°\text{C}$. Whereas Fig. 2(b) reports the $\omega$ dependence of $G'$ and $G''$ in the whole studied frequency domain for four different temperatures just near $T_p$. In this logarithmic representation, the increase of both moduli can be described, in according to the related power law, by a linear behavior as a function of $\omega$. The two continuous lines have a slope $0.72 \pm 0.03$ (corresponding to the theoretical $\Delta$); we measure separately the corresponding loss angles ($\delta_c = \pi\Delta/2$) and the value $\Delta = 0.725 \pm 0.06$.

The exponents $k$ and $t$ have been calculated accurately by numerical methods and computer simulations giving their numerical values,[26,27] for example for static percolation it is[26] $k = 0.75 \pm 0.04$, $t = 1.94 \pm 0.1$, so that $\Delta = 0.72 \pm 0.04$, for a random superconducting network in analogy with the de Gennes model $k$ is 0.7,[28] whereas a numerical simulation, in terms of a recently introduced model based on bond-fluctuation dynamics, gives $k = 1.3 \pm 0.04$. On the other hand experimental measurements performed in many different polymer gels give results strongly dependent on the polymers nature; these observations give $k$ values ranging from 0.5 to 1.5, whereas for the elastic modulus critical exponent $t$, the values are even more scattered ($1.9 < t < 3$).[10,25,27] As it has been previously said these results must be certainly ascribed to the screening of hydrodynamic interactions or entanglement effects. A situation in which there is a partial screening of hydrodynamic interactions, especially in gelling systems in the semidiluted regime, which can be considered more realistic and explains the measured $k$ values of $\sim 0.7$. These considerations will be more clear on considering the decay of density fluctuations in gels, measured by means of the dynamic light scattering. This situation will be discussed in the next sections.

## 3.2 Dynamic Light Scattering

By using the photon correlation spectroscopy we measure the time-averaged time correlation functions

$$g_T^{(2)} = \langle I(q,0)I(q,\tau)\rangle_T / \langle I(q)\rangle_T \tag{2}$$

where $I(q, t)$ is the intensity of the light scattered at the wave vector $q$ at time $t$.[29] As it is well known such a quantity is directly related to the thermally driven density-density correlation function (or dynamic structure factor) $S(q, t) = \langle \delta\rho(q, 0)\delta\rho^*(q, \tau)\rangle$, where $\delta\rho(q, \tau)$ represents the spatial Fourier component, of wave vector $\overrightarrow{q}$, of particle concentration fluctuations. The quantity of central interest is the intermediate scattering function (or normalized, ensemble-averaged, autocorrelation function) given by: $f(q, \tau) = S(q, \tau)/S(q, 0) = S(q, \tau)/S(q)$, where $S(q) = S(q, 0)$ is the static structure factor. For ergodic systems, in which the time average $\langle\ \rangle_T$ is equivalent to the ensemble average $\langle\ \rangle_E$, $g_T^{(2)} = g_E^{(2)}$, and the standard Siegert relationship give $g_E^{(2)} = 1 + c[f(q, \tau)]^2$, $c$ is the coherence factor determined by experimental conditions.[29] For a gels the ergodicity condition holds in the sol phase (i.e. before the percolation transition) whereas in the gel phase the system becomes non-ergodic. In this latter situation the ensemble and time averages must be performed separately; the correct way is averaging over different sample positions in measurements of the long time region of the time correlation. Alternatively, it can be used a nearly diffraction-limited scattering volume so that the measured correlation function is independent of the ensemble average.[22] As our system is a gel in the diluted regime we used beyond the percolation threshold this latter experimental approach performing an average on some correlation functions, just to have a correct determination of the time correlation function baseline.

Therefore the obtained ISF give a measure of the density fluctuations relaxation and the corresponding relaxation times, quantities directly related, *via* the viscosity, with the diffusional properties of the system. These latter properties have been investigated previously for a gelling system, in which gelation is driven by a cross-linking agent, (in the pre- and post-gel regimes, and at the gel point) at different concentration conditions, and an interesting picture was obtained with scaling arguments by Martin et al.[22] In this work it is well established that the concept of "radius dependent viscosity", that well explains viscoelasticity determines, also the decay of density fluctuations. In particular, $S(q, t)$ can be calculated taking into account that the scattering of a single cluster is screened (totally or partially) by the other nearest clusters. On these bases:

$$S(q, t) = \int_1^\infty mN(m)\exp(-q^2Dt)dm \tag{3}$$

where $m$ is the cluster mass, $N(m)$ the percolation number distribution and $D = D(R)$ represents the cluster diffusion coefficient; this integral, calculated by using the same scaling arguments used for $D(R)$ in viscoelasticity, adequately describes the observed dynamics in gelling systems that occurs with a precise behavior:

(i) approaching the percolation threshold the dynamic structure factor, $S(q, t) \sim \exp(-t/\tau)^{\overline{\beta}}$ $(\beta = d_f/(d_f + 1))$, has a stretched exponential tail whose characteristic time diverges at the threshold. This "critical slowing down" is due to the clustering process, i.e. to the divergence of the average cluster size.

(ii) at the gel point $S(q, t)$ becomes a power law, $S(q, t) \sim t^{-\beta/(\nu+k)}$, and maintains such a form beyond the threshold.

(iii) a moderate dilution of critical sol-gel, the long time tail of the $S(q, t)$ changes from a power law to a stretched exponential form; in these concentration conditions the $q$ dependence of $S(q, t)$ reveals two time scales that depend on different powers of the scattered wave-vector;

(iv) at more higher dilutions the stretched exponential correlation function with two time scales crosses over a more rapidly decaying form with a single decay rate; in such a situation the stretched exponential exponent rapidly increases, on decreasing concentration, and very diluted samples show exponential decays.

The divergent time scales observed before the percolation threshold can be calculated directly from the Eq. (3) by integrating the correlation functions. Using these arguments it is easy to obtain that the average relaxation time $\langle\tau\rangle$, diverges on approaching $p_c$, as $\langle\tau\rangle \sim \varepsilon^{-k}$.[27] As previously discussed for gel viscoelasticity, the knowledge of the cluster self diffusion mechanisms was fundamental to understand the complex system dynamical behavior evidenced by the light scattering,[22] i.e. the same scaling approach used for the viscosity can explain our light scattering data in the region of the percolation. However, we must consider that we work on a system in the diluted regime (the sample volume fraction is $\phi \cong 0.1$) and viscoelastic data give a value of the exponent $k$ indicating a partial screening of the hydrodynamic interactions. Therefore for the studied system it is expected in the percolation region, approaching the threshold, that the measured decay of density correlations will show a stretched exponential form.

Figure 3 reports the measured ISFs in the temperature range 35°C $< T <$ 56.7°C. The corresponding data analysis reveals, by means of a proper data fitting, a single exponential decay in the region 35°C $< T <$ 47°C, whereas for higher temperature, approaching the percolation threshold, the ISFs have the expected stretched exponential form ($f(q,\tau) \sim \exp(-t/\tau)\bar{\beta}$); such a form is maintained also for $T > T_p$. On the contrary of concentrated gels we do not observe above the percolation the scaling law behavior. Such latter result suggests a large polydispersity for fractal clusters. We observe that the stretched exponent $\bar{\beta}$, (inset of Fig. 3) decreases, increasing the temperature, at a value of $0.62 \pm 0.03$ at the threshold temperature, after that it decreases slowly approaching the phase separation. This $\bar{\beta}$ value, measured at the percolation transition, agrees with the ones predicted and measured in other gels; in the model[22] such a quantity is directly connected with the cluster fractal dimensionality. From the data fitting we measure also the average relaxation time $\langle\tau\rangle$, reported in Fig. 4(a). In the same figure are reported for comparison the relaxation times obtained in the low temperature regime where the density fluctuations decay is represented

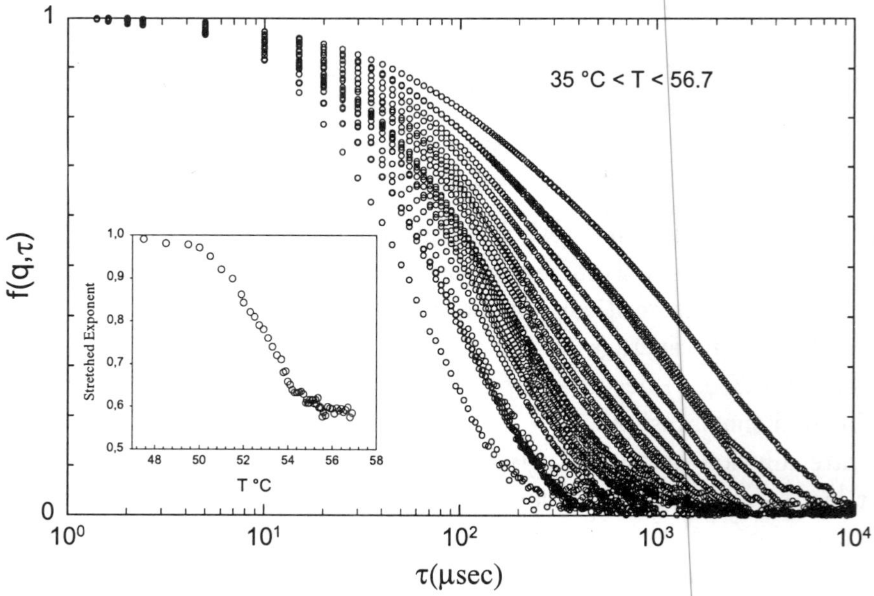


**Fig. 3** The normalized intermediate scattering functions (ISF) in the temperature range 35°C $<$ $T <$ 56.7°C. The inset reports the measured stretched exponent $\bar{\beta}$.

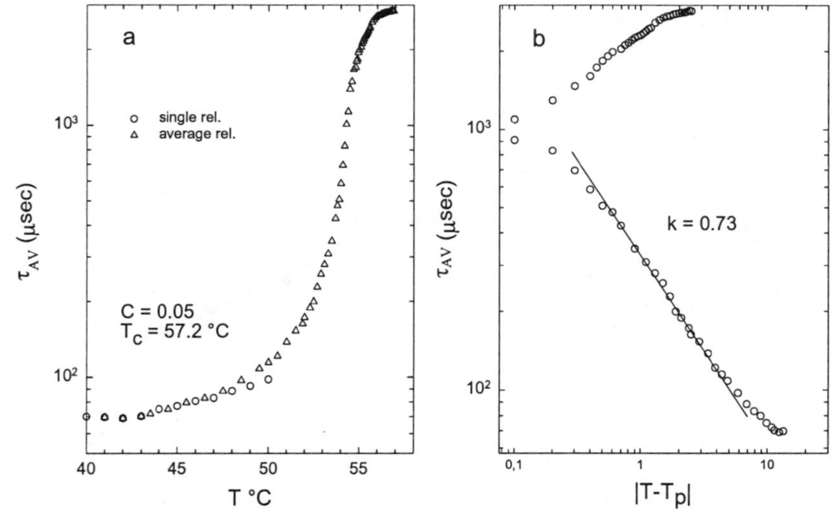


**Fig. 4** **(a)** The average relaxation times $\langle\tau\rangle$, obtained from the fitting of the ISF (Fig. 3). **(b)** $\langle\tau\rangle$ vs. $|T_P - T|$ for $T_P = 54.6°\text{C}$, we have $k = 0.73 \pm 0.04$ (slope of the straight line).

by a single exponential. From Fig. 4(b) that reports $\langle\tau\rangle$ vs. $|T_P - T|$, we measure, for $T_P = 54.6°\text{C}$, $k = 0.73 \pm 0.04$. Being $\langle\tau\rangle$ a typical macroscopic relaxation time, directly linked to the viscosity, such a result can be considered a further confirmation that the dynamics of a gelling system is dominated by a clustering process in which the screening of hydrodynamic interactions plays a primary role.

These observations have been explained (and computed) recognizing that the sol-gel transition is a connectivity divergence, and on the contrary of a thermodynamic phase transition, a divergence in the scattering intensity (critical opalescence) cannot be observed from the undiluted incipient gel. However, diluted gels present on approaching $T_p$ a large increase in the scattered intensity; such a phenomenon is due to a progressive (with concentration) polydispersity and swelling of clusters (the cluster internal structure became less and less compact) and the consequent elimination of optical screening increasing the spatial correlation length. In other words, the sol-clusters assume the same physical properties of the previous discussed Ising clusters. Therefore as it has been suggested these structures assume the same properties as the droplets studied in the Fisher's droplet model especially when percolation and critical point are located very nearest in the phase diagram.[13] In our opinion, just these mechanisms can explain the results of the elastic light scattering experiment performed in the present attractive micellar system along its critical isochore. We stress that in this experiment, by changing the temperature before reaching the critical phase separation temperature $T_c$ we cross $T_p$. The results of such experiment are shown in Fig. 5(a) where the measured scattered intensity is reported as a function of the reduced temperature $\varepsilon = |T - T_c|/T_c$.

### 3.3 Elastic Light Scattering

For a micellar system it is customary to write the scattered intensity in terms of the interparticle structure factor $S(q)$ as $I_s(q) = KS(q)$, where $K$ is a constant factor depending on the scattering geometry, the wavelength of the incident light, the particle structure factor and the sample refractive index ($n$), whereas $q = (4\pi n/\lambda)\sin(\vartheta/2)$ is the scattering wave-vector. Approaching the critical point, the particle correlation function obeys the

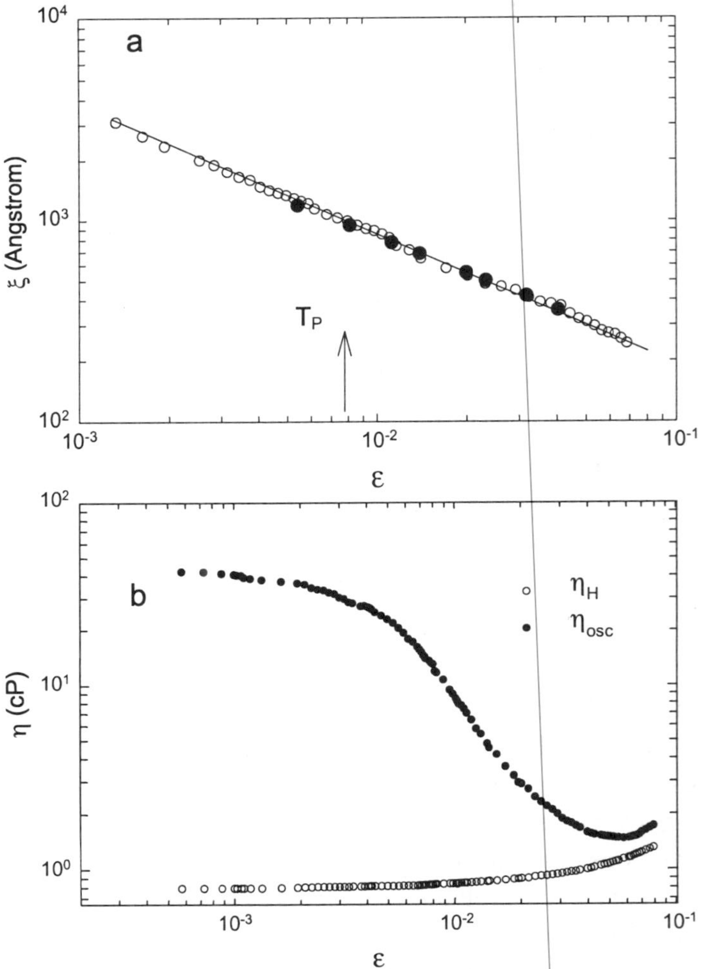


**Fig. 5** **(a)** A log-log representation of the correlation length $\xi$ vs. the reduced temperature $\varepsilon$; the use of the Ornstein-Zernike-Fisher form gives: $\gamma = 1.22 \pm 0.04$ ($\nu = 0.61 \pm 0.02$) and $\xi_0 = 61 \pm 0.5$ Å, for $T_c = 330.32$. **(b)** A log-log representation of the low shear viscosity $\eta_{osc}$ vs. $\varepsilon$ (for comparison the hydrodynamic contribution to the viscosity $\eta_H$ is reported).

Ornstein-Zernike-Fisher form, $G(q\xi)$ so that the structure factor is given by:[14]

$$S(q) = \rho k_B T \chi_T G(q\xi) = \frac{\rho k_B T \chi_0 \xi_0^2 \varepsilon^{-\gamma}}{(1 + q^2\xi^2)^{1-\eta/2}} \tag{4}$$

where $\chi_T = \chi_0 \varepsilon^{-\gamma}$ is the osmotic compressibility, $\rho$ is the particle density, $\eta$ is the Fisher exponent ($\eta = 0.0315$) and $\xi = \xi_0 \varepsilon^{-\nu}$. The fitting of the experimental data, Fig. 5(a), by using Eq. (1) gives: $\gamma = 1.22 \pm 0.04$, $\nu = 0.61 \pm 0.02$, $T_c = 330.32$ and $\xi_0 = 61 \pm 0.5$ Å. These critical indices values are in agreement with the ones predicted by the theory ($\nu = 0.6$ and $\gamma = 2\nu$). Such a result, obtained in a dilute gelling system can be considered as a confirmation for the critical system structure of the suggestions coming out from the work of Coniglio et al.[12,13]

The situation could be well different in the critical region for the transport coefficients, like viscosity and diffusion coefficients. We have studied viscoelasticity at the percolation threshold obtaining an exponent $k \sim 0.72$, by considering that also near the percolation threshold holds the result that the mean percolating cluster size diverges as the susceptibility ($\xi_p \equiv \xi$). We can recalculate this exponent from the form $k = \nu(d_f + 2 - d)$ by assuming $\nu = 0.6$, instead its value for percolation ($\nu = 0.88$), obtaining $k \sim 0.9$. Such a value considering also the complexity of the screening of hydrodynamic interaction effects can be considered in agreement with the value measured experimentally at the percolation threshold from $\eta_{osc}$ and $\langle\tau\rangle$.

As it is well known the anomalous behavior of the shear viscosity near the critical point has been a subject of great interest.[30–32] The critical exponent for the viscosity of critical molecular systems is now accepted to be very small,[33,35] so that the major problem has been the determination of the background viscosity, that is, the viscosity that would have been measured in the absence of the long ranged critical correlations. Therefore, in the treatment of dynamical critical phenomena it is customary to separate the transport coefficients, such as the concentration conductivity $\alpha$ and the shear viscosity $\eta$, into background contributions $\alpha_B$ and $\eta_B$ and a non local ($q-$dependent) critical part $\alpha_c$ and $\eta_c$.[33] One has a set of coupled integrals which relate $\alpha_c$ and $\eta_c$ to the wave vector $q$ through the osmotic susceptibility $\chi(q)$. The shear viscosity can be written as the product of the $\eta_B(T)$ to a power-law divergence[34] $\eta(T) = \eta_B(T)(Q_0\xi)^\phi$, where $\phi$ is an universal critical exponent and $Q_0$ a system dependent amplitude given in terms of two different contributions: the Debye cutoff $q_D$, and a contribution related with the diffusivity background $q_C$, as $Q_0 = 2q_Dq_C/e^{4/3}(q_D + q_C)$. Considering the divergence in $\xi$, the viscosity is:

$$\eta(T) = [\eta_B(T)(Q_0\xi_0)^\phi]\varepsilon^{-\nu\phi} = [\eta_B(T)(Q_0\xi_0)^\phi]\varepsilon^{-y} \tag{5}$$

where $y$ is the viscosity critical index, mode-mode coupling equations yield $\phi = y/\nu$ and $\phi = 8/15\pi^2$ whereas refined theoretical calculations suggest $\phi = 0.062 \pm 0.005$[35] and $y \simeq 0.032$.[33] Equation (5) well focused the physics of transport coefficients in critical phenomena in a molecular system that is characterized by a net separation between critical and fluid dependent quantities (square brackets includes only these ones). On this respect, the two characteristic lengths $Q_0$ and $\xi_0$ are relevant because they determine a precise crossover from single-particle to critical behavior, the crossover temperature $T_x$ is found to be such that $Q_0\xi(T_x) \simeq 1$; such a situation has been observed in critical microemulsions.[36] By considering the current exponent values the critical anomaly of the shear viscosity is very weak (the accessible enhancement is of the order of $20 - 30\%$). It has been also proposed, by mode coupling[37] and renormalization group[38] calculations that the shear rate or the frequency can affect $\eta_c$ near $T_c$, i.e. a non-Newtonian viscous behavior sets. In such a case shear thinning can originate a "levelling off" of the measured viscosity as a function of the temperature on approaching the critical point. However, such a situation has been never properly tested with viscometers at very extremely small shear rates and stresses.

Recently, starting from the results of some experiments that give some indication that for macromolecular systems the critical divergence of the shear viscosity is more pronounced than for molecular system,[39] it has been suggested an argument of relevant interest for viscosity in critical phenomena.[40] It has been proposed that, due to the long range character of the hydrodynamic interaction that characterizes Brownian systems, the critical divergence of the shear viscosity in macromolecules it is completely different from that of molecular systems. On the contrary of the weak divergence observed in atomic and molecular systems colloidal polymers are characterized by a strong viscosity divergence; more precisely their zero shear viscosity diverges as strong as the correlation length ($\eta(T) \sim \xi$). In addition it has been suggested that such novel behavior can be observed only under the condition that the Peclet number system is $Pe < 1$ (i.e when shear thinning effects became ineffective). In such a model the critical viscosity of a colloidal system has been written as: $\eta(T) = \eta_B(T) + \eta_C(T)$; on the contrary of Eq. (5) $\eta_C(T)$ is an *additive* critical contribution, diverging as $\eta_C(T) \sim \xi$, to the background $\eta_B(T)$. An experiment on grafted colloidal particles (interacting via a deplection mechanism), by using the volume fraction, $\phi$, as order parameter, seems to confirm these suggestions giving a viscosity critical exponent equal to $1.04 \pm 0.06$ and a $\xi$ exponent equals to $1.05 \pm 0.04$, respectively.[40]

As our viscosity measurements have been performed in an experimental condition suitable to test these latter suggestions we have plotted the obtained $\eta_{osc}$ vs. $\varepsilon$, Fig. 5(b) (in the same graph we have reported also the solvent viscosity $D_2O$); we can discern only the viscosity divergence due to the percolation phenomenon, approaching the critical point there is evidence only of a round-off in the measured data. Therefore we do not have evidence of the strong divergence proposed for colloidal macromolecules, on the contrary the critical viscosity divergence, certainly present, in according to the well known findings of the mode coupling and the renormalization group theories it is very weak. By considering the cited results on microemulsion systems (a macromolecular Baxter system[36]) it certainly holds the physics represented by Eq. (5).

## 4. CONCLUSIONS

Light scattering data and the frequency scaling behavior of the shear moduli confirm for the studied system, an attractive copolymer micellar system, a percolation transition coexisting with the critical one. As a result, static light scattering data are sensitive only to the critical process, present findings seem in agreement with the suggestion of theoretical studies on critical polymer solutions,[12] for which clusters generated by the percolation process behave, in a diluted gelling system, like critical droplets. More precisely, these polydisperse and not compact clusters (Ising clusters) have the same properties as the droplets studied in the Fisher's droplet model,[14] i.e. at the critical point their connectedness length can be assumed as the diverging correlation length ($\xi_p \equiv \xi$) and the mean cluster size diverges as the susceptibility.

Completely different is the situation for transport or dynamical properties, like the viscosity or the average relaxation time. We observe at the percolation threshold a well defined viscoelasticity, we measure also the exponent for the percolating viscosity that has a value $k \approx 0.75$. Taking into account a model developed in terms of scaling arguments (radius dependent (fractal) viscosity) physically due to the screening of hydrodynamic interaction and entaglement effects we showed that such an exponent value well represents the true structure proposed for the system. Approaching the gelation, the sol-phase clusters are polydisperse, diluted with an internal structure dominated by swelling. Dynamic light scattering data confirm by means of the stretched exponential form of the ISF of the density correlation functions and from the measured values of the stretched exponent, the validity of such an approach. In addition, the averaged relaxation time, measured from the ISF, and directly linked at $k$, confirms for such an exponent the same value measured from the viscosity and the frequency dependence of the loss and elastic moduli, respectively.

These dynamical measurements, viscoelasticity and light scattering, suggest the idea that dynamics, also for a system (like the one studied in the present experiment) in which percolation and critical phenomena coexist, is strongly determined (or influenced) by percolation clustering. For dynamical quantities seem that the main role is certainly due to the strong background contribution.

In summary, our data suggest that there are only two possibilities to describe the percolation clustering effect in transport properties: (i) percolation affects these as a background contribution completely independent from the critical phenomenon so that it can be considered in Eq. (5) as an independent contribution in $\eta_B$; (ii) alternatively there is a possible interplay between the characteristic system diverging length and the two characteristic system lengths $Q_0$ and $\xi_0$.

The first possibility corresponds also to a situation in which $T_p$ and $T_c$ are very far one from the other, so that percolation and critical phase separation act almost independently. The second case could occur when the two critical temperatures are so closed. That seems to be the actual situation as unambiguously indicated by elastic light scattering. In such a case, the onset of the simultaneous divergence of $\xi_p$ and $\xi$ can originate a sort of interference between the concentration fluctuations and bond connectivity in determining the properties of the transport parameters (viscosity). As an extreme consequence of this it is possible that for the system dynamics, $\varepsilon$ gradually looses the role of a scaling variable with a breakdown of the universality. We conclude, pointing out that these latter phenomena typical of complex fluids deserve additional studies from both the experimental and theoretical points of view.

Finally we observe that our viscoelasticity measurements on approaching the critical point, performed at a very low strain ($Pe < 1$) when shear thinning effects are certainly ineffective, disagree with the hypothesis that Brownian interaction can originate a very strong viscosity divergence. We stress that the general findings of mode coupling and renormalization groups theory still remain valid also for a system characterized by clustering processes like the ones studied here.

## ACKNOWLEDGMENTS

We are happy to contribute this manuscript to this special issue celebrating the $60^{\text{th}}$ birthday of Professor Antonio Coniglio who has significantly contributed to the statistical physics. In particular, we report an experimental study on research arguments that have involved for many years its interest. The research at MIT is supported by a grant from Materials Science Division of US DOE. The research in Italy is supported by the PRA-INFM project.

Permanent address: Francesco Mallamace, Dipartimento di Fisica, Università di Messina, and Istituto Nazionale per la Fisica della Materia Sezione di Messina, Vill. S. Agata C.P. 55, 98166

## REFERENCES

1. See e.g., F. Mallamace and H. E. Stanley (eds.), *The Physics of Complex Systems* (IOP, Amsterdam, 1997).
2. S. H. Chen, J. S. Huang and P. Tartaglia (eds.), *Structure and Dynamics of Strongly Interacting Colloids and Supramolecular Aggregates in Solutions* (Kluwer, Dordrecht, 1992).
3. H. E. Stanley and N. Ostrowsky (eds.), *On Growth and Form* (Martinus Nijhoff Publishers, Dordrecht, 1986).
4. H. Verduin and J. K. G. Dhont, *J. Colloid Interface Sci.* **172**, 425 (1995); E. Bartsch, M. Antonietti, W. Shupp and H. Sillescu, *J. Chem. Phys.* **97**, 3950 (1992).
5. J. Bergenholtz and M. Fuchs, *Phys. Rev.* **E59**, 5706 (1999); L. Fabbian, W. Gœtze, F. Sciortino, P. Tartaglia and F. Thiery, **E59**, R1347 (1999).
6. R. J. Baxter, *J. Chem. Phys.* **49**, 2770 (1968).
7. S. V. G. Menon, C. Manohar and K. Rao Srinivasa, *J. Chem. Phys.* **95**, 9186 (1991); B. Barboy, *J. Chem. Phys.* **61**, 3194 (1974); Y. C. Chiew and E. D. Glandt, *J. Phys.* **A16**, 2599 (1983).
8. A. Coniglio, U. De Angelis and A. Forlani, *J. Phys.* **A10**, 1123 (1977).
9. S. H. Chen, C. Y. Ku and Y. C. Liu, in Ref. 1, p. 243.
10. L. Lobry, N. Micali, F. Mallamace, C. Liao and S. H. Chen, *Phys. Rev.* **E60**, 7076 (1999).
11. T. Tanaka, G. Swislow and I. Ohmine, *Phys. Rev. Lett.* **42**, 1156 (1979).
12. A. Coniglio, H. E. Stanley and W. Klein, *Phys. Rev. Lett.* **42**, 518 (1979).
13. A. Coniglio and W. Klein, *J. Phys.* **A13**, 2775 (1980).

14. M. E. Fisher, *J. Math. Phys.* **5**, 944 (1964); M. E. Fisher and R. Burford, *Phys. Rev.* **156**, 583 (1967).
15. Y. Liu, S. H. Chen and J. S. Huang, *Macromolecules* **31**, 2236 (1998).
16. P.-G. de Gennes, *J. Phys. (France) Lett.* **37**, L1 (1976).
17. F. Mallamace, P. Gambadauro, N. Micali, P. Tartaglia, C. Liao and S. H. Chen, *Phys. Rev. Lett.* **84**, 5431 (2000).
18. M. Adam, A. Coniglio and D. Stauffer, *Adv. Polymer Sci.* **44**, 103 (1982).
19. P. J. Flory, *Principles of Polymer Chemistry* (Cornell University Press, Ithaca, 1953).
20. M. Doi and S. F. Edwards, *The Theory of Polymer Dynamics* (Clarendon Press, Oxford, 1986).
21. A. Aharony and D. Stauffer, *Introduction to Percolation Theory* (Taylor and Francis, London, 1994).
22. J. E. Martin, J. P. Wilconox and J. Odinek, *Phys. Rev.* **A43**, 858 (1991); J. E. Martin and J. P. Wilconox, *Phys. Rev.* **A39**, 252 (1989); J. E. Martin, A. Douglas and J. P. Wilconox, *Phys. Rev. Lett.* **61**, 2620 (1988).
23. P. E. Rouse, *J. Chem. Phys.* **21**, 1272 (1953).
24. B. H. Zimm, *J. Chem. Phys.* **24**, 269 (1956).
25. D. Durand, M. Delsanti, M. Adam and J. M. Luck, *Europhys. Lett.* **3**, 297 (1987); M. Adam, M. Delsanti, J. P. Munch and D. Durand, *Phys. Rev. Lett.* **61**, 706 (1988).
26. H. J. Herrmann, B. Derrida and J. Vannimeus, *Phys. Rev.* **B30**, 4080 (1984). Y. Kantor and I. Webman, *Phys. Rev. Lett.* **56**, 2501 (1986).
27. E. Del Gado, L. de Arcangelis and A. Coniglio, *Eur. Phys. J.* **E2**, 359 (2000).
28. J. Kerttesz, *J. Phys.* **A16**, L471 (1983).
29. B. J. Berne and R. Pecora, *Dynamic Light Scattering* (Wiley, New York, 1976).
30. M. Fixman, *J. Chem. Phys.* **36**, 310 (1962); *J. Chem. Phys.* **47**, 2808 (1967); *Adv. Chem. Phys.* **6**, 175 (1964).
31. J. M. Deutch and R. Zwanzig, *J. Chem. Phys.* **46**, 1612 (1966).
32. K. Kawasaki, *Ann. Phys.* **61**, 1 (1970).
33. J. V. Sengers, in *Critical Phenomena*, ed. M. S. Green (Academic, New York, 1971); H. C. Burstyn, J. V. Sengers, J. K. Bhattacharjee and R. A. Ferrell, *Phys. Rev.* **A28**, 1567 (1983); J. K. Bhattacharjee and R. A. Ferrell, *Phys. Rev.* **A28**, 2363 (1983).
34. T. Ohta and K. Kawasaki, *Prog. Theor. Phys.* **55**, 1384 (1976); T. Ohta, *J. Phys.* **C10**, 791 (1977).
35. P. C. Hohenberg and B. I. Halperin, *Rev. Mod. Phys.* **49**, 435 (1977).
36. J. Rouch, A. Safouane, P. Tartaglia and S. H. Chen, *J. Chem. Phys.* **90**, 3756 (1989).
37. D. W. Oxtoby, *J. Chem. Phys.* **62**, 1463 (1975).
38. A. Onuki, *Physica* **A140**, 204 (1986); A. Onuki and K. Kawasaki, *Phys. Lett.* **A75**, 485 (1980).
39. C. Ishimoto and T. Tanaka, *Phys. Rev. Lett.* **39**, 474 (1977); K. Gruner, S. Habib and S. C. Greer, *Macromolecules* **23**, 510 (1990).
40. J. K. G. Dhont, *J. Chem. Phys.* **103**, 7072 (1995); *Phys. Rev. Lett.* **77**, 5304 (1995).

# THERMALLY DILUTED ISING SYSTEMS

MANUEL I. MARQUÉS and JULIO A. GONZALO
*Depto. de Física de Materiales C-IV,*
*Universidad Atónoma de Madrid, Madrid, Spain*
JORGE ÍÑIGUEZ
*Depto. de Física Aplicada II,*
*Universidad del Pais Vasco, Bilbao, Spain*

**Abstract**

In this paper finite size scaling techniques are used to study the universality class of thermally diluted Ising systems, in which the realization of the disposition of magnetic atoms and vacancies is taken from the local distribution of spins in the pure original Ising model at criticality. The critical temperature, the critical exponents and therefore the universality class of these thermally diluted Ising systems depart markedly from the ones of short range correlated disordered systems. This result is in agreement with theoretical predictions previously made by Weinrib and Halperin for systems with long range correlated disorder.

## 1. INTRODUCTION

During last decades the systems with quenched randomness have been intensively studied.[1] The Harris criterion[2] predicts that weak dilution does not change the character of the critical behavior near second order phase transitions for systems of dimension $d$ with specific heat exponent lower than zero in the pure case (the so called P systems), $\alpha_{pure} < 0 \Longrightarrow \nu_{pure} > 2/d$, being $\nu$ the correlation length critical exponent. This criterion has been supported by renormalization group (RG),[3–5] and scaling analyses.[6] The effect of strong dilution was studied by Chayes et al.[7] For $\alpha_{pure} > 0$ (the so called R systems), the Ising 3D case for example, the system fixed point flows from a pure (undiluted) fixed point towards a new stable fixed point[3–8] at which $\alpha_{random} < 0$. Recently Ballesteros et al. have used the Monte Carlo approach to study the diluted Ising systems in two,[9] three[10] and four dimensions.[11] The existence of a new universality class for the random diluted Ising system (RDIS), different from that of the pure Ising model and independent of the average density of occupied spin states ($p$), is proved, using an infinite volume extrapolation technique[10] based upon the leading correction to scaling. The critical exponents obtained this way could be compared with the experimental critical exponents for a random disposition of vacancies in diluted magnetic systems.[12]

In all cases previously mentioned frozen disorder was always produced in a random way, that is, vacancies were distributed throughout the lattice **randomly**. Real systems, however, can be realized with other kinds of disorder, where the vacancy locations are correlated. In particular, long range correlation (LRC) has been found in X-ray and Neutron Critical Scattering experiments in systems undergoing magnetic and structural phase transitions.[13,14] This effect has been modeled by assuming a spatial distribution of critical temperatures obeying a power law $g(x) \sim x^{-a}$ for large separations $x$.[15] In general these systems behave in a way very different from RDIS, since systems with randomly distributed impurities may be considered as the limit case of short range correlated (SRC) distributions of the vacancies. The basic approach to the critical phenomena of LRC systems was established by Weinrib and Halperin[16] almost two decades ago. They found that the Harris criterion can be extended for these cases, showing that for $a < d$, the disorder is irrelevant if $a\nu_{pure} - 2 > 0$, and that in the case of relevant disorder a new universality class (and a new fixed point) with correlation length exponent $\nu = 2/a$, and a specific heat exponent $\alpha = 2(a-d)/a$ appears. In contrast, if $a > d$, the usual Harris criterion for SRC systems is recovered. LRC disorder has been studied also by the Monte Carlo approach.[17] In this case a correlation function $g(x) = x^{-a}$ with $a = 2$ (defects consisting in randomly oriented lines of magnetic vacancies inside a three dimensional Ising system) confirmed the theoretical predictions of Weinrib and Halperin.

In the present paper we will study Ising three dimensional systems where the long range correlated dilution has been introduced as a thermal order-disorder distribution of vacancies in equilibrium, governed by a characteristic ordering temperature ($\theta$), in a way similar to the thermal disordering in a binary alloy of magnetic (spins) and non-magnetic atoms (vacancies)[18] (see Fig. 1). This idea has been also applied in percolation problems.[19]

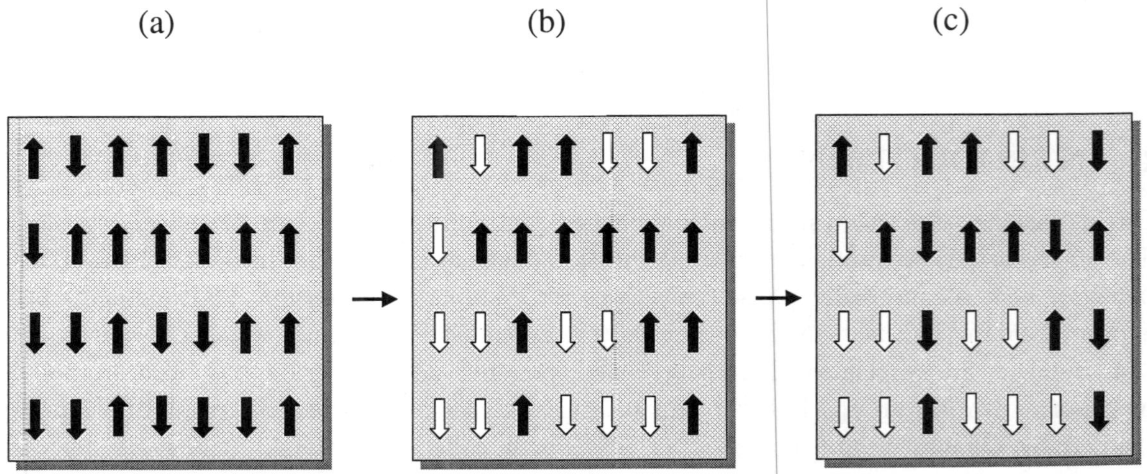


**Fig. 1** Steps to follow in order to perform a thermal dilution. **(a)** The system is thermalized to an ordering temperature ($\theta$). **(b)** One kind of spins (up or down) are turned into vacancies. **(c)** Vacancies are quenched and the system is thermalized to another temperature $T$.

Depending on the value of the ordering temperature we may distinguish clearly between:

(i) **Critically thermally diluted** Ising system realizations (TDIS), in which the quenched randomness is produced by considering a ferromagnetic Ising system at ($\theta = T_c^{3D}$): after thermalization, the spins of the dominant type (concentration $c \geq 0.5$) are taken as the locations of the magnetic atoms, and the rest are taken as the magnetic vacancies. The structure of the realization is fixed thereafter for all temperatures at which the magnetic interactions are subsequently investigated. In this case there is no relation between thermally diluted Ising systems and randomly diluted Ising systems (see Fig. 2).

(ii) **Hypercritical** Ising system realizations (obtained with $\theta \gg T_c^{3D}$). These systems are equivalent to the randomly diluted Ising systems (RDIS) with vacancy probability $p = 0.5$, resulting in $c \approx 0.5$. (see Fig. 3)

(iii) **Hipocritical** Ising system realizations (obtained with $\theta \ll T_c^{3D}$). These systems are equivalent to the randomly diluted Ising systems with vacancy probability near to $p = 1$, resulting in $c \approx 1$ (see Fig. 4).

So, if this ordering temperature $\theta$ determining the particular realization is high enough, the equilibrium thermal disorder will be very similar to the random (short range correlated) disorder of the usual previous investigations. On the other hand, if $\theta$ happens to coincide with the characteristic magnetic critical temperature ($T_c^{3D} = 4.511617$) of the undiluted system, we will have vacancies in randomly located **points**, but with a long range correlated distribution. [Note that the situation differs markedly from that of previously studied LRC systems, in which **lines** or **planes** of vacancies were considered]. The correlation of our TDIS is given by a value $a = 2 - \eta_{pure}$, where $\eta_{pure}$ is the correlation function exponent for the pure system. Since $d = 3$ and $\eta_{pure} = 0.03$ for the three dimensional Ising system, we have a long range correlated disorder with $a = 1.97 < 3 = d$. So we are in the case where

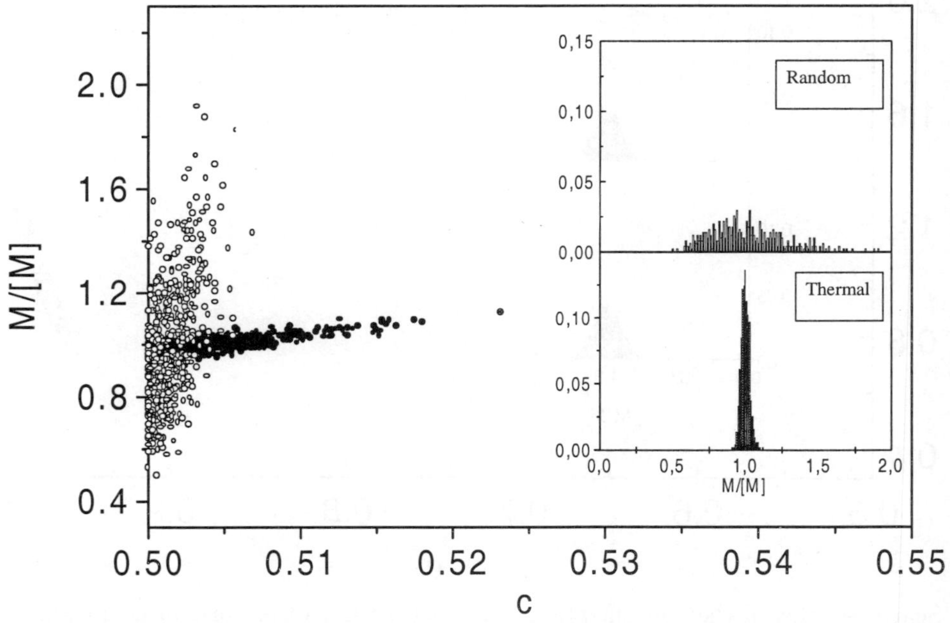


**Fig. 2** Scattered plot of the concentration of spins ($c$) versus the normalized magnetization per spin. Results are shown for the random case with $p = 0.5$ (white) and the thermal case with ($\theta = T_c^{3D}$) (black). In all cases magnetization is measured at the critical temperature of the random system. Inset shows the histograms for the values of the magnetization.

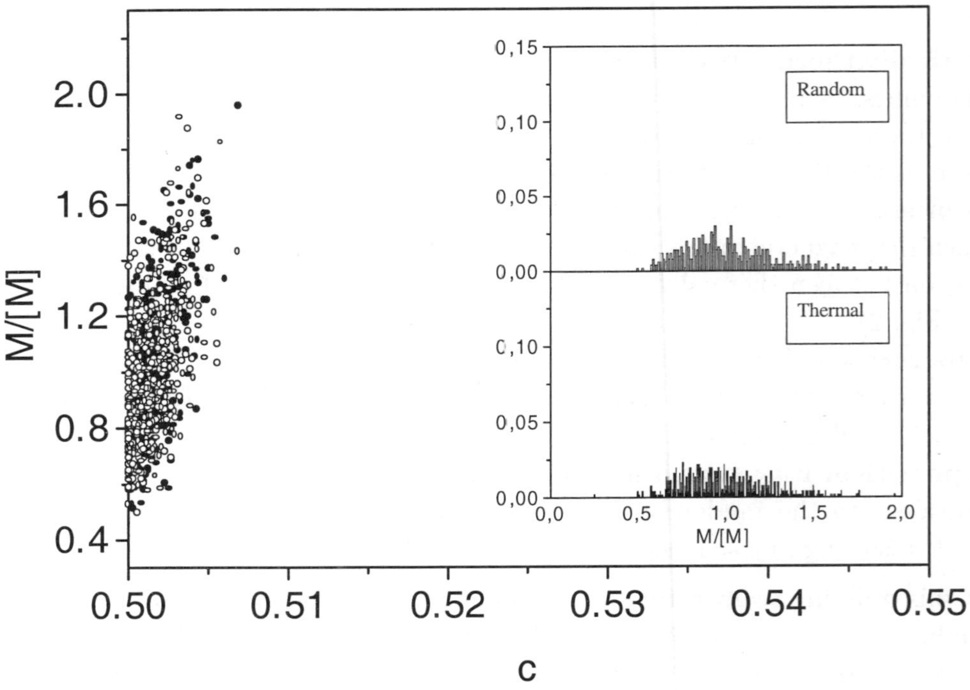


**Fig. 3** Scattered plot of the concentration of spins ($c$) versus the normalized magnetization per spin. Results are shown for the random case with $p = 0.5$ (white) and the thermal case with ($\theta = 1000$) (black). In all cases magnetization is measured at the critical temperature of the random system. Inset shows the histograms for the values of the magnetization.

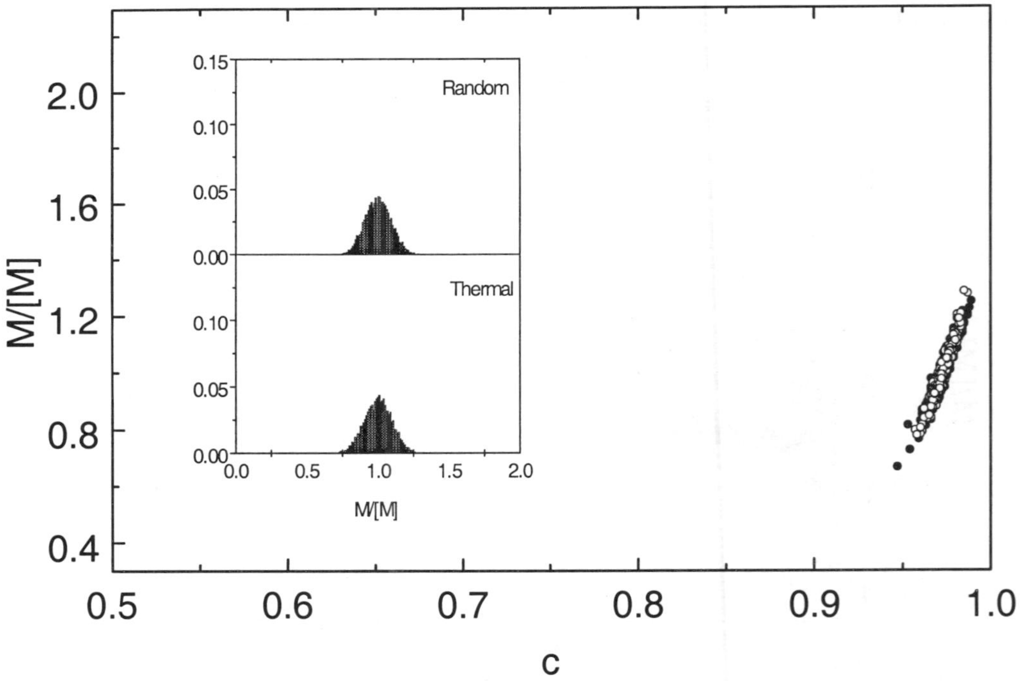


**Fig. 4** Scattered plot of the concentration of spins ($c$) versus the normalized magnetization per spin. Results are shown for the random case with $p = 0.97$ (white) and the thermal case with ($\theta = 3$) (black). In all cases magnetization is measured at the critical temperature of the random system. Inset shows the histograms for the values of the magnetization.

LRC disorder is relevant and we should detect a change of universality class with respect to the SRC case (following Weinrib and Halperin we expect for the thermally diluted Ising system $\nu \approx 1$ and $\alpha \approx -1$). Details about the construction of these thermally diluted Ising systems (TDIS) can be found in Ref. 18. In the present work we study the critical behavior and the university class of three dimensional TDIS at criticality using the Monte Carlo approach. We will compare our results with the critical behavior of the RDIS.

The structure of the paper is as follows: In Sec. 2 we study the dependence of the critical temperature (and of the self-averaging at criticality) with the size of the system for both TDIS and RDIS. Once the critical point is determined, we investigate whether or not TDIS and RDIS belong to the same universality class. In order to proceed, we study the critical behavior of both kinds of systems by applying finite size scaling techniques (Sec. 3) and by using the effective-exponent approach (Sec. 4). A summary of the main results and concluding remarks are given in Sec. 5.

## 2. TRANSITION TEMPERATURE

For a hypercubic sample of linear dimension $L$ and number of sites $N = L^d$, any observable singular property $X$ has different values for the different random realizations of the disorder, corresponding to the same dilution probability $p$ (grand-canonical constraint). This means that X behaves as a stochastic variable with average $\bar{X}$ (in the following, the bar indicates average over subsequent realizations of the dilution and the brackets indicate MC average). The variance would then be $(\Delta X)^2$, and the normalized square width, correspondingly:

$$R_X = (\Delta X)^2/\bar{X}^2 \tag{1}$$

A system is said to exhibit self-averaging (SA) if $R_X \to 0$ as $L \to \infty$. If the system is away from criticality, $L \gg \xi$ (being $\xi$ the correlation length). The central limit theorem indicates that strong SA must be expected in this case. However, the behavior of a ferromagnet at criticality (with $\xi \gg L$) is not so obvious. This point has been studied recently for short range correlated quenched disorder. Aharony and Harris (AH), using a renormalization group analysis in $d = 4 - \varepsilon$ dimensions, proved the expectation of a rigorous absence of self-averaging in critically random ferromagnets.[20] More recently, Monte Carlo simulations were used to investigate the self-averaging in critically disordered magnetic systems.[10,21,22] The absence of self-averaging was confirmed. The normalized square width $R_X$ is an universal quantity affected just by correction to scaling terms. LRC diluted systems are expected to have different critical exponents and different normalized square widths with respect to those of the usual randomly disordered systems studied previously.

We perform Monte Carlo calculations of the magnetization and the susceptibility ($\chi = (\langle M^2 \rangle - \langle M \rangle^2)/T$) per spin at different temperatures for different realizations of TDIS, and for randomly diluted systems with $p = 0.5$ (restricting to $c > 0.5$) using in both cases the Wolff[23] single cluster algorithm[24] with periodic boundary conditions, on lattices of different sizes $L = 10, 20, 40, 60, 80, 100$. Results for susceptibility vs. temperature are shown in Fig. 5.

Note how due to the existence of randomness the susceptibility points do not collapse into a single curve, since each realization has a different value of the critical temperature and of the concentration. This is even more clear for small values of L and for the critically thermal case (white points). Figure 5 indicates that the critical temperature of the TDIS clearly differs from that of the RDIS, and also than the effect of the dilution on the lack of self-averaging seems to be stronger in the thermally diluted Ising case. The critical temperature

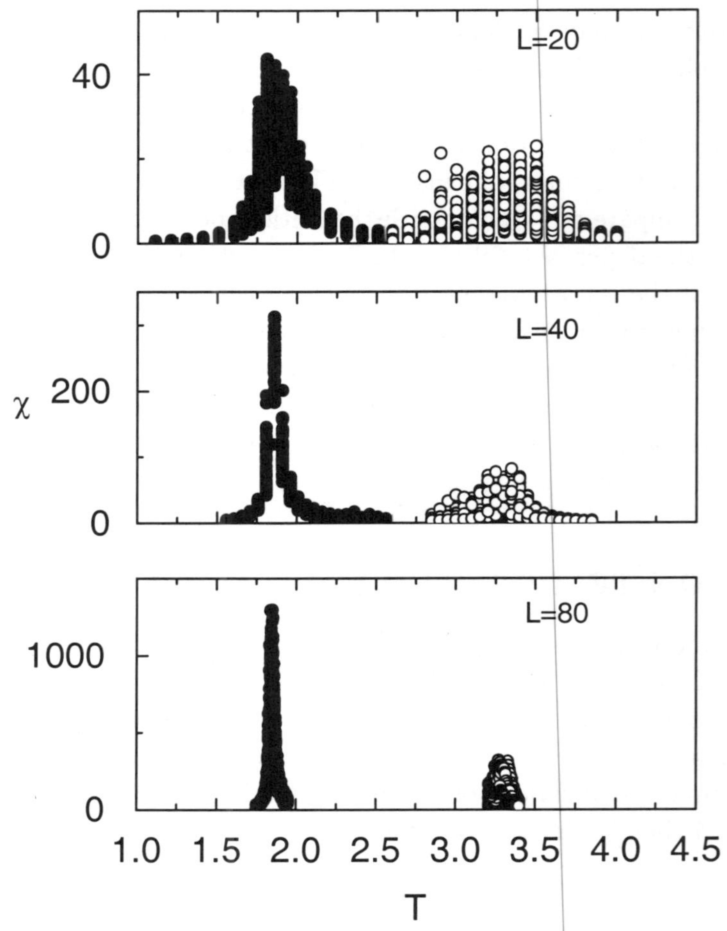


**Fig. 5** Susceptibility $\chi$ vs. temperature $T$ for random dilution realizations ($p = 0.5$) (black) and critical thermal dilution realizations (white). The size of the systems under consideration is given by $L = 20, 40, 80$.

may be obtained at the point where the normalized square width for the susceptibility, $R_\chi$, reaches its maximum. A different value of the critical temperature does not imply, of course, a different universality class. However from the dependence of the critical temperature with the length of the system we expect to detect a change in universality between TDIS and RDIS following the scaling relation:

$$T_c(L) = T_c(\infty) + AL^{-1/\nu} \tag{2}$$

being $\nu$ the critical exponent associated with the specific system's correlation length. This critical exponent has been determined by means of Monte Carlo data for the random case by Ballesteros et al.[10] They found a value $\nu_{random} = 0.683$. On the other hand the result by Weinrib and Halperin[16] indicates that the critical exponent expected for the thermal case should be $\nu_{thermal} = 2/a = 1.015$. Figure 6 represents the dependence of the critical temperature with respect to the length of the systems for random and thermal dilutions.

In both cases a fit to Eq. (2) has been performed for both values $\nu = \nu_{random} = 0.683$ (continuous line) and $\nu = \nu_{thermal} = 1.015$ (dashed line). Note how the thermal data fit better Eq. (2) for $T_c(L)$ using $\nu_{thermal}$, indicating a possible change in universality class with respect to the random case. [If we fit the data leaving all parameters free, we find $\nu_{random} \approx 0.7$ and $\nu_{thermal} \approx 1.2$, which are very near the expected results]. The extrapolated values of critical temperatures for infinite systems obtained this way are $T_c^{random}(\infty) = 1.845 \pm 0.003$

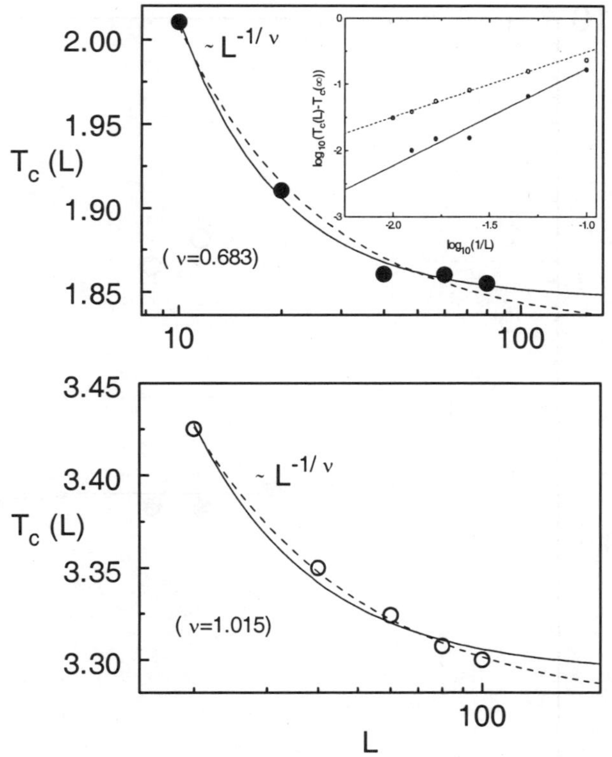


**Fig. 6** Semi-log representation of critical temperature $T_c$ vs. length $L$ for random dilution ($p = 0.5$) (black) and for thermal dilution (white). The continuous line is the fit obtained using $\nu = 0.6837$ (short range correlated random exponent) and dashed line the fit obtained using $\nu = 1.015$ (long range correlated exponent, a = 1.97) in either case. Note the diference on slopes in the log-log inset.

(close to the values previously obtained by Ballesteros et al.[10]) and $T_c^{thermal}(\infty) = 3.269 \pm 0.002$ (clearly different from $T_c(\infty)$ for the SRC case). Incidentally $\nu_{thermal}$ can be compared with $\nu$ for the observed sharp component in neutron scattering line shapes, which is around 1.3 for Tb.[14] This point deserves more careful analysis and will be taken up elsewhere.

Once the critical temperatures are known we can perform simulations for the magnetization and the susceptibility at criticality for several realizations of thermal and random diluted systems in an effort to determine the value of $R_\chi$, the normalized square width for the susceptibility. We went up to 500 realizations for $L = 10, 20, 40$ and up to 200 realizations for $L = 60, 80$. Results are shown in Fig. 7.

The arrow represents the (concentration independent) $R_\chi$ value obtained by Ballesteros et al.[10] The straight continuous line represents the average value obtained for random dilution data, and the straight dashed line gives the average value obtained for thermal dilution data. Note that the TDIS presents a lack of self-averaging around one order of magnitude larger than the RDIS. We have already presented a similar analysis for both kinds of dilution,[18] but only at the critical temperature **characteristic of the random system**. Our results are not precise enough to specify accurately the evolution of the normalized square width as a function of L governed by corrections to scaling terms. However the average we obtain for $R_\chi^{random} = 0.155$ is close to the value previously reported[10] by means of infinite volume extrapolations. For the thermal case we obtain $R_\chi^{thermal} = 1.19$, about one order of magnitude larger than for random dilution. In this case an evolution of $R_\chi^{thermal}$ vs. $L$ given by correction to scaling terms may be also expected, but according to Weinrib and Halperin[16] the analysis would be even more complicated, due to the fact that the long range correlated disordered systems present complex oscillating corrections to scaling.

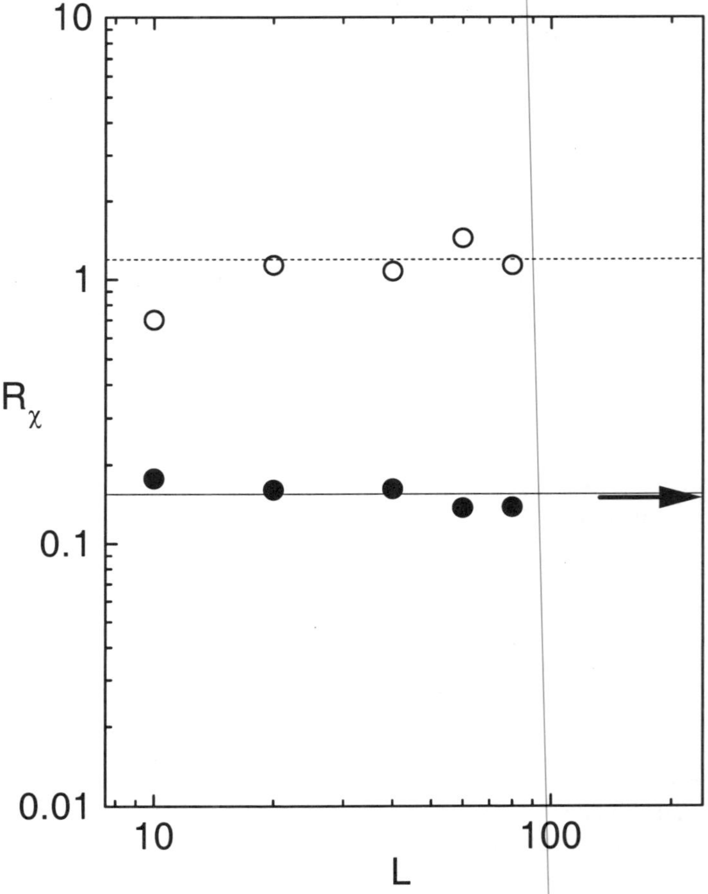


**Fig. 7** Log-log plot of the normalized square width for the susceptibility at criticality $R_\chi$ vs. length $L$ for random dilution ($p = 0.5$) (black) and thermal dilution (white).

## 3. CRITICAL BEHAVIOR OF THERMALLY DILUTED SYSTEMS

The dispersion in concentration and magnetization at criticality between the different realizations is shown at a glance in scattered plots as in Ref. 18. Each point in Fig. 8 represents a single realization with magnetization at criticality ($M$) and concentration ($c$). Note that in both cases (TDIS and RDIS), the dispersion on the magnetization and on the concentration decreases with L, but this is more clearly shown in the thermal case. Figure 8 shows clearly the difference in behavior between the **random** and the **thermal** cases, at least up to the values of $L$ considered.

From Fig. 8, we can extract averaged values for the magnetization and the inverse susceptibility, $\bar{M}$ and $\overline{\chi^{-1}}$ (with $\chi = \langle M^2 \rangle$ at the critical point) for the TDIS. Both averaged values are expected to fit the following scaling laws at criticality:

$$\overline{M(L)} \propto L^{-\beta/\nu} \tag{3}$$

$$\overline{\chi^{-1}(L)} \propto L^{-\gamma/\nu} . \tag{4}$$

Considering $(1/\nu)_{thermal} = a/2 = 0.985$, we can obtain from our data the values of $\beta_{thermal}$ and $\gamma_{thermal}$ (see f.i. the fitting for the inverse of the susceptibility in Fig. 9). We get $\beta_{thermal} = 0.56 \pm 0.05$ and $\gamma_{thermal} = 1.91 \pm 0.06$, very close to the predicted values by

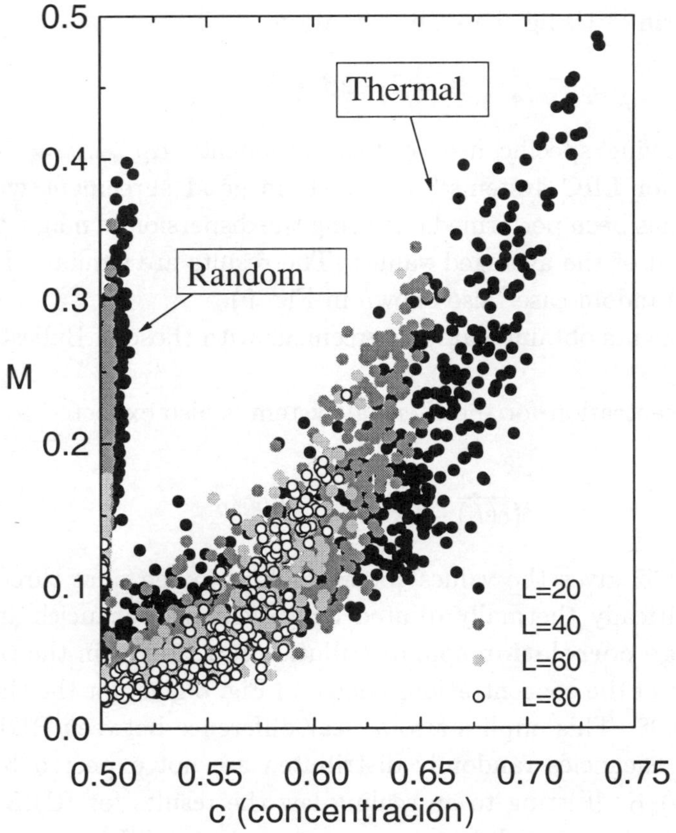


**Fig. 8** Scattered plot of magnetization $M$ at the critical temperature vs. concentration $c$ for the realizations considered of the random ($p = 0.5$) and the thermal dilutions ($L = 20, 40, 60, 80$).

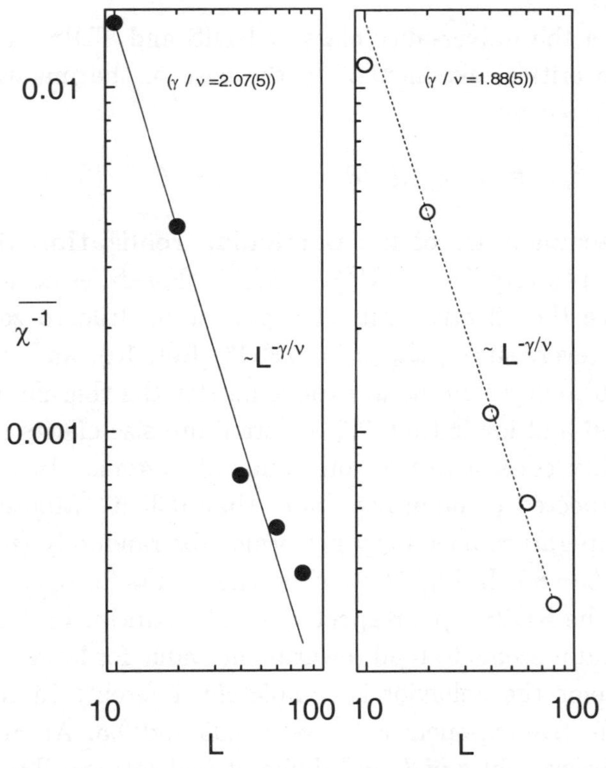


**Fig. 9** Average value of the inverse of the susceptibility $\overline{\chi^{-1}(L)}$ vs. the length of the system $L$ for the random dilution ($p = 0.5$) case (black) and the thermal dilution case (white). Continuous line and dashed line indicate fits to the random and the thermal cases, respectively.

Weinrib and Halperin.[16] Using the scaling relation:

$$\alpha = 2 - 2\beta - \gamma \tag{5}$$

we obtain the following specific heat critical exponent: $\alpha_{thermal} = -1 \pm 0.1$. Weinrib and Halperin give for LRC systems[16] $\alpha = -1$, in good agreement with our result. An analogous analysis has been performed but using the dispersion in magnetization and inverse susceptibility instead of the averaged values. The results are similar. The same study has been made for the random case (also shown in Fig. 9).

The critical exponents obtained are in agreement with those of Ballesteros et al.[10] within our error bars.

The average concentration for the thermal system is also expected to show a scaling law behavior given by:

$$(\overline{c(L)} - 0.5) \propto L^{-(\beta/\nu)_{3D}} \tag{6}$$

where $(\beta/\nu)_{3D} \simeq 0.52$ gives the values corresponding to the pure three-dimensional Ising case, because in critically thermally diluted Ising systems, vacancies are distributed with the same **long range correlation** spin distribution function as in the **pure** case.[18] The fit to the average value of the concentrations shown in Fig. 8 give, for the thermal case, a value $(\beta/\nu)_{3D} \simeq 0.55 \pm 0.08$. This implies also a clear difference between RDIS and TDIS, since Ising systems with vacancies randomly distributed are not expected to follow an scaling behavior with $(\beta/\nu)_{3D}$. [Fitting to an scaling law the results for RDIS gives an exponent around 1.4, which implies a much faster convergence to $c = 0.5$.]

## 4. EFFECTIVE CRITICAL EXPONENTS

The difference between the universality class of RDIS and TDIS can be detected also by means of the effective critical exponents. In the case of the magnetization the effective critical exponent is defined by:

$$\beta_{eff} \equiv \partial \log(M)/\partial \log(t) \quad (t = T_c^i - T) \tag{7}$$

with $T_c^i$ the critical temperature of the **particular realization** (i) characterized by a maximum of the susceptibility ($\chi = \langle M^2 \rangle - \langle M \rangle^2$). For $L \to \infty$ and $t \to 0$, $\beta_{eff} = \beta$. Finite size effects force the effective critical exponent to drop to zero before the critical value is attained. However, since $\beta_{random} = 0.3546$ (Ref. 10), and $\beta_{thermal}$ is expected to be around 0.5 (Ref. 16), effective critical exponents (for the thermal case) may rise to values greater than 0.3546 and lower than 0.5, before finite size effects appear, indicating the difference of universality class between both kinds of systems. In the RDIS the effective critical exponent is expected to be always lower than 0.3546. Monte Carlo simulations of magnetization vs. temperature have been performed for randomly ($p = 0.5$) and thermally diluted systems, with $L = 80$. In Fig. 10 we show the results for $\beta_{eff}$ vs. $\log(t)$ for two samples of the TDIS and the RDIS type, respectively. The **random** effective critical exponent is always under 0.35 and it seems to tend towards this value for large enough $L$, as expected, but for **thermal** systems the behavior is completely different: In the figure the value of the critical thermal effective exponent is between 0.35 and 0.5. An analogous investigation has been done for different values of $L$ and different realizations. The same effect has been found for lower $L$ values. However, we may note that in TDIS the effective critical exponent arrives at the maximum in a very different way depending on the realization. The reason

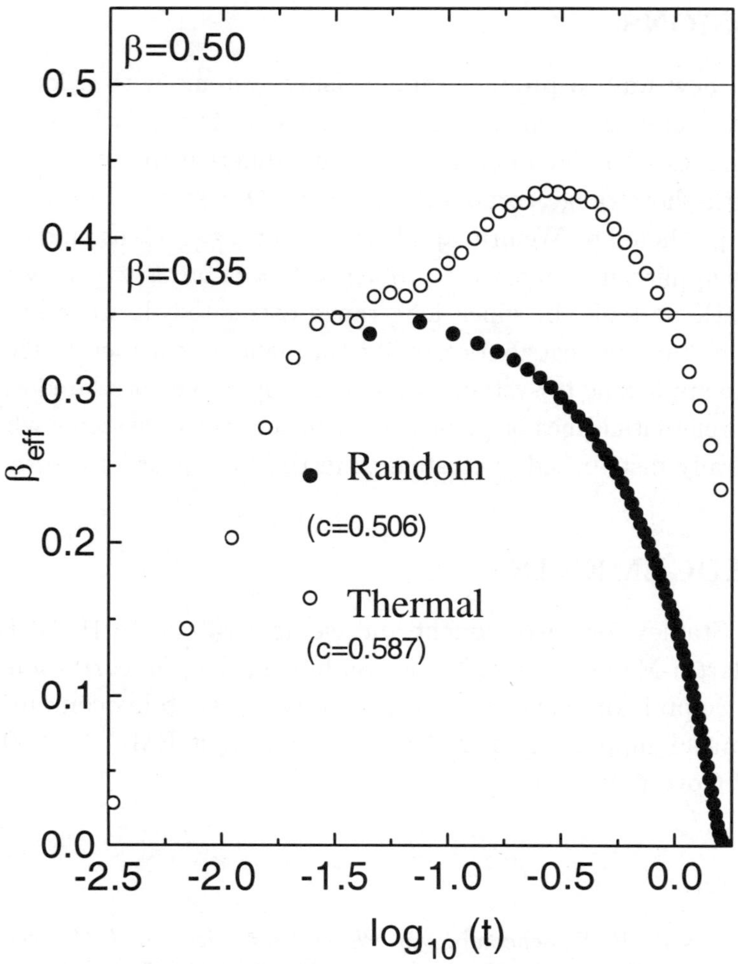


**Fig. 10** Effective critical magnetization exponent $\beta_{eff}$ vs. $\log(t)$, with $t = T_c^i - T$, for the thermal dilution case (white) and the random dilution ($p = 0.5$) case (black). The realizations considered correspond to $L = 80$.

is twofold: (1) the different **disposition** of the vacancies in each particular realization, and (2) the large differences in **concentration** for the thermal case (the rise of $\beta_{eff}$ towards the expected diluted universality value should be faster for c closer to 0.5). It may be noted that as the size ($L$) of the sample increases the possibility of local inhomogeneities in the TDIS realizations increases, giving rise to such phenomena as pseudo double peaks in the susceptibility, reduced values for the overall critical exponents, etc. However, it is important to remark than in all realizations investigated the $\beta_{eff}$ values of the TDIS (before finite size effects take over), have been clearly superior to the $\beta$ for the RDIS.

In Fig. 10 the effective critical exponent for the thermal case seems to produce a crossover towards $\beta = 0.35$ as the temperature approaches the critical point, before finite size effects finally appear. In principle this might be influenced by the indetermination in the measurement of the particular critical temperature of the realization. However, this crossover may point out that different length scales might be important at different distance from the critical point. In principle characteristic lengths such as the size of the fixed vacancies and spin clusters, the size of the system itself and the thermal spin fluctuations might all play a role. The possibility that for $L \to \infty$ and $c = 0.5$ the apparent crossover could disappear all together should not be excluded, entailing no crossover from critical thermal to random universality class.

## 5. CONCLUSIONS

To summarize, a new way to produce diluted Ising systems with a long range correlated distribution of vacancies has been analyzed by means of Monte Carlo calculations and finite size scaling techniques. We find an universality class different from that reported for diluted Ising systems with short range correlated disorder. Our systems may be included in the universality class predicted by Weinrib and Halperin for $a \approx 2$. This kind of thermal disorder had been already applied in percolation problems, but as far as we know it had never been applied to magnetic systems, in which long range correlated disorder has been previously produced mostly by random distribution of lines or planes of vacancies. The present dilution procedure may be applicable to systems where the long range correlated disorder is not due to dislocations, preferential lines or planes of vacancies, but to systems where the vacancies (points) are critically distributed in clusters as in the case of order-disorder systems.

## ACKNOWLEDGEMENTS

We thank H. E. Stanley for encourament and we are grateful to H. G. Ballesteros, L. A. Fernández, V. Martín-Mayor, and A. Muñoz Sudupe for helpful correspondence. We thank P. A. Serena for helpful comments and for generous access to his computing facilities. We acknowledge financial support from DGCyT through grant BMF 2000-0032 and from the Basque Regional Government (J. I.).

## REFERENCES

1. For a review, see R. B. Stinchcombe, in *Phase Transitions and Critical Phenomena*, eds. C. Domb and J. L. Lebowitz (Academic, New York, 1983), Vol. 7.
2. A. B. Harris, *J. Phys.* **C7**, 1671 (1974).
3. T. C. Lubensky and A. B. Harris, in *Magnetism and Magnetic Materials*, eds. C. D. Graham, G. H. Lander and J. J. Rhyne, AIP Conf. Proc. No. 24 (AIP, New York, 1975), p. 99; A. B. Harris and T. C. Lubensky, *Phys. Rev. Lett.* **33**, 1540 (1974).
4. T. C. Lubensky, *Phys. Rev.* **B11**, 3573 (1975).
5. G. Grinstein and A. Luther, *Phys. Rev.* **B13**, 1329 (1976).
6. A. Aharony, in *Phase Transitions and Critical Phenomena*, eds. C. Domb and M. S. Green (Academic, New York, 1976), Vol. 6, p. 357.
7. J. T. Chayes, L. Chayes, D. S. Fisher and T. Spencer, *Phys. Rev. Lett.* **57**, 2999 (1986).
8. W. Kinzel and E. Domany, *Phys. Rev.* **B23**, 3421 (1981); D. Andelman and A. N. Berker, *Phys. Rev.* **B29**, 2630 (1984).
9. H. G. Ballesteros, L. A. Fernández, V. Martín-Mayor, A. Muñoz Sudupe, G. Parisi and J. J. Ruiz-Lorenzo, *J. Phys.* **A30**, 8379 (1997).
10. H. G. Ballesteros, L. A. Fernández, V. Martín-Mayor, A. Muñoz Sudupe, G. Parisi and J. J. Ruiz-Lorenzo, *Phys. Rev.* **B58**, 2740 (1998).
11. H. G. Ballesteros, L. A. Fernández, V. Martín-Mayor, A. Muñoz Sudupe, G. Parisi and J. J. Ruiz-Lorenzo, *Nucl. Phys.* **B512**[FS], 681 (1998).
12. R. Folk, Yu. Holovatch and T. Yavors'kii, *cond-mat/9909121.*
13. T. R. Thurston, G. Helgesen, D. Gibbs, J. P. Hill, B. D. Gaulin and G. Shirane, *Phys. Rev. Lett.* **70**, 3151 (1993); T. R. Thurson, G. Helgesen, J. P. Hill, D. Gibbs, B. D. Gaulin and P. J. Simpson, *Phys. Rev.* **B49**, 15730 (1994).
14. P. M. Gehring, K. Hirota, C. F. Majkrzak and G. Shirane, *Phys. Rev. Lett.* **71**, 1087 (1993); K. Hirota, G. Shirane, P. M. Gehring and C. F. Majkrzak, *Phys. Rev.* **B49**, 11967 (1994).
15. M. Altarelli, M. D. Núñez-Regueiro and M. Papoular, *Phys. Rev. Lett.* **74**, 3840 (1995).
16. A. Weinrib and B. I. Halperin, *Phys. Rev.* **B27**, 413 (1983).
17. H. G. Ballesteros and G. Parisi, *cond-mat/9903230.*

18. M. I. Marqués and J. A. Gonzalo, *Phys. Rev.* **E60**, 2394 (1999); M. I. Marqués, J. A. Gonzalo and J. Íñiguez, *Phys. Rev.* **E62**, 191 (2000).
19. W. Klein, H. E. Stanley, P. J. Reynolds and A. Coniglio, *Phys. Rev. Lett.* **41**, 1145 (1978).
20. A. Aharony and A. B. Harris, *Phys. Rev. Lett.* **77**, 3700 (1996).
21. S. Wiseman and E. Domany, *Phys. Rev. Lett.* **81**, 22 (1998).
22. A. Aharony, A. B. Harris and S. Wiseman, *Phys. Rev. Lett.* **81**, 252 (1998).
23. U. Wolff, *Phys. Rev. Lett.* **62**, 361 (1989).
24. S. Wang and R. H. Swendsen, *Physica (Amsterdam)* **167A**, 565 (1990).

# CRITICAL FLUCTUATIONS IN THE BREAKDOWN OF DISORDERED SYSTEMS

ALBERTO PETRI
*Istituto di Acustica "O.M. Corbino", Consiglio Nazionale delle Ricerche,*
*via del Fosso del Cavaliere 100, 00135 Rome, Italy*
*Dipartimento di Fisica e Istituto Nazionale di Fisica della Materia,*
*Università La Sapienza, P.le A. Moro 2, 00185 Rome, Italy*

**Abstract**

In this paper some critical aspects of the behaviour of breaking lattices subject to slow driving forces are briefly reviewed. In particular, fluctuations in the response to external solicitations are discussed.

## 1. INTRODUCTION

Antonio Coniglio's contribution to the understanding of rupture phenomena in disordered systems is twofold. On one hand he has contributed indirectly to the field, by furnishing fundamental knowledge indispensable for dealing with dilute networks, like e.g. the exponent for the scaling of the density of cutting bonds in a percolating lattice.[1] On the other hand he has also pointed out important properties of the response of these systems, such as the multifractality of currents in diluted[2] and two component[3] resistor networks. These studies have continued in the observation, investigation and explanation of a large quantity of scaling properties in breaking lattices,[4,5] and the great amount of work done in this field has shown that scale invariance is a fundamental property of rupture and breakdown phenomena of disordered systems.[4,5] Indeed, disorder plays a fundamental role in generating this kind of behaviour.[6]

Scale invariance has been found to be also a more general characteristic of the dynamical response of these systems to the variation of some external parameter, and also a characteristic of the fluctuations of their response. For this reason, the adjective "critical" is often used, in analogy with the fluctuations of thermal systems at the critical point. In the following we shall mainly focus on this aspect of the breakdown phenomena and will consider in particular those situations in which an external solicitation (driving) is varied slowly with respect to the characteristic relaxation times of the system. After briefly reviewing some experimental results (Sec. 2), we shall describe how some of the observed features are reproduced by lattice models (Sec. 3), and what some current points of view on this subject are (Sec. 4).

## 2. SOME EXPERIMENTAL FACTS

Initial discoveries about the critical response of a fracturing medium are probably due to Mogi.[7] With the aim of verifying the validity of the Gutenberg Richter law also at scales much smaller than those involved in an earthquake, he designed and performed some original experiments where a controlled pressure was applied to disks made of a mixture of resin and hard grains. He observed that, by increasing pressure, elastic waves were released by some localised region within the sample. By recording the maximum wave amplitude of each series Mogi was able to show that the relation between the observed amplitudes and their frequency of occurrence was of the algebraic type, the one known as the Ishimoto and Iida's law in the seismologic field. These findings, besides to show in a quantitative way that earthquakes and fractures have common features, demonstrated the intrinsic critical nature and the importance of disorder in the response of a medium to the external solicitations.

Laboratory experiments can well reproduce not only the critical features observed in the energy release distibution by earthquakes, but also those observed in the time distribution. In 1968 Scholz succeeded in reproducing Omori's power law describing the number of aftershocks observed in a time $t$ after a main event.[8] By analysing the acoustic emission from a fracturing basalt rock he also coonfirmed the validity of the Gutenberg-Richter law.[9] In subsequent investigations Hirata[10] shewed that Omori's law holds also for generic microfracturing processes, at least for times large enough after the main event.

Thanks to the advances in technology and informatics, experiments in this fields continue to bring new results. By collecting and analyzing acoustic emission signals from concrete-like samples, we have shown[11] that power laws not only describes the frequency distribution of maximum amplitudes in the aftershock series, but also the amplitude distribution of the entire time series, as well as the time lags distribution between consecutive meaningful events of acoustic emission. We have found in particular that the energy (proportional to the squared amplitude) is distributed according to $N(E) \simeq E^{-\delta}$ with $\delta$ around 1.3, while for time lags $P(t) \simeq t^{-\zeta}$ with $\zeta$ approx 1.6.[12] In addition, lacking of characteristic scales in the fluctuations of response has been brought into evidence by measuring the power spectrum and the autocorrelation of the time series. The power spectrum has been found of the $1/f^{\gamma}$ form, with $\gamma \simeq 0.6$.[11,12]

In more recent experiments Ciliberto and coworkers[13,14] found $\delta \simeq 1.25$ for the energy scaling and conjectured its universality, whereas $\zeta$ was found to depend on the applied external stress. Acoustic emission recorded during pressurization of spherical tanks yielded still more evidences of the critical nature of breakdown phenomena and revealed the presence of logarithmic oscillations in the power laws.[15] Finally, Maes and coworkers[16] observed scale invariance in acoustic emission amplitudes, time lags, and spatial distance between consecutive events also in a cellular glass, where they found $\zeta = 1.3$, but $\delta = 2$.

Another important evidence of the critical response of disordered media concerns the roughness of the fracture surfaces.[17] After Mandelbrot and coworkers measured self-affinity properties in the fracture surface of some steels, similarity in the exponents (about 0.8) characterizing many different materials has been pointed out.[18,19] Further experiments have shown later that also 0.5 is observed.[20] According to some evidences lower values seem to characterize slow producing cracks and roughness at small scales, whereas higher values are related to large scales and should be associated with fast cracks.[21,30]

## 3. LATTICE MODELS

Existence of critical fluctuations in the response of model systems was firstly observed[22,23] in the fibre bundle model,[24] where an external applied stress is evenly shared by a stretched bundle of fibers. The elastic modulus is the same for all the fibers, but each one can stand a different, finite, amount of stress. The system is initially unloaded, then stress is applied in order to break the weaker fiber of the bundle. The excess stress is shared by the other fibres, that therefore become more prone to break, and when some fibre exceeds its own failure threshold it also breaks, so stress is redistributed again. When there no more fibres break, stress is increased again up to break the weakest of the survived fibers, and so on until all the fibres of the bundle are broken. Each breaking process is carried on at constant stress. By exact calculation Hansen and Hammer were able to show that as the applied stress is increased from zero to the global failure value, there is a probability $P(s) \propto s^{-\alpha}$ that $s$ fibers break in correspondence of the same value of applied stress. The exponent was found to be $\alpha = 5/2$ and largely independent of the statistical distribution of fiber strengths. Hansen and Hammer also observed power law distributions for "avalanches" of broken fuses in the numerical simulation of a square resistor network where, in analogy with he fiber bundle model, each fuse posseses the same conductance, but can stand different maximum current. When a fuse finds itself above its own threshold value of rupture, it breaks and the excess current is shared by the other fuses. It may happen therefore to some other of them to burn in turn. They found numerically $\alpha = 2.7$ for this system, very close to $5/2$.

Presence of similar behaviour was later observed in system with vector elasticity by numerical simulations of the Born model on a triangular lattice.[25] In this model sites interact via the potential

$$V_{ij} = (\alpha - \beta)[(\vec{u}_i - \vec{u}_j) \cdot \vec{r}_{ij}]^2 + \beta[\vec{u}_i - \vec{u}_j]^2, \tag{1}$$

where $\vec{u}_i$ is the displacement vector of site $i$ from equilibrium, $\vec{r}_{ij}$ is the unit vector between the initial equilibrium position of sites $i$ and $j$, and $\alpha$ and $\beta$ are force constants. In the case of Ref. 25 avalanches were also triggered by a corrosion mechanisms, according to which the bonds neighbouring a broken bond are weakened. Such a mechanism enhances the critical properties of the system,[26] and the related exponent was found to be $\alpha = 2.0$, the same found in the experiments on the cellular glass.[16]

The power law distributions mentioned above are computed by considering all the avalanches occurring during the whole life of the system. One can also consider what happens for a given value of the external solicitation (stress, strain, current) $\sigma$. For the fibre bundle it has been shown that the following scaling form holds:[23]

$$P(s, \sigma) = f\left(\frac{s}{s_0}\right) s^{-\tau} \tag{2}$$

where

$$s_0 \approx (\sigma_c - \sigma)^{-\kappa}, \tag{3}$$

and $\tau = 1.5$ and $\kappa = 1$; $\sigma_c$ represents the critical value of stress at which the network definitively tears. Thus, if one cumulates all the avalanches from the beginning, $\sigma = 0$, to the end of the life of the system, $\sigma = \sigma_c$, one observes

$$P(s) = s^{-\alpha} \tag{4}$$

with $\alpha = \tau + 1/\kappa = 2.5$. Zapperi et al.[31] succeeded to show that within the effective medium approximation the fuse network observes the same scaling of the fibre bundle model, and the ensemble averaged burst size $\langle s \rangle$ diverges at criticality as

$$\langle s \rangle \approx (\sigma_c - \sigma)^{-\gamma} \tag{5}$$

with $\gamma = 1/2$. This results have been confirmed numerically both on square fuse networks[31] and on elastic networks with central and bond bending potentials.[32]

A way for quantifying the vicinity of a system to some critical point is to define a branching ratio $\rho$ for the process of bond failure as the probability that the breaking of a bond gives rise to the breaking of another bond. In analogy with other processes[33] at the critical point $\rho = 1$. We have evaluated $\rho$ for the Born potential on a triangular lattice[26] by computing the average, over many realizations, of the number of broken bonds at a given value of the external solicitation (stress or strain) $\langle s \rangle$. This quantity is related to $\rho$ by:

$$\rho = \frac{\langle s \rangle - 1}{\langle s \rangle}, \tag{6}$$

and therefore

$$\rho \approx 1 - (\sigma_c - \sigma)^{-\gamma}. \tag{7}$$

Numerical simulations show that different types of solicitation produce different behaviours, as expected both from experiments and models.[4] By computing $\rho$ for the Born model we have shown in particular that breaking the bonds at constant strain makes the system less critical than breaking them at constant stress, and that the presence of corrosion mechanisms leads more quickly close to the criticality.

It must be stressed that the scale invariance for the breaking of bonds in model systems cannot be directly assimilated to that of acoustic emission. In fact, the latter carries an energy content that is not included in the broken bonds counting. Self-similarity of the energy bursts in lattice models only holds at constant sollicitation[25] or when special conditions are imposed to the system.[34]

Self-affine properties of the fracture surface are rather well reproduced by model simulations. In particular, numerical results suggest that small self-affinity exponents (around 0.5) are related to slowly moving cracks, whilst exponents close to 0.8 are associated with fast cracks.[35,36] Some different exponents, like 0.42 have been also found for slowly developping cracks in two dimensions,[37] but also if simulations well reproduce generally the experimental observations, there is no a comprehensive theory for all these behaviours (for a recent review on dynamic fractures see Ref. 38.

## 4. POINTS OF VIEW

A well coherent description of breakdown phenomena and of their critical properties is lacking. One possibility is to relate them to the Self Organized Criticality (SOC).[27] In this picture the SOC state corresponds to the final breakdown. Another tempting possibility is that of exploiting the analogies with critical phenomena for setting up a thermodynamic-like description of the fracturing process. Within this stream Zapperi et al.[31] suggested to choose the elastic moduli (or the conductance, for the fuse network) as the order parameter, drawing an analogy with first order transitions, since the order parameter suffers a jump at the final breakdown. Criticality, absent in usual first order transitions, would originate in this case from the presence of long range interactions, that in different situations have been shown to give rise to fractal fluctuations of the nucleation process.[28] However the mean field picture, from which this interpretation derives, does not describe all the features of the breaking process in euclidean lattices: perhaps the most relevant shortcoming is that the theory produces wrong predictions for the scaling of the breaking stress distribution with the system size.[29] Moreover, Sornette[39,40] and da Silveira[41,42] have found that the choice of different disorder distributions can select between continuous and discontinuous transition, with a tricritical point separating the two behaviours in the phase diagram of the system. In addition, continuous breakdown transition can be observed also when constant strain instead of constant stress driving is applied to the lattice.[43]

In conclusion we may say that in spite of the observation of critical fluctuations in many breakdown phenomena, the physical mechanisms at their origin are still far from clear and that it seems necessary to take into account both irreversibility and non self-averaging in order to come to a satisfactory explanation of these complicated non-equilibrium processes.

## REFERENCES

1. A. Coniglio, *J. Phys.* **A10**, L211 (1977).
2. L. de Arcangelis, S. Redner and A. Coniglio, *Phys. Rev.* **B31**, 4725 (1985).
3. L. de Arcangelis and A. Coniglio, *J. Stat. Phys.* **48**, 935 (1987).
4. H. J. Herrmann and S. Roux, *Statistical Models for the Fracture of Disordered Media* (North-Holland, Amsterdam, 1990), and references therein.
5. B. K. Chakrabarti and L. G. Benguigui, *Statistical Physics of Fracture and Breakdown in Disordered Systems* (Clarendon Press, Oxford, 1997), and references. therein.
6. B. Kahng, G. G. Batrouni, S. Redner, L. de Arcangelis and H. J. Herrmann, *Phys. Rev.* **B37**, 7625 (1988).
7. K. Mogi, *Tectonophysics* **5**, 35 (1967), and references therein.
8. C. H. Scholz, *Bull. Seismol. Soc. Am.* **58**, 1117 (1968).
9. C. H. Scholz, *Bull. Seismol. Soc. Am.* **58**, 399 (1968).
10. T. Hirata, *J. Geophys. Res.* **92**, 6215 (1987).
11. A. Petri, G. Paparo, A. Vespignani, A. Alippi and M. Costantini, *Phys. Rev. Lett.* **73**, 3423 (1994).
12. A. Vespignani, A. Alippi, M. Costantini, G. Paparo and A. Petri, *Fractals* **3**, 839 (1995).
13. A. Guarino, L. Bellon and S. Ciliberto, *Phys. Rev. Lett.* **79**, 3202 (1997).
14. Check *dove e' uscito dopo cond mat*
15. D. Sornette and V. J. Andersen, *Eur. Phys. J.* **B1**, 353 (1998).
16. C. Maes, A. van Moffaert, H. Frederix and H. Staruven, *Phys. Rev.* **B57**, 4987 (1998).
17. B. B. Mandelbrot, D. E. Passoja and A. J. Paullay, *Nature* **308**, 721 (1984).
18. E. Bouchaud, G. Lapasset and J. Planès, *Europhys. Lett.* **13**, 73 (1990).
19. K. J. Maloy, A. Hansen, E. L. Hinrichsen and S. Roux, *Phys. Rev. Lett.* **68**, 213 (1992).
20. V. Y. Milman, R. Blumenfeld, N. A. Stelmashenko and R. C. Ball, *Phys. Rev. Lett.* **71**, 204 (1993).
21. P. McAnulty, L. V. Meisel and P. J. Cote, *Phys. Rev.* **A46**, 3523 (1992).
22. D. Sornette, *J. Phys A: Math. Gen.* **22**, L243 (1989).
23. C. Hemmer and A. Hansen, *J. Appl. Mech.* **59**, 909 (1992).
24. H. E. Daniels, *Proc. R. Soc. London* **A183**, 405 (1945).
25. G. Caldarelli, F. di Tolla and A. Petri, *Phys. Rev. Lett.* **77**, 2503 (1996).

26. G. Caldarelli, C. Castellano and A. Petri, *Physica* **A270**, 15 (1999).
27. P. Bak, C. Tang and K. Wiesenfeld, *Phys. Rev. Lett.* **59**, 381 (1987).
28. T. S. Ray and W. Klein *J. Stat. Phys.* **61**, 891 (1990).
29. G. Caldarelli and A. Petri, *Phys. Rev. Lett.* **83**, 1483 (1999).
30. E. Bouchaud, *J. Phys C: Cond. Matt.* **9**, 4319 (1997).
31. S. Zapperi, P. Ray, H. E. Stanley and A. Vespignani, *Phys. Rev. Lett.* **78**, 1408 (1997).
32. M. Acharaya, P. Ray and B. K. Chakrabarty, *Phys. Rev.* **E53**, 140 (1996).
33. O. Kinouchi, S. T. R. Pinho and C. P. C. Prado, *Phys. Rev.* **E58**, 3997 (1998).
34. S. Zapperi, A. Vespignani and H. E. Stanley, *Nature* **388**, 658 (1997).
35. B. Chakrabarti and B. K. Chakrabarti, *Physica* **A270**, 21 (1999).
36. A. Parisi, G. Caldarelli and L. Pietronero, *Europhys. Lett.* **52**, 304 (2000).
37. A. Nakano, R. K. Kalia and P. Vashista *Phys. Rev. Lett.* **75**, 3138 (1995).
38. J. Fineberg and M. Marder, *Phys. Rep.* **313**, 1 (1999).
39. J. V. Andersen, D. Sornette and K. Lung, *Phys. Rev. Lett.* **78**, 2140 (1997).
40. *completare sorn2*
41. R. da Silveira, *Phys. Rev. Lett.* **80**, 3157 (1998).
42. R. da Silveira, *Am. J. Phys.* **67**, 1177 (1999).
43. G. Caldarelli, C. Castellano and A. Petri, *Phil. Mag.* **B79** (1999).

# CRITICAL FLUCTUATIONS IN 2D XY MAGNETS

M. SELLITTO and P. C. W. HOLDSWORTH
*Laboratoire de Physique, École Normale Supérieure de Lyon*
*46 Allée d'Italie, F-69364 Lyon Cedex 07, France*

**Abstract**

We give a review of the properties of low-temperature magnetic fluctuations in a finite-size 2D XY system. The behavior of such a simple model closely resembles that of several complex systems, notably the spectrum of power fluctuations in enclosed turbulent flow and interface width in surface growth problems. It suggests moreover new ideas in the study of aging phenomena in disordered systems.

## 1. INTRODUCTION

Fluctuations in macroscopic equilibrium systems are often assumed to be of Gaussian form. This assumption is nearly always true if the system in question can be divided into statistically independent elements, as dictated by the central limit theorem. However, in correlated systems, where this is not the case, there is no general reason to expect the central limit theorem to apply. The fluctuations spectrum can then take on a multitude of different forms. The most studied correlated systems are critical equilibrium systems. Here, the breakdown of Landau theory and the occurrence of a non-analytic fixed point for the free energy in a renormalization group flow, tells one that the fluctuations will not be Gaussian.[1] Long-range correlations are a feature not only of critical phenomena but also of systems driven into a non-equilibrium stationary state,[2] and therefore non-Gaussian universal distribution are generically expected also in this case. It has been observed that fluctuation spectrum of dissipated power in a confined turbulent flow is surprisingly similar to the spectrum of low-temperature magnetic fluctuations in the 2D XY model,[3] and that of other correlated systems driven into a non-equilibrium stationary states.[4] In the following, we briefly review some results recently obtained in the study of the 2D XY model which were mainly stimulated by these empirical observations.

## 2. FLUCTUATION STATISTICS

The XY model is perhaps the simplest possible system in which one can study critical phenomena. The critical phase that exists below the Kosterlitz-Thouless-Berezinskii transition temperature $T_{\rm KTB}$, is an attractive subject of investigation from both an analytic and a numerical point of view. Its physics is entirely captured by the harmonic spin-wave approximation with the result that many calculations can be performed microscopically from Gaussian integration, without the need to use renormalisation techniques, or the scaling hypothesis. The only price one pays for this simplicity is a critical system with a single independent exponent and the scaling relations satisfied through weak scaling only. From a numerical point of view simulation results near a single, isolated, critical point are often complicated by a shift in the effective critical temperature by an amount scaling to zero as a power-law of the system size, making it unclear exactly which temperature should be studied. In two dimensions, as the system is critical over the whole of the low temperature phase there are no such technical problems. These factors make the 2D XY model an ideal system with which to study the effects of critical correlations. The 2D XY model is defined by the Hamiltonian

$$H = -J \sum_{\langle i,j \rangle} \cos(\theta_i - \theta_j)\,, \tag{1}$$

where the angles $\theta_i$ refer to the orientation of classical spins $\mathbf{s}_i$ confined in a plane and where the sum is over nearest neighbour spins. In the following we consider a square lattice of size $N = L^2$, with periodic boundaries. The scalar instantaneous magnetisation is

$$m = \frac{1}{N} \sum_{i=1}^{N} \cos(\theta_i - \bar{\theta})\,, \tag{2}$$

where $\bar{\theta} = \tan^{-1}(\sum_i \sin\theta_i / \sum_i \cos\theta_i)$ is the instantaneous magnetisation vector direction. In Refs. 4 and 5, the following expression for the probability density function (PDF) of magnetization fluctuations has been developed:

$$\sigma P(m) = \int_{-\infty}^{\infty} \frac{dx}{2\pi} \exp\left\{ ix\,\frac{m - \langle m \rangle}{\sigma} - i\sqrt{\frac{g_1^2}{2g_2}}\,x - \frac{1}{2}\sum_{\mathbf{q}\neq \mathbf{0}} \ln\left[\mathbf{1} - i\sqrt{\frac{2}{g_2}}\frac{G(\mathbf{q})}{N}\,x\right]\right\}, \tag{3}$$

where $G(\mathbf{q}) = 1/(4 - 2\cos q_x - 2\cos q_y)$, $g_k = \sum_{\mathbf{q}\neq 0} G(\mathbf{q})^k / N^k$, and $\sigma = \langle m^2 \rangle - \langle m \rangle^2$. The above formula has been tested with extensive numerical simulations of the 2D XY model with full cosine interaction, Eq. (1), for different values of temperature and system size. In addition, a comparison has been done with microcanonical molecular dynamics (MD) simulations to check the possible dependence on the statistical ensemble or the underlying microscopic dynamics. In Fig. 1(a) the simulation results are compared with the numerical integration of Eq. (3), performed with a fast Fourier transform algorithm. The theoretical curve is clearly in extremely good agreement with the numerical data. The curve is non-Gaussian and asymmetric, with what appears to be an exponential tail for fluctuations below the mean, and with a much more rapid fall off in amplitude, for fluctuations above the mean. As mentioned in the introduction, the PDF for the fluctuations of power dissipation

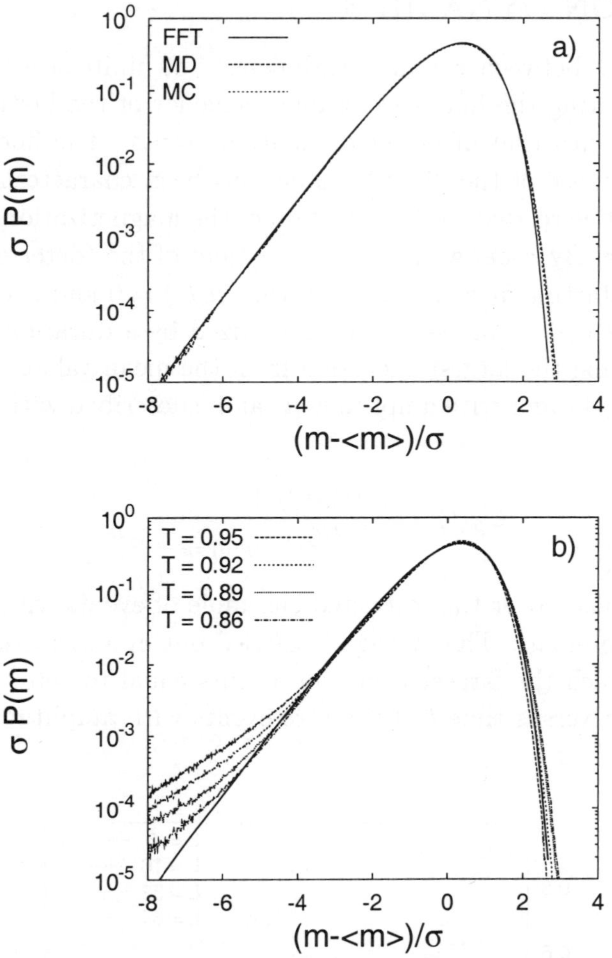


**Fig. 1** **(a)** The probability density function of magnetization fluctuations in the 2D XY model, as obtained from a fast Fourier transform (FFT) of Eq. (3), Montecarlo simulation for a system at temperature $T = 0.1$ (MC), and molecular dynamic for a system with energy $E = 0.756$ (MD). The size of the system is $N = 32^2$. **(b)** Vortex contribution to the PDF for temperature $T$ around $T_{\rm BKT} \simeq 0.89$. The continuous line curve is the FFT of Eq. (3).

in enclosed turbulent flow[3] is, within experimental error, of the same functional form as that shown in Fig. 1(a). In this case "universal" means that the suitably normalised PDF does not depend on the driving force (the Reynolds number). Several other complex correlated systems seems to share the same PDF.[4]

Deviation from the distribution shown in Fig. 1(a) can be observed in several ways. The simplest way consists of introducing a second correlation length in the problem, e.g. through an external magnetic field. Even in this case the analytical calculation can be carried out with Gaussian integration. The new family of universal distributions that appears approaches a Gaussian in the limit of infinite magnetic field.[6] Another interesting source of deviation from the behaviour observed in Fig. 1(a) is expected to occur at temperature close to $T_{\rm KTB}$, when free vortices become relevant variables. Fig. 1(b) show indeed that when the temperature is slightly below or above $T_{\rm BKT}$, the exponential tail of the distribution changes dramatically, large deviation becoming even more probable. The nature of such a change seems to have again an universal form, which is now induced by the appearance of a new critical correlation length due to the vortex unbinding. Interestingly, a similar broken-line tail also appears in the fluctuations spectrum of interface width in the 2D Kardar-Parisi-Zhang equation. A detailed investigation of this point will be given in a future work.

## 3. FLUCTUATION DYNAMICS

The statistical analogy between confined turbulence and finite-size 2D XY can be made more precise by comparing the finite-size scaling behavior of the lifetime and frequency of magnetization events with that of power dissipation events. The fluctuation dynamics in the low-temperature phase of the 2D XY model has been characterized by introducing a natural notion of event as related to the deviation of the magnetization from its mean value, $\mu(t) = (m(t) - \langle m \rangle)/\sigma$. By *event* we mean any segment of the (deterministic or stochastic) trajectory of motion starting at a time $t_1$ such that $\mu(t_1) = 0$ and ending at the next time $t_2$ such that $\mu(t_2) = 0$ again. An event is characterized by a duration $\Delta t = t_2 - t_1$ and an amplitude $\mu$ representing the largest excursion from the mean value.

The shape of a typical event with amplitude $\mu$ can be described with a coherence function defined as:

$$C_\mu(t) = \left. \frac{\langle \mu(t)\mu(0) \rangle}{\langle \mu^2(0) \rangle} \right|_{\mu(0)=\mu} \tag{4}$$

where the average is taken over the restricted ensemble of events with a fixed amplitude $\mu$ (within a small $\pm\delta\mu$ interval). The average is carried out in a synchronous way, that is by setting the time at which the largest excursion occurs equal to zero. In Fig. 2(a) we plot the coherence function versus time $t$ of the rare events with amplitude $\mu = -4$ for several

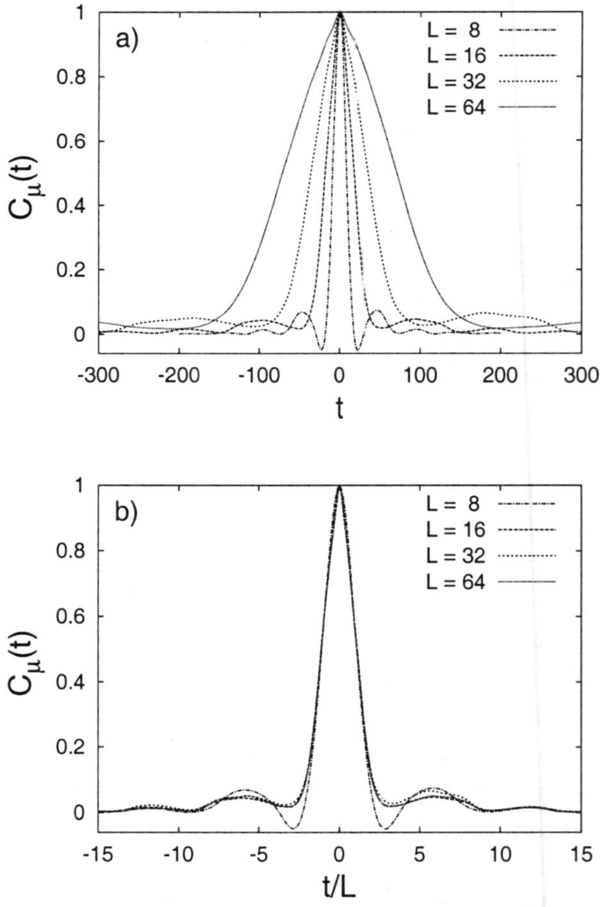


**Fig. 2** (**a**) Coherence function versus time $t$, for events with amplitude $\mu = -4$ in systems of size $N = L^2$. (**b**) The same function is plotted to the right versus the scaling variable $t/L$.

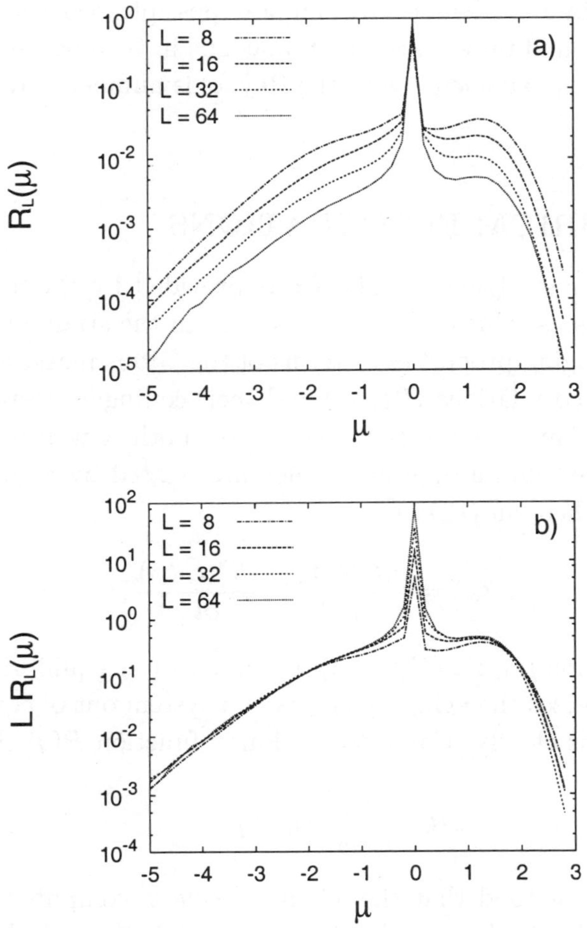


**Fig. 3** **(a)** Rate function $R_L(\mu)$ of events with amplitude $\mu$ for system with size $N = L^2$. **(b)** The same function rescaled with system size is plotted to the left.

system size $L$. The coherence function shows a fast (non-exponential in general) decay with an oscillatory damped behavior at later time which is more pronounced for the rarest events and the smallest system size ($N = 8^2$). In Fig. 2(b) we plot the same function versus $t/L$. We can see that the scaling of the coherence function of events with fixed amplitude $\mu$ is very good. A discrepancy is observed for the smallest system size, while for larger systems the agreement is excellent over the full range of value of $t$. The scaling immediately suggests that the characteristic relaxation time $\tau_L(\mu)$ of the events with amplitude $\mu$ is of the form $\tau_L(\mu) \sim L^z$ with a dynamic exponent $z = 1$. A similar result is obtained from the stochastic (Monte Carlo) dynamics. In this case however the dynamic exponent is $z = 2$. To better characterize the statistics of rare events we have also computed the rate function $R_L(\mu)$ of events with amplitude $\mu$. In Fig. 3(a) we show such a function for different system sizes. Events with $\mu < -2$ appear to be exponentially more common that their longer counterpart with $\mu > 2$. It turns out that the rate function scales linearly with $L$ (see Fig. 3(b). We therefore find that the two time scales associated with the characteristic event lifetime and the inverse of their frequency, diverge with size of the system as one would expect for a critical system, and that their ratio is independent of the system size. The same kind of dynamic finite-size scaling analysis has been experimentally carried out for the frequency and lifetime of events associated with a given power dissipation events in a confined turbulent flow. It turns out that even in this case the two characteristic time scales diverge with Reynolds number ($Re$) while their ratio is approximately constant with logarithmic correction in $Re$. This is the simplest possible scenario compatible with the universality of the PDF, and provides a direct analogue in turbulence of the critical slowing down in dynamic critical phenomena, with the Reynolds number playing the role of system size in critical systems.

## 4. NON-EQUILIBRIUM FLUCTUATIONS

A further motivation for studying the 2D XY is provided by the apparent analogy with the behavior of spin glasses. One of the main concern in the study of glassy systems is the characterization of the aging properties in terms of the Gibbs measure. In non-equilibrium dynamics the temporal correlations $C(t, t_w)$ and their conjugate response $R(t, t_w)$ are not time translational invariance, but rather depend on both a waiting time $t_w$ and $t$, the total time. In a suitable infinite-time limit, they are related by a generalized form of the fluctuation-dissipation-theorem (FDT)[7]

$$R(t, t_w) = \frac{x(C(t, t_w))}{k_B T} \frac{\partial C(t, t_w)}{\partial t_w} . \tag{5}$$

The fluctuation-dissipation ratio, $x(C(t, t_w))$, is equal to 1 at equilibrium and its departure from this value characterizes the aging dynamics of a system out of equilibrium. For a class of systems the factor $x$ is closely related to the Parisi function $P(q)$ through the relation[8]

$$x(C) = \int_0^C dq P(q) . \tag{6}$$

In relation (6), it is understood that the left hand side is computed in the limit of large waiting times after the thermodynamic limit is taken, while the right hand side is computed in an infinite system at equilibrium. One may wonder whether an appropriate generalization of (6) would allow one to predict the value of physical observables in the aging regime at a given finite time, starting from a finite equilibrated system of a suitable size. The analysis of the non-equilibrium dynamics of the 2D XY model suggests the conjecture that in critical systems the FDT violations on a finite-time scale $t_w$ are governed by the Parisi function $P(q, L)$ of a system of finite-size $L$ such that $L = \xi(t_w)$, where $\xi(t_w)$ is the dynamic correlation length.[9] This amounts to a finite-time, finite-size generalization of Eq. (6):

$$x(C(t, t_w)) = \int_0^{C(t,t_w)} dq P(q, \xi(t_w)) . \tag{7}$$

In the 2D XY model the relevant two-time dynamical observables are the following

$$C(t, t_w) = \frac{1}{N} \sum_{i=1}^{N} \langle \cos[\theta_i(t) - \theta_i(t_w)] \rangle , \quad \chi(t, t_w) = \frac{1}{Nh} \sum_{i=1}^{N} \overline{\langle \mathbf{h}_i \cdot \mathbf{s}_i \rangle} , \tag{8}$$

The integrated response, $\chi(t, t_w)$, is computed by considering a *random* perturbation of the Hamiltonian $\Delta H = -\sum_i \mathbf{h}_i \cdot \mathbf{s}_i$, where the magnetic field is a two-dimensional vector whose components are independently drawn from a bimodal distribution $\pm h$. The overlap between two equilibrium configurations $a$ and $b$ is defined as

$$q^{ab} = \frac{1}{N} \sum_{i=1}^{N} \cos(\theta_i^a - \theta_i^b) . \tag{9}$$

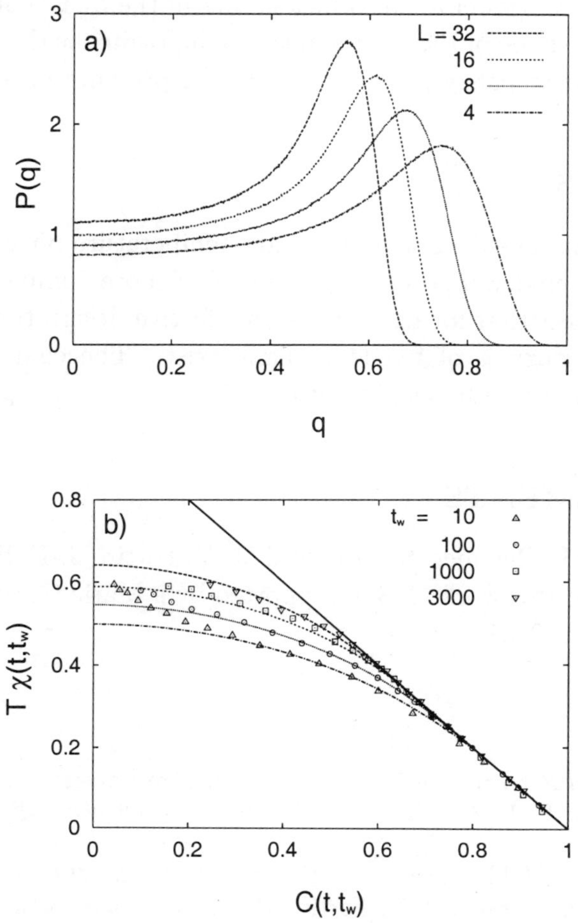


**Fig. 4** **(a)** The Parisi function $P(q,\, L)$ for the 2D XY model of size $N = L^2$ at $T = 0.7$. **(b)** Comparison between the finite-time non-equilibrium dynamics and the finite-size equilibrium behavior. The symbols represent the dynamic data at different waiting times $t_w$, while the linespoints are the static data obtained from the $P(q,\, L)$.

With this definition of the overlap one obtains a distribution reminiscent of the Parisi function for spin glasses, see Fig. 4(a). In this case, the tail in $P(q,\, L)$ is not the result of critical fluctuations but rather the product of the diffusive motion of the magnetization vector around the perimeter of a circle. While this seems rather trivial it points out that the symmetry breaking in the XY model provides a very simple analogue for the breaking of replica symmetry in a mean field spin glass: the XY magnet contains an infinity of pure states which are connected by a line of constant free energy around the circle and so in principle the system can change pure states without jumping free-energy barriers. Symmetry breaking means that the diffusion constant around the circle goes to zero in the thermodynamic limit and the system never leaves its chosen pure state. In Fig. 4(b) we compare the non-equilibrium dynamic data with the curves obtained through the Eq. (7). The agreement between the dynamic and static data is excellent. The small systematic deviations appearing on the left hand side of the parametric plot correspond to the limit where the total time $t$ is much greater than $t_w$, which is most easily seen when $t_w$ is small. In the regime where $C(t,\, t_w)$ is small and hence $\xi(t) \gg \xi(t_w)$, one can no longer expect the Eq. (7) to be satisfied. These results strongly supports the conjecture that the non-equilibrium dynamics in a critical system can be described in terms of the static distribution of a finite system of linear size $L$ such that $L = \xi(t_w)$. However, at much lower temperature the two sets of curve coincide only in the quasi-equilibrium regime, and although the qualitative shape of the curves in the aging regime is the same for the two sets, there is a quantitative discrepancy probably due to the presence of topological defects.[9]

## 5. CONCLUSIONS

The examples we have discussed show that the low-temperature phase of the 2D XY model can be a useful tool for understanding the behavior of more complex correlated systems. These observations suggest that for these cases an effective description in terms of suitably defined non-interacting degrees of freedom might exist. The characterization of such a description remains however a challenging problem.

## ACKNOWLEDGEMENTS

It is a pleasure to thank L. Berthier, S. Bramwell, J.-Y. Fortin, J.-F. Pinton and B. Portelli, for their contributions to this work. MS is supported by a Marie Curie fellowship of the EU (contract ERBFMBICT983561).

## REFERENCES

1. J. L. Cardy, *Scaling and Renormalization Group* (Cambridge University Press, Cambridge, 1996).
2. B. Schmittmann and R. K. P. Zia, *Statistical Mechanics of Driven Diffusive Systems* (Springer-Verlag, Berlin, 1995).
3. S. T. Bramwell, P. C. W. Holdsworth and J.-F. Pintona, *Nature* **396**, 552 (1998).
4. S. T. Bramwell, K. Christensen, J.-Y. Fortin, P. C. W. Holdsworth, H. J. Jensen, S. Lise, J. López, M. Nicodemi, J.-F. Pinton and M. Sellitto, *Phys. Rev. Lett.* **84**, 3744 (2000).
5. S. T. Bramwell, J.-Y. Fortin, P. C. W. Holdsworth, S. Peysson, B. Portelli, J.-F. Pinton and M. Sellitto, cond-mat/0008093, to appear in *Phys. Rev. E.*
6. B. Portelli, P. C. W. Holdsworth, M. Sellitto and S. T. Bramwell, *cond-mat/0102283.*
7. J.-P. Bouchaud, L. F. Cugliandolo, J. Kurchan and M. Mézard in *Spin Glasses and Random Fields*, ed. A. P. Young (World Scientific, Singapore, 1998).
8. S. Franz, M. Mézard, G. Parisi and L. Peliti, *Phys. Rev. Lett.* **81**, 1758 (1998).
9. L. Berthier, P. C. W. Holdsworth and M. Sellitto, *J. Phys.* **A34**, 1805 (2001).

# Part II

# Slow Dynamics

# COMPACTION OF GRANULAR MATTER: A SHORT REVIEW, AND THE RANDOM TETRIS MODEL

A. BARRAT
*Laboratoire de Physique Théorique,**
*Bâtiment 210, Université de Paris-Sud,*
*91405 Orsay Cedex, France*

V. LORETO
*Università degli Studi di Roma "La Sapienza",*
*Dipartimento di Fisica and INFM Unità di Roma 1,*
*P.le A. Moro 5, 00185 Rome, Italy*

**Abstract**

We present a short review of experimental and theoretical aspects of granular compaction, and we discuss in more details the behaviour of the so-called Random Tetris Model, a model of particles diffusing on a lattice, subject to gravity and geometrical constraints. We show how this model reproduces the experimental phenomenology, e.g. slow relaxation, irreversible/reversible cycles, memory effects. The study of the density profiles allows to interpret these results.

## 1. INTRODUCTION

Granular matter can exist in many very different states, all of which are currently the object of a large interest.[1] Since thermal energy is negligible with respect to gravitational or kinetic energy, its behaviour depends on the injected energy: strongly vibrated granular materials are studied as granular gases, for which the kinetic energy is much larger than the gravitational one; vibrated layers of grains display instabilities and pattern formation; dense granular matter can be studied at rest, and in particular many open problems concern the

---

*Unité Mixte de Recherche UMR 8627.

transmission of forces. We are here concerned with the slow dynamics of dense, compacting granular media, which has long been recognized to bear similarities with the aging of glasses[2] (indeed, compaction is slowed down by geometrical frustration and the need of collective rearrangements).

After a short description of recent experiments in granular compaction, we shall briefly summarize some theoretical approaches in Sec. 3, and focus in Sec. 4 on a particular lattice model, showing how it reproduces the experimental phenomenology.

## 2. EXPERIMENTS ON COMPACTION

The present interest in the study of the compaction of granular matter has largely been triggered by recent, detailed experiments performed in Chicago.[3] The Chicago group investigated systematically the compaction of a model granular matter (namely monodisperse spherical particles) confined in a tall vertical tube and subject to a series of taps, of intensity $\Gamma$ (ratio of the peak acceleration during a tap to the gravitational acceleration). The main results can be summarized as follows:

- starting from a low, reproducible value of the density $\rho_0$, the system evolves towards a more compact state; the evolution of the density can be fitted by an inverse logarithmic relaxation: $\rho(t) = \rho_\infty - \frac{\rho_\infty - \rho_0}{1+B\ln(1+t/\tau)}$, where $\rho_\infty$, $B$ and $\tau$ are fit parameters;
- at constant forcing $\Gamma$, the evolution is faster and yields more compact systems for larger $\Gamma$ (i.e. the asymptotic density at constant $\Gamma$ is an increasing function of $\Gamma$);
- at not too large forcing, the density is locally higher near the top of the column;
- if the granular column, starting from low density, is shaken with a slowly increasing forcing, the density increases monotonically; when the sequence is reversed, however, i.e. when the forcing is decreased, the density *keeps increasing*: the curve of density versus shaking first obtained is therefore *irreversible*; on the contrary, the second curve obtained by decreasing $\Gamma$ is retraced by increasing again $\Gamma$, showing that its reversible character. The two curves merge above a certain $\Gamma^*$.
- the density obtained by decreasing $\Gamma$ from a large value down to a small $\Gamma_0$ is much larger than the asymptotic density obtained at constant $\Gamma_0$.

New experiments, similar in spirit to the standard temperature-cycling experiments in spin-glasses,[4] were later realized by Josserand et al.:[5] the effect of a large, abrupt change in the tapping acceleration (whereas for the irreversible-reversible cycles, this acceleration was changed very slowly) during the compaction was shown to be opposite at short times to what could be expected from the long-time behaviour at constant $\Gamma$ (i.e. higher compaction rate for larger $\Gamma$); indeed, a sudden increase in $\Gamma$ leads to a decompaction, and a sudden decrease yields a temporary increase in the compaction rate. This effect was called "short-term memory", since the usual compaction rate is then recovered after a while.

Let us finally mention that, while the tapping procedure is a relatively violent way of exciting the system, an alternative experimental setup was proposed in:[6] a parallelepiped box full of beads was submitted to a slow horizontal shear of amplitude $\theta_{max}$. The same phenomenology was observed: slow compaction, faster for larger shear amplitudes, and, for sudden changes in shear amplitudes, jumps in density proportional to the change, and opposite in sign (i.e. decompaction for an increase of $\theta_{max}$, sudden increase for a decrease of $\theta_{max}$) are obtained.

## 3. THEORETICAL APPROACHES

Many theoretical attempts have been made to understand the rich phenomenology of the compaction of granular media described above. These attempts explored several paths, and we briefly summarize some of them.

Mehta et al. have simulated compaction on a microscopic level, using non-sequential moves in which particles are allowed to move and settle simultaneously:[7] this allows for cooperative rearrangements. However this approach predicted a two-timescales relaxation (individual grain motion and collective processes) that was not in accord with later experimental results;[3] this discrepancy could be explained by the fact that experiments were made in a regime of weaker vibrations.[8]

The "Parking Lot Model" (see e.g. Ref. 9) was later proposed by the Chicago experimentalists as a simple adsorption-desorption model describing the kinetics of densification of a given, horizontal layer of material: the desorption process describes the ejection of particles from the layer due to the tapping, while the adsorption corresponds to the settling down of the particles due to gravity. No diffusion is allowed within a layer, and particles cannot overlap. This model is phenomenological, but has the advantage of allowing for an analytical treatment in one dimension (the physical layer is two-dimensional, but similar results are expected in one and two dimensions).[9] In particular, a temporal regime where the density varies with an inverse logarithmic law is obtained.

A very ambitious approach has been put forward by S. Edwards and co-workers,[10] aiming at a thermodynamical description of dense granular matter by proposing an equivalent of the microcanonical ensemble: macroscopic quantities in a jammed situation should be obtained by a flat average over all blocked configurations (i.e. in which every grain is unable to move) of given volume, energy, etc.... The strong assumption is here that all blocked configurations are treated as equivalent and have the same weight in the measure. This approach, based on the idea of describing granular material with *a small number of parameters*, introduced the concept of compactivity, the analog of temperature for usual thermodynamics. Recent progresses in this direction were reported in Barrat et al.: Edwards' measure was constructed for the Kob-Andersen model (see below), and was shown to correctly predict the outcome of the aging dynamics of this model. It is however important to bear in mind that: (i) such an approach assumes that the system under consideration is spatially homogeneous, which is not always the case and (ii) the relevance of it has been checked in a few cases, but does not yet follow from any first principle.

Simple phenomenological models have been introduced and studied in Javier Bray et al.[13] In this approach one describes the dynamics of a granular medium in terms of a Markov process formulated by means of a master equation for the conditional probability of finding the system in a given configuration at time $t$, given that it was in another configuration at time $t'$.

Phenomenological models already used for glasses and spin-glasses have also been used:[6,12] they correspond to a tentative description of the phase space of the system in terms of a complicated, rough landscape, as is now quite usual in the study of aging dynamics. Using such models implies moreover the introduction of a "temperature", related to the external forcing, and pushes forward the analogy with glasses and spin-glasses, together with the idea that granular matter deserves a thermodynamical description.

Finally, let us mention some microscopic lattice models: although real grains or beads are clearly not sitting on a lattice, the idea is to define simple models that allow for a detailed study (in particular numerical) while retaining the rich phenomenology of granular media. In all cases, gravity is implemented by the choice of a tilted square (in two dimensions) or cubic (in three dimensions) lattice, of a preferred direction and of different probabilities to move upwards or downwards: the control parameter, similar to the experimental $\Gamma$, will be the ratio $x = p_{up}/p_{down}$ of these probabilities. Moreover, the particles are confined in a box with a rigid plane at the bottom, and occupy the sites with single-occupancy.

- The Ising Frustrated Lattice Gas (IFLG)[14] is directly inspired from spin-glasses: the particles can be in two possible states described by a variable $S_i = \pm 1$; they diffuse with the restriction that two neighboring particles $i$, $j$ must satisfy the constraints $S_i S_j = \epsilon_{ij}$ where the $\epsilon_{ij} = \pm 1$ are quenched random numbers that introduce frustration, and slow down the dynamics at high density: the compaction proceeds with the inverse logarithmic law seen in experiments. Irreversible/reversible curves are also obtained by varying $x$.
- In the Tetris model,[15] the constraints are only geometrical, and no quenched disorder is present (in this sense, it seems closer to real hard particles): the particles are elongated with two possible (orthogonal) orientations; geometrical frustration is implied by the fact that two particles with the same orientation cannot occupy neighboring sites in this direction. As for the IFLG, this model reproduces the main experimental phenomenology. Note that this model has two antiferromagnetic ground states, and that the dynamics can be described in terms of a coarsening process[16] and therefore has $\rho_\infty = 1$. The generalization, in which particles can have random sizes and forms, is the Random Tetris Model (RTM),[17] which has been studied in detail in Barrat and Loreto;[18] this generalization removes the pathology of having only two ground states; a description of the results follows in Sec. 4.
- The Kob-Andersen model[19] was first studied in the context of Mode-Coupling theories as a finite dimensional model exhibiting a divergence of relaxation time at a finite value of the control parameter (here the density). It is defined as a lattice gas on a three dimensional lattice, and the particles diffuse with the following restriction: a particle can move to a neighboring empty site, only if it has less than $m$ neighbours in the initial and in the final position (usually $m = 4$). Though very schematic, it has then been shown to reproduce rather well several aspects of granular compaction.[20]

## 4. THE RANDOM TETRIS MODEL

As for the Tetris model, the essential ingredient of the RTM[17] is the geometrical frustration, due to excluded volume effects: the interactions are not spatially quenched but are determined in a self-consistent way by the local arrangements of the particles, which do not allow for superpositions.

On the chosen tilted lattice (to implement the existence of a preferential direction in the system, and thus the gravity) each particle can be schematized as a cross with four arms (the number of arms is equal to the coordination number of the lattice) of different lengths, chosen in a random way. The dynamics consists in a diffusion, with probabilities $p_{up}$ (resp. $p_{down} = 1 - p_{up}$) to go up (resp. down), constrained by the particles geometry, and by a closed boundary at the bottom (lateral boundary conditions can be closed or open, and various aspect ratios can be used, without changing qualitatively the results). The control parameter is the ratio $x = p_{up}/p_{down} < 1$.

The system is initialized by inserting particles (one at a time) at the top of the system, and let them fall down in an oriented random walk on the lattice, until they reach a stable position. This yields initial configurations of a reproducible density. Various processes can then be considered: compaction at constant $x$, cycles in $x$, sudden change in $x$. We refer to Barrat and Loreto[18] for the detailed results.

Several quantities are monitored during the dynamics: (i) the bulk density, is the mean value of the density in the lower 25% or 50% of the box; we also measure the mean height of the particles, and the total potential energy; (ii) the density profile $p(j, t)$, which gives the value of the density (averaged over horizontal layers) as a function of the height $j$ at time $t$. It allows to characterize the inhomogeneities along the preferred direction; (iii) the correlation functions of the local density or of the potential energy allow to study the aging properties of the system.

## 4.1 Constant Forcing

At constant $x$, the data of Fig. 1(a) show the evolution of the density measured in the lower 50% of the system (these curves can be fitted by an inverse logarithmic law); an optimal value of $x$ exists and depends on the way the density is measured (lower 25% or 50%); this can be understood looking at the density profiles [Fig. 1(b)]: the width of the interface is larger for higher shaking amplitudes, which is quite intuitive; moreover, a denser layer forms just under the interface. This layer is denser for weaker forcing, and therefore blocks the compaction process (the particles under the interface need free volume to rearrange and compactify). Thus, at larger forcing the compaction of the bulk is more efficient since the interface is looser and particles can diffuse more easily. For very strong shaking, however, the system is very loose and the interface becomes very wide, invading the "bulk": this explains the existence of an optimal shaking amplitude, which moreover depends on the definition of the bulk. It is important to notice, looking at Fig. 1(b), how various shaking amplitudes affect various regions of the system: weak shaking allows to compactify the interface but not the bulk, strong shaking has the opposite effect.

During the compaction, the aging of the system can also be studied through the two-times correlation functions $C(t_w + t,\, t_w)$ between times $t_w$ and $t_w + t$. The usual aging behaviour is observed, with a first decay for $t \ll t_w$, approaching a quasi-equilibrium curve where time-translation invariance is respected (i.e., for $t \ll t_w$, $C(t + t_w,\, t_w)$ approaches a curve depending only on $t$) and, at $t \gg t_w$, a second decay, dependent on $t_w$, corresponding to aging. The correlation function displays a $\ln(t)/\ln(t_w)$ behaviour, in contrast with the Parking Lot Model which predicts a $t/t_w$ behaviour.

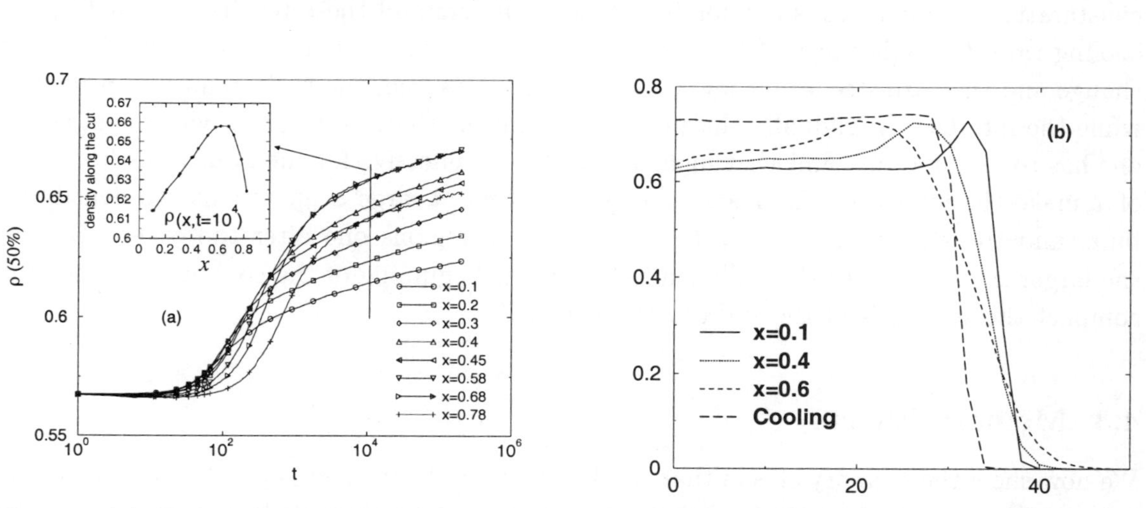


**Fig. 1** **(a)** Time-evolution of the bulk density (lower 50% of the system) for several shaking amplitudes. The insets show the density at time $t = 10^4$ as a function of $x$. **(b)** Density profiles obtained at large times ($10^6$ Monte-Carlo sweeps) in various cases: constant $x = 0.1$, 0.4, 0.6 or after a "cooling" from $x_I = 0.9$ to $x_F = 0.1$ with a cooling rate of $10^{-6}$.

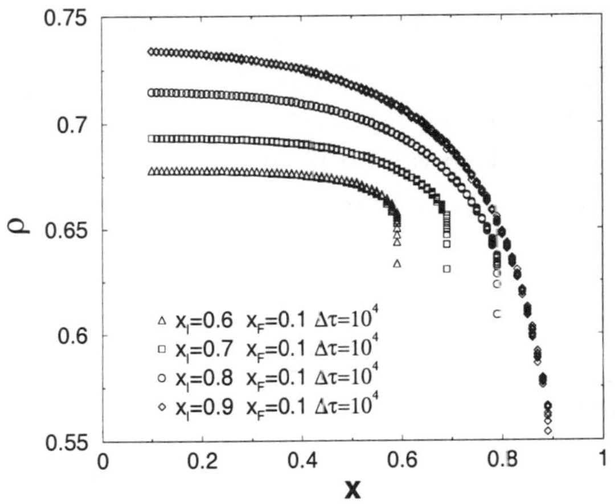


**Fig. 2** Evolution of the density during the cooling procedure, starting from various $x_I$: higher densities are obtained for larger $x_I$.

## 4.2 Cycles

It was shown in experiments that the way to obtain a very dense system is to slowly increase then decrease the forcing. The increasing part is in fact useless, since structures formed at low $x$ are destroyed by the shaking at large $x$. We therefore show in Fig. 2 the evolution of the density versus $x$ during the decrease of $x$ ("cooling", by similarity with a slow cooling for thermal systems), starting from various values $x_I$. We have in fact[18] investigated various procedures, with various initial and final values $x_I$ and $x_F$, and various "cooling rates". The outcome is that higher densities are obtained for slower procedures and for larger $x_I$ (with a saturation: increasing $x_I$ above 0.9 does not allow to increase further the asymptotic density).

The density profile obtained with $x_I = 0.9$, $x_F = 0.1$ and a slow cooling, shown in Fig. 1(b), is impressively different from the profiles at constant $x$: the system is very homogeneously dense, with a sharp interface. Moreover, the correlation functions do not display aging: the system is in a stationary state. A look at the density profiles and the constatations made at constant forcing allow to understand these results: at fixed $x_I$ and cooling rate, the bulk parts of the profiles are identical for various $x_F$: only the interfaces change and they are steeper for lower $x_F$: when $x$ is lowered, the bulk retains its properties while the interface is gradually sharpened. This means that, in order to better compactify, one has to take into account that high values of $x$ are effective for the bulk while low values of $x$ make the interface denser and steeper. Starting at large $x$ allows thus to densify the bulk, and the slow decrease of $x$ to compactify all regions one after the other. Besides, the larger $x_I$, the deeper the bulk is affected at the beginning of the cooling, and the more compact the system is at the end of the cooling.

## 4.3 Memory Effects

We now show the memory effects that can be observed during compaction when the forcing is abruptly changed at a certain time $t_w$ from $x_1$ to $x_2$, and then set back to $x_1$ at $t_w + \Delta\tau$.

The first effect is the short-term memory observed in experiments,[5] and shown in Fig. 3 for the RTM. It is symmetric, in the sense that it always occurs in a way which is counterintuitive with respect to the effect expected on the basis of the behaviour at constant $x$:

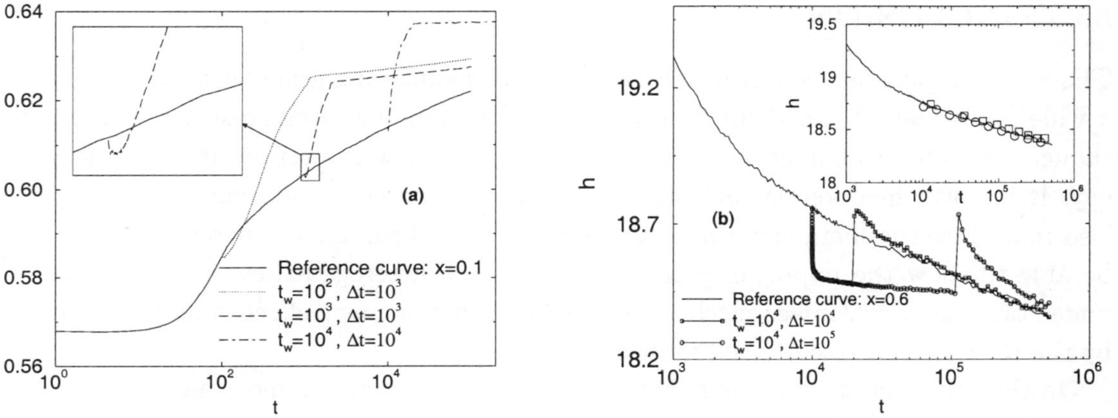


**Fig. 3** Effect of a sudden and large change in the forcing amplitude; **(a)** bulk density, for $x_2 > x_1$ ($x_1 = 0.1$, $x_2 = 0.4$): at short times a decompaction is observed (the inset is a zoom); **(b)** mean height $h$ versus time, for $x_2 < x_1$ and large $t_w$ ($x_1 = 0.6$, $x_2 = 0.3$, $t_w = 10^4$, and $\Delta t = 10^4$ and $10^5$): at short times the rate of compaction is suddenly increased, but then goes down. As $x$ is again increased, $h$ goes back to its value at $t_w$. The long time memory effect is illustrated in the inset: the symbols, corresponding to the data for $\Delta t = 10^4$ and $\Delta t = 10^5$, are shifted by $\Delta t$ and they coincide with the reference curve.

if $x_2 > x_1$, a decompaction occurs at short times while, if $x_2 < x_1$, the rate of compaction first increases. After a transient, the memory is lost, and the rate of compaction crosses over to the one observed at constant forcing $x_2$. The other memory effect occurs only in the case $x_2 < x_1$, and is therefore asymmetric in the variation of the forcing: for large $t_w$, the system is already quite compact at $t_w$, and evolves very slowly during $\Delta t$; for $t \geq t_w + \Delta t$ the compaction curves can be translated and superimposed to the reference one, as shown in the inset of Fig. 3(b): the system has kept memory of its state at $t_w$.[a]

Once more, it is straightforward to interpret these results using the density profiles: when $x$ is abruptly lowered, the first effect is that the particles tend to go down, and the interface becomes more compact. Therefore the density first increases with respect to the unperturbed case. At larger times however, the evolution is slowed down by the creation of a dense layer at the interface, which blocks the bulk rearrangements needed for the compaction. After $t_w + \Delta t$, the increase in the forcing allows to suppress the dense layer, and the compaction can become again fast. Moreover, if $t_w$ is large enough, the bulk of the system is already quite compact, and therefore the smaller value of the forcing during $\Delta t$ leads to a compaction of the interface but the bulk almost does not evolve. At $t_w + \Delta t$, the forcing is again increased: the relaxation of the interface being fast, this leads the system back to its state at $t_w$.

In the symmetric case $x_2 > x_1$, the first effect after $t_w$ is a decompaction, especially at the interface. The fact that the interface is less compact then allows for a much better compaction of the bulk. At $t = t_w + \Delta t$, the bulk has been deeply modified, so the system cannot have any memory of its configuration at $t_w$.

It is worth mentioning how memory effects have been put in evidence also in the context of the "Parking Lot Model".[21]

---

[a]If, on the other hand, $t_w$ is small, the system is not very compact, able to evolve a lot during $\Delta t$, and does not display memory for times larger than $t_w + \Delta t$.

## 5. CONCLUSION

The very rich phenomenology displayed by dense, compacting granular media is still raising a wide experimental and theoretical interest. The approach of microscopic lattice models, though restricted to numerical simulations, has been shown to reproduce this phenomenology. It is worth mentioning that other models, for example phase-space based models, can also reproduce the main experimental features. The advantage of microscopic models is to be able to show the importance of the study of the heterogeneities, and therefore of the understanding of real space mechanisms. A long time memory effect is moreover predicted by the RTM in the case of aged granular media.

On the other hand, it is clear that approaches of a more thermodynamical nature, like the one proposed by Edwards and co-workers, are necessary and should be pushed forward.

New experiments would be welcome to understand the nature of local rearrangements, and e.g. to test the predictions of the various models for the correlation functions. Moreover, the study e.g. of diffusion and mobility of tracer particles within driven granular media could allow for a detection of a dynamical temperature and would be an important step towards a fundamental understanding of the compaction of granular media.

## ACKNOWLEDGMENTS

It is a pleasure to thank C. Josserand, E. Trizac and P. Viot for useful discussions.

## REFERENCES

1. See e.g. H. M. Jaeger and S. R. Nagel, *Science* **255**, 1523 (1992); *Granular Matter: An Interdisciplinary Approach*, ed. A. Mehta (Springer-Verlag, New York, 1994); H. M. Jaeger, S. R. Nagel and R. P. Behringer, *Rev. Mod. Phys.* **68**, 1259 (1996); *Proceedings of the NATO Advanced Study Institute on Physics of Dry Granular Media*, eds. H. J. Herrmann et al. (Kluwer Academic Publishers, Netherlands, 1998).
2. See, for example Chapter 7 of: L. C. E. Struik, *Physical Ageing in Amorphous Polymers and Other Materials* (Elsevier, Houston, 1978).
3. J. B. Knight, C. G. Fandrich, C. N. Lau, H. M. Jaeger and S. R. Nagel, *Phys. Rev.* **E51**, 3957 (1995); E. R. Nowak, J. B. Knight, M. Povinelli, H. M. Jaeger and S. R. Nagel, *Powder Technol.* **94**, 79 (1997); E. R. Nowak, J. B. Knight, E. Ben-Naim, H. M. Jaeger and S. R. Nagel, *Phys. Rev.* **E57**, 1971 (1998); H. M. Jaeger, in *Physics of Dry Granular Media*, eds. H. J. Herrmann, J.-P. Hovi and S. Luding (Kluwer Academic, Dordrecht, The Netherlands, 1998), pp. 553–583.
4. See e.g. E. Vincent, J. Hammann, M. Ocio, J.-P. Bouchaud and L. F. Cugliandolo, in *Spin Glasses and Random Field*, ed. A. P. Young (World Scientific, 1997), and references therein.
5. C. Josserand, A. Tkachenko, D. M. Mueth and H. M. Jaeger, *Phys. Rev. Lett.* **85**, 3632 (2000).
6. S. Luding, M. Nicolas and O. Pouliquen, *cond-mat/0003172* and in *Compaction of Soils, Granulates and Powders*, eds. D. Kolymbas and W. Fellin (A. A. Balkema, Rotterdam, 2000), p. 241; M. Nicolas, P. Duru and O. Pouliquen, *Eur. Phys. J.* **E3**, 309 (2000).
7. A. Mehta, *Physica* **A186**, 121 (1992); G. C. Barker and A. Mehta, *Phys. Rev.* **E47**, 184 (1993).
8. D. Head and G. J. Rodgers, *J. Phys.* **A31**, 107 (1998).
9. P. L. Krapivsky and E. Ben-Naim, *J. Chem. Phys.* **100**, 6678 (1994); X. Jin, G. Tarjus and J. Talbot, *J. Phys.* **A27**, L195 (1994); J. Talbot, G. Tarjus and P. Viot, *Phys. Rev.* **E61**, 5429 (2000).
10. S. F. Edwards, in *Current Trends in the Physics of Materials* (Italian Physical Society and North Holland, Amsterdam, 1990); S. F. Edwards and R. B. S. Oakeshott, *Physica* **A157**, 1080 (1989); S. F. Edwards and C. C. Mounfield, *Physica* **A210**, 279 (1994); S. F. Edwards and D. V. Grinev, *Phys. Rev.* **E58**, 4758 (1999).

11. A. Barrat, J. Kurchan, V. Loreto and M. Sellitto, *Phys. Rev. Lett.* **85**, 5034 (2000); A. Barrat, J. Kurchan, V. Loreto and M. Sellitto, *Phys. Rev.* **E**, preprint cond-mat/0011492 (2001).
12. D. Head, *Phys. Rev.* **E62**, 2439 (2000).
13. J. Javier Brey, A. Prados and B. Sanchez-Rey, *Phys. Rev.* **E60**, 5685 (1999); *Physica* **A275**, 310 (2000).
14. M. Nicodemi, A. Coniglio and H. J. Herrmann, *Phys. Rev.* **E55**, 3962 (1997).
15. E. Caglioti, V. Loreto, H. J. Herrmann and M. Nicodemi, *Phys. Rev. Lett.* **79**, 1575 (1997).
16. A. Baldassarri, S. Krishanmurthy, V. Loreto and S. Roux, *Coarsening and Slow-Dynamics in Granular Compaction*, preprint (2001).
17. E. Caglioti, S. Krishnamurthy and V. Loreto, *Random Tetris Model*, unpublished (1999).
18. A. Barrat and V. Loreto, *J. Phys.* **A33**, 4401 (2000); *Europhys. Lett.* **53**, 297 (2001).
19. W. Kob and H. C. Andersen, *Phys. Rev.* **E48**, 4364 (1993).
20. M. Sellitto and J. J. Arenzon, *Phys. Rev.* **E62**, 7793 (2000).
21. J. Talbot, G. Tarjus and P. Viot, *Aging and Response Properties in the Parking-Lot Model*, cond-mat/0008183.

# WHY CONDUCTIVITY DECREASES WITH PRESSURE IN ION-DOPED POLYMERS

JOHN T. BENDLER and JOHN J. FONTANELLA
*Physics Department, U.S. Naval Academy,*
*Annapolis, MD 21402-5026, USA*
MICHAEL F. SHLESINGER
*Physical Sciences Division, Office of Naval Research,*
*800 N. Quincy St., Arlington, VA 22217, USA*

**Abstract**

The relaxation time scale in glassy materials is derived within a model of anomalous defect diffusion. The effect of the defects on ion-doped polymeric glasses is to produce a stretched exponential waiting time distribution for ion jumps. The characteristic time scale for ion jumps is connected to the temperature and pressure dependent concentration of mobile defects. The resultant expression for ionic conductivity is compared with experimental results for the polymer electrolyte poly(propylene glycol) (PPG) containing $LiCF_3SO_3$.

## 1. INTRODUCTION

In 1889, Arrhenius introduced the concept of an activation energy together with a law for the related characteristic relaxation time

$$\tau_A = \nu_0^{-1} e^{\Delta/kT} \tag{1}$$

where $k$ is Boltzmann's constant and $T$ is the absolute temperature in degrees Kelvin. The Arrhenius law was later derived by Kramers in terms of the trajectory of a particle successfully crossing an energy barrier of height, $\Delta$, with an attempt frequency of $\nu_0$. However the Arrhenius law is typically not valid for most glass-forming materials. A corresponding law for these materials was proposed by Vogel[1] in 1921, for viscosity in the form

$$\eta(T) = \eta_0^{(T-T_1)/(T-T_0)}. \tag{2}$$

Vogel used this empirical fit for viscosity experiments on mercury, water, and oils. If viscosity is proportional to a time scale $\tau_\nu$, Vogel's law can be written equivalently and more transparently as,

$$\tau_\nu = A_\nu e^{\frac{B_\nu}{T-T_0}} . \tag{3}$$

This later form was proposed by Fulcher[2] in 1925, and Tammann and Hesse[3] in 1926. Today this is called the Vogel law or the VFT law or some other combination of the above. The interpretation of the parameters $A_\nu$, $B_\nu$ and $T_0$ is not so straightforward. First, one notices that $T_0$ is a special temperature where the time scale diverges. The temperature $T_0$ is typically well below $T_g$, the glass transition temperature of the material, i.e. the relaxation dynamics are focused on $T_0$ and not on $T_g$. There have been several attempts to derive the Vogel law or alternative laws. Previously, we had derived the equation for the temperature dependence of the relaxation time at zero pressure[4-6]

$$\tau_{DD} = A_{DD} e^{\frac{B_{DD}}{(T-T_c)^{3/2}}} \tag{4}$$

where $A_{DD}$, $B_{DD}$ and $T_c$ are constants. The derivation of Eq. (4) starts by initially arriving at a stretched exponential relaxation time distribution within a defect diffusion (DD) model.[4-6] The relaxation time scale is the mean time of the stretched exponential law. This time scale depends on the (temperature and pressure dependent) mobile defect concentration. In this model, the defects carry free volume and can unfreeze the parts of the glass that they visit. With enough defects the glass is not frozen. As the temperature is lowered, defects cluster, and the number of mobile defects decreases. As this occurs, the material becomes more viscous (rigidity begins to set in). At $T_g$, the defect concentration decreases to the point where rigidity percolates, and the glassy state is formed. Relaxation, however, is still occurring. A phase transition in the number of mobile defects, as the temperature is lowered towards $T_c$, creates the behavior characterized by Eq. (4).

Equation (4) has consistently been as good as or better than the Vogel law (3) for describing ionic conductivity, dielectric relaxation and viscosity data for glass-forming materials.[7-9] The Vogel law often fails to fit data adequately near $T_g$ and an Arrhenius law is sometimes used near $T_g$, in conjunction with a Vogel law in the region above $T_g$. Equation (4) provides a consistently better fit throughout both regions.

In this paper it is shown how Eq. (4) can be extended to include the effect of pressure. It will be demonstrated that the resultant generalized Vogel-type law successfully describes measurements of the pressure dependence of the ionic conductivity in a polymer electrolyte.

## 2. THEORY

Consider a glass-forming material possessing a concentration $c$ of defects, where $c_m$ of these are mobile. Ion hopping motion at an ion occupied site occurs, at time $t$, due to the flux of defects, $F(t)$, into that site. To calculate $F(t)$ we employ a waiting time probability density, $\psi(t)$, for holding a defect for a time $t$, in between its jumps. This incorporates randomness into the defect motion. The relaxation function $\phi(t)$ is given by $\phi(t) = e^{-F(t)}$. The defect flux into a site, within a time $t$, is $c_m N(t)$ where $N(t)$ is the number of distinct sites a random walking defect visits within a time $t$. The relaxation law is,[4-6].

$$\phi(t) = e^{-c_m N(t)}. \tag{5}$$

For a typical Brownian motion type random walk, $N(t)$ is proportional to $t$ and exponential relaxation occurs. Slower than exponential relaxation, i.e. stretched exponential relaxation is another possible relaxation law, i.e.

$$\phi(t) = e^{-\lambda c_m t^\beta} \equiv e^{-(t/\tau)^\beta} \tag{6}$$

where $\beta < 1$ and $\lambda$ is a constant and on the RHS $\tau = \tau_{DD} = (\lambda c_m)^{-1/\beta} = c_m^{-1/\beta}\tau_0$. This is the typical behavior exhibited by glassy materials. There are several ways in which one can arrive at Eq. (6). If one relates a time scale to overcoming a free energy barrier as follows: $t = t_0 e^{(\Delta - TS)/kT}$ then the distributions of energy barriers, $\Delta$, entropies, $S$, and prefactors, $t_0$, can each generate a distribution $\psi(t)$ of waiting times between jumps. Consider first that only $S$ is a random variable and the distribution of environments produces $f(S) = S_0^{-1}e^{-S/S_0}$. Then $\psi(t)dt = f(S)dS$ leads at long times to $\psi(t) \approx 1/t^{-1-\beta}$ with $\beta = S_0/k$. When $k > S_0$, $\beta < 1$. For this case, for a single defect, the mean waiting time between jumps, $\int_0^\infty t\psi(t)dt$, is infinite. If, instead, one places all the randomness in the energy term a similar stretched exponential decay arises, but with the exponent being temperature dependent. In general, both mechanisms can be expected.

Since only the defects are mobile in the model, as the temperature is lowered, the defects cluster (or correlate their motion) to lower the system entropy. We now make the assumption that single defects, of concentration $c_1$, are more mobile than a cluster of defects. We therefore replace $c_m$ in Eq. (6) by $c_1$. To have a single (isolated) defect at a site, one must first have a defect there with probability $c$ and also have all of the $z$ neighbor sites within its correlation volume unoccupied, i.e.

$$c_1 = c(1-c)^z \tag{7}$$

with $z = (\xi/d)^3$, where $\xi$ is the defect correlation length, and $d$ is the mean defect-defect spacing. In a mean field lattice gas model, the correlation length $\xi$ between the defects grows near and above a critical temperature $T_c$ as $\xi(T) \approx L(T_c/T - T_c)^{1/2}$ where $L$ is a constant and $T_c$ is the temperature at which single defects disappear and below which only defect clusters exist. With increasing pressure, the defect-defect spacing $d$ is assumed to decrease isotropically as $d^3 = d_0^3(1 - \delta(T,\, P))$ where $1 - \delta(T,\, P) = V(T,P)/V(T,\,0)$ is the fractional volume change of the material as pressure increases and $d_0$ is the mean lattice spacing separation at zero pressure. The time scale in the stretched exponential can now be expressed as,

$$\tau \approx c_1^{-1/\beta}\tau_0 = c^{-1/\beta}\tau_0 e^{\frac{BT_c^{3/2}}{(T-T_c)^{3/2}(1-\delta(T,P))}} \tag{8}$$

where $B = -(L/d_o)^3 \ln(1-c)/\beta$. This represents a new relaxation law that is Vogel-like, but with a 3/2 temperature exponent and the inclusion of pressure effects. Note that $T_c$ is a function of $P$. In an ion-containing polymeric glass-forming material, as described by our model, ion transport is controlled by the defects and it is assumed that the time scale for ionic conductivity is related to that for diffusion. An individual defect has an infinite waiting time between jumps, but the ion, hit by a flux of defects, has the stretched exponential waiting time distribution. All the temporal moments of the stretched exponential are finite, including the first moment, which we label as $\tau_1$. Since the diffusion constant is of the form $D = l^2/(6\tau_1)$, employing the Nernst-Einstein relation $\sigma = q^2 nD/kT$, where $q$ is the charge on an ion and $n$ is the ion concentration

$$\sigma(T,\, P) = \frac{q^2 n l^2 c^{1/\beta}}{6kT\tau_0} e^{-\frac{BT_c^{3/2}}{(T-T_c)^{3/2}(1-\delta)}}\,. \tag{9}$$

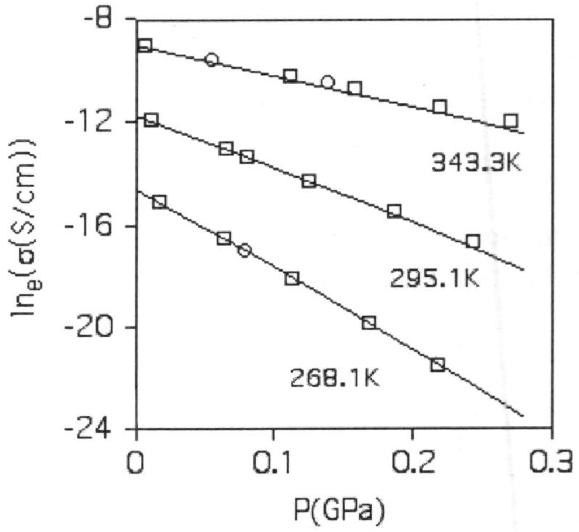


**Fig. 1** Electrical conductivity vs. pressure for PPG:$LiCF_3SO_3$ vs. temperature. The points are experimental. The squares represent decreasing pressure and the circles are increasing pressure. The lines are predicted by the generalized Vogel equation.

Basically, as the pressure is increased, defects are pushed closer together and become more clustered leaving fewer single (mobile) defects. This will decrease the defect flux, increasing the time scale $\tau$ for ion hopping and thus decreasing the conductivity $\sigma$.

## 3. COMPARISON WITH EXPERIMENT

For the purpose of comparing the theory of the ionic conductivity with experiment, Eq. (9) is written as follows:

$$\sigma(T,\, P) = \frac{A_\sigma}{T(1-\delta)}\, e^{-\frac{BT_c^{3/2}}{(T-T_c)^{3/2}(1-\delta)}}\,. \tag{10}$$

where $A_\sigma$ and $B$ are constants determined from a best fit to the data. The $\delta$ in the denominator of the pre-exponential accounts for the increase in ion concentration as pressure increases. Other terms in the pre-exponential may be pressure dependent but are taken to be constant in the present paper. It is assumed that the volume of the material changes with pressure according to:

$$1 - \delta = 1 - \chi(T)P + f(T)P^2 + g(T)P^3. \tag{11}$$

This form is chosen as it is supported by the $PVT$ data of Zoller and Walsh[10]. Finally, the critical temperature is assumed to be be pressure dependent.

To summarize the results for ionic conductivity vs. pressure, the theoretical values for $\ln(\sigma)$ vs. pressure at three temperatures were calculated using Eq. (10). The theoretical and experimental values are shown in Fig. 1. The figure shows good agreement between theory and experiment. The investigation of the first and second pressure derivatives of the conductivity will be reported elsewhere.

## 4. CONCLUSIONS

A defect diffusion model has been developed that is capable of describing the temperature and pressure variation of several dynamical processes in glass-forming materials. The basic principle of the model is that ionic conduction occurs when a defect encounters an ion. Each of the parameters in the theory has a clear, physical interpretation. There is an underlying temperature that represents the temperature below which no defects are mobile. There is a dimensionless constant in the exponent that depends upon the correlation length, separation and concentration of the defects. The pre-exponential involves a product of a characteristic relaxation time (taken to be about that for a lattice vibration) and the defect concentration. Comparison of theory and experiment reveals an excellent representation of the data over a wide range of temperatures and pressures. This model can be extended to cover dielectric relaxation and viscosity, as well.[11]

## ACKNOWLEDGMENTS

It is a true pleasure to dedicate this work to Antonio Coniglio in honor of his 60th birthday. Antonio represents what is good about physicists and the profession of physics. In addition to brilliant work and lectures in statistical mechanics, and his wonderful physical and mathematical insight, Antonio always raises the level of fellowship and spirit of any scientific meeting and endeavor. It is an honor to be his friend and colleague. This work was supported in part by National Science Foundation Grant DMR-9815957 and the Office of Naval Research.

## REFERENCES

1. H. Vogel, *Phys. Z.* **22**, 645 (1921).
2. G. S. Fulcher, *J. Am. Ceram. Soc.* **8**(6), 339 (1925).
3. G. Tammann and W. Hesse, *Z. Anorg. Allmeg. Chem.* **156**, 256 (1926).
4. J. T. Bendler and M. F. Shlesinger, *Nucl. Phys.* **A5**, 82 (1988).
5. J. T. Bendler and M. F. Shlesinger, *J. Mol. Liq.* **36**, 37 (1987).
6. J. T. Bendler and M. F. Shlesinger, *J. Stat. Phys.* **53**, 531 (1988).
7. J. J. Fontanella, M. C. Wintersgill, C. S. Coughlin, P. Mazaud and S. G. Greenbaum, *J. Polym. Sci.: Part B: Polym. Phys.* **29**, 747 (1991).
8. J. J. Fontanella, M. C. Wintersgill and J. J. Immel, *J. Chem. Phys.* **110**, 5392 (1999).
9. J. J. Fontanella, *J. Chem. Phys.* **111**, 7103 (1999).
10. P. Zoller and D. J. Walsh, *Standard Pressure-Volume-Temperature Data for Polymers* (Technomic Publishing Co., Inc., Lancaster, PA, 1995).
11. J. T. Bendler, J. J. Fontanella and M. F. Shlesinger, *Phys. Rev. Lett.*, to appear (2001).

# DYNAMICAL NON-LINEAR SUSCEPTIBILITY OF THE QUENCHED AND ANNEALED FRUSTRATED LATTICE GAS MODELS

ANTONIO DE CANDIA, ANNALISA FIERRO and ANTONIO CONIGLIO
*Università di Napoli "Federico II",*
*INFM, Unità di Napoli*
*Complesso Universitario di Monte S. Angelo,*
*I-80126, Napoli, Italy*

**Abstract**

In this paper we study the 3D frustrated lattice gas model in the quenched and annealed versions. In the first case, the dynamical non-linear susceptibility grows monotonically as a function of time, until reaching a *plateau* that corresponds to the static value. The static non-linear susceptibility diverges at some density, signaling the presence of a thermodynamical transition. In the annealed version, where the disorder is allowed to evolve in time with a suitable kinetic constraint, the thermodynamics of the model is trivial, and the static non-linear susceptibility does not show any singularity. Nevertheless, the model shows a maximum in the dynamical non-linear susceptibility at a characteristic value of the time. Approaching the density corresponding to the singularity of the quenched model, both the maximum and the characteristic time diverge. We conclude that the critical behavior of the dynamical susceptibility in the annealed model is related to the divergence of the static susceptibility in the quenched case. This suggests a similar mechanism also in supercooled glass-forming liquids, where an analogous behavior in the dynamical non-linear susceptibility is observed.

## 1. INTRODUCTION

There has been a long debate on whether the glass transition corresponds to a real thermodynamical transition, or it is a purely dynamical phenomenon. Two outstanding theories that give a different answer to this question are the Adam and Gibbs entropy theory[1] and the mode-coupling theory.[2] In the last years, a unifying picture has been proposed, which states that there exist two transition temperatures in supercooled liquids: a higher one corresponding to the purely dynamical transition of mode-coupling theory, which in real liquids does not give rise to the structural arrest thanks to the so-called "hopping processes"; and a lower one corresponding to a thermodynamical transition of the Adam and Gibbs kind, with the vanishing of the configurational entropy and the consequent structural arrest of the liquid.

This picture was inspired mainly by the analogy with a class of mean field spin glass models, the so-called $p$-spin models, that show exactly this behavior[3] (apart from the absence of hopping processes), and have become a sort of paradigmatic model for the glass transition. The $p$-spin model is formed by $N$ Ising spins $S_i$, with the Hamiltonian

$$H = \sum_{i_1 < \cdots < i_p} J_{i_1 \ldots i_p} S_{i_1} \ldots S_{i_p} \tag{1}$$

where the sum is extended to all the groups of $p$ spins out of $N$, with $p \geq 3$, and $J_{i_1 \ldots i_p}$ are quenched random Gaussian variables, with zero mean and variance proportional to $N^{1-p}$. The first purely dynamical transition of the $p$-spin model, at the temperature $T_c$, is characterized by the fact that the configuration space becomes disconnected into an exponentially large number of "ergodic components", $\mathcal{N} \sim \exp(N\Sigma)$, where $\Sigma$ is the so-called configurational entropy. Below $T_c$, the system gets trapped in a single ergodic component and ergodicity is broken. The second transition at the temperature $T_s$ is characterized by the vanishing of the configurational entropy $\Sigma$, and corresponds to a real thermodynamical transition.

The analogy between structural glasses and $p$-spin models was pushed forward by the "effective potential" theory of Franz and Parisi.[4] In this theory one introduces a measure of the similarity between two configuration of the system, the overlap $q$. The effective potential $V(q)$ is the free energy of two systems that are constrained to have an overlap $q$. Both in $p$-spin models, and in structural liquids in the hypernetted chain approximation, the effective potential $V(q)$ develops a secondary minimum below the temperature $T_c$. This signals breaking of ergodicity: while physical quantities calculated in the primary minimum represent averages computed with the Boltzmann weight, the same quantities computed in the secondary minimum represent averages computed only within a single ergodic component. Now the non-linear susceptibility $\chi = N(\langle q^2 \rangle - \langle q \rangle^2)$, evaluated for $T < T_c$ in the secondary minimum, that is within a single ergodic component, diverges when $T_c$ is approached from below.

Donati et al.[5] have recently defined a dynamical non-linear susceptibility, and evaluated it both in the mean-field spherical $p$-spin model, and in a Lennard-Jones binary mixture by molecular dynamics simulation. It is defined by $\chi(t) = N(\langle q(t)^2 \rangle - \langle q(t) \rangle^2)$, where $q(t)$ is the time dependent self-overlap, that is the overlap between an equilibrium configuration at time $t'$, and the same configuration evolved at time $t' + t$. While the static susceptibility inside a single ergodic component can be evaluated only for $T < T_c$, where ergodicity is broken (at least in mean-field) and single ergodic components do exist, the dynamical one can be evaluated also for $T > T_c$, and it shows a characteristic behavior, with a maximum $\chi(t^*)$ at some time $t^*$, followed by a decay to the equilibrium value. The maximum diverges when $T_c$ is approached from above, together with the characteristic time $t^*$.

In this paper, we calculate the dynamical non-linear susceptibility in the 3D frustrated lattice gas, with quenched and annealed interactions.[6] The frustrated lattice gas was introduced as a simple lattice model of a glass-forming liquid.[7] Each site carries two kinds of variables, a lattice gas variable $n_i = 0, 1$, which represent the presence or absence of a particle on the $i$th site, and an Ising spin variable $S_i$, which represent an internal degree of freedom of the particle, such as for example the orientation of a non-symmetrical molecule. The Hamiltonian of the model is

$$H = J \sum_{\langle ij \rangle} (1 - \varepsilon_{ij} S_i S_j) n_i n_j - \mu \sum_i n_i \,, \tag{2}$$

where $\varepsilon_{ij} = \pm 1$. In the limit $J \to \infty$, the first term of the Hamiltonian implies that two nearest neighbor sites can be simultaneously occupied by two particles only if their spin variables satisfy the constraint $\varepsilon_{ij} S_i S_j = 1$. Therefore, if we identify the variables $S_i$ with the orientation of a non-symmetrical molecule, this condition means that two molecules can be near only if their relative orientation is appropriate. Another possible interpretation of the model is that $n_i$ represent the presence or absence of a particle in the $i$th cell, while $S_i$ represent the position of the particle inside the cell. If the size of the particles is of the order of that of the cells or greater, then not all the pairs of positions $S_i$ and $S_j$ inside two neighboring cells are allowed, due to the excluded volume.

Being constituted essentially by diffusing particles, this model is suited to study quantities like the diffusion coefficient, or the density autocorrelation functions, that are usually important in the study of liquids. Indeed, the model has proven to reproduce fairly well many features of supercooled glass-forming liquids, as for example the "cage effect". At low temperature and high density, the model shows a two step relaxation in the self correlation function

$$\phi^s_{\mathbf{q}}(t) = \langle e^{-i\mathbf{q}(\mathbf{r}(t'+t)-\mathbf{r}(t))} \rangle \tag{3}$$

and in the mean square displacement

$$\Delta(t) = \langle (\mathbf{r}(t'+t) - \mathbf{r}(t))^2 \rangle \tag{4}$$

In this paper we calculate, by means of Monte Carlo simulations, the dynamical non-linear susceptibility, both in the case where variables $\varepsilon_{ij}$ are quenched, that is they are extracted randomly for each sample and do not evolve in time, and in the case where they are annealed, that is they are allowed to evolve in time. We will see that one observes two different behaviors: while in the first case the dynamical susceptibility grows monotonically until reaching the static value, which is divergent at some density $\rho_c$, in the second case it shows a maximum at a finite time, which seems to diverge when approaching $\rho_c$, and decays at long times to the equilibrium value, which does not have any singularity.

## 2. THE QUENCHED MODEL

In this section we study the frustrated lattice gas model (2) in the case of quenched interactions, that is when the variables $\varepsilon_{ij}$ are extracted randomly for each sample, and do not evolve in time. In the limit $\mu \to \infty$ all sites are occupied, and the model reproduces the Ising spin glass. In the other limit $J \to \infty$ the model describes a frustrated lattice gas with properties tipical of "frustrated" liquids.

The variables $\varepsilon_{ij}$ represent the disordered environment in which the particles move. It is clear that in real liquids the environment in which a particle move is constituted by the position of all the other particles, and the disorder in some sense is "self-generated". Here instead we separate the variables that represent the environment, $\varepsilon_{ij}$, from the variables

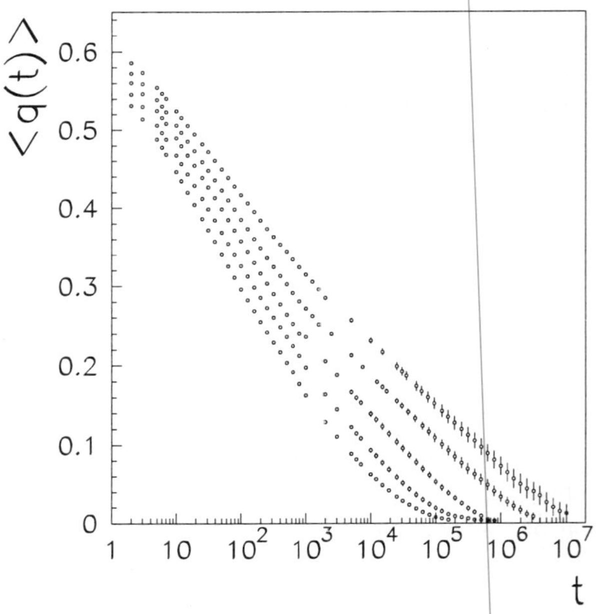


**Fig. 1** Relaxation functions of the self-overlap in the quenched model, for a system of size $16^3$ and densities $\rho = 0.58$, $0.59$, $0.60$, $0.61$, $0.62$.

that represent the position of the particles, $n_i$ and $S_i$. Furthermore, while in real liquids the environment changes in time as soon as the system undergoes a structural rearrangement, in our model the variables $\varepsilon_{ij}$ are quenched and thus fixed in time. Nevertheless, we expect that this model reproduces the properties of the liquid at very low temperature, where the disordered environment evolves so slowly that it can be considered as quenched. From the thermodynamic point of view, it has been shown[8] that there exists some density $\rho_c \simeq 0.62$ where the system has a thermodynamical transition, signaled by the divergence of the static non-linear susceptibility

$$\chi_{SG} = \frac{1}{N}\sum_{ij}[\langle S_i n_i S_j n_j\rangle^2] \tag{5}$$

where the average $\langle\cdots\rangle$ is over the Boltzmann measure, while the average $[\cdots]$ is over the disorder configurations $\{\varepsilon_{ij}\}$.

Here we show the results for the relaxation of the self-overlap, which is defined as

$$q(t) = \frac{1}{N}\sum_i S_i(t')n_i(t')S_i(t'+t)n_i(t'+t) \tag{6}$$

and for the dynamical susceptibility

$$\chi(t) = N[\langle q(t)^2\rangle - \langle q(t)\rangle^2] \tag{7}$$

where the average $\langle\cdots\rangle$ is over the reference time $t'$.

In Fig. 1 we show the relaxation functions $\langle q(t)\rangle$ for a system size $16^3$, and for densities between $\rho = 0.58$ and $0.62$. Each curve is obtained averaging over a time interval between $10^6$ and $10^7$ Monte Carlo steps, and over 16 runs with different realizations of the disorder, the initial conditions and the thermal history. The error bars are evaluated as the mean standard deviation of the average over the 16 runs. The long time tail of the functions can be well fitted by a stretched exponential form

$$\langle q(t)\rangle \propto \exp\left\{-(t/\tau)^\beta\right\} \tag{8}$$

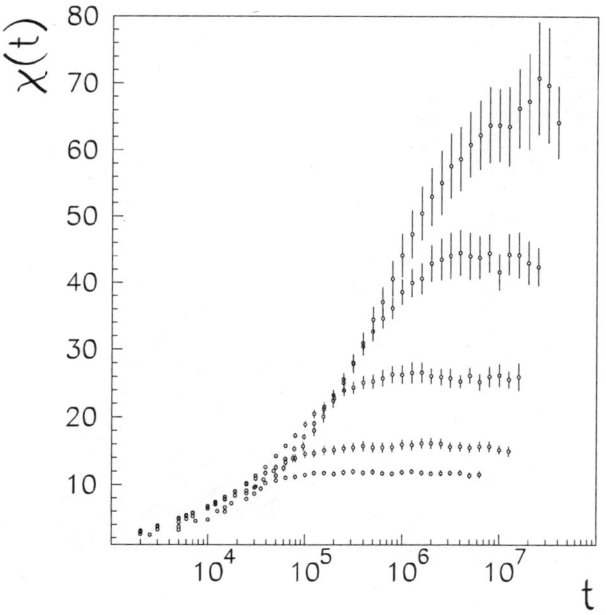


**Fig. 2** Dynamical susceptibility in the quenched model, for the same system size and densities of Fig. 1.

with an exponent $\beta$ strongly dependent from the density, which tends to very low values $\beta \simeq 0.2$ at high density.

In Fig. 2 it is shown the dynamical susceptibility $\chi(t)$ for the same system size and densities of Fig. 1. Note that $\chi(t)$ grows monotonically, and has no maximum at finite time. The asymptotic value $\chi(\infty)$ corresponds to the static susceptibility (5), and therefore has a divergence at a density $\rho_c \simeq 0.62$.

## 3. THE ANNEALED MODEL

Here we study the frustrated lattice gas model (2) in the case of annealed interactions, that is when the variables $\varepsilon_{ij}$ are allowed to evolve in time. In this case, the partition function of the model can be easily evaluated. Summing over all the configurations of the $\varepsilon_{ij}$ we obtain, up to an irrelevant factor,

$$Z = \sum_{\{n_i, S_i, \varepsilon_{ij}\}} \exp(-\beta H) \propto \sum_{\{n_i\}} \exp(-H_{\text{eff}}) \tag{9}$$

where the effective Hamiltonian $H_{\text{eff}}$ is given by

$$H_{\text{eff}} = K \sum_{\langle ij \rangle} n_i n_j - \beta\mu \sum_i n_i \tag{10}$$

with $K = \log(2/(1+e^{-2\beta J}))$. Therefore the static properties of the model are equal to those of a lattice gas with a repulsion between nearest neighbor particles, and with no correlation between spins, that is $\langle S_i S_j \rangle = \delta_{ij}$. One can verify that the intensity $K$ of the repulsive interaction is always lower than the value that corresponds to the critical point of the lattice gas, so we can conclude that the model (2) in the case of annealed interactions does not present any thermodynamical transition. In the following we will always take $J = \infty$.

We have performed a Monte Carlo simulation of the model, by allowing the interactions $\varepsilon_{ij}$ to evolve with a kinetic constraint. Namely, an interaction can change its value only if the sites $i$ and $j$, and all of their nearest neighbors, are empty. The variables $\varepsilon_{ij}$ play

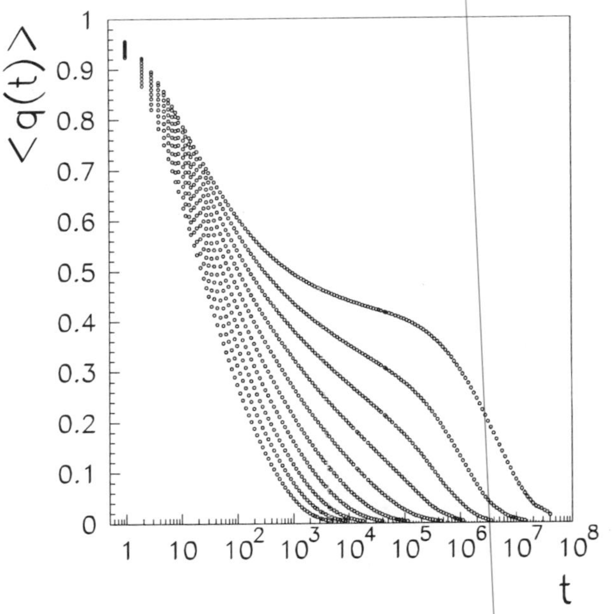


**Fig. 3** Relaxation functions of the self-overlap in the annealed model, for a system of size $16^3$ and densities $\rho = 0.52$, 0.53, 0.54, 0.55, 0.56, 0.57, 0.58, 0.59, 0.60, 0.61.

the role of a "disordered environment" in which the particles move. Therefore, the kinetic constraint represents the condition that, for the local environment to undergo a structural rearrangement, it is necessary that a fluctuation brings the local density below a certain value, leaving enough space to realize the rearrangement otherwise hindered. We expect that, as the temperature decreases and the density increases, the local environment represented by the variables $\varepsilon_{ij}$ changes so slowly, that it can be considered as a "self-induced quenched disorder" on the shorter time scale on which the particles diffuse.

We start the simulation by generating an equilibrium configuration at a given density. In order to do this, we simulate the model without any kinetic constraint. In this case we can thermalize the system in a very short time even at the highest density considered. Once an equilibrium configuration is obtained, we perform a diffusive dynamics for the particles, while the interactions evolve with the kinetic constraint described above.

During the simulation we evaluated the overlap relaxation function $\langle q(t) \rangle$ and the dynamical non-linear susceptibility $\chi(t)$, see Eqs. (6) and (7). Note that, as the density grows, the relaxation time gets longer and longer, until at a density $\rho \simeq 0.63$ it gets longer than the observation time allowed by our CPU time resources, which is about $10^8$ Monte Carlo steps for a system size $16^3$.

In Fig. 3 we show the relaxation functions of the self-overlap, for system size $16^3$ and for various densities between $\rho = 0.52$ and 0.61. Each curve is obtained averaging over a time interval of $10^6 - 10^8$ Monte Carlo steps, and over 16 runs with different initial conditions and thermal histories. The error bars are evaluated as the mean standard deviation of the average over the 16 runs.

Observe that for high density the relaxation functions $\langle q(t) \rangle$ clearly develop a two-step relaxation, signaling the existence of two well separated time scales in the system. We interpret the first short time decay of the relaxation functions as due to the motion of the particles in the environment made by the interactions $\varepsilon_{ij}$, which on this time scale appear as frozen, while the second decay is due to the rearrangement of the environment, and final relaxation to equilibrium. The long time tail of the relaxation functions is well fitted

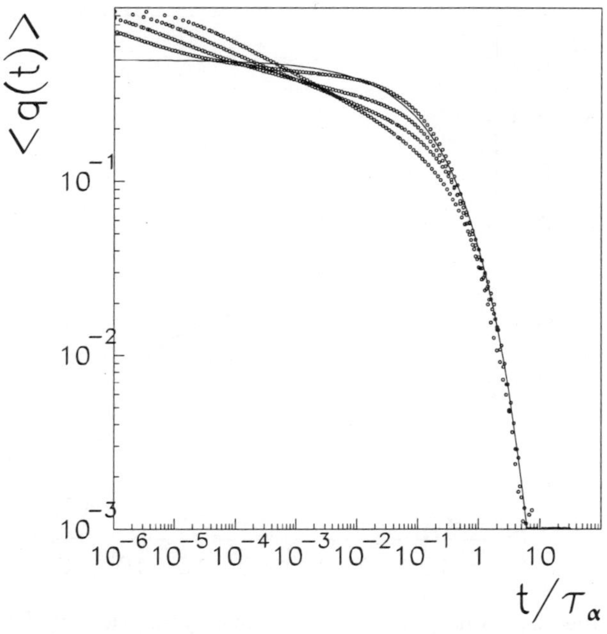


**Fig. 4** Time temperature superposition principle for the relaxation functions of the self-overlap, for densities $\rho = 0.58, 0.59, 0.60, 0.61$. The fitting function is a stretched exponential with exponent $\beta = 0.5$.

by a stretched exponential form (8), where the exponent $\beta$ depends very weakly from the temperature (it is constant within the errors) and is between $\beta = 0.4$ and 0.6.

In Fig. 4 we show the time-temperature superposition of the relaxation functions, for densities between $\rho = 0.58$ and 0.61. The fitting function is a stretched exponential with exponent $\beta = 0.5$.

We tried to fit the intermediate part, corresponding to the *plateau*, of the relaxation function of the self-overlap for density $\rho = 0.61$, with the function predicted by the mode-coupling theory of the density fluctuations in a supercooled liquid. This function can be written in a simplified form as

$$\langle q(t) \rangle = f + At^{-a} - Bt^{b} \tag{11}$$

where f is the value of the *plateau*, the first power law is the short time decay to the plateau, and the second is the von Schweidler law that should describe the departure from the *plateau* in the liquid phase. The mode-coupling theory predicts that the exponents $0 < a < \frac{1}{2}$ and $0 < b \leq 1$ are related by the equation

$$\frac{\Gamma^2(1-a)}{\Gamma(1-2a)} = \frac{\Gamma^2(1+b)}{\Gamma(1+2b)} = \lambda\,, \tag{12}$$

where $\frac{1}{2} \leq \lambda < 1$ is the so-called exponent parameter. We made a fit using as fitting parameters $f$, $A$, $B$ and $\lambda$, while $a$ and $b$ were given from $\lambda$ by the relation (12). The result is shown in Fig. 5, where the full line is the fitting curve with $a = 0.339 \pm 0.002$ and $b = 0.69 \pm 0.01$.

In Fig. 6 we show the dynamical non-linear susceptibility for the same size and densities of Fig. 3. It shows the same behavior observed in the mean field $p$-spin model at the dynamical transition, and in molecular dynamics simulations of a Lennard-Jones binary mixture,[5] namely a maximum $\chi(t^*)$ at some characteristic time $t^*$, followed by a decay for very long times to the equilibrium value, that in the annealed model is easily calculated,

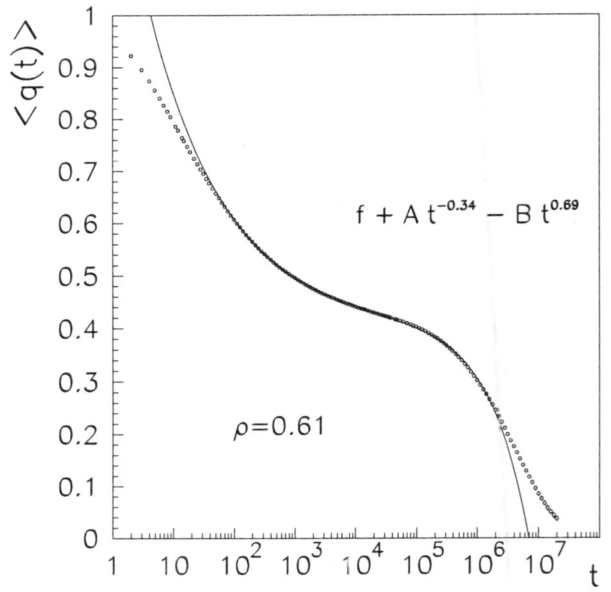


**Fig. 5** Fit of the intermediate time part of the relaxation function of the self-overlap, for density $\rho = 0.61$, with the fitting function $f + At^{-a} - Bt^{b}$, where the fitting parameters are $f$, $A$, $B$ and $\lambda$, and $a$ and $b$ are given by the relation (12).

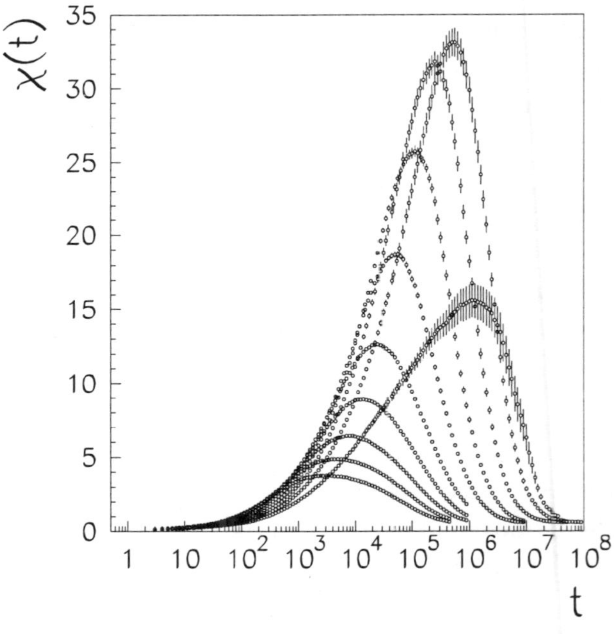


**Fig. 6** Dynamical susceptibility in the annealed model, for the same system size and densities of Fig. 3.

$\chi(\infty) = \rho^2$. When the density grows, and approaches the critical point of the quenched model, the maximum $\chi(t^*)$ seems to diverge, together with the characteristic time $t^*$. For the highest density the maximum decreases, but if one performs the averages over longer and longer observation times, the maximum increases at an extremely slow rate. It is therefore possible that the last curve is strongly affected by a too short observation time.

We obtain that the maximum $\chi(t^*)$ as a function of the density can be fitted quite well (taking out the last three points) by the power law $\chi(t^*) \propto (\rho_c - \rho)^{-\alpha}$, with $\rho_c = 0.66 \pm 0.01$

and $\alpha = 3.6 \pm 0.2$. At very long times $\chi(t)$ decays to the equilibrium value, which is simply $\chi(\infty) = \rho^2$.

## 4. CONCLUSIONS

We have evaluated the dynamical non-linear sysceptibility of the frustrated lattice gas model, both in the quenched and annealed version. In the first case, the interactions $\varepsilon_{ij}$, that represent the disordered environment in which particles move, are frozen. The presence of frozen disorder causes the presence of a thermodynamical transition, with a divergence of the static non-linear susceptibility. When the interactions are allowed to evolve in time, the thermodynamical transition disappears, and the static susceptibility is regular at any densities. However, the critical behavior shows up in the dynamical susceptibility, which presents a maximum at a finite value of the time, which diverges at some value of the density. This can be due to the fact that the dynamical constraint makes the dynamics of the interactions $\varepsilon_{ij}$ slower and slower when the density is increased, and therefore at short enough time the environment in which particles move appears as frozen disorder.

We suggest that a similar mechanism might be relevant also for supercooled liquids, where a similar behavior in the dynamical non-linear susceptibility is observed. Indeed, if one introduces a "quenched disorder" in a supercooled liquid by coupling it to a reference configuration, that is to a frozen configuration of the same system, one finds a diverging static non-linear susceptibility. However, this means that we have treated all the degrees of freedom of the liquid, that is the $3N$ coordinates of the particles, on the same footing. It would be interesting, inspired by the example of the frustrated lattice gas, to separate the "environment" degrees of freedom of the liquid, represented in our model by the variables $\varepsilon_{ij}$, and to treat only them as quenched.

## ACKNOWLEDGMENTS

We acknowledge support by the TMR network contract ERBFMRXCT980183, INFM PRA-HOP 99, MURST PRIN-2000, and INFM Progetto Calcolo Parallelo.

## REFERENCES

1. G. Adam and J. H. Gibbs, *J. Chem. Phys.* **43**, 139 (1965).
2. W. Götze, in *Liquids, Freezing and the Glass Transition*, eds. J. P. Hansen, D. Levesque and J. Zinn-Justin (North Holland, 1990).
3. T. R. Kirkpatrick and D. Thirumalai, *Phys. Rev.* **B36**, 5388 (1987).
4. S. Franz and G. Parisi, *Physica* **A261**, 317 (1998).
5. C. Donati, S. Franz, S. C. Glotzer and G. Parisi, *cond-mat/9905433.*
6. A. Fierro, A. de Candia and A. Coniglio, *Phys. Rev.* **E62**, 7715 (2000).
7. A. Coniglio, *Proceedings of the International School on the Physics of Complex Systems* (Varenna, 1996); M. Nicodemi and A. Coniglio, *J. Phys. Lett.* **A30**, L187 (1996).
8. A. de Candia and A. Coniglio, in preparation.

# LACK OF EQUILIBRATION IN A MODEL FOR CONTINUOUSLY SUPERCOOLED LIQUIDS

F. CORBERI
*Istituto Nazionale per la Fisica della Materia, Unità di Salerno and Dipartimento di Fisica, Università di Salerno, 84081 Baronissi (Salerno), Italy*

A. CONIGLIO
*Dipartimento di Scienze Fisiche, Università di Napoli Federico II and Istituto Nazionale di Fisica della Materia, Unità di Napoli Monte S. Angelo — Via Cintia, 80126 Napoli, Italy*

**Abstract**

We consider the dynamics of supercooled fluids subject to a continuous quenching procedure, with cooling rate $r = dT(t)/dt$. The analysis is carried out analytically in the framework of a mean field schematic model recently introduced.[1] We show the existence of a glass temperature $T_g(r)$ below which the system falls out of equilibrium. $T_g(r)$ approaches logarithmically in $r$ the ideal glass temperature $T_0 = \lim_{r \to 0} T_g(r)$, where the relaxation time diverges *á la* Vogel-Fulcher, similarly to some experimental observations. Well above $T_g(r)$ a *simple fluid* behavior is observed. As $T_g$ is approached from above a characteristic wave vector $k_d$ divides an high momenta equilibrated region, where a *fluid*-like behavior is obeyed, from the non thermalized modes with $k < k_d$, for which time translational invariance lacks. Below $T_g$ the system is found in a globally off-equilibrium glassy state characterized by a logarithmic decay of the density fluctuations and aging. The two time correlator decays as an enhanced power law.

## 1. INTRODUCTION

Glassy systems, such as fluids supercooled below the melting temperature, are characterized by a slow structural rearrangement due to a complex cooperative dynamics caused by configurational restrictions. This results in a strong divergence of the equilibrium relaxation time $\tau$ as the ideal glass temperature $T_0$ is approached from above. When a fluid is supercooled with a finite cooling rate $r$, namely the temperature of the thermal bath $T$ is lowered as $dT/dt = -r(T)$, it follows the equilibrium curve down to a characteristic temperature $T_g(r)$, when its relaxation time $\tau$ becomes comparable to the timescale $r^{-1}$ of the temperature variations. In experiments[2] $T_g(r)$ is observed to depend logarithmically on the cooling rate $r$. By still decreasing the temperature the system goes out of equilibrium, since the rapid increase of $\tau$ does not allow instantaneous equilibration with the bath. In this region molecular rearrangement still occur but on timescales much larger than $r^{-1}$ and become slower and slower as time goes on: The system ages. The structure recovery in this region is characterized by a slow decay of the density toward its equilibrium value,[3] which is consistent with a logarithmic behavior. This feature, together with the particular dependence of $T_g$ on $r$, cannot be explained by most successful theoretical approaches to glassy kinetics such as the mode coupling theory and are not nowadays clearly understood.

In this article we study the evolution of a cooled system in the framework of a model[1] recently introduced for the description of the dynamics close to the glass transition. The approach is simple enough to be handled analytically, allowing explicit calculations at a mean field level. We show the existence of a dynamical transition temperature $T_g(r) \simeq T_0\{1 + [-a \ln r - c]^{-1}\}$ ($a$, $c$ = constants), $T_0$ being the ideal glass temperature where the relaxation time diverges *á la* Vogel-Fulcher, which separates two regions with different dynamical properties. Well above $T_g(r)$ the system behaves as a *simple fluid* and physical observables attain their equilibrium value at the current temperature (see, for instance, the particle squared density fluctuations $S(T)$ behavior in Fig. 1). As $T_g(r)$ is approached

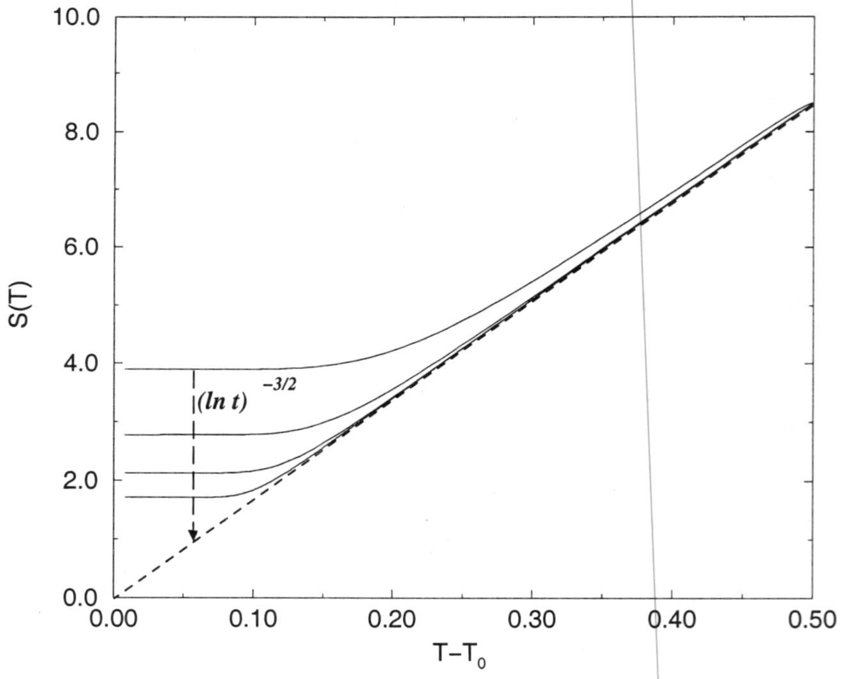


**Fig. 1** The behavior of the average squared density fluctuations $S$ is shown as the temperature is gradually lowered with $r = 10^{-5}$, $10^{-4}$, $10^{-3}$, $10^{-2}$ (continuous lines from bottom to top). This lines start deviating from the bold dashed line representing the equilibrium behavior at a characteristic temperature $T_g$ which depends logarithmically on $r$. Below $T_g$ the relaxation process becomes extremely slow, resulting in an apparent constancy of $S$. Observation of $S$ on much longer timescales reveals that $S \simeq (\ln t)^{-3/2}$, as schematically depicted in the figure.

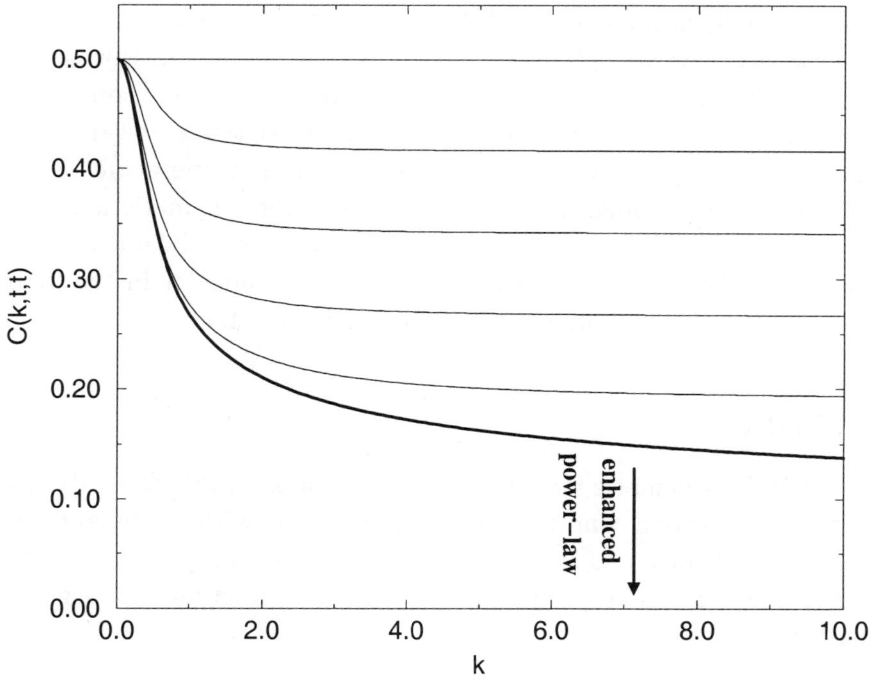


**Fig. 2** The structure factor $C(\mathbf{k}, t, t)$ is plotted against $k$ at different times ($t = 1$, 168, 318, 468, 618, 968 from top to bottom), for a cooling procedure with $r = 5 \cdot 10^{-4}$. In equilibrium at a constant temperature $T$ one would have $C = T$. This is actually the case at the very beginning ($t = 1$). As the quenching proceeds one still have $C(\mathbf{k}, t, t) \simeq T(t)$ for large momenta ($k > k_d(t)$) but not for $k < k_d(t)$. $k_d(t)$ moves to the right until, at $t \simeq 968$ it hits the ultraviolet cutoff, $T$ crosses $T_g$ and the glassy phase is entered. Below $T_g$ the relaxation process becomes extremely slow so that subsequent curves apparently collapse onto the curve $C_{pin}(\mathbf{k}) = C(\mathbf{k}, t = 968)$ (bold line). On much larger timescales (not shown in the picture) this curve decays as an enhanced power-law, as indicated.

a characteristic momentum $k_d$ separates the wave vector region with $k > k_d$, behaving as a fluid, from the out of equilibrium modes with $k < k_d$ for which time translational invariance is lost (see the behavior of the structure factor $C(\mathbf{k}, t, t)$ in Fig. 2). By lowering the temperature $k_d$ increases and the contribution of the out of equilibrium modes become progressively more important until, at the glass temperature $T_g$, $k_d$ becomes comparable with the high momentum cutoff of the model. Below $T_g$, since $k < k_d$ for all momenta, the system is found in a globally out of equilibrium glassy state. Here the evolution of the density fluctuations $\sqrt{S(t)}$ is shown to proceed logarithmically slow, as $(\ln t)^{-3/4}$. In this regime the system ages and the two time correlator decays as an enhanced power law $C(\mathbf{k}, t', t) \sim \exp\{-b([\ln(t)]^\delta - [\ln(t')]^\delta)k^2\}$, with $\delta = 6/d$.

The relation between the response function $\chi(\mathbf{k}, t', t)$ to an external perturbation and the autocorrelation $C(\mathbf{k}, t', t)$ of the unperturbed system can also be considered. Whenever the fluid is nearby equilibrium, for $T \gg T_g$, the fluctuation dissipation theorem is obeyed: $\chi$ and $C$ are proportional through the temperature $T$ of the reservoir. This applies even if the bath temperature gradually changes, provided the system is well into the fluid phase. On the other hand, when the glass temperature is approached, still being $T > T_g$, the fluid falls gradually out of equilibrium and the fluctuation dissipation theorem starts to be violated. Interestingly, however, $\chi$ and $C$ are still proportional through a constant $\Theta$ which depends on the time $t'$ elapsed after the perturbation, suggesting its interpretation as an effective inner temperature. In the presence of a reservoir at a fixed temperature $T$, $\Theta$ approaches $T$ from above. How fast this happens depends on the scale considered, since

$\Theta$ is a function in momentum space such that large wave vectors thermalize before than smaller ones. The situation with $T_0 < T < T_g$ is also interesting. In this case it is possible to show that for sufficiently large momenta $\Theta$ remains pinned to $T_g$ on the experimental typical timescale $r^{-1}$. Eventually $\Theta$ evolves, on much larger times, towards $T$. In the limit $r \to 0$, $\Theta$ converges for large wave vectors to the ideal glass temperature $T_0$.

This paper is organized as follows: In Sec. 2 we recall the schematic model for quenched fluids and discuss its applicability to the case of a progressively quenched system at a constant pressure. In Sec. 3 the model is studied and its behavior in the fluid and glassy phase is described. Conclusive remarks are discussed is Sec. 4.

## 2. THE MODEL

Let us briefly recall the schematic model[1] for the description of the glassy dynamics in order to generalize it to the case of a continuous cooling procedure with a constant pressure. In its original form the model considers the variable $\rho(\mathbf{r}, t)$ representing a coarse grained particle density with average $\bar{\rho}(t)$. The evolution of the density fluctuations is governed by the diffusion equation

$$\frac{\partial}{\partial t}[\rho(\mathbf{r}, t) - \bar{\rho}(t)] = \nabla \cdot \left[M(\rho)\nabla \frac{\delta}{\delta \rho} F\{\rho\}\right] + \eta(\mathbf{r}, t) \tag{1}$$

where $F\{\rho\}$ is the free energy. In the following we restrict the analyses to the simple form $F\{\rho\} = \rho^2$ since the spirit of the present approach, as discussed below, is to reproduce some basic phenomenology of the glass transition with minimal ingredients. $\eta(\mathbf{r}, t)$ is a Gaussianly distributed random field, representing thermal noise, with expectations $\langle \eta(\mathbf{r}, t)\rangle = 0$, $\langle \eta(\mathbf{r}, t)\eta(\mathbf{r}', t')\rangle = -2T\nabla \cdot \{M(\rho)\nabla[\delta(\mathbf{r}-\mathbf{r}')\delta(t-t')]\}$, where $\langle \cdots \rangle$ means ensemble averages and $T$ is measured in units of the Boltzmann constant. We consider the mobility

$$M(\rho) = e^{[\rho(\mathbf{r},t)-1]^{-1}} e^{-v(T)}. \tag{2}$$

This form of $M$ is assumed on phenomenological grounds because it provides[1] the expected Vogel-Fulcher divergence of the equilibrium relaxation time; a similar form has been also derived in different context such as the free volume theories of the glass transition or car parking problems.[5] The temperature dependence in Eq. (2) can be assumed to be of the Arrhenius type, namely $v(T) = v/T$ ($v =$ const.).[6]

The analytic approach to the model[1] amounts to the substitution $M(\rho) \to D(t)$ in Eq. (1), where $D(t) = \langle M(\rho)\rangle$ does not depend on space due to homogeneity. Introducing the density fluctuation field $\phi(\mathbf{r}, t) = \rho(\mathbf{r}, t) - \bar{\rho}(t)$ Eq. (1) becomes

$$\frac{\partial \phi(\mathbf{r}, t)}{\partial t} = D(t)\nabla^2 \phi(\mathbf{r}, t) + \eta(\mathbf{r}, t)\,. \tag{3}$$

The thermal noise appearing in Eq. (3) is Gaussian with zero mean and $\langle \eta(\mathbf{r}, t)\eta(\mathbf{r}', t')\rangle = -2TD(t)\nabla^2[\delta(\mathbf{r} - \mathbf{r}')\delta(t - t')]$.

The model described insofar aims to describe the dynamical feature characterizing a supercooled fluid at a simple level. It is schematic in spirit and, in order to be generic, lives aside as much system specific details as possible. The basic assumption is that a good deal of the complex behavior of glassy systems is of an essentially dynamical origin and can be encoded into the conventional diffusion equation (1) by means of a suitably chosen particle mobility $M(\rho)$. Additional features provided by the actual form of the free energy of realistic systems are disregarded here; the form of $F$ we adopt is as simple as possible and, in particular, regular when crossing the glass temperature, which turns out to be a purely dynamical effect. Previous studies of this model[1] considered the case of a sudden quench at time $t = 0$ from an high temperature $T_i$ to a lower temperature state at $T_f < T_i$ characterized by the equilibrium density $\bar{\rho}_f$. The dynamical evolution of the system was then studied from $t = 0$ onwards with fixed $T = T_f$ and $\bar{\rho} = \bar{\rho}_f$. Here, instead, we are interested in the case of a system progressively cooled with a certain rate $r$ at the constant pressure $P$. The density of the system increases correspondingly during the process and an equation for $\bar{\rho}$ is needed. In this article we make the simple assumption that the average density $\bar{\rho}$ obeys the equilibrium equation of state of a non-interacting gas with a finite co-volume

$$\bar{\rho}^{-1}(T) = \rho_{cp}^{-1} + \frac{T}{P} \tag{4}$$

where $\rho_{cp} > 1$ is a random close packing density and $P$ is the pressure. We expect this assumption to be appropriate for the non-interacting system we are dealing with at least for $T > T_g$. For $T < T_g$, since the system falls out of equilibrium, a departure of $\bar{\rho}$ from Eq. (4) is expected. An analyses of the effects of the non-equilibration of the average density will be discussed in a forecoming article. Equations (3) and (4) will be used to describe the cooling of a system, with rate $r(T)$, from an initial equilibrium state at $T = T_i$. The results of the numerical integration of the model which are presented in Figs. 1 and 2 refer to $\rho_{cp} = 1.1$, $P = 6 \cdot 10^{-2}$. The cooling procedure is such that $T = T_i - rt$ ($r =$ constant), with $T_i = 0.5$ and $\bar{\rho}(T_i) = 10^{-1}$.

Transforming Eq. (3) into reciprocal space, one obtains the following form for the two time correlator $C(\mathbf{k}, t', t) = \langle \phi(\mathbf{k}, t)\phi(-\mathbf{k}, t')\rangle$, $(t \geq t')$

$$\begin{aligned} C(\mathbf{k}, t', t) &= T(t')e^{-[R^2(t)-R^2(t')]k^2} - e^{-[R^2(t)+R^2(t')]k^2} \int_0^{t'} ds e^{2R^2(s)k^2} r[T(s)] && (5) \\ &= C_0(\mathbf{k}, t', t) + C_1(\mathbf{k}, t', t) && (6) \end{aligned}$$

where

$$R^2(t) = \int_0^t D(s)ds \tag{7}$$

and $C_0$ and $C_1$ correspond to the first and second terms in Eq. (5). $D(t)$ is given by

$$D(t) = \mathcal{N}^{-1}(t) \int_0^1 M(\rho) e^{-(\rho-\bar{\rho})^2/[2S(t)]} d\rho\,. \tag{8}$$

$\mathcal{N}(t) = \int_0^1 e^{-(\rho-\bar{\rho})^2/[2S(t)]}$ being a normalization factor. The mean square density fluctuations $S(t) = \langle \phi^2 \rangle$, which in the present approximation represents also the free energy (per particle), can be computed through the structure factor as

$$S(t) = (2\pi)^{-d} \int_{k<\Lambda} C(\mathbf{k}, t, t) d\mathbf{k}\,. \tag{9}$$

Equations (5), (7), (8) and (9), together with the equation for $\bar{\rho}$ (4) are a close set of equations governing the evolution of the mean field model. In the following these equations will be studied analytically. A comparison with the outcome of a numerical integration of the model is presented in Figs. 1 and 2.

## 3. STUDY OF THE MODEL

From Eq. (6) one sees that the correlator can be decomposed into the sum of two terms. In a thermalized state with a constant temperature bath $C_1$ vanishes whereas, when the system is not in equilibrium, due to a temperature change, both $C_0$ and $C_1$ contribute to the correlation. The ratio $\varepsilon(\mathbf{k}, t') = C_1/C_0$ can therefore be a measure of the degree of non-equilibration of the system playing the role of a *distance* from the thermalized state. This can be made more precise by computing the relation between the integrated response function to a small perturbing field $h(\mathbf{k}, t')$ applied at time $t'$, $\chi(\mathbf{k}, t', t) \equiv \int_{t'}^{t} d\tau \frac{\delta\langle\rho(\mathbf{k}, t)\rangle}{\delta h(-\mathbf{k}, \tau)}$ and $C(\mathbf{k}, t', t)$

$$\chi(\mathbf{k}, t', t) = 1 - \frac{C(\mathbf{k}, t', t)}{\Theta(\mathbf{k}, t')} \tag{10}$$

where

$$\Theta(\mathbf{k}, t') = T(t')[1 + \varepsilon(\mathbf{k}, t')] \,. \tag{11}$$

In an equilibrium system at the constant temperature $T(t) = T(t') = T$, since $C \equiv C_0$, one has $\Theta \equiv T$ so that the fluctuation dissipation theorem is obeyed. Then in the parametric plot of $\chi$ versus $C$ one obtains a straight line of slope $-T^{-1}$. Let us consider the system very rapidly quenched at time $t = 0$ to the final temperature $T > T_0$. From Eq. (10) one has that if $\chi$ is plotted against $C$ for a fixed $t'$ and a certain $\mathbf{k}$ we still find a single straight line of slope $-\Theta^{-1}$, because $\Theta$ does not depend on $t$. This suggests to regard $\Theta$ as an effective inner temperature[7] and $\epsilon$ as the *distance* from complete thermalization at $\Theta = T$. Computing $\Theta$ for times $t' > 0$, since the system is out of equilibrium, one finds that it approaches $T$ from above. This features of $\Theta$ can be compared with recent investigations of the response of supercooled liquids in molecular dynamic simulations. A similar behavior with an *internal* temperature depending on $t'$ is observed[8] for the *inter-basin* response, namely for large $t - t'$, of a Lennard Jones binary mixture. For small $t - t'$ the response proves the *intra-basin* vibrational dynamics, in equilibrium at the bath temperature. Therefore for this system a broken line with two definite slopes is observed. In the present model the fast-equilibrating *intra-basin* modes are not taken into account from the beginning and for this reason a plot with a single straight line is recovered. Interestingly, a fluctuation-dissipation ratio depending on $t'$ has also been found in models for granular media[9] and in experiments on structural glasses.[10] Moreover $\varepsilon$, and so $\Theta$, have also an interesting structure in momentum space. Notice, in fact, that $\varepsilon$ is a function of the wave vector and large momenta thermalize better then smaller ones. This is because from Eqs. (5) and (6) one has $\lim_{|k|\to 0} \varepsilon = [T_i/T]-1$, which is larger then 1, while generally $\lim_{|k|\to\infty} \varepsilon = 0$. The more general case of a continuously varying temperature can be discussed along the same lines. With this set up $\Theta$ closely follows $T(t')$ well inside the fluid phase while it manages to approach $T^{(}t')$, which is itself $t'$-dependent, in out-of-equilibrium situations. Again $\Theta$ depends on the smaller time $t'$ alone and a single line is observed in the plot of $\chi$ versus $C$. The structure in momentum space is also preserved with $\lim_{|k|\to 0} \varepsilon = [T_i/T(t')]-1$ and $\lim_{|k|\to\infty} \varepsilon = 0$. If $T(t')$ crosses the glass temperature $T_g$ at some time $t_g$ the system goes globally out of equilibrium and, in particular, the structure factor $C(\mathbf{k}, t, t)$ remains pinned for long times to the valued assumed at the time $t_g$ when $T_g$ was crossed, while a further evolution is only observed on much larger timescales. Moreover, for sufficiently large momenta one has $C(\mathbf{k}, t, t) \simeq T(t)$ for all temperatures down to $T_g$. This features of $C(\mathbf{k}, t, t)$ will be better explained below (Sec. 3.2) and are summarized in Fig. 2. Given that $\Theta$ can also be cast in the form $\Theta(\mathbf{k}, t') = C(\mathbf{k}, t', t')$ one has that, for large wave vectors, the effective temperature remains itself pinned to the value $\Theta = T_g$ on timescales $t' - t_g$ of the order of $r^{-1}$. Only on much larger times a further evolution of $\Theta$ towards the temperature of the reservoir is observed. Besides, in the vanishing cooling rate limit, $r \to 0$, for large momenta $\Theta$ follows the bath temperature down to the ideal glass temperature. Once $T_0$ is met, the system freezes and $\Theta$ remains fixed at $T_0$.

From the previous considerations it is clear that a typical wave vector $k_d$ exists, defined through $\varepsilon(k_d, t') \simeq 1$, which separates a region of large momenta where $\varepsilon < 1$, from the small $k$ sector where $\varepsilon > 1$. For $k > k_d$ the system behaves essentially as in equilibrium at the current temperature, whereas for $k < k_d$ it exhibits a different pattern, owing to non-equilibrium effects. We define $T_g$ as the temperature at which the structure factor $C(\mathbf{k}, t, t)$ deviates sensibly from $C_0(\mathbf{k}, t, t)$ for all momenta, namely $\varepsilon(\Lambda, t) \simeq 1$. In order to determine $T_g$ we estimate $\varepsilon$ for large $t'$ evaluating the integral defining $C_1$ by saddle point techniques

$$\varepsilon(\mathbf{k}, t') = \frac{r(t')}{2k^2 D(t')}[1 - e^{-2k^2 D(t')t'}]. \tag{12}$$

Using the determination (12) we obtain $T_g \simeq r(t')/[2\Lambda^2 D(T_g)]$; here we have replaced the value $D(t)$ of the average mobility at the time when $T_g$ is reached with its equilibrium value $D_{eq}(T_g)$, since this is appropriate for $T > T_g$. In the small cooling rate $T_g(r)$ approaches $T_0$ so that $\bar{\rho}(T)$ can be expanded as $\bar{\rho}(T) \simeq 1 + (d\bar{\rho}/dT)_{T_0}(T - T_0)$. With this expression, and letting $S \simeq 0$ in Eq. (8), as appropriate to low temperatures, one has

$$T_g(r) \simeq T_0 \left[1 + \frac{1}{-a \ln r - c}\right] \tag{13}$$

where $a = -T_0 d\bar{\rho}/dT|_{T_0}$ and $c = av/T_0 + T_0 d\bar{\rho}/dT|_{T_0} \ln(2\Lambda^2 T_0)$ are constant. An analogous logarithmic dependence of $T_g$ on the cooling rate is also observed in experiments.[2] Notice that the temperature $T_g(r)$ is of an essentially dynamical origin, since the equilibrium measure is a regular function along $T_g(r)$. Moreover the mobility $M(\rho)$ itself is regular and finite at $T_g(r)$ (it vanishes at $T_0 < T_g(r)$). This shows that the dynamical glass transition at $T_g$ is neither a thermodynamic transition neither a complete freezing, the kinetics proceeds below $T_g$, but simply the point at which the relaxation times of the system meet the experimental timescale.

In the following we will discuss separately the behavior of the model in the regimes above and below $T_g$.

## 3.1 Fluid Phase ($T > T_g$)

For $T \gg T_g(r)$ the characteristic wave vector $k_d$ is small. Therefore extensive one time variables which are obtained as a convolution over momenta do not deviate appreciably from their equilibrium value at the current temperature $T(t)$. This is shown in Fig. 1 where it is observed that $S(t)$ initially follows the equilibrium line. One also has $R^2(t) \simeq D_{eq}(T)t$. Non-extensive quantities which depend on the wave vector, on the other hand, behave differently depending on the particular momentum considered. We consider one or two time quantities separately. For $k > k_d$ one time observables attain their equilibrium value at $T(t)$, whereas smaller momenta are out of equilibrium. As an example the behavior of the structure factor is shown in Fig. 2. In equilibrium one has $C(\mathbf{k}, t, t) \equiv T$. Figure 2 shows that indeed $C(\mathbf{k}, t, t) = C_0(\mathbf{k}, t) = T(t)$ for $k \gg k_d$ while, for $k < k_d$ it is out of equilibrium and behaves differently. For $|k| \to 0$, $C(\mathbf{k}, t, t)$ remains at its initial value $T_i$ since, as already observed, $\varepsilon(\mathbf{k}, t) \simeq [T_i/T(t)] - 1$ in this limit.

Next we discuss the behavior of two time non extensive quantities, such as $C(\mathbf{k}, t', t)$. On time scales $t - t'$ smaller than the temperature variation the mobility $D(t)$ can be approximated by a constant value: One has $R^2(t) - R^2(t') \simeq D_{eq}(T)(t - t')$.

For $k > k_d$ the two time correlator is well approximated by $C_0(\mathbf{k}, t', t) \simeq T(t') \exp(t - t')k^2/\tau[T(t')]$ which decays exponentially over a characteristic time $\tau[T(t')] = D^{-1}(T')$. Notice that time translational invariance is also obeyed in this time domain for $k > k_d$. The characteristic time $\tau$ can be evaluated as follows: If the current temperature is sufficiently low, thermal fluctuations can be approximatively neglected. Then, expanding $\bar{\rho}$ around $T_0$ and letting $S \simeq 0$ in Eq. (8), we obtain $\tau \simeq M^{-1}\{1 + [d\bar{\rho}/dT]_{T_0}[T' - T_0]\}$, which diverges at $T_0$. With the form (2) one has $\tau \simeq \exp[g/(T' - T_0)]$, where $g^{-1} = -d\bar{\rho}/dT|_{T_0}$ is a constant, namely a Vogel-Fulcher divergence at $T_0$, as observed in experiments.[11]

For $k < k_d$, $C$ still decays exponentially as a function of $t$ but, due to the relevant contribution of $C_1$ time translational invariance is not fulfilled. These behaviors are expected for $T \gg T_g$; for temperatures closer to $T_g$, $R^2(t)$ grows slower then $D_{eq}(T)t$ so that the decay of $C$ slows down too.

If one keeps cooling the system with a finite $r$ until the glass temperature $T_g$ is met, since $k_d$ gets larger and larger, all the wave vectors fall progressively more and more out of equilibrium until the systems globally enters the glassy phase at $T \simeq T_g$. Let us consider also a different situation, which is often implemented experimentally, where the cooling procedure is halted at some time $t^*$. Then for $t' > t^*$ Eq. (5) becomes

$$C(\mathbf{k}, t', t) = C_0^*(\mathbf{k}, t', t)[1 - Ae^{-2R^2(t')k^2}] \tag{14}$$

where $C_0^*$ is $C_0$ evaluated at the constant temperature $T = T(t^*)$ and the constant $A$ is given by $A = \int_0^{t^*} ds \exp\{2R^2(s)k^2\} r[T(s)]$. Then if $T > T_g$ one has $C \simeq C_0^* \simeq \exp(t - t')k^2/\tau(T)$ for sufficiently large $t'$ so that time translational invariance is recovered at all momenta.

### 3.2 Glassy phase ($T < T_g$)

At $T \simeq T_g(r)$, $k_d$ becomes comparable with the momentum cutoff $\Lambda$. At this point all the observables deviate sensibly from the equilibrium values at the current temperature $T_g$. This is true also for the average density $\bar{\rho}$; the validity of Eq. (4) is questionable in this case and a more general approach considering also the dynamical equation obeyed by $\bar{\rho}$ is, in principle, needed. In the spirit of a schematic description of the glassy kynetics pursued by this approach, however, we will still retain Eq. (4) considering it as a first rough and improvable approximation, hopefully capable to give some hints for the study of more realistic systems. With this specification, by lowering the temperature below $T_g$ the inner relaxation process described by this model becomes very slow so that on the timescale of the thermal variations $r^{-1}$ the system pins: All the quantities remain fixed at the value they assumed at $T_g$. This is shown in Figs. 1 and 2. Notice that the limiting form $C_{pin}(\mathbf{k})$ of the structure factor is of a purely dynamical origin since, given the non-interacting system we are considering, the equilibrium states are disordered with a constant structure factor $C_0(k) = T$.

Even in this glassy state, however, the system is still evolving, although on huge timescales, because $M(\rho)$ is finite. In order to study this slow kinetics let us consider the physically realistic case of a system cooled down to $T < T_g$ and successively kept at $T$ from $t^*$ onward. In this situation the kinetics of the model is similar to the previously studied case[1] of a sudden deep quench with a constant density. In this case it was shown that, $S(t)$ decays slowly as $(\ln t)^{-3/2}$, as represented schematically in Fig. 1. The structure recovery in real supercooled fluids is also characterized by a logarithmic relaxation of the density. The correlator decays as an enhanced power law $C(\mathbf{k}, t', t) \sim \exp\{-b([\ln(t)]^\delta - [\ln(t')]^\delta)k^2\}$. In this low temperature region time translational invariance is lacking on all lengthscales and the system ages globally before crossing over to a stationary time translational state for extremely long times.

## 4. CONCLUSIONS

In this article we have considered the dynamics of a continuously cooled glassy system in the framework of a simple mean field approach where explicit calculations can be carried out. The model is based on a diffusion equation for the density fluctuations characterized by a fast vanishing mobility at the freezing density $\bar{\rho} = 1$. In the present approach the equilibrium behavior of a non-interacting gas for the average density $\bar{\rho}$ is assumed. This simple hypotesis is expected to be appropriate in the fluid phase well above $T_g$ because, as discussed, global quantities which do not depend on space position follow their equilibrium behavior. However this feature of $\bar{\rho}$ is expected to progressively break down as the transition temperature $T_g$ is approached. In this case the average density of real systems[11] evolves itself slowly and, in particular, $\bar{\rho} < \bar{\rho}_{eq}$. This effect could cause a further slow down of the kinetics of this model, if properly taken into account, and a more complex dynamical behavior. Despite its schematic nature, however, this approach allows one recover some phenomenology observed in real systems, like the functional dependence of the glass temperature on the cooling rate and the logarithmic decay of $S$ in the supercooled phase. Moreover the model makes new predictions regarding the behavior of the correlation functions that could be tested experimentally. In this approach the parameter $\varepsilon$ tunes the degree of non-equilibration of the system and is responsible for the violation of the fluctuation dissipation theorem. $\varepsilon$ is simply given by the ratio between $C_0$ and $C_1$, namely the *equilibrium* and *non-equilibrium* contribution to the autocorrelation function, providing a clear physical interpretation of the interplay between the thermalized modes and the off-equilibrium ones. Whether a similar conceptual scheme survives when the mean field approximation we have considered in detail is released, and to which extent it can be shared by more realistic models for glassy kinetics is matter of future investigation.

## ACKNOWLEDGMENTS

We acknowledge support by the TMR network contract ERBFMRXCT980183 by INFM PRA-HOP 99 and by MURST PRIN-2000.

## REFERENCES

1. F. Corberi, M. Nicodemi, M. Piccioni and A. Coniglio, *Phys. Rev. Lett.* **83**, 5054 (1999).
2. J. Zarzycky, *Glasses and the Vitreous State* (Cambridge University Press, 1991).
3. A. J. Kovacs, *Fortschr. Hochpolym.-Forsch.* **3**, 394 (1964).
4. F. L. Hill, *An Introduction to Statistical Thermodynamics* (Addison-Wesley Publishing Company, 1960).
5. G. Adam and J. H. Gibbs, *J. Chem. Phys.* **43**, 139 (1965); E. Ben-Naim, J. B. Knight, E. R. Nowak, H. M. Jaeger and S. R. Nagel, *Physica* **D123**, 380 (1998).

6. E. L. Houston, J. W. Cahn and J. E. Hilliard, *Acta Metall.* **14**, 1685 (1966).
7. L. F. Cugliandolo, J. Kurchan and L. Peliti, *Phys. Rev.* **E55**, 3898 (1997).
8. F. Sciortino and P. Tartaglia, *cond-mat/0007208.*
9. M. Nicodemi, *Phys. Rev. Lett.* **82**, 3734 (1999).
10. T. S. Grigera and N. E. Israeloff, *Phys. Rev. Lett.* **83**, 5038 (1999).
11. C. A. Angell, *Science* **267**, 1924 (1995); M. D. Ediger, C. A. Angell and S. R. Nagel, *J. Phys. Chem.* **100**, 13200 (1996).

# EFFECTS OF AN IMPOSED FLOW ON PHASE-SEPARATING BINARY MIXTURES

F. CORBERI
*Istituto Nazionale per la Fisica della Materia, Unità di Salerno*
*Dipartimento di Fisica, Università di Salerno,*
*84081 Baronissi (Salerno), Italy*

G. GONNELLA
*Istituto Nazionale per la Fisica della Materia, Unità di Bari*
*Dipartimento di Fisica, Università di Bari*
*Istituto Nazionale di Fisica Nucleare, Sezione di Bari,*
*via Amendola 173, 70126 Bari, Italy*

A. LAMURA
*Institut für Festkörperforschung, Forschungszentrum Jülich,*
*52425 Jülich, Germany*

**Abstract**

We study the phase separation of a binary mixture in uniform shear flow in the framework of the continuum convection-diffusion equation based on a Ginzburg-Landau free energy. This equation is solved both numerically and in the context of large-$N$ approximation. Our results show the existence of domains with two typical sizes, whose relative abundance changes in time. As a consequence log-time periodic oscillations are observed in the behavior of most thermodynamic observables.

## 1. INTRODUCTION

When a binary mixture is suddenly quenched from a disordered initial state to a coexistence region below the critical temperature, the two components segregate and form domains which grow with time. In the unsheared case a single length scale $R(t)$, which measures the average size of domains, characterizes the kinetics of phase separation. This length grows with the power law $R(t) \sim t^\alpha \sim t^{1/3}$ in a purely diffusive regime.[1] The application of a shear flow greatly affects the phase separation process. A large anisotropy is observed in typical patterns of domains which appear elongated in the direction of the flow.[2] Previous numerical studies confirm these observations.[3–5] These studies, however, were carried out on rather small systems so that an accurate resolution of the spatial properties, which is usually inferred from the knowledge of the structure factor, was not yet available.

In this article we investigate the segregation process both by numerical simulations of large scale systems in order to compute at a fine level of resolution the structure factor and in the context of a self-consistent approximation which allows to overcome any possible finite-size effect of the simulations. We show that two typical lengths in each spatial direction characterize the phase separation process. Domains are stretched and broken cyclically producing an alternate prevalence of thin and thick domains. As a consequence of this mechanism all the physical observables are modulated by oscillations which are periodic on a logarithmic time scale.

This paper is organized as follows. In Sec. 2 we introduce the model. In Sec. 3 we present the results of the numerical simulations. Section 4 is devoted to the analysis of the behavior of the model in the large-$N$ approximation. Finally we summarize and draw our conclusions.

## 2. THE MODEL

In the following we describe a binary mixture by means of a model with a coupling between a diffusive field $\varphi$, representing the concentration difference between the two components of the mixture, and an applied velocity field. This approach neglects hydrodynamic effects. For weakly sheared polymer blends with large polymerization index and similar mechanical properties of the two species the present model is expected to be satisfactory in a preasymptotic time domain when velocity fluctuations are small.[1] When hydrodynamic effects become important, instead, the full dynamical model[6] where the velocity is governed by the Navier-Stokes equation must be considered.

The mixture, which we assume to be symmetric in the two components, is described by the Langevin equation

$$\frac{\partial \varphi}{\partial t} + \nabla \cdot (\varphi \mathbf{v}) = \nabla^2 \frac{\delta \mathcal{F}}{\delta \varphi} + \eta \tag{1}$$

where $\mathbf{v}$ is the external velocity field describing plane shear flow with profile $\mathbf{v} = \gamma y \mathbf{e}_x$,[2] $\gamma$ and $\mathbf{e}_x$ being, respectively, the shear rate and the unit vector in the $x$ direction. $\eta$ is a Gaussian white noise, representing the effects of thermal fluctuations with zero mean and correlations that, according to the fluctuation-dissipation theorem, are given by

$$\langle \eta(\mathbf{r}, t) \eta(\mathbf{r}', t') \rangle = -2T\nabla^2 \delta(\mathbf{r} - \mathbf{r}')\delta(t - t') \tag{2}$$

where $T$ is the temperature of the mixture, and $\langle \cdots \rangle$ denotes the ensemble average. The term $\nabla \cdot (\varphi \mathbf{v})$ describes the advection of the field $\varphi$ by the velocity and $\nabla^2(\delta\mathcal{F}/\delta\varphi)$ takes into account the diffusive transport of $\varphi$. The equilibrium free-energy can be chosen as

$$\mathcal{F}\{\varphi\} = \int d\mathbf{r} \left\{ -\frac{1}{2}\varphi^2 + \frac{1}{4}\varphi^4 + \frac{1}{2}|\nabla\varphi|^2 \right\} \tag{3}$$

in the ordered phase.

The main observable for the study of the growth kinetics is the structure factor defined as

$$C(\mathbf{k},\, t) = \langle \varphi(\mathbf{k},\, t)\varphi(-\mathbf{k},\, t)\rangle \tag{4}$$

where $\varphi(\mathbf{k},\, t)$ are the Fourier components of $\varphi$. From the knowledge of the structure factor the average sizes of domains in different directions can be computed as

$$R_x(t) = \frac{\int d\mathbf{k} C(\mathbf{k},\, t)}{\int d\mathbf{k}|k_x|C(\mathbf{k},\, t)} \tag{5}$$

and analogously for the other directions. Of experimental interest are also the rheological quantities. We consider the excess viscosity which can be defined as[2]

$$\Delta\eta(t) = -\frac{1}{\gamma}\int \frac{d\mathbf{k}}{(2\pi)^d} k_x k_y C(\mathbf{k},\, t) \tag{6}$$

where $d$ denotes the number of spatial dimensions.

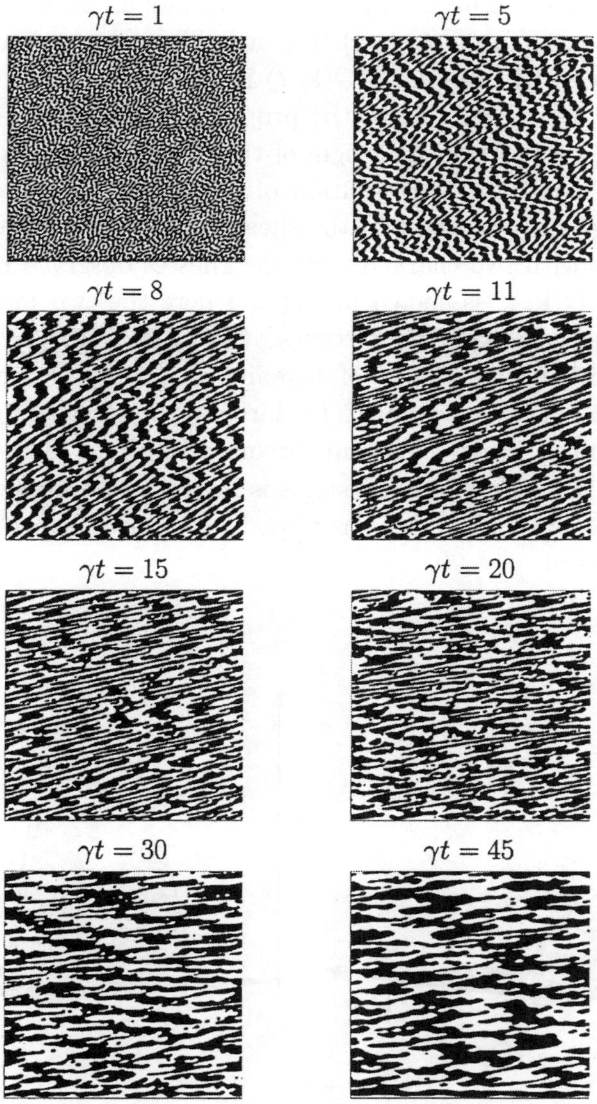


**Fig. 1** Configurations of a portion of 512 × 512 sites of the whole lattice are shown at different values of the shear strain $\gamma t$ at $T = 0$. The $x$ axis is in the horizontal direction.

## 3. NUMERICAL SIMULATIONS

The most straightforward way of studying Eq. (1) is to simulate it numerically. We have used a first-order Euler discretization scheme. Periodic boundary conditions have been adopted in the $x$ and $z$ directions; Lees-Edwards boundary conditions[7] were used in the $y$ direction. These boundary conditions require the identification of a point at $(x, 0, z)$ with one located at $(x + \gamma L \Delta t, L, z)$, where $L$ is the size of the lattice and $\Delta t$ is the time discretization interval. Simulations were run using lattices of size $L = 4096$ in $d = 2$ and $L = 256$ in $d = 3$. Results are presented here for the two-dimensional case at temperature $T = 0$.[8]

A sequence of configurations at different values of the strain $\gamma t$ is shown in Fig. 1. After an early time, when well defined interfaces are forming, the pattern of domains starts to be distorted for $\gamma t > 1$. The growth is faster in the flow direction and domains assume the typical striplike shape aligned with the flow direction. As the elongation of the domains increases, non-uniformities appear in the system: Regions with domains of different thickness can be clearly observed at $\gamma t = 11$ and $\gamma t = 20$.

An analysis of length scales present in the system can be done by studying the behavior of the structure factor. At the beginning $C(\mathbf{k}, t)$ exhibits an almost circular shape. Then shear-induced anisotropy deforms $C(\mathbf{k}, t)$ into an elliptical pattern. The profile of the structure factor changes with time until $C(\mathbf{k}, t)$ is separated in two distinct foils, each of them characterized by two peaks. Since the property $C(\mathbf{k}, t) = C(-\mathbf{k}, t)$ holds, the two foils are symmetric with respect to the origin of the $k$-space and it is sufficient to consider only the two peaks of one foil. The position of each peak identifies a couple of typical lengths, one in the flow and the other in the shear $(y)$ direction. This corresponds to the observation of domains with two characteristic thicknesses observed in Fig. 1. The relative height of the peaks of $C(\mathbf{k}, t)$ is shown in Fig. 2 where the two maxima in each foil are observed to dominate alternatively at the times $\gamma t = 11$ and $\gamma t = 20$.

The competition between two kinds of domains is a cooperative phenomenon. In a situation like that at $\gamma t = 11$, the peak with the larger $k_y$ dominates, describing a prevalence of stretched thin domains. When the strain becomes larger, a cascade of ruptures occurs in those regions of the network where the stress is higher and elastic energy is released. At this point the thick domains, which have not yet been broken, prevale and the other peak of $C(\mathbf{k}, t)$ dominates, as at $\gamma t = 20$.

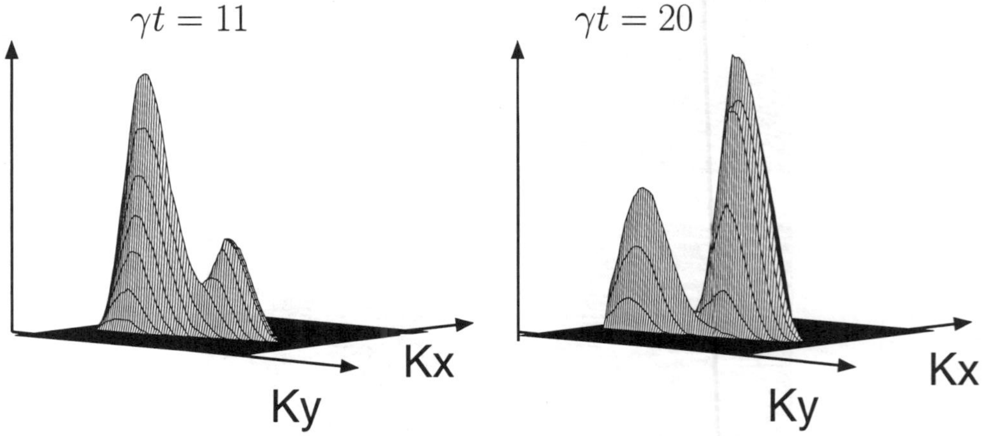


**Fig. 2** Three-dimensional plot of the structure factor at $\gamma t = 11, 20$. Only one foil of $C(\mathbf{k}, t)$ is shown (see the text for details).

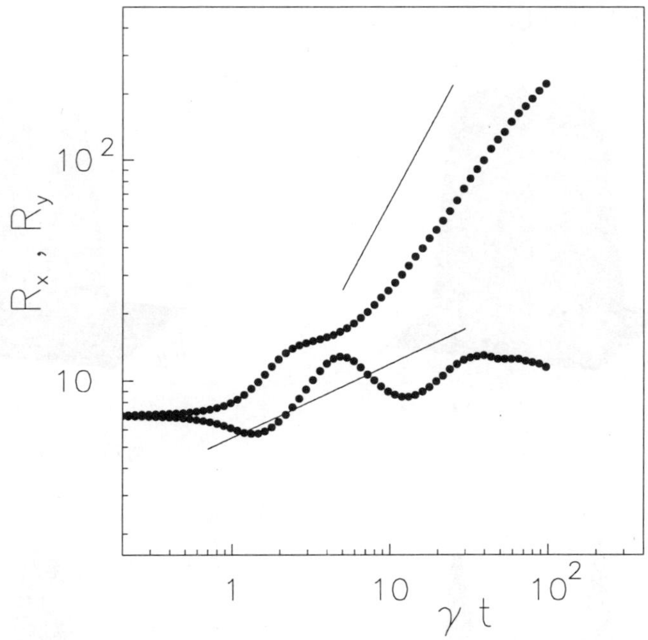


**Fig. 3** Evolution of the average domains sizes in the shear (lower curve) and flow (upper curve) directions. The straight lines have slopes 1/3 and 4/3.

This dynamics affects the behavior of the average size of domains $R_x(t)$, $R_y(t)$. Their behavior is shown in Fig. 3. Due to the alternative dominance of the peaks of $C(\mathbf{k}, t)$, $R_x$ and $R_y$ increase with amplitudes modulated by an oscillation in time. Using a generalization of the renormalization group scheme for the unsheared case[9] we have shown that $R_x \sim \gamma t^{4/3}$ and $R_y \sim t^{1/3}$. The simulations cannot give evidence of such a scaling regime because finite-size effects[8] prevent to follow the alternate predominance of the two peaks on sufficiently long times. The present result indicates that the growth in the shear direction is not affected by the flow, while in the $x$ direction the convective term in Eq. (1) increases the growth exponent of 1 with respect to the unsheared case. Similar results have been obtained for temperatures in the range $0 \leq T \leq 5$ and in $d = 3$ at $T = 0$.[10]

## 4. A SELF-CONSISTENT APPROXIMATION

Another possible way of studying the present model is in the context of the large-$N$ approximation.[11,12] In this context the model is generalized to a vectorial order parameter with an arbitrary number $N$ of components and the limit $N \to \infty$ is taken. This special limit is known to provide a mean-field picture of the phase-separation process which can give an insight of the phenomenon at a semi-quantitative level.[13] The governing equation for the structure factor in this limit reads

$$\frac{\partial C(\mathbf{k}, t)}{\partial t} - \gamma k_x \frac{\partial C(\mathbf{k}, t)}{\partial k_y} = -k^2[k^2 + S(t) - 1]C(\mathbf{k}, t) + k^2 T \tag{7}$$

where $S(t)$ is obtained through the self-consistent prescription

$$S(t) = \int_{|\mathbf{k}|<q} \frac{d\mathbf{k}}{(2\pi)^d} C(\mathbf{k}, t) \tag{8}$$

and $q$ is a phenomenological cutoff.

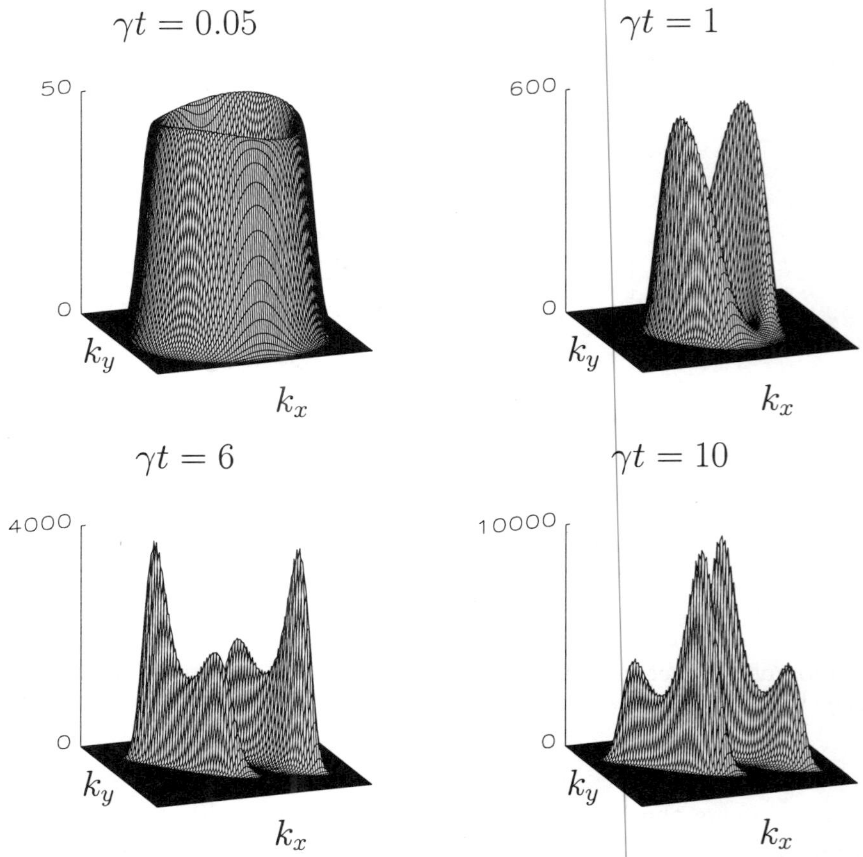


**Fig. 4** The structure factor at consecutive times for $\gamma = 0.001$ in the large-$N$ approximation. The scales on the $k_x$ and $k_y$ axes have been enlarged differently for a better view.

Using a scaling ansatz[14] we found that

$$R_x \sim \gamma t^{5/4} \tag{9}$$

and

$$R_\perp \sim t^{1/4} \tag{10}$$

in the direction of the flow and perpendicular to it, respectively. The excess viscosity behaves as

$$\Delta\eta \sim \gamma^{-2} t^{-3/2}. \tag{11}$$

The self-consistency condition (8) has been worked out explicitly in the long-time domain.[15] It is found that the model has a multiscaling symmetry and that the asymptotic behaviors (9)–(11) have logarithmic corrections.

In Fig. 4 the solution of Eq. (7) in $d = 2$ at $T = 0$ is plotted for different values of the shear strain $\gamma t$. The picture resembles the one outlined in the previous section but the numerical integration of Eqs. (7) and (8) is much less demanding than Eq. (1) and the evolution of the structure factor can be neatly obtained over much longer timescales. The evolution of $C(\mathbf{k}, t)$ shows that an alternate prevalence of the two peaks on each foil continues in time. As a consequence $R_x$ and $R_y$ grow in time with a power law behavior, $R_x \sim \gamma t^{5/4}$ and $R_y \sim t^{1/4}$, decorated by oscillations which are found to be periodic in the logarithm

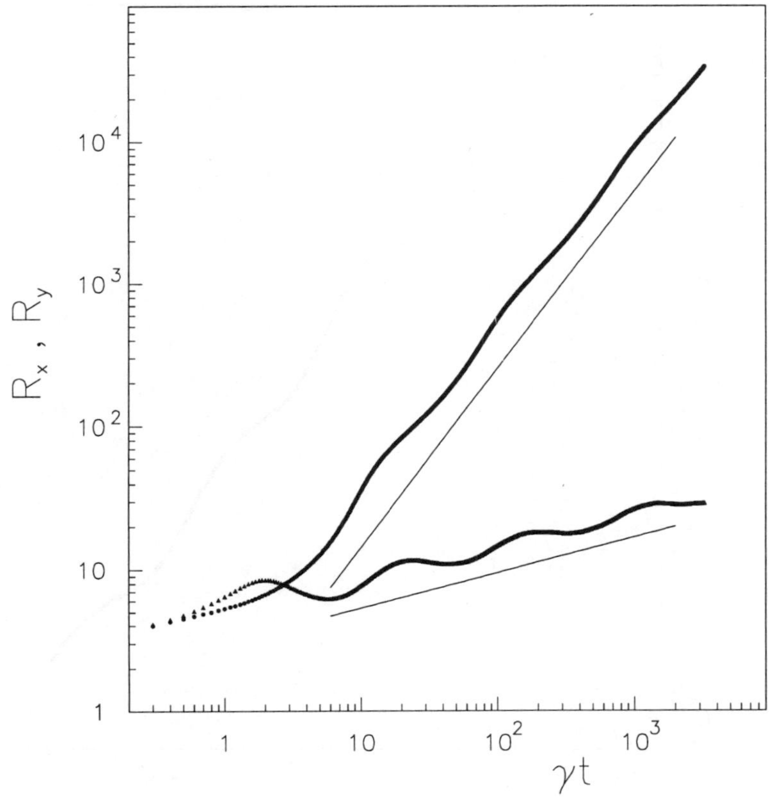


**Fig. 5** Evolution of the average domains size in the shear (lower curve) and flow (upper curve) directions. The straight lines have slopes 1/4 and 5/4.

of time, as shown in Fig. 5. The growth exponent 1/4 is expected for systems with a continuous symmetry ($N > 1$) without shear and corresponds to the diffusive exponent 1/3 characteristic of the scalar model. Again a difference 1 between the two growth exponents is found.

Stretching of domains requires work against surface tension and burst of domains dissipate energy resulting in an increase $\Delta\eta$ of the viscosity.[2] The time behavior of $\Delta\eta$ is shown in Fig. 6. Starting from zero, $\Delta\eta$ grows up to a global maximum, then it relaxes to smaller values oscillating. The decay is consistent with the expected power-law behavior (11). The oscillations are related to the dynamics of the peaks of $C(\mathbf{k}, t)$. Relative maxima of $\Delta\eta$ occur when the domains are maximally stretched (see Figs. (4) and (5) at $\gamma t = 6$). Later, when the domains are deformed to such an extent that they start to break up, $\Delta\eta$ decreases and more isotropic domains are formed. Then a minimum of $\Delta\eta$ is observed. The process of storing elastic energy with consequent dissipation through cascades of ruptures reproduces periodically in time with a characteristic frequency. Similar log-time periodic oscillations have been observed in solid materials subject to an external strain.[16] In three dimensions we proved that the growth of domains is not affected by the external shear in the directions perpendicular to the flow and the growth exponent 5/4 is still found to characterize the growth in the $x$ direction.[17]

## 5. CONCLUSIONS

In the theoretical framework above outlined we have shown and accounted for many of the most significative features of the phase separation of sheared binary mixtures. Moreover,

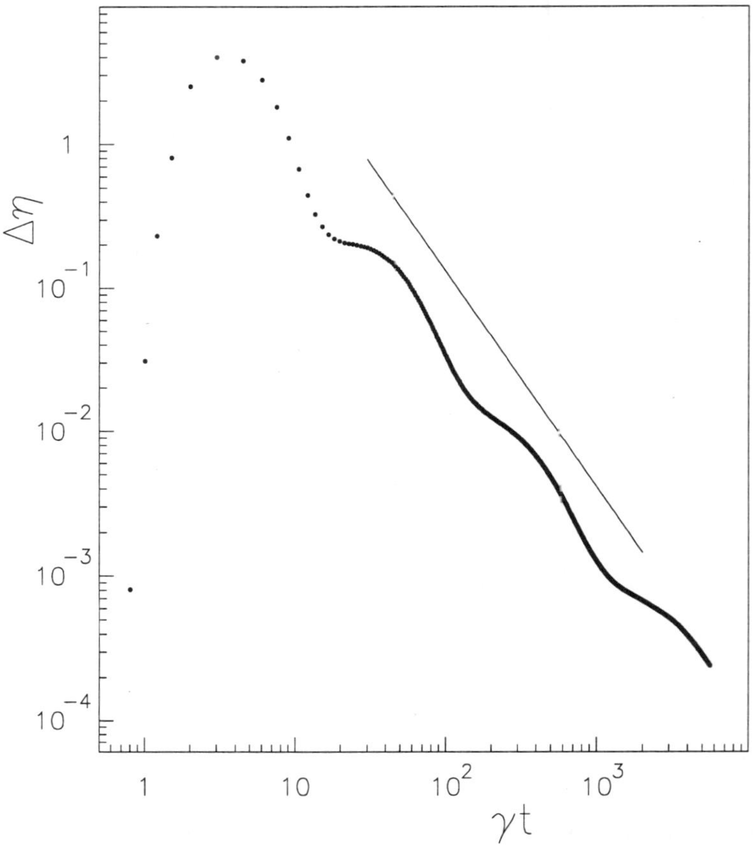


**Fig. 6** The excess viscosity as a function of the shear strain $\gamma t$. The slope of the straight line is $-3/2$.

additional predictions are allowed. We could connect the alternating dominance of the peaks of the structure factor and the overshoots in the excess viscosity. A four-peaked structure factor was observed in shallow quenching of polymer solutions,[18] together with a double overshoot of the excess viscosity. The results of our study strongly suggest that the interplay between the four peaks of the structure factor, with their alternate prevalence, gives rise to an oscillatory phenomenon which reflects itself on the main observables. On these bases our conjecture is that the experimental double overshoot can be interpreted as the first part of an oscillatory pattern superimposed on the global trend of the excess viscosity.

Our simulations indicate that domains of two characteristic sizes in each direction alternatively prevail, because the thicker are thinned by the strain and the thinner are thickened after multiple ruptures in the network. This is reflected on the properties of the structure factor. These results have been confirmed, at a semiquantitative level, in the context of the large-$N$ approximation, where the observed oscillations are periodic in the logarithm of time. The logarithmic periodicity of these overshoots in time is one of the relevant predictions of the theory to be tested in future experiments on longer time scales.

## ACKNOWLEDGMENTS

F. C. and G. G. acknowledge support by the TMR network contract ERBFMRXCT980183, by INFM PRA-HOP 99 and by MURST PRIN-2000.

## REFERENCES


1. See e.g., A. J. Bray, *Adv. Phys.* **43**, 357 (1994).
2. A. Onuki, *J. Phys. Condens. Matter* **9**, 6119 (1997).
3. T. Ohta, H. Nozaki and M. Doi, *Phys. Lett.* **A145**, 304 (1990); *J. Chem. Phys.* **93**, 2664 (1990).
4. D. H. Rothman, *Europhys. Lett.* **14**, 337 (1991).
5. Z. Shou and A. Chakrabarti, *Phys. Rev.* **E61**, R2200 (2000).
6. For simulations with hydrodynamics, see M. E. Cates, V. M. Kendon, P. Bladon and J. C. Desplat, *Faraday Discuss.* **112**, 1 (1999); A. J. Wagner and J. M. Yeomans, *Phys. Rev.* **E59**, 4366 (1999); A. Lamura and G. Gonnella, e-print cond-mat/0005518, in press in *Physica A*.
7. A. W. Lees and S. F. Edwards, *J. Phys.* **C5**, 1921 (1972).
8. F. Corberi, G. Gonnella and A. Lamura, *Phys. Rev. Lett.* **83**, 4057 (1999).
9. A. J. Bray, *Phys. Rev.* **B41**, 6724 (1990).
10. F. Corberi, G. Gonnella and A. Lamura, *Phys. Rev.* **E62**, 8064 (2000).
11. G. F. Mazenko and M. Zannetti, *Phys. Rev. Lett.* **53**, 2106 (1984); *Phys. Rev.* **B32**, 4565 (1985).
12. G. Pätzold and K. Dawson, *Phys. Rev.* **E54**, 1669 (1996).
13. A. Coniglio, P. Ruggiero and M. Zannetti, *Phys. Rev.* **E50**, 1046 (1994).
14. F. Corberi, G. Gonnella and A. Lamura, *Phys. Rev. Lett.* **81**, 3852 (1998).
15. N. P. Rapapa and A. J. Bray, *Phys. Rev. Lett.* **83**, 3856 (1999).
16. D. Sornette, *Phys. Rep.* **297**, 239 (1998).
17. F. Corberi, G. Gonnella and A. Lamura, *Phys. Rev.* **E61**, 6621 (2000).
18. K. Migler, C. Liu and D. J. Pine, *Macromolecules* **29**, 1422 (1996).

# FAST RELAXATION TIME IN A SPIN MODEL WITH GLASSY BEHAVIOR

GIANCARLO FRANZESE
*Center for Polymer Studies and Department of Physics, Boston University*
*590 Commonwealth Avenue, Boston, MA 02215, USA*

**Abstract**

We consider a frustrated spin model with a glassy dynamics characterized by a slow component and a fast component in the relaxation process. The slow process involves variables with critical behavior at finite temperature $T_p$ and has a global character like the (structural) $\alpha$-relaxation of glasses. The fast process has a more local character and can be associated to the $\beta$-relaxation of glasses. At temperature $T > T_p$ the fast relaxation follows the non-Arrhenius behavior of the slow variables. At $T \lesssim T_p$ the fast variables have an Arrhenius behavior, resembling the $\alpha - \beta$ bifurcation of fragile glasses. The model allows us to analyze the relation between the dynamics and the thermodynamics.

## 1. INTRODUCTION

A recent review[1] has pointed out some of the leading questions in the theoretical study of glassforming liquids. Following the fruitful path of theoretical studies on spin models with glassy behavior,[2] here we present results on a frustrated spin model that give an insight on some of these questions, showing relations between the dynamics and the statics of the model.

We consider a model, introduced by Coniglio and coworkers,[3–8] with two kinds of variables, coupled to each other. One of them has a finite temperature transition and a slow relaxation, while the other has a transition only at zero temperature in two dimensions (2D) and has a fast relaxation at finite temperature. The model has been show[6,8] to have glassy behavior and the slow and fast components of the relaxation can be related to the

$\alpha$ and $\beta$ processes, respectively, of glasses. Here we show numerical results resembling the *$\alpha$-$\beta$ bifurcation process*[1] observed in experiments on fragile liquids.[9,10]

In the following we give an introduction to the problem. In Sec. 2 we introduce the spin model, reviewing some results on its phase diagram and on its glassy behavior. In Sec. 3 we show some numerical results about the bifurcation in this model and the connection with the thermodynamics. In Sec. 4 we give the conclusions and the perspectives for future work.

## 1.1 Glasses and the $\alpha$-$\beta$ Bifurcation

The majority of liquids can form a glass if cooled at high enough rate. The glass state is experimentally defined by measuring macroscopic quantities, like the diffusion coefficient or the viscosity. The calorimetric glass transition temperature $T_g$ is defined as the temperature at which the viscosity reaches a value of $10^{12}$ sPa and depends on the cooling rate.

The glass transition is considered as a dynamic off-equilibrium phenomenon, but a large amount of work has recently shown that the dynamics of glassy systems is strictly related to its static properties, i.e. to its thermodynamics.[11] The main focus of these studies is the slow component of the dynamics. One of the aim of this paper is to show that also the fast processes can be related to the thermodynamics of the system.

Following the Angell's classification,[12] glassforming liquids are divided in 'strong' and 'fragile'. Strong liquids are characterized by a relaxation time $\tau(T)$ that increases with decreasing temperature $T$ in a way well described by an Arrhenius law

$$\tau(T) = \tau_\infty \exp[A/(k_B T)] \tag{1}$$

where $\tau_\infty$ is interpreted as a characteristic microscopic relaxation time at infinite $T$ and $A$ is an activation energy (energy barrier) for global rearrangements ($k_B$ is the Boltzmann constant). A strong glassformer is, for example, $SiO_2$.

Fragile liquids are defined as those whose $\tau(T)$ shows a large departure from the Arrhenius law (non-Arrhenius behavior) for $T > T_g$. For $T < T_g$ the relaxation time is well described by an Arrhenius law. Example of fragile glassformers are glycerol and orto-therphenyl. A function widely used to fit data for fragile liquids is the Vogel-Tamman-Fulcher (VTF) law

$$\tau(T) = \tau_\infty \exp[B/(T - T_0)] \tag{2}$$

where $\tau_\infty$ is a high-$T$ characteristic time, $B$ and $T_0$ are fitting constants. Since $T_0 < T_g$ the diverging behavior is never reached.

An alternative non-Arrhenius behavior is derived in the Mode Coupling Theory (MCT)[13] for $T > T_g$, in which the relaxation time increases as a power law of the temperature,

$$\tau(T) \simeq (T - T_{MCT})^{-\gamma}, \tag{3}$$

where $T_{MCT}$ is the MCT transition temperature (with $T_{MCT} < T_g$) and $\gamma$ is a parameter.

The non-Arrhenius behavior can be explained, in a consistent way with the MCT,[11] also by the Adam-Gibbs theory of the excluded volume, in which the relaxation time is

$$\tau = \tau_\infty \exp(C/TS_c) \tag{4}$$

where $\tau_\infty \simeq 10^{-14}$s, $C$ is a constant and $S_c$ is the entropy difference between liquid and crystal – 'excess' entropy. Since $S_c$ is, usually, constant below $T_g$, the low-$T$ Arrhenius behavior is recovered.

This phenomenology concerns the structural $\alpha$-relaxation, i.e. the process described by the long-time part of the relaxation functions of the system. For high $T$, the relaxation functions, like the density-density correlation function, have a simple exponential behavior ('simple liquid' regime)

$$f(t) = f_0 \exp(t/\tau)\,, \tag{5}$$

with $f_0$ and $\tau$ depending on $T$ and where $t$ is the time.

Decreasing $T$ the relaxation functions usually show a 'two-steps' behavior where a fast relaxation process, associated to the first step, is followed by a slow relaxation process, the second step. The second step is the $\alpha$-relaxation and is associated to a macroscopic process involving non-local degrees of freedom (as the global reorganization of the system).[1] Below a characteristic temperature $T^* > T_g$, the second step is well approximated in most cases by the Kohlrausch-Williams-Watts (KWW) stretched exponential function

$$f_{KWW}(t) = f_0 \exp[(t/\tau)^\beta]\,, \tag{6}$$

where $f_0$, $\tau$ and $\beta$ depend on $T$. The KWW function recovers the simple exponential for $\beta = 1$, with the parameter $\beta$ describing the departure from exponentiality. The fit of glassy relaxation functions, usually, shows $\beta$ decreasing with the temperature, with $\beta < 1$ for $T < T^*$. In many cases, a better approximation of the second step of the relaxation function is given by using the Ogielski[14] stretched exponential function

$$f_O(t) = t^{-x} f_{KWW}(t)\,, \tag{7}$$

with $x$ depending on $T$.

The first step (or $\beta$-relaxation) is usually associated to any microscopic process[1] occurring at very short time scale. In the MCT for atomic liquids it is associated to the fast diffusion of mobile particles inside 'cages' of immobile particles.[13] In molecular liquids it can be associated also to the relaxation of internal degrees of freedom.[15]

The $\beta$-relaxation presents some non-universal features in fragile liquids,[9] and sometimes in intermediate liquids,[10] that are not well understood. The $\beta$-relaxation time follows the non-Arrhenius $\alpha$-relaxation behavior at high $T$ and shows a crossover (the $\alpha-\beta$ bifurcation) to an Arrhenius behavior at a temperature $T_{\alpha-\beta} \simeq T_{MCT}$.[1] To have an insight on this phenomenon and its connection with other static and dynamic transitions, we consider the spin model presented in the next section.

## 2. THE SPIN MODEL

We consider a lattice model[3] with, on each lattice site, a Potts variable[16] $\sigma_i$, with and integer number $s$ of states ($\sigma_i = 1, \ldots, s$), coupled to an Ising spin $S_i$ with two states ($S_i = \pm 1$). The model can be considered as a schematic representation of structural glasses, such as dense molecular glasses, plastic crystal, or orto-therphenyl at low temperature, with orientational degrees of freedom frustrated by geometrical hindrance between non-spherical molecules. The orientational degrees of freedom are represented by the Potts variables and the frustration is modeled by means of ferro/antiferromagnetic interactions for the Ising spins. The model is defined by the Hamiltonian

$$H_s\{\tau_i, \varepsilon_{i,j}\} = -2sJ \sum_{\langle i,j \rangle} \delta_{\varepsilon_{i,j}\tau_i,\tau_j} \tag{8}$$

where $\tau_i \equiv S_i\sigma_i = \pm 1, \pm 2, \ldots, \pm s$ has $q \equiv 2s$ states, the sum is extended over all the nearest neighbor (n.n.) sites, $J$ is the strength of interaction, $\varepsilon_{i,j} = \pm 1$ is a quenched variable that represents the sign of the ferro/antiferromagnetic interaction, $\delta_{n,m} = 0,\ 1$ is a Kronecker delta.

In the original formulation,[3] the Hamiltonian shows a clearer separations between the two ($\sigma_i$ and $S_i$) coupled variables:

$$H_s\{S_i,\, \sigma_i,\, \varepsilon_{i,j}\} = -sJ \sum_{\langle i,j \rangle} \delta_{\sigma_i,\sigma_j}(\varepsilon_{i,j} S_i S_j + 1)\,. \tag{9}$$

Equation (9) shows that the Ising spins interactions are diluted by a ferromagnetic Potts model, i.e. two n.n. Ising spins can interact only when the corresponding n.n. Potts variables are in the same state (or orientation).

The model has been proposed in two versions. In the first version, the interaction signs $\varepsilon_{i,j}$ are randomly assigned, giving rise to frustration and disorder. This model can be considered the generalization of an Ising spin glass (SG) — with two states — to a model with $2s$ states — the Potts SG (PSG).[4]

In the other version, the interactions signs are assigned in a deterministic way and there is an odd number of $\varepsilon_{i,j} = -1$ (antiferromagnetic interaction) on each lattice cell. In this way the model has frustration and *no* disorder and is the generalization of the Ising fully frustrated (IFF) model[5] to a $2s$-states Potts fully frustrated (PFF) model[8] [Fig. 1(a)]. In the following we will review some results on the PSG and the PFF model.

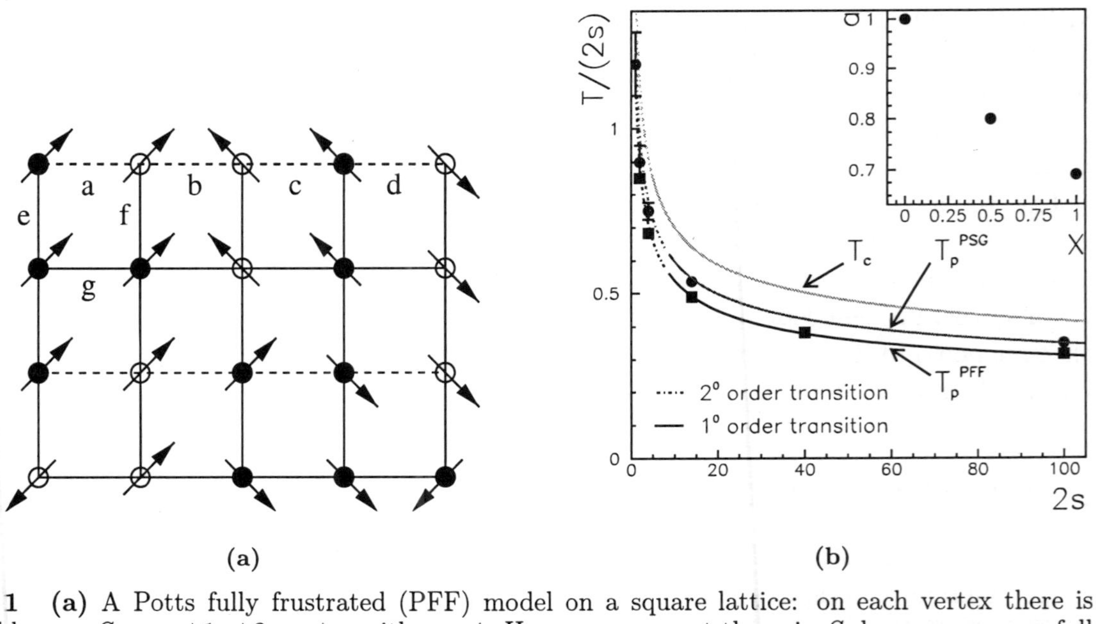


(a) (b)

**Fig. 1** **(a)** A Potts fully frustrated (PFF) model on a square lattice: on each vertex there is variable $\tau_i \equiv S_i\sigma_i = \pm 1, \pm 2 \ldots \pm s$ with $s = 4$. Here we represent the spin $S_i$ by an open or a full dot (respectively positive and negative, for example) and the orientation $\sigma_i$ by an arrow pointing in four different directions. Ferromagnetic (antiferromagnetic) interactions are represented by full (dotted) lines. **(b)** The analytic value of the transition temperature (in 2D) $T_c \equiv T_p(q,\, 0)$, with $q = 2s$, of the ferromagnetic Potts model (upper line), corresponding to $X = 0$, is given together with the 2D Monte Carlo data for the PSG model (circles), corresponding to $X = 0.5$, with transition at $T_p^{PSG}$ and for the PFF model (squares), corresponding to $X = 1$, with transition at $T_p^{PFF}$. The data are fitted with Eq. (10) with $T_p^{PSG} \equiv T_p(q,\, 0.5)$ and $T_p^{PFF} \equiv T_p(q,\, 1)$. The parameters $a(X)$ are shown in the inset. Errors are smaller than symbols size.

## 2.1 The Thermodynamic Transitions

Both the PFF and the PSG model have two thermodynamic transitions, one associated to the variables $\sigma_i$ and one to the spins $S_i$. At zero temperature (in 2D) there is a transition for the Ising spins $S_i$. The transition is in the IFF universality class for the PFF model, and in the Ising SG class for the PSG model. At finite temperature $T_p$ it has been shown, by means of Monte Carlo (MC) simulations[4,8] and analytic approaches,[3] that the Potts variables undergo a transition in the universality class of a $s$-state Potts ferromagnetic model [Fig. 1(b)].

It is possible to extend numerically[8] the analytic expression of the ferromagnetic Potts transition temperature[16] to the PSG and the PFF case, by introducing the fraction $X$ of elementary frustrated cell in the lattice, with the ferromagnetic Potts model corresponding to $X = 0$, the PSG model to $X = 0.5$ and the PFF model to $X = 1$. The generalized relation is

$$\frac{k_B T_p(q, X)}{a(X)qJ} = \frac{1}{\ln\left[1 + \sqrt{a(X)q}\right]} \tag{10}$$

where $q = 2s$ and the parameter $a(X)$ is reported in Fig. 1(b). Note that in 3D the transition of the Ising spins is expected at finite $T < T_p$.

## 2.2 The Glassy Behavior and the Connection with the Thermodynamics

It has been shown[6,5,8] that the PSG and the PFF models have a glassy behavior. For example in the PFF with $q \equiv 2s = 2, 4$ in 2D and 3D the autocorrelation function

$$f_A(t, T) = \frac{\langle A(t, T)A(0, T)\rangle - \langle A(T)\rangle^2}{\langle A(0, T)^2\rangle - \langle A(T)\rangle^2}, \tag{11}$$

for a global quantity $A$ (like the total energy $E$ or the Potts order parameters $M$) shows the 'two-step' behavior of glasses, with a non-exponential second step [upper panel in Fig. 2(a)].

These global correlation functions can be considered as measures of the structural relaxation process. Their relaxation times [$\tau_0^E$ and $\tau_0^M$ in Fig. 2(b)] diverge, in the thermodynamic limit, at the ordering Potts transition temperature $T_p(q, 1)$ in Eq. (10).[8] For $T > T_p$ this diverging relaxation time resembles the $\alpha$-relaxation process occurring for $T > T_g$ in fragile liquids.[a]

In the PSG case the onset $T^*$ of non-exponentiality coincides with the Griffiths temperature corresponding[6,17] to the transition temperature $T_c(q) \equiv T_p(q, 0)$ of the ferromagnetic $q$-states Potts model.[b] This finding generalizes the results for the Ising SG.[14,18] In the Ising

[a]For $T < T_p$, these correlation times decrease with $T$ [e.g. for $M$ in lower panel in Fig. 2(a)], because they are proportional to $\xi^z$, where the correlation length $\xi$ is finite for $T \neq T_p$ [$z$ is the quantity-dependent dynamical exponent, approximated by the exponents in Fig. 2(b)]. In real glasses, instead, the relaxation time is always increasing for decreasing $T$. A relaxation time with a more appropriate behavior could be the one associated to the overlap between Potts configurations visited at different times. We will propose an 'extension' of the model, in the last section, that should give a monotonic slow relaxation time.

[b]In disordered systems it is possible to show that for finite external field a free energy (Griffiths) singularity arises.[17] In the limit of external field going to zero, the temperature at which this singularity occurs goes to the transition temperature $T_c$ of the corresponding system with no disorder, and the singularity vanishes. The case with no disorder, corresponding to the $q$-states PSG model, is the $X = 0$ case (i.e. the ferromagnetic $q$-states Potts model) and is $T_c \equiv T_p(q, 0)$.

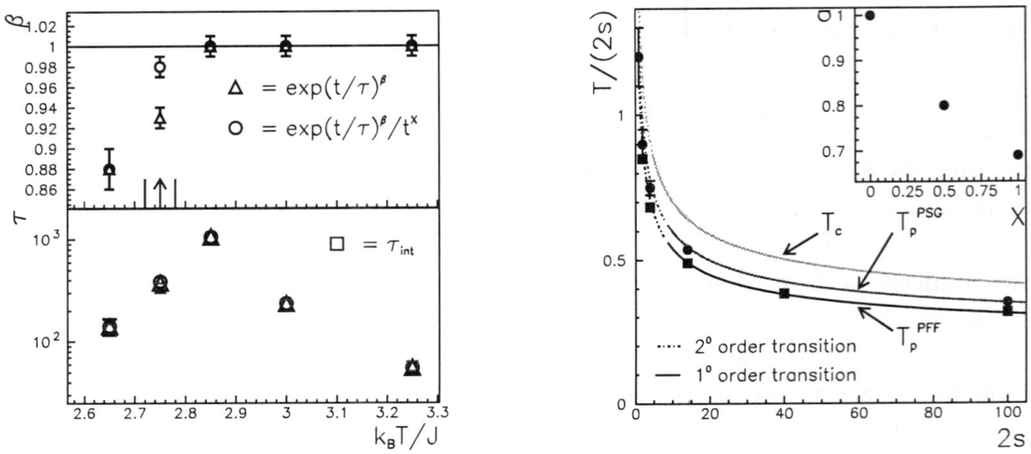


(a) (b)

**Fig. 2** The slow relaxation in the PFF model with $q \equiv 2s = 4$ in 2D. **(a)** The fitting parameter $\beta$ (upper panel) and $\tau$ (lower panel) for the second step of the correlation function of the Potts order parameter $M$ (extrapolated to the thermodynamic limit). The circles are the parameters for the Ogielski stretched exponential form in Eq. (7), the triangles for KWW stretched exponential form in Eq. (6). The arrow shows the estimate of $T_p(4, 1)$ in Eq. (10). The lower panel includes also data for the integral correlation time $\tau_{int} = \lim_{t_{max}\to\infty}[\frac{1}{2} + \sum_{t=0}^{t_{max}} f_M(t)]$. Where not shown, the errors are smaller that the symbols size. **(b)** The log-log plot of the autocorrelation time $\tau_0$, for the energy density $E$ (squares) and the Potts order parameter $M$ (circles) at the finite-size transition temperature $T_p(L)$ as function of the size $L$ = 20, 24, 30, 40, 50. The autocorrelation time $\tau_0$ is defined as the time (in unit of Monte Carlo steps) at which $f_M(\tau_0, T_p(L)) = 0.4$ and $f_E(\tau_0, T_p(L)) = 0.3$. Here $T_p(L) \to T_p(4, 1)$ for $L \to \infty$. The positive slopes $z_M$ and $z_E$ show that these times diverge in the thermodynamic limit $L \to \infty$.

SG, indeed, the non-exponential behavior for $T < T_c$ was related[19] to the presence of unfrustrated regions, due to the randomness, with a size probability distribution that decreases exponentially. At $T_c$ the ferromagnetic-like correlation length is equal to the characteristic size of each region, giving rise to a non-Gaussian distribution of relaxation times, that is responsible for the non-exponentiality of the global relaxation time.

Due to the lack of randomness, in the PFF model this explanation is not valid (the Griffiths temperature is not defined in this case). Indeed, for the PFF model, the onset of non-exponential relaxation corresponds to the Potts transition temperature, as has been shown numerically[5,8] in 2D for $q = 2, 4$ [upper panel in Fig. 2(a)] and[5] in 3D for $q = 2$. In these cases the previous argument can be extended considering that the ferromagnetic-like correlation length is now associated to the Potts variables, ordering at $T_p$.

Summarizing, the results suggest that if the system is disordered ($X \neq 1$), then $T^*$ corresponds to the Potts transition temperature for $X = 0$ (i.e. the Griffiths temperature $T_c$); if the system is fully frustrated ($X = 1$), $T^*$ corresponds to the Potts transition temperature for $X = 1$ (there is no Griffiths temperature in this case). Note that the transition is vanishing for $X \neq 1$ (because the Griffiths temperature marks a transition that disappears for vanishing external field) and for the case $X = 1$ with $q = 2$ (in this case the transition is defined only as a percolation transition, because there are no orientational states). The transition is actually present for the cases with $X = 1$ and $q > 2$.

Hence the onset of the non-exponentiality is marked by a temperature related, at least in this model, to a thermodynamic transition. The transition, in this case, is due to the Potts (orientational) variables, but, in a more general case, could be associated to any global (structural) ordering. This kind of ordering could be, in principle, due to some internal degrees of freedom non easily detected in real experiments.

## 3. THE $\beta$-RELAXATION

Both PSG and PFF models in 2D show[6] a non-exponential behavior also for the normalized correlation functions $f_\chi(t)$ of the time-dependent nonlinear susceptibility

$$\chi_{SG}(t) = \frac{1}{N}\overline{\left\langle\left[\sum_{i=1}^{N} S_i(t+t_0)S_i(t_0)\right]^2\right\rangle} \tag{12}$$

(with $N$ total number of spins, $t_0$ equilibration time, $\chi_{SG}(0) = N$) where the angular brackets stand for the thermal average and the bar stands for the average over the disorder in the PSG case, and is absent in the PFF case. Note that $f_\chi(t)$ depends explicitly only on the Ising spins.

In the PSG case, the onset $T^*$ of non-exponentiality for $f_\chi$ corresponds to the Griffiths temperature $T_c$ (Fig. 3). In the PFF case it has been shown[6] that $T^*$ corresponds to the Potts transition temperature. These results seem to have the same interpretation as those for the correlation functions of the quantities depending explicitly on the Potts variables (previous section).

The difference in this case is that the Ising spins have no thermodynamic transition at $T_p$. Therefore their correlation time $\tau$, associated to $f_\chi(t)$, does not diverge at $T_p$ [lower panel in Fig. 3(b)] and, hence, describes a fast process, respect to the slow dynamics of the Potts variables. This fast $\tau$ can be considered as a measure of the time needed by the Ising spins to minimize the energy locally, on the diluted lattice given by the clusters of ordered Potts variables. This fast (local) relaxation corresponds to the $\beta$-process.

The intriguing result in this case is that this correlation time $\tau$ has a non-Arrhenius behavior for $T > T_p$ and an Arrhenius behavior for $T < T_p$ [Figs. 3(b) and 4]. This

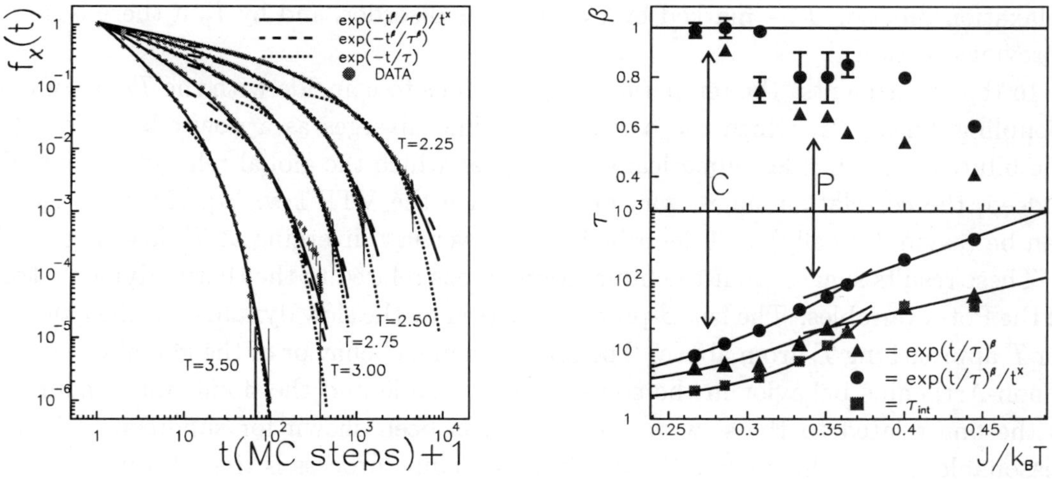


(a) (b)

**Fig. 3** The $\beta$-relaxation in the PSG model in 2D for $q \equiv 2s = 4$. **(a)** The normalized correlation function $f_\chi(t)$ of the nonlinear susceptibility: temperatures are in units of $J/k_B$; times in units of MC steps. The data are fitted with the forms in Eqs. (5)–(7). The Ogielski form in Eq. (7) is the only able to fit the data on four decades. Where not shown the error bars are smaller then the symbols size. **(b)** The fitting parameters $\beta$ and $\tau$ for the stretched exponential forms in Eqs. (6) and (7). For the values of $x$, see Ref. 6. The onset $T^*$ of the stretched exponential corresponds to the Griffiths temperature $T_c$ (marked with $C$) above the Potts transition temperature $T_p$ (marked with $P$). The lower panel includes also data for the integral correlation time $\tau_{int} = \lim_{t_{max}\to\infty}[\frac{1}{2} + \sum_{t=0}^{t_{max}} f_\chi(t)]$. The lines are only guides for the eyes.

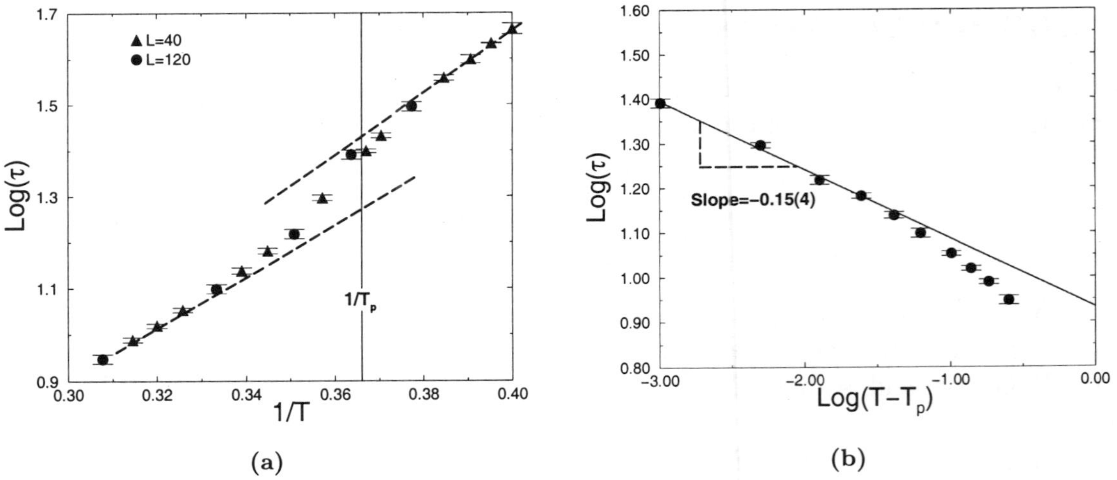


**Fig. 4** The $\beta$-relaxation in the PFF model in 2D with $q = 4$. **(a)** The logarithm of the correlation time $\tau$, associated to $f_X$, for two finite systems with linear sizes $L = 40$, 120, plotted against the inverse temperature ($T$ is in units of $J/k_B$). The finite size effect is evident only near the Potts transition temperature $T_p$ (marked by a vertical line), where the Potts correlation length reach the system size. An Arrhenius behavior in this plot is represented by a straight line. **(b)** The data for $T > T_p$ plotted against the logarithm of $T - T_p$ to verify the power law predicted by MCT. The power law is asymptotically satisfied and the slope is an approximate estimate of the power exponent. In both panels the lines are only guides for the eyes.

result resembles the not well understood $\alpha - \beta$ bifurcation, presented in the introduction, that is seen at a characteristic temperature $T_{\alpha-\beta}$ in some fragile liquids[9] and intermediate liquids.[10]

Moreover the crossover, from non-Arrhenius to Arrhenius, occurs at $T_p$ in both models (with or without disorder). Therefore $T_{\alpha-\beta}$ is separated by the onset of non-exponential relaxation function $T^*$ – marked by $T_c$ if there is disorder, and by $T_p$ if there is no disorder (previous section).

In the experiments[1] the temperature $T_{\alpha-\beta}$ seems to coincide with the $T_{MCT}$ of the Mode Coupling Theory, at which the $\alpha$-relaxation time diverges as a power law, Eq. (3). Here the bifurcation coincides numerically with $T_p$, at which the global relaxation time diverges. Indeed, the correlation time $\tau$ can be fitted with the VTF law, Eq. (2), with $T_0 = T_p$, or can be (asymptotically) well described by a power law diverging at $T_p$ [e.g. Fig. 4(b)].

These results suggest that the bifurcation here is related to the thermodynamic transition of the Potts variables. The fast $\beta$-process is driven by the slow dynamics of the global process for $T$ approaching $T_p$ from above. The non-Arrhenius behavior of the global process induces a non-Arrhenius behavior in the coupled fast variable and the dominant dynamic process is the one related to the slow variable. As has been shown for supercooled liquids,[11] is reasonable that in this regime the dominant dynamic process is the relaxation to the lower accessible energy, at the given $T$, following the instability directions, in the energy landscape, of the visited states, i.e. the activated processes play no role.

For the supercooled liquid case[11] it has been shown that $T_{MCT}$ corresponds to the temperature below which the average number of instability directions for the visited configuration goes to zero. From preliminary results, the same situation appears to be valid here at $T_p$.

Finally, below $T_p$ the system is exploring one of the basins of attraction of the Potts variables $\sigma_i$ and the dynamics of the fast Ising spins is no longer coupled to that of the $\sigma_i$. The dynamics of the fast variables become activated (Arrhenius) consistently with the lack of instability directions. Indeed, in this case the energy cannot be lowered by following an instability direction, but only by doing an activated process. This is analogous of what has been found for the global relaxation in supercooled liquids,[11] but in our case the focus is on the local $\beta$-processes.

Therefore the results here suggest that the $\alpha - \beta$ bifurcation can be related to the thermodynamics of the systems, and that the Potts transition temperature $T_p$ of these systems plays the role of the $T_{MCT}$ of supercooled liquids.

## 4. CONCLUSION AND PERSPECTIVES

The glassy spin models considered here, the PSG and the PFF model, are characterized by the presence of two coupled variables, whose dynamics decouple when the relaxation time of the slow variable diverges. This happens at the temperature $T_p$ where the slow variable has a thermodynamic phase transition.

The picture that can be derived is the following. Due to the symmetry breaking of the slow variable for $T \leq T_p$, the system is attracted in one of the basin of the energy landscape of the slow variable. Inside this basin, the dynamics of the system is mainly due to the fast variable and the dominant dynamical mechanism is the activated process, as consequence of the lack of instability directions for the visited states.

Macroscopically, this is shown by the $\alpha - \beta$ bifurcation that can be seen in some fragile and intermediate glassforming liquids,[1,9] i.e. by the crossover of the $\beta$-relaxation time from a non-Arrhenius behavior to an Arrhenius law at a temperature $T_{\alpha-\beta}$.

The experiments suggest that $T_{\alpha-\beta} \simeq T_{MCT}$, consistently with the analysis reported here. Indeed, in our models the bifurcation occurs at the temperature which plays the role of $T_{MCT}$, i.e. $T_p$, where the slow process has a diverging relaxation time.

Therefore the $\alpha$-$\beta$ bifurcation can been related to the thermodynamics, as well as the onset $T^*$ of non-exponential relaxation. What we can learn from the 'toy model' studied here is that, in cases in which there is disorder, $T^*$ and $T_{\alpha-\beta}$ do not coincide. The first, indeed, corresponds to the Griffiths temperature $T_c$ – associated to a real transition only for a non-zero external field coupled to the slow variable – and the second to the temperature $T_p$, where the slow relaxation time diverges. The two temperatures coincides only in the particular case of a frustrated system with no disorder.

An open question is the effect of the $T$-dependence of the global correlation length $\xi$. An interesting case is the one in which $\xi$ does not decrease below the thermodynamic transition temperature. A model with such characteristic is the $XY$ model, undergoing the Kosterlitz-Thouless-Berezinskii[20] transition. The resulting Hamiltonian will be

$$H\{S_i, \phi_i, \varepsilon_{i,j}\} = -J \sum_{\langle i,j \rangle} \cos(\phi_i - \phi_j - A_{i,j})(\varepsilon_{i,j} S_i S_j + 1) \tag{13}$$

where $A_{i,j}$ are constants depending on the gauge (or the external field) and $\phi_i \in [0, 2\pi)$ are a more realistic representation of the continuous orientational variables.

## ACKNOWLEDGMENTS

This work is consequence of many years of collaboration with Antonio Coniglio, during my studies in Naples where I benefited from his advise and his friendship. It has been an honor for me to be his student. I wish to thank C. Austen Angell for bringing to my attention the $\alpha$-$\beta$ bifurcation in a very instructive private discussion. I am in debt with S. Franz,

G. Parisi, S. Mossa, E. La Nave, G. Ruocco, A. Scala, F. Sciortino, H. E. Stanley, F. Starr for enlightening discussions and precious advises.

## REFERENCES


1. C. A. Angell, *J. Phys.* **C12**, 6463 (2000).
2. See for example, *Spin Glasses and Random Fields*, ed. A. P. Young (World Scientific, Singapore, 1998).
3. A. Coniglio, F. di Liberto, G. Monroy and F. Peruggi, *Phys. Rev.* **B44**, 12605 (1991); U. Pezzella and A. Coniglio, *Physica* **A237**, 353 (1997); F. di Liberto and F. Peruggi, *Physica* **A248**, 273 (1998).
4. G. Franzese and A. Coniglio, *Phys. Rev.* **E58**, 2753 (1998).
5. G. Franzese, A. Fierro, A. De Candia and A. Coniglio, *Physica* **A257**, 376 (1998); A. Fierro, G. Franzese, A. de Candia and A. Coniglio, *Phys. Rev.* **E59**, 60 (1999).
6. G. Franzese and A. Coniglio, *Phys. Rev.* **E59**, 6409 (1999).
7. G. Franzese and A. Coniglio, *Philos. Mag.* **B79**, 1807 (1999).
8. G. Franzese, *Phys. Rev.* **E61**, 6383 (2000).
9. G. P. Johari and M. Goldstein, *J. Chem. Phys.* **53**, 2372 (1970); G. P. Johari and M. Goldstein, *J. Chem. Phys.* **55**, 42459 (1971).
10. F. Stickel, E. W. Fisher and R. Richert, *J. Chem. Phys.* **102**, 6251 (1995); C. Hansen, F. Stickel, T. Berger, R. Richert and E. W. Fisher, *J. Chem. Phys.* **107**, 1086 (1997); H. Wagner and R. Richert, *J. Phys.* **C103**, 4071 (1999).
11. E. La Nave, A. Scala A, F. W. Starr, F. Sciortino and H. E. Stanley *Phys. Rev. Lett.* **84**, 4605 (2000); A. Scala, F. W. Starr, E. La Nave, F. Sciortino and H. E. Stanley, *Nature* **406**, 166 (2000); L. Angelani, R. Di Leonardo, G. Ruocco, A. Scala and F. Sciortino, *Phys. Rev. Lett.* **85**, 5356 (2000); S. Satry, *Nature* **409**, 164 (2001); L. Angelani, R. Di Leonardo, G. Parisi and G. Ruocco, *preprint cond-mat/0011519.*
12. C. A. Angell, in *Relaxation in Complex Systems*, eds. K. L. Ngai and G. B. Wright (US Department of Commerce, Springfield, 1985); K. Ito, C. T. Moynihan and C. A. Angell, *Nature* **398**, 492 (1999).
13. W. Götze, in *Liquids, Freezing and Glass Transition*, eds. J. P. Hansen, D. Levesque and J. Zinn-Justin (North-Holland, Amsterdam, 1991); W. Götze and L. Sjögren, *Rep. Prog. Phys.* **55**, 241 (1992).
14. A. T. Ogielski, *Phys. Rev.* **B32**, 7384 (1985).
15. G. Monaco, S. Caponi, R. Di Leonardo, D. Fioretto and G. Ruocco, *Phys. Rev.* **E62**, 612 (2000); S. Mossa et al., unpublished.
16. F. Wu, *Rev. Mod. Phys.* **54**, 235 (1982).
17. R. B. Griffiths, *Phys. Rev. Lett.* **23**, 17 (1969).
18. W. L. McMillan, *Phys. Rev.* **B28**, 5216 (1983).
19. M. Randeria, J. P. Sethna and R. G. Palmer, *Phys. Rev. Lett.* **57**, 245 (1986); A. Bray, *Phys Rev. Lett.* **59**, 586 (1987); F. Cesi, C. Maes and F. Martinelli, *Commun. Math. Phys.* **188**, 135 (1997).
20. V. L. Berezinskii, *Zh. Eksp. Teor. Fiz.* **59**, 207 (1970); *Sov. Phys. JETP* **32**, 493 (1971); J. M. Kosterlitz and D. J. Thouless, *J. Phys.* **C6**, 1181 (1973);

# MOLECULAR-DYNAMICS STUDIES OF BIATOMIC SUPERCOOLED LIQUIDS: INTERMITTENCY, STICK-SLIP TRANSITION AND THE BREAKDOWN OF THE STOKES-EINSTEIN LAWS

CHRISTIANO DE MICHELE
*Dipartimento di Scienze Fisiche, Universitá di Napoli "Federico II",*
*and INFM, UdR Napoli,*
*V. Cinthia, 45, I-80100 Naples, Italy*
*demichel@na.infn.it*

DINO LEPORINI
*Dipartimento di Fisica "Enrico Fermi", Universitá di Pisa,*
*and INFM, UdR Pisa,*
*V. F. Buonarroti 2, Pisa, I-56127, Italy*
*dino.leporini@df.unipi.it*

## Abstract

The transport and the relaxation properties of a biatomic supercooled liquid are studied by molecular-dynamics methods. Both translational and rotational jumps are evidenced. At lower temperatures their waiting-time distributions decay as a power law at short times. The Stokes-Einstein relation (SE) breaks down at a temperature which is close to the onset of the intermittency. A precursor effect of the SE breakdown is observed as an apparent stick-slip transition. The breakdown of Debye-Stokes-Einstein law for rotational motion is also observed. On cooling, the changes of the rotational correlation time $\tau_1$ and the translational diffusion coefficient at low temperatures are fitted by power laws over more than three and four orders of magnitude, respectively. A less impressive agreement is found for $\tau_l$ with $l = 2 - 4$ and the rotational diffusion coefficient.

## 1. INTRODUCTION

It is well known that, on approaching the glass transition temperature $T_g$ from above, diffusion coefficients and relaxation times exhibit remarkable changes of several orders of magnitude which are under intense experimental, theoretical and numerical investigation.[1,2] In the high-temperature regime the changes usually track the shear viscosity $\eta$ in the sense that, if $X$ denotes the diffusion coefficient or the inverse of a relaxation time, the product $X\eta/T$ is nearly temperature-independent. In particular, both the Stokes-Einstein, $D \approx kT/6\pi\eta a$, and the Debye-Stokes-Einstein laws, $D_r \approx kT/\eta a^3$ are found to work nicely, $D$, $D_r$ and $a$ being the translational and the rotational diffusion coefficients and the molecular radius, respectively. Differently, in deeply supercooled regimes there is wide evidence that the product *increases on cooling* evidencing the breakdown of the hydrodynamic behavior at molecular level and the decoupling by the viscous flow.[3–11]

The decoupling between microscopic time scales and the viscous flow is a signature of the heterogeneous dynamics which develops close to the glass transition, i.e. a spatial distribution of transport and relaxation properties.[12–15] Crossover temperatures to that regime are broadly located around 1.2 $T_g$, i.e. in the region where the critical temperature $T_c$ predicted by the mode-coupling theory of the glass transition (MCT) is found.[3,16]

During the last years molecular-dynamics simulations (MD) proved to be a powerful tool to investigate supercooled liquids (for a recent review see ref. 17). Most MD studies

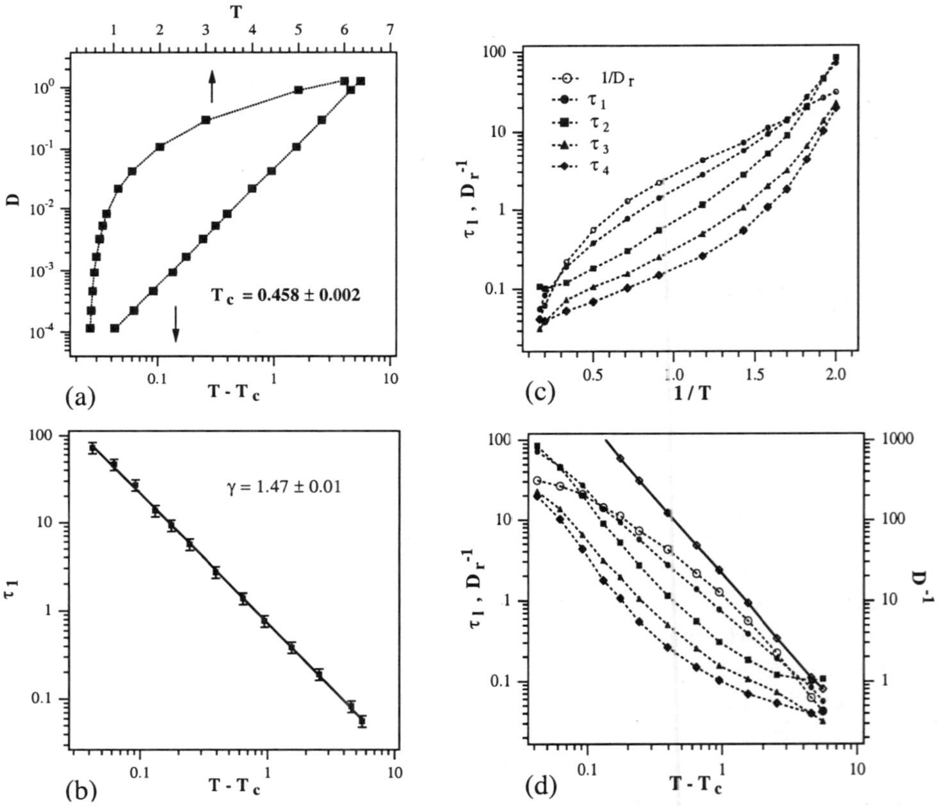


**Fig. 1** **(a)** Temperature dependence of the translational diffusion coefficient $D$. The dashed line is a fit with the power law Eq. (3) with $\gamma_D = 1.93 \pm 0.02$, $T_c = 0.458 \pm 0.002$ and $C_D = 0.0481 \pm 0.0004$. **(b)** MCT scaling analysis of $\tau_1$. $T_c = 0.458$. **(c)** Arrhenius plot and **(d)** MCT scaling analysis of the rotational correlation times $\tau_l$, $l = 1-4$ and the rotational diffusion coefficient $D_r$. $T_c = 0.458$. The dashed lines are guides for the eyes. The translational diffusion constant $D$ and the best fit are also shown for comparison (open diamonds).

confirmed that the decoupling is due to dynamic heterogeneities.[18–23] In fact, "active"[20] or "mobile"[23] regions which largely contribute to set the macroscopic average value have been identified. In such regions hopping processes, enhancing the transport with respect to the hydrodynamic behavior, have been evidenced.[18,20] The occurrence of jumps in glasses has been reported several times in the recent past.[24–27]

The paper discusses the decoupling phenomena in viscous liquids in the framework of recent MD studies on molecular supercooled liquids.[28,29] In Sec. 2 the model is presented. The results are discussed in Sec. 3. The conclusions are summarized in Sec. 4.

## 2. THE MODEL

The system under study is a model molecular liquid of rigid dumbbells.[30–32] The atoms A and B of each molecule have mass $m$ and are spaced by $d$. Atoms on different molecules interact via the Lennard-Jones potential:

$$V_{\alpha\beta}(r) = 4\varepsilon_{\alpha\beta}[(\sigma_{\alpha\beta}/r)^{12} - (\sigma_{\alpha\beta}/r)^{6}], \quad \alpha, \beta \in \{A, B\}. \tag{1}$$

The model parameters in reduced units are: $\sigma_{AA} = \sigma_{AB} = 1.0$, $\sigma_{BB} = 0.95$, $\varepsilon_{AA} = \varepsilon_{AB} = 1.0$, $\varepsilon_{BB} = 0.95$, $d = 0.5$, $m_A = m_B = m = 1.0$. The sample has $N = N_{at}/2 = 1000$ molecules which are accommodated in a cubic box with periodic boundary conditions. Further details on the simulations may be found elsewhere.[28] We examined the isobar at $P = 1.5$ by equilibrating the sample under isothermal-isobaric conditions and then collecting the data by a production run in microcanonical conditions. The temperatures we investigated are $T = 6$, 5, 3, 2, 1.4, 1.1, 0.85, 0.70, 0.632, 0.588, 0.549, 0.52, 0.5.

## 3. RESULTS AND DISCUSSION

### 3.1 Diffusion Coefficients and Rotational Correlation Times

#### 3.1.1 *Translational diffusion coefficient*

The translational diffusion coefficient $D$ is evaluated by the Einstein relation[33]:

$$D = \lim_{t\to\infty} \frac{R(t)}{6t} \tag{2}$$

where $R(t)$ is the mean squared displacement of center of mass at time $t$. In Fig. 1(a) the temperature dependence of $D$ is shown and fitted by the power-law

$$D = C_D(T - T_c)^{\gamma_D}. \tag{3}$$

Theoretical justification of Eq. (3) is provided by MCT[16]. In particular, the ideal MCT predicts the inequality $\gamma_D > 1.5$. The best fit values are $\gamma_D = 1.93 \pm 0.02$ and $T_c = 0.458 \pm 0.002$, $C_D = 0.0481 \pm 0.0004$.

#### 3.1.2 *Rotational diffusion coefficient and correlation times*

The rotational diffusion coefficient may be evaluated as[30,31] :

$$D_r = \lim_{t\to\infty} \frac{R_r}{4t} \tag{4}$$

where $R_r$ is the mean square value of the angular displacement:

$$\phi_i(t) - \phi_i(0) = \Delta\phi_i(t) = \int_0^t \omega_i(t')dt'. \quad (5)$$

The rotational correlation times $\tau_l$ are defined as the area below the rotational correlation functions.[33] $l$ is the rank of the involved Legendre polynomial. Figures 1(c) and (d) present the temperature dependence of $\tau_l$, $l = 1-4$ and $D_r$. It is seen that a wide region exists where the above quantities exhibit approximately the same Arrhenius behavior (about $0.7 < T < 2$). At lower temperatures the apparent activation energy of the rotational correlation times increase. In particular, $\tau_1$ becomes shorter than $\tau_2$ and a similar crossover is anticipated between $\tau_3$ and $\tau_4$ at temperatures just below 0.5. Differently, the rotational diffusion coefficient $D_r$ exhibits the same activation energy for $T < 0.9$. This behavior extends below the critical temperature $T_c$.[30]

Figures 1(b) and (d) show the MCT analysis of the rotational diffusion and the correlation times. According to MCT, $\tau_l$, $D_r^{-1} \propto (T - T_c)^{-\gamma}$.[34,35] Figure 1(b) proves that $\tau_1$ complies with MCT scaling over more than three orders of magnitude. Deviations are seen for $l > 1$.

## 3.2 Waiting-Time Distributions

At low temperatures a fraction of the overall biatomic molecules moves by jumps of finite size.[28,29] The waiting time between two jumps is a random quantity. The related distributions for both the translational and rotational jumps are presented here. They are denoted as $\psi(t)$ and $\psi_{rot}(t)$, respectively.

### 3.2.1 *Translational jumps*

In the present study a molecule jumps at time $t$ if the displacement between $t$ and $t + \Delta t^*$ ($\Delta t^* = 24$) exceeds $\sqrt{\Delta R^*} = \sigma_{AA}/2 = 0.5$. Further details on the procedure to detect jump events are provided elsewhere.[28]

Figures 2(a) and (b) show the waiting-time distribution $\psi(t)$ at different temperatures. At high temperature $\psi(t)$ is exponential. On cooling, the exponential decay is replaced at short times by a slowly-decaying regime. We fitted the decay by the function:

$$\psi(t) = [\Gamma(\xi)\tau^{\xi}]^{-1} t^{\xi-1} e^{-t/\tau} \quad 0 < \xi \leq 1. \quad (6)$$

The choice is motivated by the remark that in glassy systems rearrangements are rare events due to the constraints hampering the structural relaxation. It is believed that intermittent behaviour in particle motion develops on cooling.[22,36–38] A signature of the intermittence is the power law decay of $\psi(t)$ (for 2D liquids see Ref. 27) and related quantities such as the first-passage time distribution.[22] The exponent $\xi$ of Eq. (6) has a simple interpretation. If a dot on the time axis marks each relaxation event (a jump), the fractal dimension of the set of dots is $\xi$. For $\xi < 1$, it is found $\psi(t) \propto t^{\xi-1}$ at short times.[36,39]

The best fits at $T = 0.5, 0.549$ and $0.632$ are shown in Fig. 2(a) and (b). The increase of temperature results in a weak increase of the exponent $\xi$ and a more marked decrease of $\tau$. Interestingly, $\psi(t)$ exhibits small but reproducible deviations from Eq. (6) at $T = 0.5$. They suggest that the long-time decay is *faster* than the exponential one. If the exponential decay is replaced by a Gaussian one, the fit improves quite a lot and the $\xi$ exponent changes from 0.49 to 0.45.

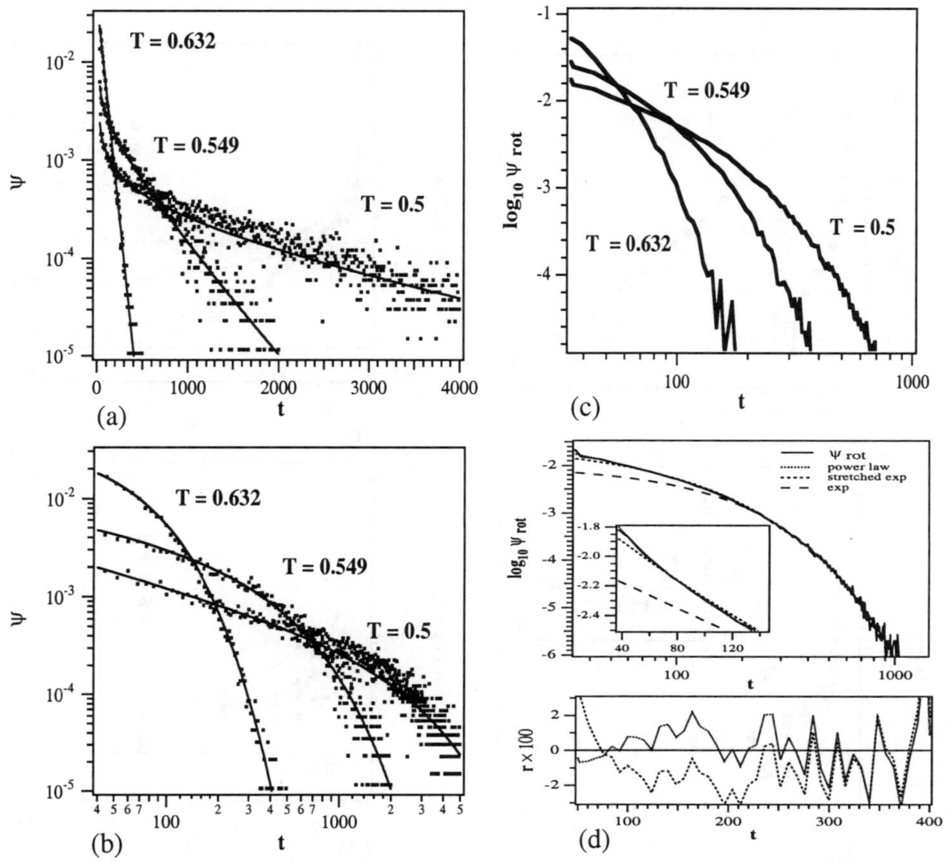


**Fig. 2** Long-time (**a**) and short-time (**b**) behavior of the waiting-time distribution $\psi(t)$ at different temperatures. The superimposed lines are best fits with Eq. (6) and $\xi = 0.49$, $\tau = 2550$ ($T = 0.5$), $\xi = 0.63$, $\tau = 420$ ($T = 0.549$) and $\xi = 1$, $\tau = 49$ ($T = 0.632$). bf (c) The rotational waiting-time distribution $\psi_{rot}$ at different temperatures. (**d**) Top: best fits of $\psi_{rot}$ at $T = 0.5$ with the exponential ($\tau = 102 \pm 3$), the stretched exponential ($\beta = 0.78 \pm 0.02$ and $\tau = 63 \pm 4$) and the truncated power law Eq. (6) ($\xi = 0.34 \pm 0.04$ and $\tau = 125 \pm 3$). The insert is a magnification of the short-time region. Notice that the fit with the power law is virtually superimposed to $\psi_{rot}$. Bottom: residuals of the fits by the truncated power-law and the stretched exponential.

### 3.2.2 *Rotational jumps*

The distribution $\psi_{rot}(t)$ of the waiting-time, namely the residence time in one angular site of the unit vector $\mathbf{u}_i$ being parallel to the axis of the $i$th molecule, offers a simple way to characterize rotational jumps. A jump of the $i$th molecule is detected at $t_0$ if the angle between $\mathbf{u}_i(t_0)$ and $\mathbf{u}_i(t_0 + \Delta t^*)$ is larger than 100° with $\Delta t^* = 24$. Other details are identical to the translational case.

Figure 2(c) shows $\psi_{rot}(t)$ at different temperatures. At $T = 0.632$ is virtually exponential whereas at lower temperatures the short-time behavior is different. In Fig. 2(d) $\psi_{rot}(t)$ at $T = 0.5$ is compared to the truncated power law Eq. (6), the stretched ($\exp[-(t/\tau)^\beta]$) and the usual exponential functions. The better agreement of Eq. (6) at short times is appreciated by the residuals.

It is worth noting that it was found that the time needed to complete the translational jumps exhibits a distribution.[28,29] The absence of a similar distribution for the rotational jumps points to a larger freedom of the reorientation process.

It is an important conclusion of the present study that intermittency has been evidenced at short times in both the translational and the rotational jump motion.

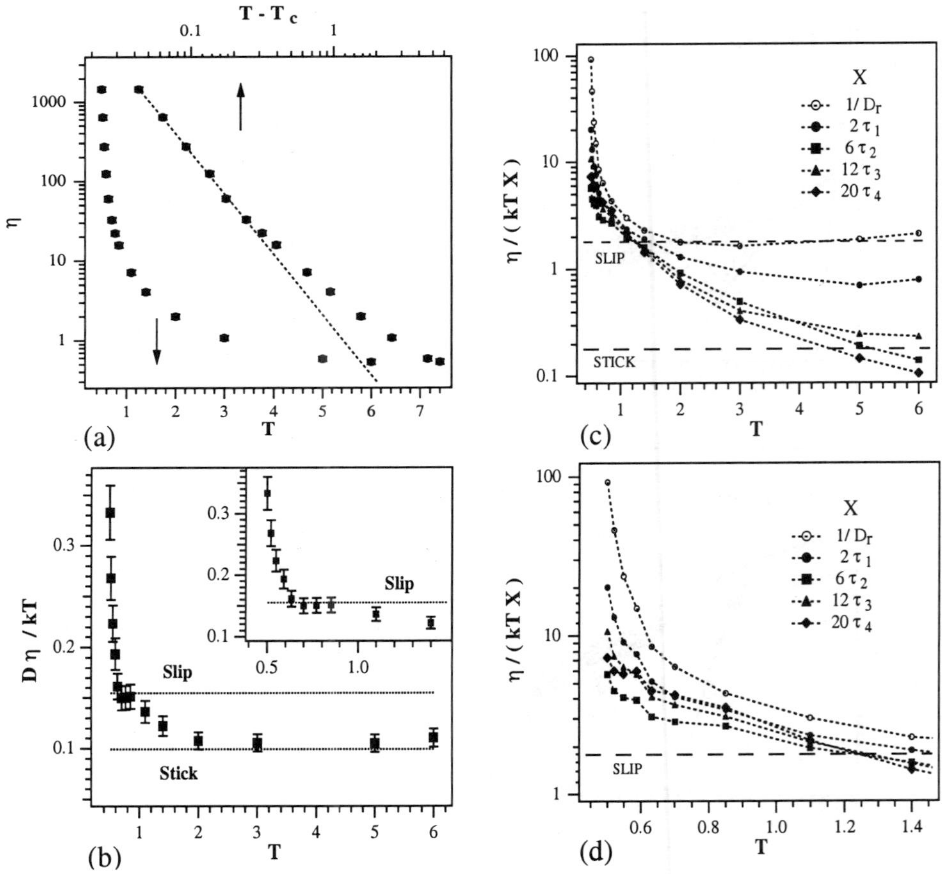


**Fig. 3** **(a)** Temperature dependence of the shear viscosity. The slope of the superimposed line is $\gamma_\eta = -2.20 \pm 0.03$. **(b)** Temperature dependence of the ratio $D\eta/kT$. Dashed lines are the SE predictions for prolate ellipsoids with semiaxis $b = 0.69$ and $c = 0.46$ and either stick or slip BC. Magnification of the supercooled region is shown by the inset. **(c)** Plots of the quantity $\eta/XkT$ with $X = D_r^{-1}, l(l+1)\tau_l$. **(d)** Magnification of the the supercooled regime. The dashed lines in (c) and (d) are the DSE predictions for stick and slip BC. In the region $1 < T < 2$ the ratio $\eta/XkT$ depends little on $X$ due to diffusional behavior.

## 3.3 Breakdown of Stokes-Einstein and Debye-Stokes-Einstein Laws

This section is devoted to discuss the decoupling of the single-particle dynamics from the viscosity which occurs at low temperatures.

### 3.3.1 *The shear viscosity*

The system under study exhibits a dramatic increase of the viscosity on cooling.[28,29] Figure 3(a) shows that the changes are conveniently described by a power law analogous to Eq. (3) in a temperature range smaller than the one of the translational diffusion coefficient [see Fig. 1(a)].

### 3.3.2 *Breakdown of the Stokes-Einstein law*

Several experimental[3,4,7,9] and numerical[18,19,22,24,40,41] works evidenced a decoupling of the translational diffusion and the viscosity on approaching the glass transition. Typically, the decoupling occurs around $T_c$.[3] MD investigated the issue in one- and two-components *atomic* systems. It is therefore of interest to examine the present molecular system from that respect.

The decoupling manifests as an enhancement of the translational diffusion $D$ with respect to the prediction of the Stokes-Einstein relation (SE) which reads[42]

$$D = \frac{kT}{\mu\eta} \tag{7}$$

$\mu$ is a constant that depends on both the molecule geometry and the boundary conditions (BC). For a sphere of radius $a$, $\mu$ equals $6\pi a$ and $4\pi a$ for stick and slip BC, respectively. The cases of uniaxial ellipsoids with both stick and slip BC were worked out.[42,43]

In Fig. 3(b) we plot the ratio $D\eta/kT$ as a function of temperature. At higher temperatures the ratio levels off at about $0.105 \pm 0.007$. On cooling, there is first a mild change followed by a steep increase below $T = 0.632 = 1.38T_c$. We remind that SE predicts a temperature-independent ratio. It is worth noting that the SE law breaks down at $T \sim 0.632$ below which intermittent jump-motion becomes apparent [see Fig. 2(b)].

It is believed that the SE failure is a signature of the heterogeneous dynamics of supercooled liquids.[12–15] Alternative views are provided by frustrated lattice gas models[40] and the "energy landscape" picture.[2,41,44–46]

Around $T = 0.77$ a plateau at $0.151 \pm 0.01$ is reached. The diatomic molecule under study may be sketched as a prolate ellipsoid with semiaxis $b = 3/4$ and $c = 1/2$. SE predicts that the corresponding ratio $D\eta/kT$ for stick and slip BC is equal to 0.091 and 0.1415, respectively. The values compare well to the high- and low-T plateau in Fig. 3(b). By setting $b = 0.69$ and $c = 2/3b$ the agreement is improved with $D\eta/kT = 0.098$ and 0.154 for stick and slip BC, respectively. The above analysis provides reasonable evidence of a precursors effect of the SE breakdown which manifests itself as an apparent stick-slip transition. A similar crossover has been observed in colloidal suspensions.[47]

### 3.3.3 *Breakdown of the Debye-Stokes-Einstein Law*

For large Brownian particles the reorientation in a liquid occurs via a series of small angular steps, i.e. it is diffusive. Hydrodynamics predicts that the diffusion manifests a strong coupling to the viscosity $\eta$ which is accounted for by the Debye-Stokes-Einstein law (DSE). For biaxial ellipsoids it takes the form[48]

$$D_i = \frac{kT}{\mu_i \eta}, \quad i = x,\, y,\, z \tag{8}$$

$D_{x,y,z}$ are the principal values of the diffusion tensor, $k$ is the Boltzmann constant. The coefficients $\mu_i$ depend on the geometry and BC. For a sphere with stick BC $\mu_{x,y,z} = 6v$, $v$ being the volume of the sphere. The cases of uniaxial ellipsoids with stick and slip BC were worked out.[42,48,49]

If the viscosity is not high ($\eta < 1 Poise$) DSE works nicely even at a molecular level. At higher viscosities DSE is found to overestimate the rotational correlation times of tracers in supercooled liquids by time-resolved fluorescence and Electron Spin Resonance (ESR) studies.[50–52] In this decoupling region ESR evidenced that the tracer rotates by jump motion.[53] On the other hand, photobleaching[7] and NMR[4], studies found only small deviations from DSE even close to $T_g$.

We have studied the coupling between the rotational relaxation and the viscosity. From this respect we also considered the popular alternative form of Eq. (8) which is written in terms of rotational correlation times by assuming that the reorientation is diffusive, i.e. it occurs by small angular steps. It is well suited for comparison with the experiments which do not usually provide direct access to the rotational diffusion coefficients. In the present case the alternative form is obtained by replacing $D_r^{-1}$ with the product $l(l+1)\tau_l$ in Eq. 8.

In Fig. 3(c) we plot the ratio $\eta/XkT$ with $X = D_r^{-1}$, $l(l+1)\tau_l$ with $l = 1-4$ together with DSE predictions for both stick and slip BC.

At high temperatures if $X = l(l+1)\tau_l$ with $l > 1$ the ratio roughly approaches the value expected for stick BC. For $X = D_r^{-1}$ DSE with slip BC fits better. The $\tau_1$ case is intermediate. On cooling $\eta/XkT$ increases. For intermediate temperatures the rotational diffusion takes place since $l(l+1)\tau_l$ and $D_r$ are all close to the DSE expectation with slip BC. Notably, $D_r\eta/kT$ remains close to this value in the wide interval $2 < T < 6$.

At lower temperatures $\eta/XkT$ diverges. The stronger deviations are exhibited by $D_r$ and $\tau_{1,3}$. The results are readily intepreted. At low temperature molecules undergo 180° flips.[29] These affect the pair $\tau_{1,3}$ much more than the pair $\tau_{2,4}$. The large decoupling of $D_r$ is also understood since the latter is mainly affected by fast dynamics.

## 4. CONCLUSIONS

The transport and the relaxation properties of a biatomic supercooled liquid on the isobar $P = 1.5$ have been studied. The results point out that at low temperature a fraction of the overall molecules performs jump motion with intermittent behavior. The resulting changes in the dynamics weaken the coupling with the viscosity. Interesting precursor effects to be described as apparent slip-stick transitions are observed.

## REFERENCES


1. M. Giordano, D. Leporini and M. P. Tosi (eds.) "Proceedings of the II Workshop on Non-Equilibrium Phenomena in Supercooled Fluids, Glasses and Amorphous Materials," *J. Phys.: Condens. Matter*, **11**, 10A (1999).
2. For a short review see M. D. Ediger, C. A. Angell and S. R. Nagel, *J. Phys. Chem* **100**, 13200 (1996).
3. E. Rössler, *Phys. Rev. Lett.* **65**, 1595 (1990).
4. F. Fujara, B. Geil, H. Sillescu and G. Fleischer *Z. Phys.* **B88**, 195 (1992); I. Chang, F. Fujara, B. Geil, G. Heuberger, T. Mangel and H. Sillescu *J. Non-Cryst. Solids* **172–174**, 248 (1994).
5. E. Rössler, J. Tauchert and P. Eiermann, *J. Phys. Chem.* **98**, 8173 (1994).
6. E. Rössler and P. Eiermann, *J. Chem. Phys.* **100**, 5237 (1994).
7. M. T. Cicerone, F. R. Blackburn and M. D. Ediger, *J. Chem. Phys.* **102**, 471 (1995); M. T. Cicerone and M. D. Ediger, *J. Chem. Phys.* **104**, 7210 (1996).
8. L. Andreozzi, A. Di Schino, M. Giordano and D. Leporini, *Europhys. Lett.* **38**, 669 (1997).
9. A. Voronel, E. Veliyulin, V. Sh. Machavariani, A. Kisliuk and D. Quitmann, *Phys. Rev. Lett.* **80**, 2630 (1998).
10. M. Faetti, M. Giordano, L. Pardi and D. Leporini, *Macromolecules* **32**, 1876 (1999).
11. L. Andreozzi, M. Faetti, M. Giordano and D. Leporini, *J. Phys. Chem.* **B103**, 4097 (1999).
12. J. A. Hodgdon and F. H. Stillinger, *Phys. Rev.* **E48**, 207 (1993); F. H. Stillinger and J. A. Hodgdon, *ibid.* **50**, 2064 (1994);
13. C. Z.-W. Liu and I. Oppenheim, *Phys. Rev.* **E53**, 799 (1996).
14. J. F. Douglas and D. Leporini, *J. Non-Cryst. Solids* **235–237**, 137 (1998).
15. H. Sillescu, *J. Non-Cryst. Solids* **243**, 81 (1999).
16. W. Götze, *J. Phys.: Condens. Matter* **11**, A1 (1999); H. Z. Cummins, *J. Phys.: Condens. Matter* **11**, A95 (1999).
17. W. Kob, *J. Phys.: Condens. Matter* **11**, R85 (1999).
18. D. Thirumalai and R. D. Mountain, *Phys. Rev.* **E47**, 479 (1993).

19. R. Yamamoto and A. Onuki, *Phys. Rev.* **E58**, 3515 (1998).
20. R. Yamamoto and A.Onuki, *Phys. Rev. Lett.* **81**, 4915 (1998).
21. D. Perera and P. Harrowell, *Phys. Rev. Lett.* **81**, 120 (1998).
22. P. Allegrini, J. F. Douglas and S. H. Glotzer, *Phys. Rev.* **E60**, 5714 (1999).
23. W. Kob, C. Donati, S. J. Plimpton, P. H. Poole and S. C. Glotzer, *Phys. Rev. Lett.* **79** 2827 (1997); C. Donati, J. F. Douglas, W. Kob, S. J. Plimpton, P. H. Poole and S. C. Glotzer, *Phys. Rev. Lett.* **80** 2338 (1998).
24. J.-L. Barrat, J.-N. Roux and J. P. Hansen, *Chem. Phys.* **149**, 197 (1990).
25. H. Miyagawa, Y. Hiwatari, B. Bernu and J. P. Hansen, *J. Chem. Phys.* **88**, 3878 (1988).
26. G. Wahnström, *Phys. Rev.* **A44**, 3752 (1991).
27. T. Muranaka and Y. Hiwatari, *J. Phys. Soc. Japan* **67**, 1982 (1998).
28. C. De Michele and D. Leporini, *Phys. Rev. E*, to appear (March 2001).
29. C. De Michele and D. Leporini, *Phys. Rev. E*, to appear (March 2001).
30. S. Kämmerer, W. Kob and R. Schilling, *Phys. Rev.* **E56**, 5450 (1997).
31. S. Kämmerer, W. Kob and R. Schilling, *Phys. Rev.* **E58**, 2141 (1998).
32. S. Kämmerer, W. Kob and R. Schilling, *Phys. Rev.* **E58**, 2131 (1998).
33. J.-P. Hansen and I. R. McDonald, *Theory of Simple Liquids*, 2nd ed. (Academic Press, London, 1986).
34. R. Schilling and T. Scheidsteger, *Phys. Rev.* **E56**, 2932 (1997).
35. T. Franosch, M. Fuchs, W. Götze, M. R. Mayr and A. P. Singh, *Phys. Rev.* **E56**, 5659 (1997); W. Götze, *J. Phys.: Condens. Matter* **11**, A1 (1999); H. Z. Cummins, *J. Phys.: Condens. Matter* **11**, A95 (1999).
36. L. Sjögren, *Z. Phys.* **B74**, 353 (1989).
37. J. F. Douglas and J. B. Hubbard, *Macromolecules* **24**, 3163 (1991); J. F. Douglas, *Comp. Mat. Sci.* **4**, 292 (1995).
38. T. Odagaki, *Phys. Rev. Lett.* **75**, 3701 (1995).
39. R. Hilfer and L. Anton, *Phys. Rev.* **E51**, R848 (1995).
40. M. Nicodemi and A. Coniglio, *Phys. Rev.* **E56**, R39 (1998); A. Coniglio, A. DeCandia, A. Fierro and M. Nicodemi, *J. Phys.: Condens. Matter* **11**, A167 (1999).
41. L. Angelani, G. Parisi, G. Ruocco and G. Viliani, *Phys. Rev. Lett.* **81**, 4648 (1998).
42. H. Lamb, *Hydrodynamics*, 6th ed. (Cambridge University Press, Cambridge, 1932).
43. S. Tang and G. T. Evans, *Mol. Phys.* **80**, 1443 (1993).
44. M. Goldstein, *J. Chem. Phys.* **51**, 3728 (1969).
45. S. Sastry, P. G. Debenedetti and F. H. Stillinger, *Nature* **393**, 554 (1998).
46. C. A. Angell, *J. Res. Natl. Inst. Std. Tech.* **102**, 171 (1997); C. A. Angell, B. E. Richards and V. Velikov, *J. Phys.: Condens. Matter*, **11**, A75 (1999).
47. P. N. Segre, S. P. Meeker, P. N. Pusey and W. C. K. Poon, *Phys. Rev. Lett.* **75**, 958 (1995).
48. L. D. Favro, *Phys. Rev.* **119**, 53 (1960).
49. C.-M. Hu and R. Zwanzig, *J. Chem. Phys.* **60**, 4354 (1974).
50. L. Andreozzi, A. Di Schino, M. Giordano and D. Leporini, *Europhys. Lett.* **38**, 669 (1997).
51. L. Andreozzi, M. Giordano and D. Leporini, *J. Non Cryst. Solids* **235–237**, 219 (1998).
52. M. Faetti, M. Giordano, L. Pardi and D. Leporini, *Macromolecules* **32**, 1876 (1999).
53. L. Andreozzi, F. Cianflone, C. Donati and D. Leporini, *J. Phys.: Condens. Matter* **8**, 3795 (1996)

# VORTEX MATTER OUT OF EQUILIBRIUM

MARIO NICODEMI
*Universitá di Napoli "Federico II", Dip. Scienze Fisiche,*
*INFM and INFN, 80126 Napoli, Italy*
*Imperial College, 180 Queen's Gate, London SW7 2BZ, UK*
*Mario.Nicodemi@na.infn.it*

**Abstract**

The present paper provides a short review of recent progresses in understanding out of equilibrium vortex behaviours in type II superconductors by use of a schematic coarse grained vortex model. In particular, it is possible to depict a unifying scenario for magnetic and transport properties of off-equilibrium vortex matter, ranging from the *reentrant* phase diagram, to magnetisation *loops*, "anomalous" 2-*nd peak*, logarithmic *creep*, "anomalous" *finite creep rate* for $T \to 0$, *"memory"* and *"irreversibility"* in I-V characteristics, *"rejuvenation"* and *"stiffening"* of the system response.

## 1. INTRODUCTION

In presence of an external magnetic field, vortex lines penetrates in type II superconductors. Vortices form a strongly interacting system, usually named "vortex matter", characterised by very rich physics.[1] The structural properties of vortex dynamics are very important because they crucially affects the overall system behaviour and, so, have also essential consequences in technological applications. In particular, it has been discovered that vortex matter exhibits relevant, even dominant, history dependent phenomena in magnetic as well as transport properties, such as memory in magnetisation curves along with aging in I-V characteristics (see for instance Refs. 1 to 4). These phenomena are markedly *out of equilibrium* effects, similar to those recorded in other glass formers such as supercooled liquids, polymers and random magnets,[5] and are the subject of the present review.[6]

Samples within a broad range of material parameters show these effects, suggesting they originate from few general basic properties.[1] Specifically, we consider here a schematic model to describe the features of vortex dynamics in superconductors, namely a system of repulsive particles wandering in a pinning landscape in presence of an external drive.[6] The model describes several phenomena of vortex physics, ranging from a reentrant phase diagram in the $(B, T)$ (field-temperature) plane, to the "anomalous second peak" in magnetisation loops (the "fishtail" effects), logarithmic creep and "aging" of magnetic relaxation, the "anomalous creep" at very low temperatures, "memory" and history dependent effects in vortex flow and in I-V characteristics, and many others.[6] Thus, it is possible to depict a comprehensive and unifying framework for magnetic and transport phenomena.

## 2. THE R.O.M. MODEL

A system of straight parallel vortex lines, corresponding to a magnetic field $B$ along the $z$-axis, interacts via a potential:[1] $A(r) = \frac{\phi_0^2}{2\pi\lambda'^2}[K_0(r/\lambda') - K_0(r/\xi')]$, $K_0$ being the MacDonald function, $\xi$ and $\lambda$ the correlation and penetration lengths ($\xi' = c\xi/\sqrt{2}$, $\lambda' = c\lambda$, $c = (1-B/B_{c2})^{-1/2}$). The typical high vortex densities and long $\lambda$ imply that the vortex system is strongly interacting. To make it theoretically more tractable, as proposed in Refs. 6 and 7, one can coarse grain in the $xy$-plane by introducing a square grid of lattice spacing, $l_0$, of the order of the London length, $\lambda$ (see Fig. 1). The number of vortices on the $i$th coarse grained cell, $n_i$, is an integer number smaller than $N_{c2} = B_{c2}l_0^2/\phi_0$ ($B_{c2}$ is the upper critical field and $\phi_0 = hc/2e$ is the flux quantum). The coarse grained interaction Hamiltonian is thus:[6] $\mathcal{H} = \frac{1}{2}\sum_{ij} n_i A_{ij} n_j - \frac{1}{2}\sum_i A_{ii}|n_i| - \sum_i A_i^p n_i$. The first two terms describe the repulsion between the vortices and their self energy, and the last the interaction with a random pinning background. For sake of simplicity, we consider the simplest version of $\mathcal{H}$: we choose $A_{ii} = A_0 = 1$; $A_{ij} = A_1 < A_0$ if $i$ and $j$ are nearest neighbours; $A_{ij} = 0$ otherwise; the random pinning is delta-distributed $P(A^p) = (1-p)\delta(A^p) + p\delta(A^p - A_0^p)$.[8] Particles can also be given a "charge" $s_i = \pm 1$ (corresponding to opposite direction of magnetic flux) and neighbouring particles with opposite "charge" annihilate. To control the overall system "charge density" we can add a chemical potential term $-\mu\sum_i s_i n_i$ to the above Hamiltonian (where $n_i \to n_i s_i$). In analogy with computer investigation of dynamical processes in fluids, the time evolution of the model is simulated by a Monte Carlo Kawasaki dynamics on a square lattice of size $L$ at a temperature $T$.[8] The system is periodic in the $y$-direction. The two edges parallel to the $y$-axis are in contact with a vortex reservoir, i.e. an external magnetic field, of density $N_{ext}$. Particles can enter and leave the system only through the reservoir. The above Restricted Occupancy Model (ROM) is described in full details in Ref. 6.

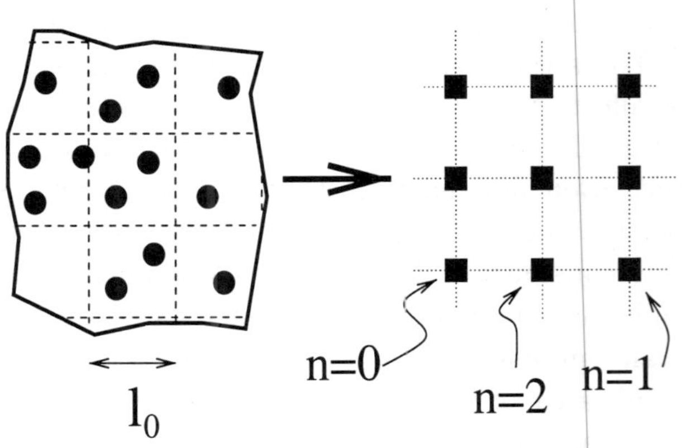


**Fig. 1** The ROM lattice model is defined by a coarse graining of the vortex system on a scale $l_0$.

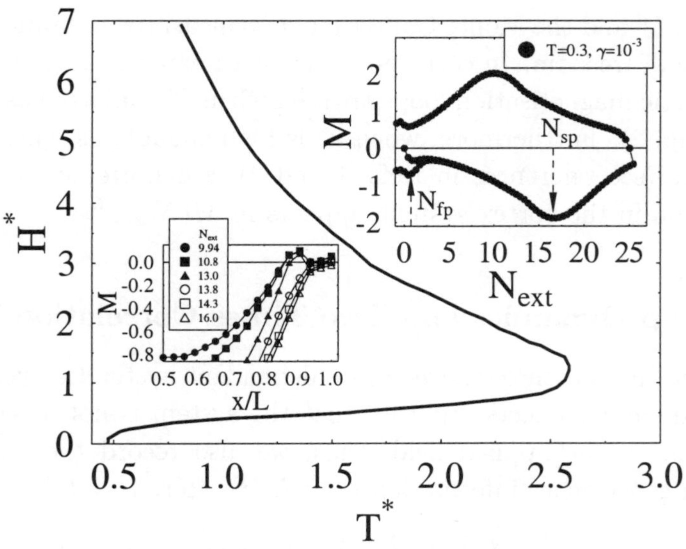


**Fig. 2 Main frame** The phase diagram of the ROM model in the plane ($H^*$, $T^*$), where $H^* = \mu/k_BT$ and $T^* = T/A_1$ are the dimensionless chemical potential of the external applied field and temperature. It is evaluated in mean field approximation for $\kappa^* = 10$ and $A_0^p = 0.0$ (bold full line). **Inset top** The magnetisation, $M$, is plotted as a function of the applied field, $N_{ext}$, while cycling the field with a sweep rate $\gamma = 1.1\ 10^{-3}$ at $T = 0.3$. $M(N_{ext})$ shows the so called "fish-tail" feature, i.e., a second peak $N_{sp}$. **Inset bottom** The Bean profile, i.e. the magnetisation $M(x)$, as a function of the transversal spatial coordinate $x/L$ ($L$ is the system linear size), for the shown values of the external field, during the above cycle.

## 2.1 The Equilibrium Phase Diagram

To understand the equilibrium properties of the ROM model we briefly consider its replica mean field theory (MF). In this approximation the equilibrium phase diagram in the field-temperature plane ($H^*$, $T^*$) (where $T^* = k_BT/A_1$ and $H^* = \mu/k_BT$) can be analytically dealt with (see Fig. 2). In absence of disorder it clearly shows a reentrant phase transition from a high temperature low density fluid phase to an ordered phase, in analogy to predictions in superconductors.[1] For moderate values of the pinning energy ($A_0^p \leq A_1$), a second order transition still takes place, which at sufficiently strong pinning is expected to become a "glassy" transition, as is seen in Random Field Ising Models.[5] For the 2D lattice we consider below, a numerical investigation is consistent with a first order transition (in limit $A^p \to 0$). In MF, the extension of the low $T$ phase shrinks by increasing $A_0^p$ (i.e. the highest critical temperature, $T_m^*$, decreases) and the higher is $\kappa^*$ the smaller the reentrant region (the parameter $\kappa^* = A_1/A_0$ can be directly related to the Ginzburg-Landau parameter $\kappa = \lambda/\xi$). These findings are in agreement with experimental results on vortex phase diagrams (see references in Refs. 1 and 2).

# 3. MAGNETIC PROPERTIES

In the following sections, we are going to discuss the dynamical behaviour of the ROM model. In our Monte Carlo simulations the system is prepared by zero field cooling and then increasing $N_{ext}$ at constant rate, $\gamma$, up to the working value. By now, we have no externally applied currents. In particular, we recorded the magnetisation

$$M(t) = N_{in}(t) - N_{ext}(t) \tag{1}$$

where $N_{in} = \sum_i n_i/L^d$ and the Monte Carlo time, $t$, is measured in units of single attempted updates per degree of freedom. In correspondence with experiments,[2] at low temperatures pronounced hysteretic magnetisation loops are seen when $M$ is plotted as a function of $N_{ext}$[6] (see upper inset Fig. 2). Furthermore, when $\kappa^*$ is high enough, as much as in experiments on superconductors (see references in Refs. 1 and 2), a definite *second peak*, related to a true phase transition in the vortex system, appears in $M(N_{ext})$.[6]

### 3.1 Aging Creep Dynamics and Two Times Correlation Functions

At low temperature the presence of sweep rate dependent hysteretic cycles, slowly relaxing magnetisation, and similar effects, indicate that the system is not at equilibrium. At the given working value of the applied field, thus, we also record the magnetic correlation function, $C$, which gives richer information than $M(t)$. Here ($t > t_w$):

$$C(t, t_w) = \langle [M(t) - M(t_w)]^2 \rangle \,. \tag{2}$$

If the temperature is not too low (say, around $T = 1.0$), the system creep is characterised by finite relaxation times and no "aging" is seen: $C(t, t_w)$ is a function of $t - t_w$. At long times, $C(t, t_w)$ is well fitted by the so called Kohlrausch-Williams-Watts law:[6]

$$C(t, t_w) \simeq C_\infty\{1 - e^{-[(t-t_w)/\tau]^\beta}\} \,. \tag{3}$$

The time scale, $\tau$, and the Kohlrausch-exponent, $\beta$, depend on the temperature $T$ (see Fig. 3) and on the overall field $N_{ext}$. The pre-asymptotic dynamics (i.e. $t - t_w \ll \tau$) is also interesting and characterised by various regimes. In particular, for not too short times, a power law is observed over several decades.

The above behaviours are recorded in a broad region at low temperatures. However, around $T = 0.5$, a steep increase of $\tau$ is found (see Fig. 3). In fact, for temperatures below $T_g \simeq 0.25$, the characteristic time gets larger than our recording window and the system definitely loses contact with equilibrium. The crossover temperature $T_g(N_{ext})$ has a physical

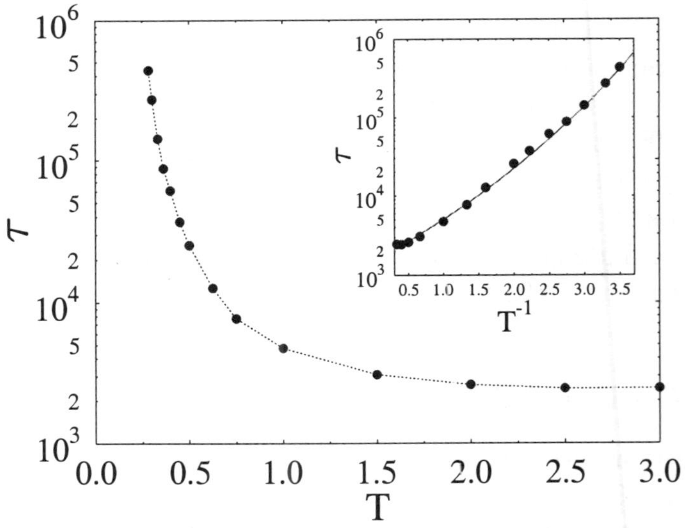


**Fig. 3** **Main frame** The system equilibration time, $\tau$, from Eq. (3), enormously grows by decreasing the temperature $T$ (here $N_{ext} = 10$). Below the crossover temperature $T_g \sim 0.25$, $\tau$ is larger than the observation window. **Inset** Close to $T_g$, $\tau$ plotted as a function of $1/T$ approximately shows a Vogel-Tamman-Fulcher or an Arrhenius behaviour [see Eq. (4)].

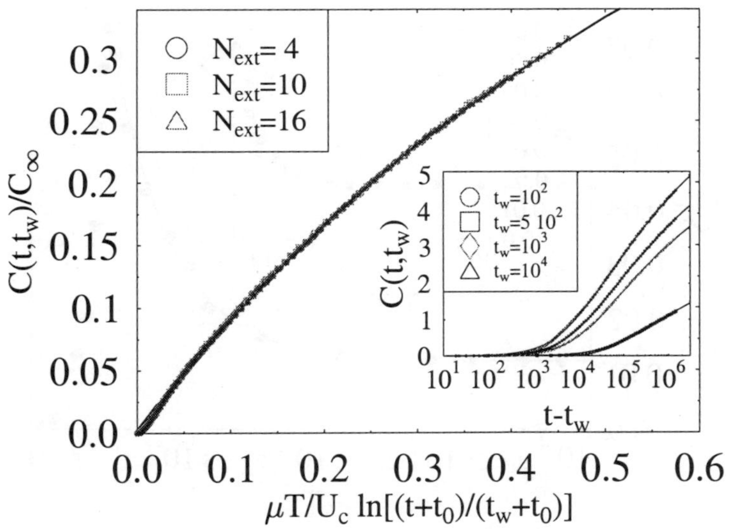


**Fig. 4** **Inset** The two-times correlation function, $C(t,\, t_w)$ (here recorded for $N_{ext} = 16$), at $T = 0.1$ shows "aging" in time. The continuous lines are logarithmic fits from Eq. (5). **Main Frame** The data from the inset are superimposed on the same master function for each of the shown $t_w$ (and for $N_{ext} = 4,\, 10,\, 16$).

meaning similar to the phenomenological definition of the so called glass transition point in supercooled liquids.[5] The presence of an underlying "ideal" glass transition point, $T_c(N_{ext})$, can be located by some fit of $\tau$ from the high $T$ regime, such as a Vogel-Tamman-Fulcher or a power law (see Fig. 3):

$$\tau = \tau_0 \exp\left(\frac{E_0}{T - T_c}\right). \tag{4}$$

Below $T_g$ relaxation times are huge and the system off equilibrium dynamics has remarkably rich "aging" properties. In the inset of Fig. 4 we show that $C(t,\, t_w)$, at $T = 0.1$, depends on both times $t$ and $t_w$: the system evolution is slower the older is its "age" $t_w$ (see Fig. 4). In the entire low $T$ region ($T < T_g$), after a short initial power law behaviour, $C(t,\, t_w)$ can be well fitted by a generalisation of a known interpolation formula, often experimentally used,[1] which now depends on the *waiting time*, $t_w$:

$$C(t,\, t_w) \simeq C_\infty \left\{1 - \left[1 + \frac{\mu T}{U_c} \ln\left(\frac{t + t_0}{t_w + t_0}\right)\right]^{-1/\mu}\right\}. \tag{5}$$

Equation (5) (in agreement with the general scenario of Ref. 10) implies that for times large enough (but smaller than the equilibration time) we have $C(t,\, t_w) \sim \mathcal{S}(t/t_w)$, showing the presence of dynamical *scaling properties.*

Aging has been experimentally observed in magnetic relaxations of vortex matter in Ref. 3. The above phenomena are common to many different systems ranging from polymers, to supercooled liquids, spin glasses, granular media.[5,12]

## 3.2 Anomalous Creep at Very Low Temperatures

The results of the previous section are very useful to understand an intriguing experimental observation[11] about vortex matter: even at very low temperatures (where activated processes should be absent) magnetic relaxation does take place. This phenomenon, previously interpreted in terms of "quantum tunnelling" of vortices,[1] is also found in the present

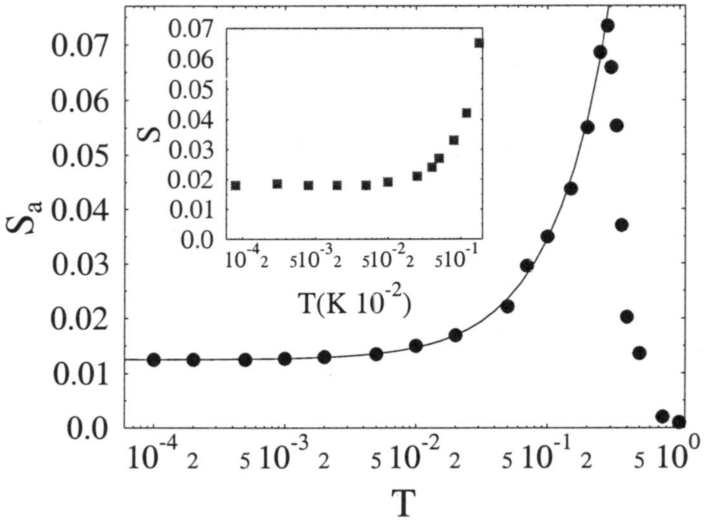


**Fig. 5** **Main frame** The creep rate, $S_a$, in the ROM model as a function of the temperature, $T$, in units of $A_0$ ($N_{ext} = 10$, $\gamma = 10^{-3}$). The continuous line is a linear fit. The decrease of $S_a$ at higher $T$ is also experimentally found. **Inset** The creep rate, $S$, in a BSCCO crystal at 880 Oe (from Aupke et al. in Ref. 11).

purely "classical" vortex model. We show that a non-zero creep rate for $T \to 0$ is to be expected in systems "aging" in their off equilibrium dynamics.

In the experiments the creep rate, $S_a$, dependence with $T$ is investigated, where

$$S_a = \left| \frac{\partial \ln(M)}{\partial \ln(t)} \right| . \tag{6}$$

When the temperature is very low, in both experiments and in our simulations, $S_a$ approaches a *finite* value, $S_a(0) > 0$, for $T \to 0$. In Fig. 5, we plot the creep rate, $S_a$, as a function of $T$ in a broad temperature range. For comparison we present experimental data in BSCCO (from Ref. 11) as inset (note that the values of $S_a$ in our model and in real samples are very similar).

We have seen that at very low $T$, the system equilibration time $\tau(T)$ diverges exponentially. In that region, the typical observation time windows, $t_{obs}$, are such that $t_{obs}/\tau \ll 1$, and the system is in the early stage of its off equilibrium relaxation from its initial state. This is schematically the origin of the flattening of $S_a$ at very low $T$.[6] Notice that, if one could observe the system for an exponentially long time, i.e. if $t_{obs}/\tau \gg 1$, then the creep rate, $S_a$, would indeed go to zero.

In the slow off equilibrium relaxation at very low temperatures no activation over barriers occurs and the system simply wanders in its very high dimensional phase space through the few channels where no energy increase is required. Experimentally, the present scenario, where off equilibrium phenomena dominate the anomalous low $T$ creep, is supported by the discovery of "aging".[2–4].

## 4. VORTEX FLOW AND I-V CHARACTERISTICS

Type II superconductors subject to an external current (which activates a Lorentz driven vortex flow) also shows strong memory and history dependent effects. Here, we propose a scenario for a broad set of these kind of phenomena found in I-V characteristics. In relation to recent experimental results,[4] we discuss in particular the nature of "memory" effects observed in the response of the system to an external drive, i.e. the I-V characteristic. The essential step is to identify the relevant time scales in the dynamics.

For a given applied field, $N_{ext}$ (below $N_{ext} = 10$), we monitor the system relaxation in presence of a drive, $I$ (due to the Lorentz force), in the $y$-direction. As in similar driven lattice gases,[9] the effect of the drive is simulated by introducing a bias in the Metropolis coupling of the system to the thermal bath: a particle can jump to a neighbouring site with a probability $\min\{1, \exp[-(\Delta\mathcal{H} - \epsilon I)/T]\}$. Here, $\Delta\mathcal{H}$ is the change in $\mathcal{H}$ after the jump and $\varepsilon = \pm 1$ for a particle trying to hop along or opposite to the direction of the drive and $\varepsilon = 0$ in orthogonal jumps. A drive $I$ generates a voltage $V$[13]:

$$V(t) = \langle v_a(t) \rangle \tag{7}$$

where $v_a(t) = \bar{v}(t)$ is an average vortex "velocity" at time $t$.[6] Here, $v(t) = \frac{1}{L}\sum_i v_i(t)$ is the instantaneous flow "velocity", $v_i(t) = \pm 1$ if the vortex $i$ at time $t$ moves along or opposite to the direction of the drive $I$ and $v_i = 0$ otherwise.

## 4.1 Memory in Driven Vortex Flow

Some experiments where the drive is cyclically changed[4] show an interesting manifestation of "memory". A drive $I$ is applied to the system and, after a time $t_1$, abruptly changed to a new value $I_1$; finally, after waiting a time $t_2$, the previous $I$ is restored and the system evolves for a further $t_3$ (see lower inset of Fig. 6). The measured $V(t)$, in the ROM model, is shown in the main panel of Fig. 6 for $T = 0.1$. A first observation is that after the switch to $I_1$ the system seems to abruptly reinitiate its relaxation approximately as if it has always been

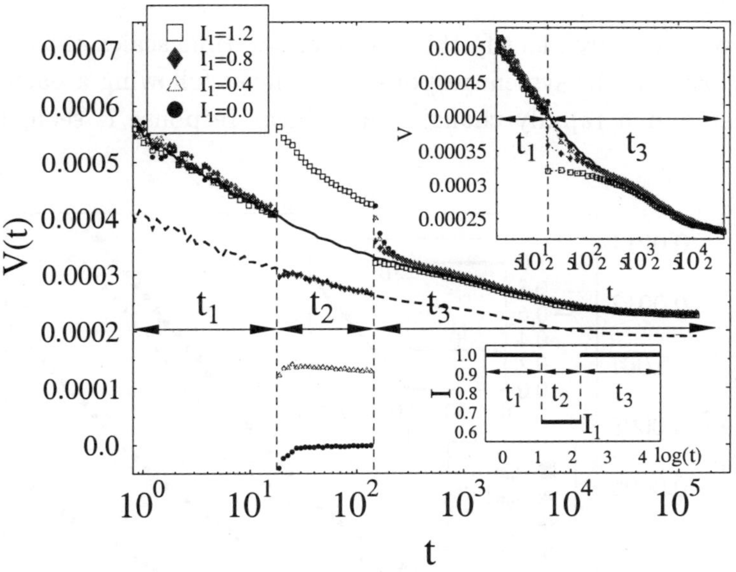


**Fig. 6** The system voltage drop, $V(t)$, is plotted as a function of time at $T = 0.1$ for an applied drive $I = 1$ in the ROM model. The lower inset shows the drive protocol used: after a time lag $t_1$, $I$ is abruptly changed to a new value $I_1$ for a time $t_2$ and finally it is set back to its previous value. When $I$ is switched to $I_1$ the system seems to *"rejuvenate"*: it suddenly restarts its relaxation along the path it would have had if $I = I_1$ at all times (consider the continuous and dashed bold curves, corresponding to $I = I_1 = 1$ and $I = I_1 = 0.8$, plotted for comparison). By restoring $I$ after $t_2$, the system shows a strong form of "memory": if $t_2$ and $I_1$ are small enough (see text) the relaxation of $V(t)$ restarts where it was at $t_1$. However, if $t_2$ and $I_1$ are too large, this is not the case, as shown in the upper inset.

at $I_1$ (see for example the dashed curve in Fig. 6), a phenomenon known as *"rejuvenation"* in thermal cycling of spin-glasses and other glassy systems.[14] The more surprising fact is, however, that for $I_1$ small enough (say $I_1 \ll I^*$, $I^*$ to be quantitatively defined below) when the value $I$ of the drive is restored the voltage relaxation seems to restart from where it was at $t_1$, i.e. where it stopped before the application of $I_1$ (see Fig. 6). Actually, if one "cuts" the evolution during $t_2$ and "glues" together those during $t_1$ and $t_3$, an *almost* perfect matching is observed (see upper inset of Fig. 6). What is happening during $t_2$ is that the system is trapped in some metastable states, but *not* completely frozen as shown by a small magnetic, as well as voltage, relaxation. These non trivial "memory" effects are experimentally found in vortex matter[4] (and glassy systems[14]). They show a sort of *"imperfect memory"*, because they disappear when the time spent at $I_1$ becomes too long or when, for a given $t_2$, $I_1$ becomes too high, as shown in the inset of Fig. 6. This will be explained below.

## 4.2 History Dependent I-V

Time dependent properties also strongly affect the current-voltage characteristic. As in real experiments on vortex matter,[4] we let the system undergo a current step of hight $I_0$ for a time $t_0$ before starting to record the I-V by ramping $I$, as sketched in the inset of Fig. 7. Figure 7 shows (for $T = 0.1$) that the I-V depends on the waiting time $t_0$. The system response is "aging": the longer $t_0$ the smaller the response, a phenomenon known as "*stiffening*" in glass formers.[5,14] The model also reproduces the experimentally found time dependence of the critical current.[4] Usually, one defines an effective critical current, $I_c^{eff}$, as the point where $V$ becomes larger than a given threshold (say $V_{thr} = 10^{-5}$ in our case): one then finds that $I_c^{eff}$ is $t_0$ and $I_0$ dependent (like in experiments[4] $I_c^{eff}$ is slowly increasing with $t_0$, see Fig. 7).

In our model we can easily identify the characteristic time scales of the driven dynamics. After applying a drive, $I$, the system response, $V$, relaxes following a pattern with two very different parts: at first a rapidly changing non-linear response is seen, later followed by

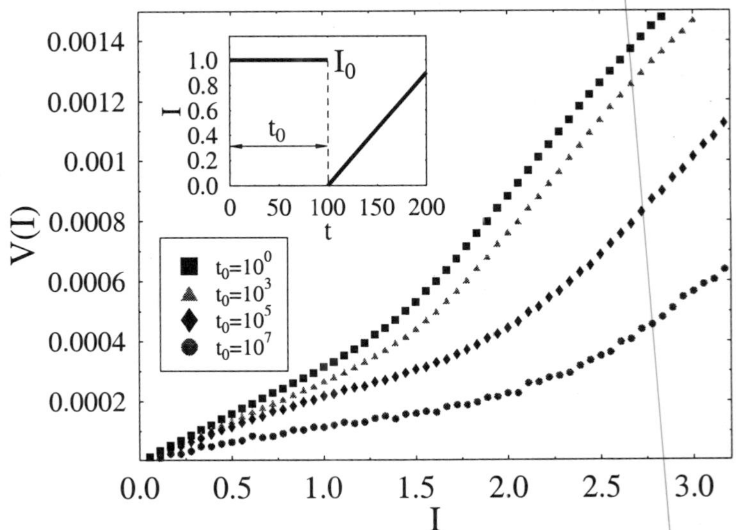


**Fig. 7** The I-V obtained at $T = 0.1$ by ramping $I$ after keeping the system in presence of a drive $I_0 = 1$ for a time $t_0$ as shown in the inset. The response, $V$, is "aging" (i.e. depends on $t_0$) and, more specifically, *stiffening*: it is smaller the longer $t_0$.

a very slow decrease towards stationarity. In agreement with experimental findings,[4] the latter has a characteristic double step structure, which asymptotically can be well fitted by stretched exponentials[15]: $V(t) \propto \exp(-t/\tau_V)^\beta$. The above long time fit defines the characteristic asymptotic scale, $\tau_V$, of relaxation. The exponent $\beta$ and $\tau_V$ are a function of $I$, $T$ and $N_{ext}$: in particular $\tau_V(I)$ decreases with $I$ and seems to approach a *finite plateau* for $I < I^*$, with $I^* \simeq O(1)$. In this sense, the presence of a drive $I$ makes the approach to stationarity faster and has an effect similar to an increase in $T$.

The above properties of $\tau_V$ are sufficient to explain the history dependent effects found in the experiments previously considered. For instance, the "imperfect memory", discussed in Fig. 6, is caused by the presence of a long, but finite, scale $\tau_V$ in the problem: for a given $I_1$ the system seems to be frozen whenever observed on times scales smaller than $\tau_V(I_1)$. Thus, if $t_2$ is short enough ($t_2 < \tau_V(I_1)$) the system preserves a strong "memory" of its state at $t_1$. The weakening of such a "memory" found for higher currents $I_1$ in Fig. 6, is also a consequence of the strong decrease of $\tau_V(I)$ with $I$. The phenomenon of "rejuvenation" (see Fig. 6) is, in turn, a consequence of the presence of the extremely fast first part of relaxation found in $V(t)$ upon applying a drive and of the above long term memory. The existence of the slow part in the $V(t)$ relaxation also affects the "stiffening" of the response in the I-V of Fig. 7, which is due to the non-stationarity of the vortex flow on scales smaller that $\tau_V$. Actually, in Fig. 7, for a given $I$ the value of $V$ on the different curves corresponds to the system being probed at different stages of its non-stationary evolution.

The origin of these time dependent properties of the driven flow, and in turn those of I-V's, traces back to the concurrent vortex creep and reorganisation of vortex domains. In fact, both with or without an external drive, the system evolves in presence of a non uniform density profile (the Bean profile, see lower inset Fig. 2) which in turn relaxes. An important discovery is that the characteristic times of voltage and magnetic relaxation are approximately proportional. This outlines that the non-stationary voltage relaxation is structurally related to the reorganisation of vortices during the creep (a fact confirmed by recent experiments[2]).

## 5. CONCLUSIONS

In conclusion, we showed that the replica mean field theory and Monte Carlo simulations of a schematic statistical mechanics lattice model[6] for vortices in type-II superconductors (a system of particles diffusing in pinning landscape) allow to depict a comprehensive framework of magnetic and transport properties experimentally observed in vortex matter. These results are also supported by some molecular dynamics simulations.[6,16]

We have seen that the model shows a reentrant phase diagram in the field-temperature plane (B, T), analogous to what observed in vortex matter. The system dynamics exhibits equilibration times which become huge around a crossover point, $T_g(N_{ext})$, and seem to diverge at a lower temperature, $T_c(N_{ext})$, where an "ideal" glass transition could be located. Around $T_g$ the system dynamics has a crossover corresponding to a change in microscopic vortex motion: from diffusive (above $T_g$) to strongly subdiffusive.[6] Related to that is a crossover from power law and Kohlrausch-Williams-Watts to logarithmic relaxation found in magnetic creep, which below $T_g$ has apparent "aging" features. The above "off equilibrium" scenario also explains the experimental finding (previously interpreted in terms of "quantum tunnelling" of vortices[1]) concerning the existence of a finite creep rate, $S_a$, when $T \to 0$.

We saw that magnetisation loops are typically found when $M$ is plotted as a function of the applied field, including a definite "second peak" when the Ginzburg-Landau parameter is not too low. The "second peak" is associated with a new phase transition in the system.

Concerning transport properties, we have shown how they are strongly related to creep. We also explained a broad set of "memory" effects in vortex flow of driven type II superconductors, such as "rejuvenation" and "stiffening" of the system response, "memory" and "irreversibility" in I-V characteristics.

The emerging unifying scenario of magnetic and transport properties in vortex physics has interesting relations with off equilibrium phenomena in other glass formers such as random magnets and supercooled liquids.

## ACKNOWLEDGMENTS

It is a pleasure to thank D. K. Jackson, G. Perkins, N. A. Lindop and H. J. Jensen for their essential contributions to this work. This work was partially supported by the TMR Network ERBFMRXCT980183, INFM-PRA(HOP) and INFM Parallel Computing Initiative.

## REFERENCES

1. G. Blatter, M. V. Feigel'man, V. B. Geshkenbein, A. I. Larkin and V. M. Vinokur, *Rev. Mod. Phys.* **66**, 1125 (1994); Y. Yeshurun, A. P. Malozemoff and A. Shaulov, *ibid.* **68**, 911 (1996); E. H. Brandt, *Rep. Prog. Phys.* **58**, 1465 (1995); L. F. Cohen and H. J. Jensen, *ibid.* **60**, 1581–1672 (1997).
2. M. Daeumling, J. M. Seuntjens and D. C. Larbalestier, *Nature* **346**, 332 (1990); Y. Yeshurun et al., *Phys. Rev.* **B49**, 1548 (1994); G. K. Perkins et al., *Phys. Rev.* **B51**, 8513 (1995); K. Ghosh et al., *Phys. Rev. Lett.* **76**, 4600 (1996); B. Khaykovich et al., *Phys. Rev. Lett.* **76**, 2555 (1996); D. Giller et al., *Phys. Rev. Lett.* **79**, 2542 (1997); S. B. Roy and P. Chaddah, *J. Phys. Cond. Mat.* **9**, L625 (1997); S. Kokkaliaris et al., *Phys. Rev. Lett.* **82**, 5116 (1999); Y. Paltiel et al., *Nature* **403**, 398 (2000); V. F. Correa, G. Nieva and F. de la Cruz, *Phys. Rev. Lett.* **83**, 5322 (1999).
3. E. L. Papadopoulou, P. Nordblad, P. Svedlindh, R. Schöneberger and R. Gross, *Phys. Rev. Lett.* **82**, 173 (1999).
4. W. Henderson, E. Y. Andrei, M. J. Higgins and S. Bhattacharya, *Phys. Rev. Lett.* **77**, 2077 (1996); Z. L. Xiao, E. Y. Andrei and M. J. Higgins, *Phys. Rev. Lett.* **83**, 1664 (1999); Z. L. Xiao, E. Y. Andrei, P. Shuk and M. Greenblatt, *ibid.* **85**, 3265 (2000).
5. M. D. Ediger, C. A. Angell and S. R. Nagel, *J. Phys. Chem.* **100**, 13200 (1996); J. P. Bouchaud, L. F. Cugliandolo, J. Kurchan and M. Mezard, in *Spin Glasses and Random Fields*, ed. A. P. Young (World Scientific, 1997).
6. D. K. Jackson, M. Nicodemi, G. Perkins, N. A. Lindop and H. J. Jensen, *Europhys. Lett.* **52**, 210 (2000); M. Nicodemi and H. J. Jensen, *J. Phys.* **A34**, L11 (2001); *Physica* **C341–348**, 1065 (2000); cond-mat/0007028; cond-mat/0006491; cond-mat/0006258; cond-mat/0101065.
7. K. E. Bassler and M. Paczuski, *Phys. Rev. Lett.* **81**, 3761 (1998).
8. Here $A_0^p = 0.3A_0$, $N_{c2} = 27$, $p = 1/2$ and, if not differently written, $A_1 = 0.28A_0$ (the $\mathcal{H}$ parameters can be related to the real material parameters[6]). $L$ is taken up to $L = 128$.
9. S. Katz, J. L. Lebowitz and H. Spohn, *Phys. Rev.* **B28**, 1655 (1983); B. Schmittmann and R. K. P. Zia, in *Phase Transition and Critical Phenomena*, eds. C. Domb and J. L. Lebowitz (Academic, 1995).
10. A. Coniglio and M. Nicodemi, *Phys. Rev.* **E59**, 2812 (1999).
11. A. C. Mota et al., *Physica* **C185–189**, 343 (1991); K. Aupke et al., *Physica* **C209**, 15 (1993); A. F. Th Hoekstra, *Phys. Rev. Lett.* **80**, 4293 (1998).
12. M. Nicodemi and A. Coniglio, *Phys. Rev. Lett.* **82**, 961 (1999); *J. Phys.* **C12**, 6601 (2000); M. Nicodemi, *Phys. Rev. Lett.* **82**, 3734 (1999); A. Coniglio, A. de Candia, A. Fierro and M. Nicodemi, *J. Phys.* **C11**, A167 (1999).

13. R. A. Hyman, M. Wallin, M. P. A. Fisher, S. M. Girvin and A. P. Young, *Phys. Rev.* **B51**, 15304 (1995).
14. F. Lefloch, J. Hamman, M. Ocio and E. Vincent, *Europhys. Lett.* **18**, 647 (1992); K. Jonason, E. Vincent, J. Hamman, J. P. Bouchaud and P. Nordblad, *Phys. Rev. Lett.* **81**, 3243 (1998).
15. At lower $T$ inverse logarithmic relaxations are found.[6]
16. H. J. Jensen, A. Brass and A. J. Berlinsky, *Phys. Rev. Lett.* **60**, 1676 (1988); F. Nori, *Science* **271**, 1373 (1996); A. B. Kolton, D. Dominguez and N. Grönbech-Jensen, *Phys. Rev. Lett.* **80**, 3061 (1999); C. Reichhardt, A. van Otterlo and G. T. Zimanyi, *Phys. Rev. Lett.* **84**, 1994 (2000).

# ON THE STATISTICAL PROPERTIES OF THE LARGE TIME ZERO TEMPERATURE DYNAMICS OF THE SK MODEL

GIORGIO PARISI
*Dipartimento di Fisica, INFM and INFN, Università La Sapienza*
*Piazzale Aldo Moro, Roma 00187*

**Abstract**

Here we study the zero temperature dynamics of the Sherrington Kirkpatrick model and we investigate the statistical properties of the configurations that are obtained in the large time limit. We find that the replica symmetry is broken (in a weak sense). We also present some general considerations on the synchronic approach to the off-equilibrium dynamics, that has motivated the present study.

## 1. INTRODUCTION

In recent years there have been many progresses in our understanding of the non-equilibrium dynamics[1,16] of the infinite Sherrington Kirkpatrick model[6,7] and of other glassy systems.[8–10] The aim of these studies is to compute the properties of the systems (e.g. the energy or the magnetization) as function of time, knowing the initial configuration at time zero.

These progresses have been done using a diachronic approach in which the evolution of the system is studied by writing closed equations for the correlation functions and response functions at different times.

An alternative synchronic approach has been put forward;[11,12] here one considers the probability distribution of the configurations of the system at a given time $t$ ($P_t(\sigma)$) and one writes it as

$$P_t(\sigma) \propto \exp(-\mathcal{H}_t(\sigma))\,. \tag{1}$$

In this approach there are two crucial steps:

- (a) The determination of the effective Hamiltonian $\mathcal{H}_t$. It is possible that in order to obtain qualitative and semiquantitative information it is not necessary to compute exactly $\mathcal{H}_t$, and an approximate knowledge is sufficient.
- (b) The computation of the statistical properties of the system at given effective Hamiltonian.

This approach may be successfully only if the effective Hamiltonian $\mathcal{H}_t$ (or a reasonable approximation to it) is not too complicated. While the diachronic approach is rather systematic, a good amount of guesswork is needed in the synchronic approach in order to chose a reasonable form of the effective Hamiltonian.

In order to get some intuition on the possible forms of the effective Hamiltonian we have studied in this note the statistical properties of the configurations that are obtained at large times in the SK model using a zero temperature dynamics. This problem has already been studied in the past[4,13,14,16–18] and it has its own interest. We have obtained some new and unexpected results, i.e. we have found that the replica symmetry is broken (in a weak sense[19]) and that the distribution of the local fields has unexpected properties.

In the second section we present some general considerations on the synchronic approach. These considerations have been the motivation of the present study, but they may be skipped by the reader interested only in the results of the paper for the SK model. In the third section we define the dynamics which we study and we recall some known results on the zero temperature solutions of the TAP equations.[20] In Sec. 4 we show the numerical results of our investigations and finally in the last section we present some tentative conclusions.

## 2. THE SYNCHRONOUS APPROACH

We consider a system whose variables satisfy some kind of stochastic or deterministic evolution equations. The probability distribution of the configurations at time zero is given:

$$P_0 \propto \exp(-\mathcal{H}_0)\,. \tag{2}$$

The quantity $\mathcal{H}_0$ plays the role of boundary condition.

In the simplest case the system is random at time zero and $\mathcal{H}_0 = 0$. Our aim is to compute $\mathcal{H}_t$ given $\mathcal{H}_0$. If we are rather lucky we can compute it exactly. This happens for example in the Gaussian SK[12] or in the spherical SK model. Before discussing the general approach it may be useful to show in details the soluble example.

### 2.1 The Gaussian Sherrington Kirkpatrick Model

In the Gaussian Sherrington Kirkpatrick[12] the Hamiltonian is

$$H = -\sum_{i,k} J_{i,k}\sigma_i\sigma_k\,, \tag{3}$$

where the variables $\sigma_i$ are real. At infinite temperature they are uncorrelated Gaussian variables with unit variance. The variables $J_{i,k}$ are Gaussian distributed and uncorrelated.

The dynamics is controlled by the Langevin equation:

$$\frac{d\sigma_i}{dt} = -\sigma_i + \beta F_i(t) + \eta_i(t)\,, \tag{4}$$

where the force is given by

$$F_i(t) = -\frac{\partial H}{\partial \sigma_i} = \sum_k J_{i,k}\sigma(t)_k\,. \tag{5}$$

The variables $\eta$ have an uncorrelated white noise distribution

$$\langle \eta_i(t)\eta_k(t')\rangle = 2\delta(t-t')\delta_{i,k}\,. \tag{6}$$

We suppose that at time zero the initial condition is simply $\sigma(0)_i = 0$. It is easy to show that the effective Hamiltonian at time $t$ is given by

$$\mathcal{H}_t(\sigma) = \sum_{i,k} \sigma_i f^t(J)_{i,k}\sigma_k\,, \tag{7}$$

where the function $f^t$ is given by

$$f^t(z) = \frac{z}{1-\exp(-tz)} \tag{8}$$

and $f^t(J)_{i,k}$ denotes the $i$, $k$ element of the matrix $f^t(J)$.

The proof of this statement may be obtained by noticing that the evolution equations become much simpler in the basis where the matrix $J$ is diagonal.[12,21] In this basis the components of the variables $\sigma$ along the eigenvectors of the matrix $J$ are uncorrelated. Their variance can be computed and in this way one obtains Eq. (8).

## 2.2 The General Approach

In the general case the system has an Hamiltonian $H_J(\sigma)$ and the evolution is described by a Langevin equation of the form in Eq. (4) or by some analogous equation for discrete systems.

We suppose that the effective Hamiltonian can be approximated as

$$\mathcal{H}_0 = g(\sigma,\, J,\, \lambda(t))\,, \tag{9}$$

where $g$ is a preassigned function which depend on $M$ variables $\lambda_\alpha(t)$ ($\alpha = 1 \ldots M$); $M$ may also be infinite. In this framework the effective Hamiltonian depends on time only through the variables $\lambda(t)$.

If we suppose to know the function $g$ (or a good approximation to it), the problem consists in finding the appropriate values of the functions $\lambda_\alpha(t)$.

There are two strategies which we can follow:

- (a) We choose a set of observables $O_\alpha(\sigma, J)$ for $\alpha = 1 \ldots M$ and we impose the validity of the equations:

  $$\frac{d\langle O_\alpha\rangle_t}{dt} = \sum_\gamma \frac{\partial\langle O_\alpha\rangle_t}{\partial\lambda_\gamma}\frac{d\lambda_\gamma}{dt}\,, \tag{10}$$

  where the l.h.s. is computed using the Langevin equations.

  If we are near to equilibrium it is convenient to write

  $$\mathcal{H}_t = \beta H + \sum_\gamma O_\gamma\lambda_\gamma(t)\,. \tag{11}$$

For small values of $\lambda_\gamma(t)$ we can linearize the equations and using the Langevin equations we get

$$\sum_\gamma \left\langle \sum_i \frac{\partial O_\alpha}{\partial s_i} \frac{\partial O_\gamma}{\partial s_i} \right\rangle \lambda_\gamma(t) = \sum_\gamma \langle O_\alpha O_\gamma \rangle_c \frac{d\lambda_\gamma(t)}{dt} \,, \tag{12}$$

where the expectation values are computed at equilibrium and $\langle\,\rangle_c$ denotes the connected expectation value.

- (b) We write down the Fokker Plank equation for the probability $P(t, \sigma)$:

$$\frac{dP(t, \sigma)}{dt} = \mathcal{L}P(t, \sigma)\,, \tag{13}$$

where $\mathcal{L}$ is the appropriate linear operator. A variational principle may be used to chose the variables $\lambda_n(t)$. If we call $\mathcal{P}(\lambda(t), \sigma)$ the probability distribution corresponding to the effective Hamiltonian 7. For example we can impose that the quantity

$$\sum_{\{\sigma\}} \left( \sum_\gamma \frac{\partial P(t)}{\partial \lambda_\gamma} \frac{d\lambda_\gamma}{dt} - \mathcal{L}P(t) \right)^2 \tag{14}$$

takes the minimum value.

In the past this point of view has been advocated in the study of turbulence and both strategies have been followed.[22,23]

In their original work Cooley and Sherrington[11] have analysed the dynamics of the SK model with Ising spins. They have taken $M = 2$. In the case of all spins equal to 1 at time zero, they have

$$g(\sigma, J, \lambda(t)) = \lambda_1(t) \sum_{i,k} J_{i,k}\sigma_i\sigma_k + \lambda_2(t) \sum_i \sigma_i \,. \tag{15}$$

Reasonable results have been obtained especially at the short times.

It is clear that the previous equation (with (M=2)) is only a first approximation. In the case of the Gaussian SK model it was proved[12] that it does not reproduce the exact result. Indeed in this last case we have seen that the correct expression (for $\sigma_i(0) = 0$) is given by

$$g(\sigma, J, \lambda(t)) = \sum_{\alpha=1,\infty} \lambda_\alpha(t) \sum_{i,k} (J^\alpha)_{i,k}\sigma_i\sigma_k \,, \tag{16}$$

where as usual $(J^\alpha)_{i,k}$ is the matrix element of the matrix $J$ to the power $\alpha$.

For the Ising case is possible that the previous formula is not adequate and that other terms may be present as

$$\sum_i \sigma_i (\sum_k J_{i,k}\sigma_k)^3. \tag{17}$$

## 3. ZERO TEMPERATURE DYNAMICS AND THE TAP EQUATIONS

Here we consider the SK Ising model with the Hamiltonian in Eq. (3), where the $J$ are random independent Gaussian variables with variance $1/N$, $N$ being the total number of Ising spins.

We are interested in studying the statistical properties of the configurations that are obtained by the dynamics at zero temperature at large times. In this case the dynamics is such to orient the spins with the effective field $h_i = \sum_k J_{i,k}\sigma_k$. In other words one set

$$\sigma_i(t+1) = \mathrm{sign}(h_i(t)) \,. \tag{18}$$

Different algorithms differs in the order in which the rule Eq. (18) is applied.

The algorithm stops when in all the sites the following zero temperature TAP equation is satisfied

$$\sigma_i(t) = \mathrm{sign}(h_i(t)) \,. \tag{19}$$

The asymptotic configuration at large times depends on the initial configuration. We are will study the ensemble where each solution of zero temperature TAP Eq. (19) is weighted with the probability of being obtained by the algorithm with a random choice of the initial configuration. In other words, we weight each solution with the size of its attraction basin.

It is known that in the large $N$ limit the energy density of the asymptotic configuration does not depend on the starting configuration with probability one and it is higher that the ground state energy. Indeed for a sequential algorithm it is about $E_S = -0.715$,[14] while the ground state energy is $E_0 = -0.7633$.

The simplest hypothesis would be that the set of configurations weighted with the attraction basin statistically coincides with the set of all the solutions of the zero temperature TAP equations with energy equal to $E_S$.

A precise computation of the statistical properties of the solutions of the zero temperature TAP equations for this value of the energy has not been done. However we can use the information we have for energies greater that $E_{RSB} = -0.672$ (at which an exact computation can be done) and for energy equal to the ground state the value.[15] The energy $E_S$ is intermediate among the two so that an educated guess can be done. The maximum overlap among two generic solutions should be 0 at $E_{RSB}$ and it should be 1 at $E_0$ so that we could guess its value around $0.6 - 0.8$. This value is purely indicative.

The probability distribution of the effective field $P(h)$ is approximatively a shifted Gaussian for energies greater that $E_{RSB}$. There are no indications that $P(0)$ should be zero for $E < E_{RSB}$.

We will see later that some these expectations are in variance with our results coming from numerical simulations. We conclude that the original hypothesis is wrong and that the generic solution of the TAP equations, weighted with its attraction basin, is not the generic solution of the TAP equations of the appropriate energy.

Before presenting the numerical results we will describe the three minimisation algorithms that we have used: the sequential algorithm, the greedy algorithm and the reluctant algorithm.

- The Sequential Algorithm.

  This is the simplest algorithm to implement. One cycle of the algorithm consists in applying the rule (18), sequentially for increasing $i$ from $i = 1$ to $i = N$. We repeat the cycle up to the moment when a solution of the TAP equation is reached. This algorithm corresponds to the zero temperature limit of an usual Monte Carlo or heath bath dynamics.

- The Greedy Algorithm.

  This is the simplest algorithm to understand analytically. One step of the algorithm consists in applying the rule (18), for that $i$ which minimises $\sigma_i h_i$. We repeat the step up to the moment at which a solution of the TAP equation is reached. This algorithm corresponds to the zero temperature limit of the Glauber dynamics.

- The Reluctant Algorithm.

  This algorithm is the opposite of the greedy algorithm. One step of the algorithm consists in applying the rule (18), for that $i$ which maximises $\sigma_i h_i$ among those $i$ such that $\sigma_i h_i$ is negative. One repeats the step up to the moment at which a solution of the TAP equation is reached. At each step the energy decreases, but it decreases of the smallest possible amount. As we shall see later this slower algorithm is the most effective in finding the configurations of smallest energies.

## 4. NUMERICAL RESULTS

### 4.1 The Sequential Algorithm

We have studied numerically the zero temperature dynamics of the SK model for systems with $N$ in the range 16–1600. We report firstly the results for sequential updating; the results for the other algorithms (the greedy and the reluctant) are not qualitatively different and they will not be discussed in detail.

For each value of $N$ many instances of the system have been generated (from 10000 at $N = 16$ to 100 at $N = 1600$). For each choice of the coupling $J$ we have followed the dynamics starting from $M$ different initial configurations.[a]

For each configuration we have recorded the energy, the distribution of the forces $h_i$. For a pair of configurations (i.e. $\sigma_i$ and $\tau_i$ for $i = 1, N$), we can define the overlap

$$q = \left| \frac{\sum_{i=1,N} \sigma_i \tau_i}{N} \right| . \tag{20}$$

We have computed the overlap among all the $\frac{M(M-1)}{2}$ pairs of the $M$ final configurations with the same couplings.

In Fig. 1 we show the expectation value of the energy density $E$ as function of $N$. The data are has been fitted as

$$E = E_S + \frac{C}{N^\alpha} , \tag{21}$$

where the values of the parameters of the fit are $E_S = -0.715$, $C = 0.25$ and $\alpha = 0.33$. The value of $E_\infty$ differs from the ground state energy (which is $-0.7633$). The value of $\alpha$ is similar to the one which is obtained for the $N$ dependence of the ground state and it is compatible with being equal to $1/3$.

The fluctuations of the energy density go to zero when $N$ goes to infinity approximately proportionally to $1/N$ (a best fit gives $1/N^{0.98}$). The energy is a self average quantity as expected.

The expectation value of $q$ as function of $N$ goes to zero approximately as $N^{-0.9}$ when $N \to \infty$ (see Fig. 2). A behaviour of the type $\frac{\ln(N)}{N}$ cannot be excluded. Also the overlap is a self-averaging quantity: in the infinite volume limit it goes to zero with probability 1.

[a]The average results do not depend on the value of $M$. In order to gain computer time, we found convenient to take a value of $M$ equal to $N^{1/2}$.

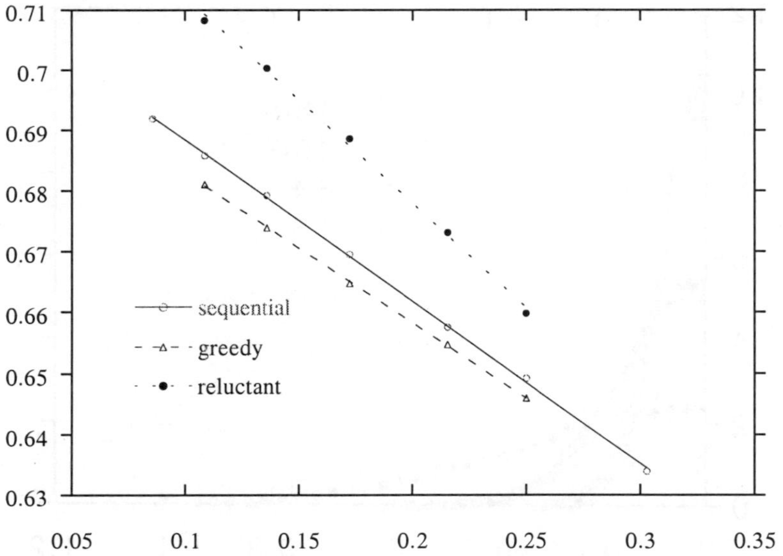


**Fig. 1** Minus the average energy as function of $N^{-1/3}$ for the sequential, greedy and reluctant algorithms. The lines are linear fits.

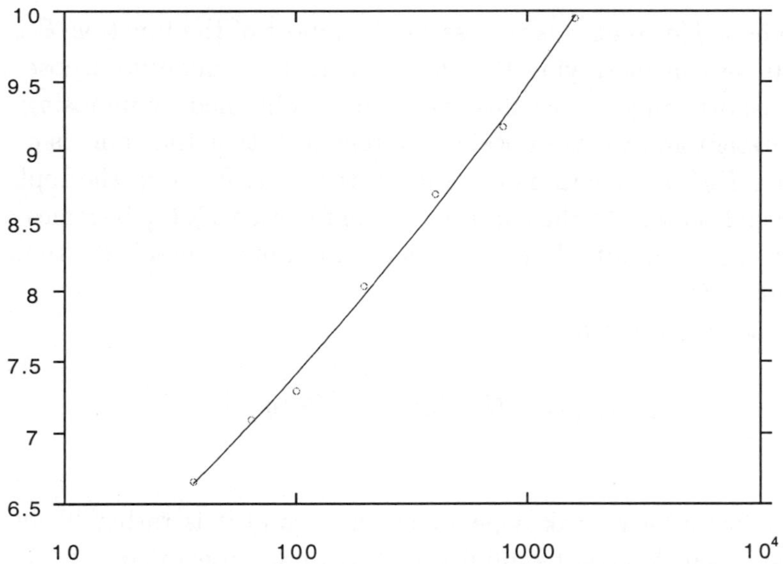


**Fig. 2** The average value of $N\langle q^2\rangle$ as function of $N$ in the case of the sequential algorithm. The line is a power law fit with exponent 0.1; a logarithmic dependance (a straight line) is also possible.

These results are in good agreement with older studies.[14,17] The surprise comes from the study of the distribution function $P(q)$. If we plot the function $P(q)$ for different sizes we do not see anything unexpected (see Fig. 3). The distribution becomes more and more concentrated at $q = 0$ when $N \to \infty$.

The crucial point is however how fast the function $P(q)$ decreases with $N$ at fixed $q$. More precisely can define the function

$$f(q) = -\lim_{N\to\infty} \frac{\ln(P_N(q))}{N}\,. \tag{22}$$

In the usual equilibrium SK model the function $f(q)$ is equal to zero in the interval $[0\text{-}q_{EA}]$ and it is different from zero for $q > q_{EA}$.

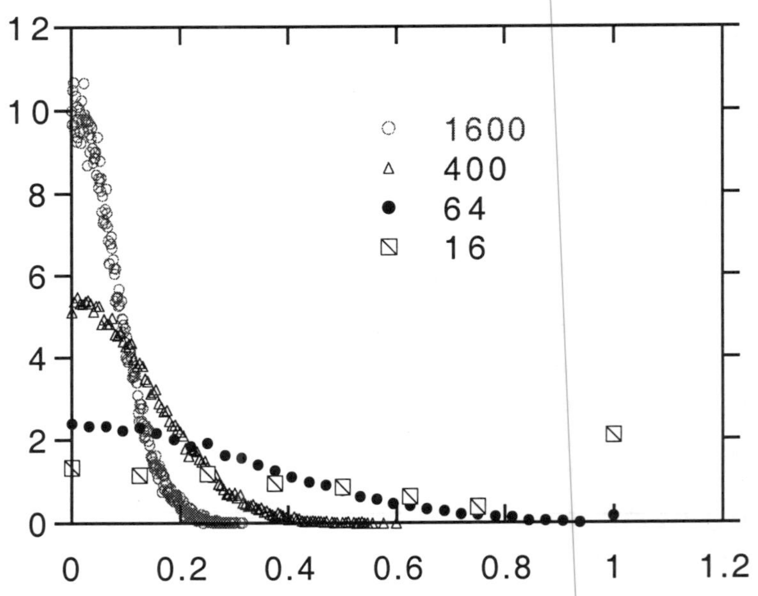


**Fig. 3** The function $P(q)$ for different values of $N$ ($N$ = 16, 64, 400, 1600) in the case of the sequential algorithm.

The region where $f(q)$ is zero is the essential support of the function $P(q)$. In the region where $f(q) = 0$ we can modifying the Hamiltonian by a quantity whose density goes to zero with $N$ (in short range model this effect can be obtained by imposing the appropriate boundary conditions) and we can obtain a system such that the value is $q$.

If the function $P_N(q)$ becomes non-trivial in the limit $N \to \infty$ the replica symmetry is broken in the usual sense. On the contrary, if the function $P_N(q)$ becomes a delta function in the limit $N$ going to infinity, but $f(q)$ is zero in a finite interval, the replica symmetry is broken in a weak sense.

In Fig. 4 we plot the quantity

$$r_N(q) = N^{-\nu} \ln(\int_q^1 dq' P_N(q')) \,, \tag{23}$$

with $\nu = 2/3$.

This quantity has a very weak dependence on $N$ and it is rather likely that it goes to a nonzero value when $N$ goes to infinity in the whole interval $0-1$. If we accept this conclusion, we find that $P_N(q)$ decreases slower than an exponential for all values of $q$ less than 1 and therefore the function $f(q)$ is always zero in the whole interval $0-1$. We cannot exclude that $r_N(q)$ diverges also for a value of $q$ less that one. In any case replica symmetry is broken in a weak sense.

The other quantity for which we obtain surprising results is the probability distribution of the fields $h_i = \sum_k J_{i,k}\sigma_k$. The function $P_N(h)$ depends weakly on $N$ (the expectation value of $h$ is the energy.

The function $P_N(h)$ seem to vanish linearly at $h = 0$. In order to show this behaviour in Fig. 5 we plot the function

$$\tilde{P}_N(h) \equiv \frac{P_N(h)}{h + \Delta} \,, \tag{24}$$

for $N = 1600$ where the constant $\Delta$ (which vanishes when $N \to \infty$) has been chosen in such a way to have a smooth behaviour at small $h$. We have found that $\Delta = 2N^{-1/2}$ is a good

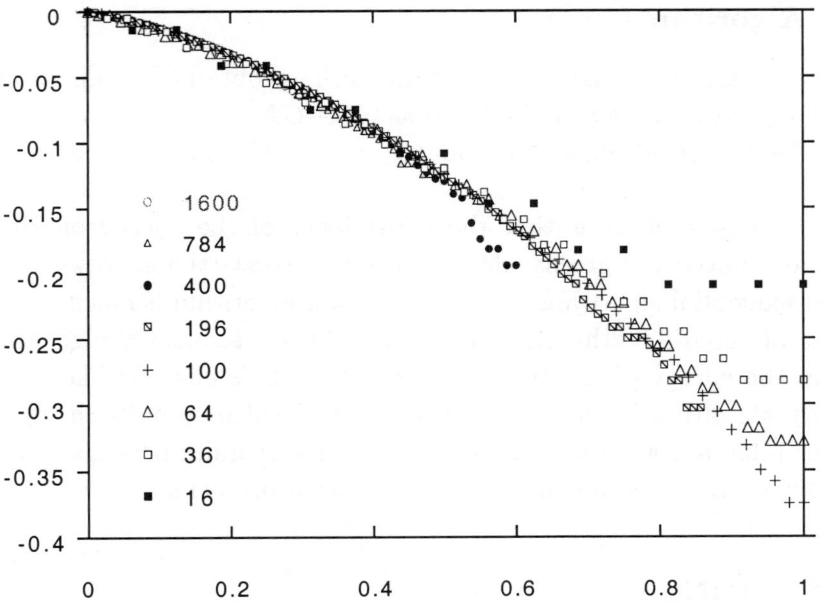


**Fig. 4** The results obtained with the sequential algorithm for the function $r_N(q)$ for $N$ and $q$ ranging from 16 to 1600 and from 0 to 1 respectively.

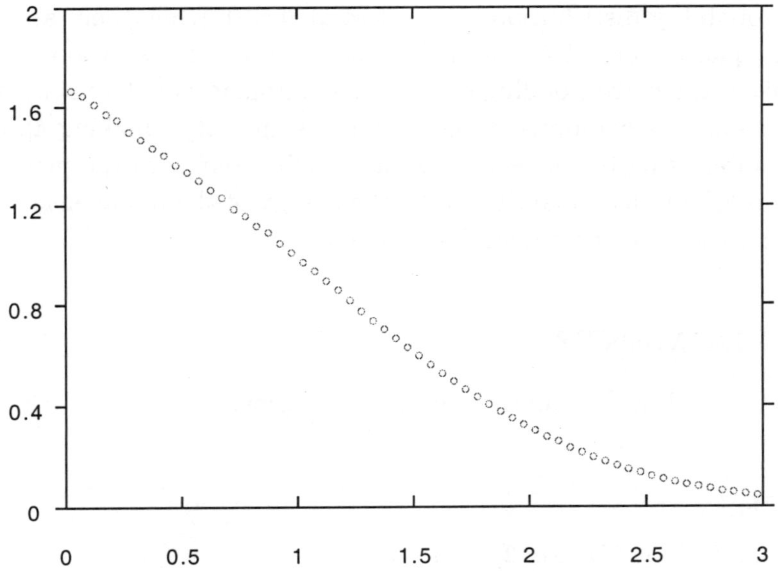


**Fig. 5** The function $\tilde{P}(h)$ in the case of the parallel algorithm for $N = 1600$.

choice. This value of $\Delta$ is natural, indeed the variable $h$ may take the values $(n + 1/2)\Delta$, with $n$ integer.

A linear behaviour of the function $P(h)$ is similar to the one observed at thermal equilibrium at zero temperature, where $P(h)$ is proportional to $h$.

We do not have a simple explanation for such behaviour. The process of minimizing the energy of a given spin has the side effect of decreasing also the energy of the other spins, pushing the distribution of $h$ far from zero. However it is unclear how to transform this observation in a quantitative prediction.

## 4.2 Other Algorithms

We have analyzed also the other dynamics ad we have obtained comparable results. We find that in all cases the energy can be fitted as $E_\infty + CN^{-1/3}$. We have that $E_\infty = -0.707$ and $C - 0.24$ for the greed algorithm and $E_\infty = -0.747$ and $C = 0.34$ for the reluctant algorithm.

The only remarkable effect is that the dependence of the energy on the algorithm is counterintuitive. Indeed the greedy algorithm is the worst (the energy is higher of about 0.007 than the sequential algorithm) and the reluctant algorithm is much better (the energy $-0.746$ is lower of about 0.03 than the that obtained using sequential algorithm and distant only 0.02 from the energy of the true ground state). This effect is likely due to the fact that the greedy algorithm is more easily trapped in local minima with high energies and that the slower reluctant algorithm avoids these traps. It quite possible that the reluctant algorithm will work much better in other minimization problems.

## 5. CONCLUSIONS

The probability distribution of the configurations obtained by the zero temperature dynamics has rather peculiar statistical properties. The results obtained are very different from those coming taking a generic solution of the TAP equations with the appropriate energy. The maximum allowed value of $q$ (from the thermodynamic point) is 1, or very near to 1. Moreover the probability distribution of $h$ is zero at $h = 0$, while there are no reasons that $P(h)$ should be equal to zero at $h = 0$ in the case if the generic solution.

It is rather likely the corresponding effective Hamiltonian is not so simple and its statistical properties should be computed using a replica symmetry breaking approach.

It would also interesting to connect the small $h$ behaviour with the power laws observed in the dependence of physical quantities (like the energy of the remanent magnetization) or the number of iterations in the sequential dynamics.

## ACKNOWLEDGMENTS

It is a pleasure to thank E. Marinari for useful discussions.

## REFERENCES

1. J.-P. Bouchaud, *J. Phys. (France)* **2**, 1705 (1992).
2. G. Parisi and F. Ritort, *J. Phys.* **A26**, 6711 (1993).
3. L. F. Cugliandolo, J. Kurchan and F. Ritort, *Phys. Rev.* **B49**, 6331 (1994).
4. M. Opper, *Phys. Rev. Lett.* **71**, 71 (1992).
5. L. F. Cugliandolo and J. Kurchan; *J. Phys.* **A27**, 5749 (1994).
6. M. Mézard, G. Parisi and M. A. Virasoro, *Spin Glass Theory and Beyond* (World Scientific, Singapore, 1987).
7. G. Parisi, *Field Theory, Disorder and Simulations* (World Scientific, Singapore, 1992).
8. S. Franz and M. Mézard, *J. Phys.* **A26**, 209 (1994).
9. L. F. Cugliandolo and J. Kurchan, *Phys. Rev. Lett.* **71**, 93 (1993), and references therein.
10. E. Marinari, G. Parisi and F. Ritort, *J. Phys. A: Math. Gen.* **27**, 7615 (1994).
11. A. C. C. Cooley and D. Sherrington, *Phys. Rev.* **E49**, 1921 (1994).
12. A. C. C. Cooley and S. Franz, cond-mat preprint 9406082.
13. G. Parisi, N. Parga and M. A. Virasoro, *J. Phys. Lett.* **45**, L1063 (1984).
14. N. Parga and G. Parisi, ICTP preprint (1985), unpublished.

15. A. J. Bray and M. A. Moore, *J. Phys.* **C12**, L441 (1979).
16. G. Ferraro, *Infinite Volume Relaxation in the Sherrington-Kirkpatrick Model*, cond-mat/9407091.
17. N. Parga, *J. Phys.* **A19**, 51 (1986).
18. S. Cabasino, E. Marinari, G. Parisi and P. Paolucci, *J. Phys. Math. Gen.* **A21**, 4201 (1988).
19. G. Parisi and M. A. Virasoro, *J. Phys.* **50**, 3317–3329 (1989).
20. J. Thouless, P. W. Anderson and R. G. Palmer, *Phil. Mag.* **35**, 137 (1977).
21. L. F. Cugliandolo, J. Kurchan and G. Parisi, *J. Phys. I (France)* **4**, 1691 (1994), and references therein.
22. R. Biferale, R. Benzi and G. Parisi, *Europhys. Lett.* **6**, 6 (1992).
23. S. Migdal, *Phys. Rev.* **E5**, 123 (1992).

# THE RELATIONSHIP BETWEEN THE SCALING PARAMETER AND RELAXATION TIME FOR NON-EXPONENTIAL RELAXATION IN DISORDERED SYSTEMS

YA. E. RYABOV and YU. FELDMAN
*School of Applied Science, The Hebrew University of Jerusalem, Givat Ram, Jerusalem, 91904, Israel*

**Abstract**

A memory function equation and scaling relationships were used for the physical interpretation of the Cole-Cole exponent. The correspondence between the relaxation time, the geometrical properties, the self-diffusion coefficient and the Cole-Cole exponent was established. Using this approach the dielectric relaxation spectra of the polymer-water mixtures and the glass transition process in the *nylon 6,6* quenched, crystalline and micro-composite samples were analyzed.

## 1. INTRODUCTION

It is well known[1] that experimental dielectric response in frequency domain for most complex systems cannot be described by a simple Debye expression with a single dielectric relaxation time. A number of empirical relationships have been proposed in order to fit the broad dielectric spectrum of these systems. In the most general way such non-Debye dielectric behavior can be described by the so-called Havriliak-Negami formula[2]

$$\varepsilon^*(\omega) = \varepsilon_\infty + \frac{\varepsilon_s - \varepsilon_\infty}{(1 + (i\omega\tau)^\alpha)^\beta} \,. \tag{1}$$

In Eq. (1) $\alpha > 0$ and $\beta > 0$, $\omega$ is the cyclic frequency, $i$ is the imaginary unit, $\varepsilon_\infty$ is the high frequency limit of the complex dielectric permittivity $\varepsilon^*(\omega)$, $\varepsilon_s$ is a static dielectric permittivity and $\tau$ is a mean relaxation time.

The parameters $\alpha$ and $\beta$ reflects low and high frequency power-law asymptotics of the Havriliak-Negami formula and are usually referred to as shape parameters of the relaxation peak. The experimental data shows that $\alpha$ and $\beta$ are strictly dependent on temperature, structure, composition and other controlled physical parameters.[3] However, until recently time the reasons underlying such dependencies were not clear.[1,3,4] The $\alpha$ and $\beta$ were discussed as the parameters of the distribution of the relaxation times or mentioned as broadening parameters without further discussion. Over the last years significant progress in this direction has been made. A lot of efforts[5–7] were undertaken in order to understand the relaxation dynamics of glass forming liquids and different polymer mixtures before the paper by G. Katana et al.[8] where the concentration fluctuation model was introduced. In the framework of this model a reasonable explanation of the $\alpha$-relaxation process in homogenous polymer mixtures was presented. Although relaxation peak broadening was discussed, no relationships have been derived for exponents $\alpha$ and $\beta$ versus structural or dynamical parameters.[8] This kind of non-exponential response is observed in numerous physical systems and actually describes not only the dielectric properties. For instance, in the works of F. Yonezwara and co-authors[9,10] the Monte Carlo simulation of random walks on two-dimensional fractal structures was carried out. It was shown there that the complex susceptibility of this process has a power-law frequency asymptotic similar to the Havriliak-Negami type. Moreover, the model coupling theory[11] ascribes to the power-law exponents universal nature by virtue of universality of the correlation functions for different dynamical variables.

There are also a large number of works where mathematical formalism of fractional calculus[12,13] has been applied to the anomalous diffusion and relaxation problems. The physical applications as well as the mathematical issues of fractional calculus were recently investigated.[14] The review by R. Metzler and J. Klafter[15] discusses anomalous diffusion in detail. In the papers by R. R. Nigmatullin and Ya. E. Ryabov[16] the relaxation equation for the Havriliak-Negami processes was obtained. Nevertheless, at this time there is no unambiguous clear understanding of the non-exponential response in complex disordered heterogeneous systems. Therefore an understanding of the relationship between the $\alpha$ and $\beta$ exponents and the physical properties of the systems is very important.

## 2. THEORY

### 2.1 Memory Function Equation

Let us consider in this paper the specific case of the Havriliak-Negami process with $\beta = 1$, and $0 < \alpha < 1$. This is the so-called Cole-Cole equation.[17] The complex susceptibility $\chi(i\omega)$ in this case reads as

$$\chi(i\omega) = \frac{\chi_0}{1 + (i\omega\tau)^\alpha}, \tag{2}$$

where $\chi_0$ is a static susceptibility. The imaginary part of the complex susceptibility (2) represents itself as a rather wide symmetric peak with a maximum situated at the $\omega = 1/\tau$. This peak narrows with increasing $\alpha$ and finally reaches Debye's shape (the usual exponential relaxation) at the limit $\alpha = 1$.

In time domain the relaxation process could be described by normalized correlation function $f(t)$. In terms of the linear response approach the $f(t)$ should obey the equation

$$\frac{df(t)}{dt} = -\int_0^t m(t-t')f(t')dt', \tag{3}$$

where $m(t)$ is the memory function, and $t$ is the time variable. In the case of dielectric relaxation, $f(t)$ can be considered as a dipole correlation function. The specific type of the memory function depends on the features of interaction between the relaxing system and the surroundings.[18] For instance, the exponential relaxation corresponds to the $m(t) \sim \delta(t)$, where $\delta(t)$ is the Dirac delta function.

The representation of any analytical function $g(t)$ in frequency domain may be obtained through the Laplace transform[19]

$$G(z) = \mathcal{L}[g(t)] = \int_0^\infty g(t)e^{-zt}dt\,, \tag{4}$$

where $z = \sigma + i\omega$ is the so-called complex frequency or Laplace parameter. Thus, Laplace transform of Eq. (3) leads to

$$zF(z) - 1 = -M(z)F(z)\,, \tag{5}$$

where $F(z)$ and $M(z)$ are Laplace images of $f(t)$ and $m(t)$. Combining (2) with (5) and taking into account the relationship between the complex susceptibility and the correlation function,[20] $\chi(z)/\chi_0 = 1 - zF(z)$, one can obtain the Laplace image of the memory function for the Cole-Cole process as follows

$$M(z) = z^{1-\alpha}\tau^{-\alpha}. \tag{6}$$

Since $0 < \alpha < 1$, the exponent in Eq. (6) $1 - \alpha > 0$ and there is no univalent analytical representation of function $M(z)$ from Eq. (6) in time domain.[19] However, it is possible to obtain the dynamic relaxation equation for the Cole-Cole process in time domain by using fractional integro-differential operators.

According to definition,[12,13] the Riemann-Liouville fractional integration $D_0^{-\nu}$ and derivation $D_0^{\gamma}$ operators are

$$D_0^{-\nu}[g(t)] = \frac{1}{\Gamma(\nu)}\int_0^t (t-t')^{\nu-1}g(t')dt', \quad 0 < \nu \leq 1\,, \tag{7}$$

$$D_0^{\gamma}[g(t)] = \frac{d}{dt}D_0^{\gamma-1}[g(t)]\,, \quad 0 < \gamma \leq 1\,, \tag{8}$$

where $\Gamma(\nu)$ is the Gamma function. Therefore, Eq. (5) with the memory function (6) in time domain can be rewritten as follows

$$\begin{aligned} \frac{df(t)}{dt} &= -\tau^{-\alpha}D_0^{1-\alpha}[f(t)] - C\,, \\ C &= D_0^{-\alpha}[f(t)]\Big|_{t=+0} \equiv 0\,. \end{aligned} \tag{9}$$

Equation (9) was already discussed elsewhere[14,16,21] as a phenomenological representation of the relaxation equation for the Cole-Cole law. Let us examine now the right-hand-side of Eq. (9) and especially the fractional derivation operator $D_0^{1-\alpha}$. The fractional differentiation and integration operators were studied very closely in the book by A. Le Méhauté and co-authors.[22] It was shown there that the fractional integration of any smooth analytical function could be considered as its averaging on a fractal set. Thus for Eq. (9), the fractional differentiation operator $D_0^{1-\alpha}$ would be considered a result of the statistical averaging of a discrete memory functions set. As such, each discrete memory function consists of a sum of delta functions, $\delta(t-t_i)$, where sequence $t_i$ constructs a fractal set with *the fractional dimension equal to the Cole-Cole exponent* $\alpha$. This kind of memory function is an intermediate situation between total memory ($m(t) = Const$) and the absence of memory ($m(t) \sim \delta(t)$). Schematically this procedure of averaging is represented in Figs. 1 and 2. In this case the resulting memory function obeys the power-law time scaling relationship with scaling exponent $\alpha$. From another point of view these interrupted discrete memory functions could be considered as a renewal process[23] with asymptotic power-law dependence.[11]

Thus, one can summarize that the Cole-Cole process would be produced by scaled interrupted interactions between the statistical reservoir and the independently relaxing units.

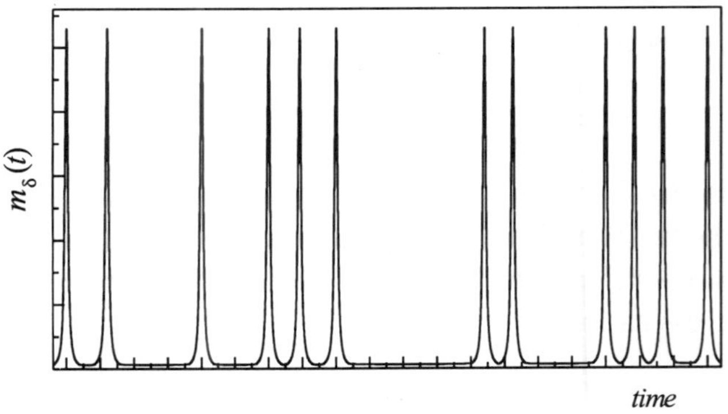


**Fig. 1** Schematic picture representing the "*interrupted*" discrete memory function $m_\delta(t)$, which consists of a sum of delta functions $\delta(t-t_i)$.

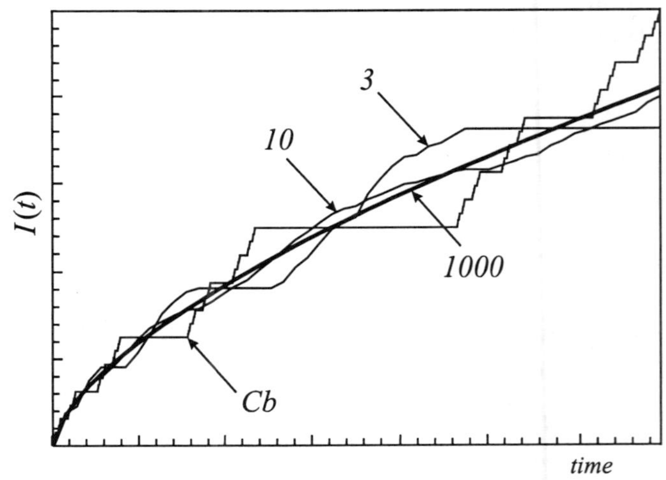


**Fig. 2** Schematic presentation of the interrupted discrete memory function "*statistical*" averaging. Here $I(t) = \int_0^t \mathcal{M}(t')dt'$, where $\mathcal{M}(t)$ is a "*statistically averaged*" memory function. The curve *Cb* corresponds to the situation when the statistical ensemble consists of only one system, and that the memory function of this system has a Cantor set distribution of $t_i$ moments. Curve 3 represents the situation when the statistical ensemble contains three systems of this kind. Curve 10 corresponds to the ten systems in the ensemble. Curve 1000 describes the statistical ensemble that consists of one thousand systems. The last one has the actual power-law behavior $I(t) \sim t^{\ln 2/\ln 3}$.

## 2.2 Scaling Relationship

In general the scaling properties of the memory function (6) can be provided by different physical causes. Moreover, there is a computer simulation proof[9,10] that the anomalous diffusion on fractal structures exhibits a Cole-Cole behavior. Therefore, one can assume that the scaling property of the memory function has its origin in the geometrical self-similarity of the investigated system. In this case the exponent $\alpha$ depends on both space and time scalings. The equation

$$\alpha = \frac{\ln(N)}{\ln(\xi)} \tag{10}$$

with time scaling parameter $\xi$ and space scaling parameter $N$ is the mathematical expression of the above hypothesis.

The time scaling parameter $\xi$ is a dimensionless characteristic time quantity of the Cole-Cole process. However, there already exists a characteristic time constant of the process — the mean relaxation time $\tau$. Thus, the scaling parameter $\xi$ and the mean relaxation time could be related to each other by

$$\xi = \frac{\tau}{\tau_0}, \tag{11}$$

where the constant $\tau_0$ is the cutoff time of the scaling in time domain.

The space scaling parameter $N$ is actually the number of points where the relaxing units are interacting with the statistical reservoir. The assumption of geometrical self-similarity of the considered system implies that this number is

$$N = G\left(\frac{R}{R_0}\right)^{d_G}, \tag{12}$$

where $d_G$ is a fractal dimension of the point set where relaxing units are interacting with the statistical reservoir, $R$ is the linear size of the volume where the motion of single relaxing unit occurs, $R_0$ is the cutoff size of the scaling in the space, and $G$ is a geometrical coefficient approximately equal to unity, which depends on the shape of the system heterogeneity. For instance, the well-known two-dimensional recurrent fractal Seprinski carpet[24] has $d_G = \ln 8/\ln 3 \approx 1.89$, $G = \sqrt{3}/4 \approx 0.43$.

Non-exponential behavior is commonly observed[9–11,25,26] for complex non-crystalline materials. The relaxation process in this type of material is usually accompanied by diffusion. Therefore, the mean relaxation time for such disordered systems is the time during wich the relaxing structural unit moved a distance of $R$. The Einstein-Smoluchowski theory[27] gives the relationship between $\tau$ and $R$ as follows:

$$R^2 = 2d_E D_s \tau\,, \tag{13}$$

where $D_s$ is the self-diffusion coefficient, and $d_E$ is the Euclidean dimension. Thus, by substitution (11), (12) and (13) into (10), one can get the relationship between the Cole-Cole parameter $\alpha$ and the mean relaxation time $\tau$ in the form

$$\alpha = \frac{d_G}{2}\frac{\ln(\tau\omega_s)}{\ln(\tau/\tau_0)}, \tag{14}$$

where

$$\omega_s = G^{2/d_G}\frac{2d_E D_s}{R_0^2}$$

is the characteristic frequency of the self-diffusion process.

# 3. EXPERIMENTAL EVIDENCE

## 3.1 Composite Material

The Cole-Cole process in particular is inherent to the glass-transition in some polymers[9–11]. The glass-transition relaxation process in these systems is due to the motion of some section of a polymer chain that is accompanied by diffusion. The dielectric spectroscopy measurements[28] were carried out for *nylon 6,6*. There were three samples: a quenched *nylon 6,6* in amorphous phase ($QN$), a sample with a presence about of 60% crystalline phase which we call here *crystalline nylon 6,6* ($CN$) and a *microcomposite crystalline nylon 6,6* sample ($MCN$) with incorporated *kevlar 29* fibers. The diameter of a fiber is 12.5 $\mu$m. They occupy about 0.1% – 0.2% of the sample volume and are practically parallel to each other and in the plane of the specimens[29]. The measurements were performed in the frequency range 20 Hz – 1 MHz and temperature interval from $-100°$C up to 240°C. The accuracy of the dielectric permittivity and losses measurements was better than 3%. The details of sample preparation, experimental conditions and data treatment are described elsewhere.[28,29] The quantitative analysis of the dielectric spectra of the glass-transition process was carried out by fitting the isothermal dielectric loss spectra according to the Havriliak-Negami formula (1). It was found from the fitting that $\beta$ was equal to 1 for the glass-transition process in all of the samples.

Figure 3 represents the dependence of the Cole-Cole-parameter $\alpha$ on the mean relaxation time $\tau$ for $QN$, $CN$ and $MCN$ samples. From this figure one can see that the presence of the crystalline phase as well as the *kevlar* fibers affects the $\tau$ dependency of $\alpha$ and slows down the relaxation process. From Fig. 3 one can see that there is a fair agreement between the experimental data and the presented theoretical description. The quantitative analysis of these dependencies was performed by using Eq. (14). The fitting results are presented in Table 1.

Let us discuss now the space fractional dimension $d_G$. It is possible to estimate the average length of a *nylon 6,6* polymer chain that is about 50 – 100 $\mu$m (each polymer chain contains about $10^5$ groups while the length of a polymer group $r_g$ is about 10Å). This length is comparable to the thickness of a sample[29] 120 – 140 $\mu$m. This means that the movement of

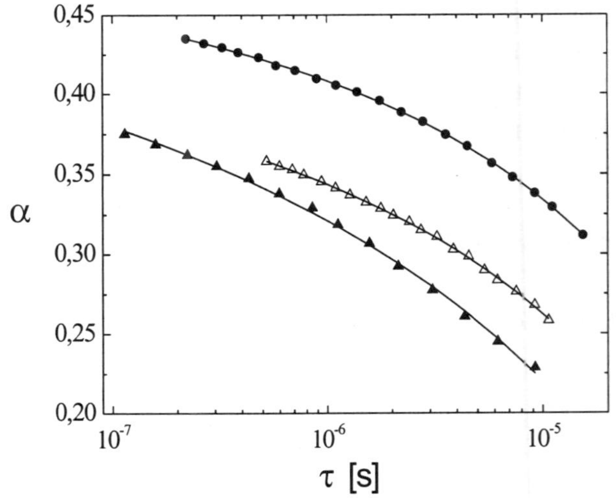


**Fig. 3** Cole-Cole $\alpha$ parameter versus relaxation time $\tau$; (•) for $QN$ data, (▲) for $CN$ data and (△) for $MCN$ data. The experimental points were obtained by temperature variations in the range 50°C – 120°C. The curves are fitted according to Eq. (14).

**Table 1 Values of the space fractional dimension $d_G$, the cutoff time of the scaling in time domain $\tau_0$ and the characteristic frequency $\omega_s$ for the samples $QN$, $CN$ and $MCN$.**

| Sample | $d_G$ | $\tau_0$ [ms] | $\omega_s$ [kHz] |
|---|---|---|---|
| *QN* | $1.12 \pm 0.01$ | $1.1 \pm 0.1$ | $5.9 \pm 0.3$ |
| *CN* | $1.20 \pm 0.05$ | $5.8 \pm 4.4$ | $9.7 \pm 1.9$ |
| *MCN* | $1.04 \pm 0.02$ | $1.5 \pm 0.4$ | $8.1 \pm 0.7$ |

the chains is most likely occurring in the plane of the sample. This fact correlates with the values of the space fractional dimension $d_G$. For all the samples $d_G \in (1, 2)$ (see Table 1). Thus, the Euclidean dimension of the space in which chain movement occurs is $d_E = 2$.

Although there is no presently unambiguous data about the mesoscale structure of the investigated samples, nevertheless it is possible to estimate the order of magnitude of some physically important quantities from the cutoff time $\tau_0$ and characteristic frequency $\omega_s$ values. It was already mentioned that the characteristic frequency $\omega_s$ is related to the self-diffusion process. For the estimation one can accept $R_0^2 \approx 10^{-16}$ m$^2$ ($R_0$ is the cube root of the volume occupied by one polymer chain), $G \approx 1$. Then, the self-diffusion coefficient evaluated by the expression

$$D_s = \omega_s R_0^2/4 \tag{15}$$

falls into the interval $10^{-14} - 10^{-13}$ m$^2$s$^{-1}$, which is usual for these kinds of polymer materials.[30]

The cutoff time $\tau_0$ is related to the maximal size of the cooperativity region $l_c$ as $l_c^2 = 4D_s\tau_0$. Thus, in the two-dimensional case the number of the polymer groups $n_g$ in the cooperativity region can be estimated by $n_g = l_c^2/r_g^2 \approx 10^2$, which is in fair agreement with the results obtained by E. Donth.[31]

In addition from Table 1 it can be seen that the presence of the crystalline phase as well as the *kevlar* fibers leads to the increase of the cutoff time $\tau_0$, which means a slowdown of the relaxation process. The presence of the *kevlar* fibers or the crystalline phase in a sample leads to the increase of the $\omega_s$ value also. This is a manifestation of a decrease in the mobile polymer chain length.

## 3.2 Polymer-Water Mixtures

Another experimental example of the $\alpha$ versus $\tau$ dependency is that of polymer water mixtures. In this article we consider the experimental data obtained by N. Shinyashiki et al.[32] They investigated the water mixtures of seven water-soluble polymers. Poly(vinyl alcohol) (*PVA*; $M_w = 77\,000$), poly(allylamine) (*PAlA*; $M_w = 10\,000$), poly(acrylic acid) (*PAA*; $M_w = 5\,000$), poly(ethylenimine) (*PEI*; $M_w = 50\,000$), poly(ethylene glycol) (*PEG*; $M_w = 8\,000$), poly(vinyl methyl ether) (*PVME*; $M_w = 90\,000$) and poly(vinylpyrrolidone) (*PVP*; $M_w = 10\,000$). The measurements of the complex dielectric permittivity were performed in the frequency range from 300 MHz up to 10 GHz and different sample compositions from 50% up to 100% of water in solution at 25°C. Note that in contrast to the previous set of experimental data the temperature was held constant.

It was shown[32] that the relaxation process in the specified frequency range reflects cooperative relaxation of water molecules while the relaxation of a polymer substance was observed

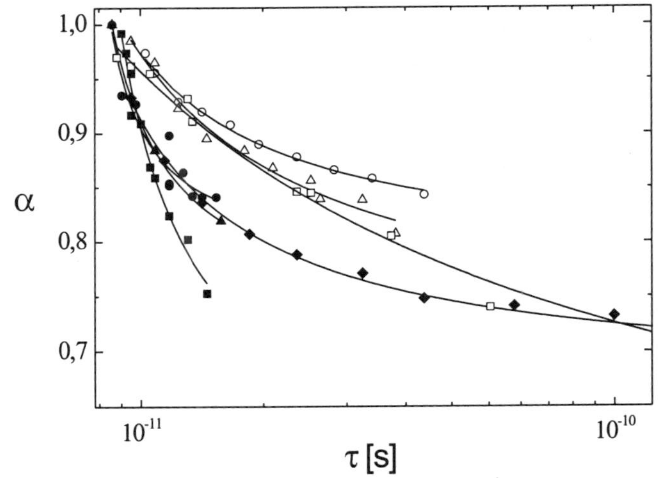


**Fig. 4** Cole-Cole $\alpha$ parameter versus relaxation time $\tau$. The full symbols correspond to the hydrophilic polymer water solutions while the open symbols correspond to the hydrophobic samples; (•) for *PVA*, (▲) for *PAlA*, (■) for *PAA*, (♦) for *PEI*, (○) for *PEG*, (△) for *PVME*, (□) for *PVP*. The experimental points were obtained by variations of water concentration. The solid lines represent the best fit according to Eq. (14).

at a longer time interval. The experimental dependencies of the Cole-Cole-exponent $\alpha$ versus the relaxation time together with fitting curves for all the samples are plotted in Fig. 4. As mentioned[32] earlier, the Fig. 4 shows a remarkable separation of the experimental curves into two groups. The hydrophilic polymer solutions *PVA*, *PAlA*, *PAA* and *PEI* are distributed within one group while the hydrophobic polymer water mixtures *PEG*, *PVME* and *PVP* are distributed into another. Note that the hydrophobic group is above the hydrophilic one.

According to the described model the Cole-Cole exponent reflects the interaction between the relaxing units and the statistical reservoir. The biggest deviation of the Cole-Cole exponent from unity corresponds to the stronger interaction. We can argue that in this case the relaxing units are the water molecules while the statistical reservoir is the polymer macromolecules. The deviation $\alpha$ from unity for the hydrophilic group (see Fig. 4) is bigger than for the hydrophobic one. This confirms the presented concept and indicates that the water-polymer interaction is more significant for the hydrophilic samples.

From this figure we can see also that for all the samples the experimental curves rise up to $\alpha = 1$ at $\tau \sim 8$ ps. This point corresponds to the zero concentration of a polymer in solution and represents the relaxation of bulk water. Increasing the polymer concentration leads to an increase of the relaxation time and to the decrease of the Cole-Cole exponent. The changing of the Cole-Cole exponent is an integral effect. When we discussed the interaction between the relaxation units and the statistical reservoir we implied that this interaction includes in itself the water-water interactions in the cooperative region, the water molecule confinement and caging by the polymer chains, *etc.* In particular, caging and confinement depends on the local structure and this is why the scaling properties become so important in that consideration.

In the framework of the presented model the size of the cooperativity region $R$ in Eq. (13) is related to the molecules that are affected by the polymer chains. These water molecules construct the dynamic cooperative structure, which is different from the structure of the bulk water. The increase of the polymer concentration leads to the increase of this co-operative region size $R$ and consequently leads to the increase of the relaxation time $\tau$. The cooperativity implies long-range space correlation. Therefore it seems that at non-zero

**Table 2 Values of the space fractional dimension $d_G$, the cutoff time of the scaling in time domain $\tau_0$, the characteristic frequency $\omega_s$ and the self-diffusion coefficient $D_s$ for all polymer-water mixtures.**

| Sample | $d_G$ | $\tau_0$ [ps] | $\omega_s \cdot 10^{-11}$ [Hz] | $D_s \cdot 10^9$ [m$^2$s$^{-1}$] |
|---|---|---|---|---|
| *PVA* | $1.56 \pm 0.09$ | $7.18 \pm 0.74$ | $1.47 \pm 0.21$ | 3.31 |
| *PAlA*[a] | 1.43 | 6.46 | 1.74 | 3.92 |
| *PAA* | $1.12 \pm 0.17$ | $6.34 \pm 0.83$ | $2.08 \pm 0.68$ | 4.68 |
| *PEI* | $1.33 \pm 0.02$ | $4.89 \pm 0.45$ | $2.67 \pm 0.40$ | 6.01 |
| *PEG* | $1.54 \pm 0.04$ | $4.45 \pm 0.74$ | $2.78 \pm 0.63$ | 6.26 |
| *PVME* | $1.38 \pm 0.10$ | $3.58 \pm 1.23$ | $4.24 \pm 2.47$ | 9.54 |
| *PVP* | $0.92 \pm 0.09$ | $0.72 \pm 0.33$ | $241 \pm 348$ | 542 |

polymer concentration all water molecules (at least the molecules which are contributing to the discussed relaxation process) are built into a cooperative structure that is different from the structure of bulk water.

Table 2 contains the fitting parameters for the water-polymer systems described above. It is well known[33] that the macroscopic dielectric relaxation time of bulk water (8.27 ps at 25°C) is about ten times larger than the microscopic relaxation time[34] of a single water molecule reorientation that corresponds to a hydrogen bond lifetime $\sim$ 0.7 ps. This fact follows from the associative structure of bulk water where the macroscopic relaxation time is reflects the cooperative relaxation process on the scale of water molecule clusters. As mentioned above, the water in the polymer-water mixture constructs the cooperative structure but is different from the bulk with another microscopic relaxation time. This parameter depends only on the dynamical properties of the cooperativity and does not depend on the polymer concentration. Different polymers build different cooperative water structures with different microscopic relaxation times. The model presented above implies that the microscopic relaxation time of water molecules is equal to the cutoff time of the scaling in time domain $\tau_0$. In the case of the hydrophilic polymers this time is longer (see Table 2).

For the most hydrophilic polymer *PVA* the strong interaction between the polymer and the water molecule results in the largest value of $\tau_0$, which is only 10% less than the macroscopic relaxation time of the bulk water. Weakening of the hydrophilic properties (or intensification of the hydrophobic properties) results in a decrease of the interaction between water and polymer and consequently in the decrease of $\tau_0$. The aqueous solution of the most hydrophobic polymer *PVP* has the smallest value of a single water molecule microscopic relaxation time which is practically equal to the microscopic relaxation time of a water molecule in the bulk (see Table 2).

The interaction between water and polymer occurs in the vicinity of the polymer chains and only the water molecules situated at the interface are affected by the interaction. The space fractal dimension $d_G$ in this case is the dimension of this interface. If a polymer chain is stretched as a line then its dimension is 1. In any other conformation the wrinkled polymer chain has a larger space fractal dimension which falls into the interval $d_G \in (1, 2)$. Thus, it is possible to say that this value of fractal dimension is the measure of polymer chain meandering. Straighter (probably more rigid) polymer chains have $d_G$ close to 1. More wrinkled (probably more flexible) polymer chains have $d_G$ close to 2. It is remarkable

[a]For this sample there is only three experimental points. This is the reason why it is impossible to determine the mean square deviation value and consequently the confidence intervals for the fitting parameters.

that according to the fitting results, the straightening of the polymer chains leads to the weakening of the hydrophilic property.

The presence of polymer in water affects not only the relaxation but the diffusion of the solvent as well. As in the previous case, we can use Eq. (15) for the self-diffusion coefficient estimation. However, in this case the scaling cutoff size in space is equal to the size of one water molecule $R_0 \approx 3$ Å. Note that the polymer affects only water molecules situated in the vicinity of the polymer chains. Thus, the estimated self-diffusion coefficient corresponds only to these water molecules and does not depend on the polymer concentration. The averaged self-diffusion coefficient estimated for the whole polymer-water mixture should be different and depend on the polymer concentration. Therefore, the coefficient presented in Table 2 does not characterize the mixture as a whole but reflects only the water-polymer interaction.

The self-diffusion coefficient for the bulk water[35] at 25°C is $2.57 \times 10^{-9}$ $m^2s^{-1}$. The presence of a polymer in the water prevents clusterization of water and relieves the diffusion. At the same time the strong interaction between polymer and water for hydrophilic samples slows down the diffusion. The clear tendency of the diffusion coefficient to increase with the reduction of hydrophility can be observed from Table 2. For the most hydrophilic sample *PVA* the self-diffusion coefficient is only 30% larger than for the bulk water while for *PVP* it is more than two orders of magnitude larger.

## 4. CONCLUSION

The paper reports the relationship between the relaxation time, the geometrical properties, the self-diffusion coefficient and the Cole-Cole exponent. The main distinguishing feature of the presented approach is a direct relationship between the Cole-Cole exponent and mean relaxation time. Therefore, this exponent depends on the temperature, the sample composition and other controlled physical parameters only through the mean relaxation time. This was confirmed by presented experimental examples (polymer composites and polymer water mixtures). Indeed, the polymer composite samples were measured at different temperatures but with constant sample structure. The polymer water mixtures were investigated at constant temperature but with different sample compositions. Nevertheless, the model describes the experimental data in the first case as well as in the second one. It is most likely that the relaxation time sets not only the frequency range of the relaxation process but determines its scaling properties also.

The cutoff quantities $\tau_0$ and $R_0$ are the borders of this scaling. They could be at high frequency and short distances, as well at the low frequency and long distance limits. For instance, in the case of polymer water mixtures these quantities are connected to the high frequency limit while for microcomposite samples the low frequency cutoff is important.

Thus, the presented model could be used for understanding the relationship between the scaling properties, structure and dynamics of the non-exponential relaxation processes in the disordered heterogeneous systems.

## ACKNOWLEDGMENTS

The authors express their sincere gratitude to Prof. G. Marom and H. Nuriel for providing the samples, Prof. S. Yagihara and Dr. N. Shinyashiki for experimental data, A. Gutina for the experiments and data treatment. One of us (Ya. E. Ryabov) thanks the Valazzi-Pikovsky Fellowship Fund for the financial support of his work at the Hebrew University.

## REFERENCES


1. C. J. F. Böttcher and P. Bordewijk, *Theory of Electric Polarization* (Elsevier, Amsterdam, 1992).
2. S. Havriliak and S. Negami, *J. Polym. Sci.* **C14**, 99 (1966).
3. A. K. Jonscher, *Dielectric Relaxation in Solids* (Chelsea Dielectric Press, London, 1983); A. K. Jonscher, *Universal Relaxation Law* (Chelsea Dielectric Press, London, 1996).
4. B. K. P. Scaife, *Principles of Dielectrics* (Oxford University Press, Oxford, 1989).
5. A. Zetsche, F. Kremer, W. Jung and H. Schulze, *Polymer* **31**, 1883 (1990).
6. G. Katana, F. Kremer, E. W. Fischer and R. Plaetschke, *Macromolecules* **26**, 3075 (1993).
7. A. Schönhals and F. Kremer, *J. Non-Cryst. Solids* **172–174**, 336 (1994).
8. G. Katana et al., *Macromolecules* **28**, 2714 (1995).
9. S. Fujiwara and F. Yonezwara, *Phys. Rev.* **E51**, 2277 (1995).
10. S. Gomi and F. Yonezwara, *Phys. Rev. Lett.* **74**, 4125 (1995).
11. W. Götze and L, Sjögren, *Rep. Prog. Phys.* **55**, 241 (1992).
12. K. Oldham and J. Spanier, *The Fractional Calculus* (Academic Press, New York, 1974).
13. K. Miller and B. Ross, *An Introduction to the Fractional Calculus and Fractional Differential Equations* (John Wiley Inc., New York, 1993).
14. *Applications of Fractional Calculus in Physics*, ed. R. Hilfer (World Scientific, London, 2000).
15. R. Metzler and J. Klafter, *Phys. Reports* **339**, 1 (2000).
16. R. R. Nigmatullin and Ya. E. Ryabov, *Phys. Solid State* **39**, 87 (1997); R. R. Nigmatullin and Ya. E. Ryabov, *Russ. Phys. J.* **40**, 314 (1997).
17. K. S. Cole and R. H. Cole, *J. Phys. Chem.* **9**, 341 (1941).
18. H. Mori, *Prog. Theor. Phys.* **33**, 423 (1965); H. Mori, *Prog. Theor. Phys.* **34**, 399 (1965).
19. J. A. Korn snd T. M. Korn, *Mathematical Handbook* (McGraw-Hill, New York, 1968).
20. R. Kubo, *J. Phys. Soc. Jpn.* **12**, 571 (1957).
21. K. Weron and A. Klauzer, *Ferroelectrics* **236**, 59 (2000).
22. A. Le Méhauté, R. Nigmatullin and L. Nivanen, *Flèches du Temps et Géométrie Fractale* (Hermez, Paris, 1998).
23. W. Feller, *An Introduction to Probability Theory and Its Applications* (John Wiley Inc., New York, 1971).
24. J. Feder, *Fractals* (Plenum Press, New York, 1988).
25. E. W. Fisher, E. Donth and W. Steffen, *Phys. Rev. Lett.* **68**, 2344 (1992).
26. J. Kakalios, R. A. Street and W. B. Jackson, *Phys. Rev. Lett.* **59**, 1037 (1987).
27. A. Einstein, *Ann. Phys.* **322**, 549 (1905); P. K. Pathria, *Statistical Mechanics* (Pergamon Press, Oxford, 1972).
28. N. Klein, G. Marom and E. Wachtel, *Polymer* **37**, 5493 (1996).
29. H. Nuriel et al., *Composites* **A31**, 69 (2000).
30. J. E. M. Snaar, P. Robyr and R. Bowtell, *Mag. Res. Im.* **16**, 587 (1998); P. Gribbon, B. C. Heng and T. E. Hardingham, *Biophys. J.* **77**, 2210 (1999); F. Joabsson, M. Nyden and K. Thuresson, *Macromolecules* **33**, 6772 (2000).
31. E. Donth, *J. Non-Cryst. Solids* **131–133**, 204 (1991).
32. N. Shinyashiki et al., *J. Phys. Chem.* **B102**, 3249 (1998).
33. U. Kaatze et al., *J. Chem. Eng. Data* **34**, 371 (1989).
34. R. Buchner, J. Barthel and J. Stauber, *Chem. Phys. Lett.* **306**, 57 (1999); C. Ronne, P. Astrand and S. R. Keiding, *Phys. Rev. Lett.* **82**, 2888 (1999); U. Kaatze, *J. Mol. Liq.* **56**, 95 (1993).
35. D. Eisenberg and W. Kauzmann, *The Structure and Properties of Water* (Clarendon Press, Oxford, 1969).

# SLOW DYNAMICS, AGING AND HISTORY-DEPENDENT EFFECTS IN THE PARKING-LOT MODEL

P. VIOT[*,†], J. TALBOT[*,‡] and G. TARJUS[*]
[*]*Laboratoire de Physique Théorique des Liquides, UMR 7600, Université Pierre et Marie Curie, 4, Place Jussieu 75252 Paris, Cedex 05, France*
[†]*Laboratoire de Physique Théorique, UMR 8627, Bat. 210, Université de Paris-Sud, 91405 ORSAY Cedex, France*
[‡]*Department of Chemistry and Biochemistry, Duquesne University, Pittsburgh, PA 15282-1530, USA*

**Abstract**

We review the properties of the Parking Lot Model and their connection with the phenomenology of vibrated granular materials. New simulation results concerning the out-of-equilibrium, aging behavior of the model are presented. We investigate in particular the relation between two-time response and correlation functions and the so-called violation of the fluctuation-dissipation theorem.

## 1. INTRODUCTION

Quite frequently, the time required for a system to reach equilibrium exceeds the experimental timescale. All measured quantities are then inherently nonequilibrium ones and depend upon the history of the system. While such history-dependent effects were initially studied in systems with quenched disorder, they can also be observed in disorder-free systems. An important example is the one-dimensional adsorption-desorption or Parking Lot Model (PLM)[1,2] that has been shown to reproduce the phenomenology of weakly vibrated granular materials.[3,4] In this model, hard rods arrive at randomly selected positions on a line at rate $k_+$. A new rod is accepted if it does not overlap with any preadsorbed rod. In addition, all rods present on the line are subject to desorption at a rate $k_-$. The existence of insertion and deletion mechanisms implies the existence of a steady (equilibrium) state characterized by a single control parameter $K = k_+/k_-$. When $K$ is sufficiently large, the saturation density is close to one, and this state is approached very slowly. Despite the simplicity of the model, it has a rich kinetic behavior. While we focus on the one-dimensional model, higher-dimensional versions can be envisaged as well.

From a purely theoretical perspective, the model has a number of features that make it worth investigating. All equilibrium properties are known exactly. With no desorption, the process is totally irreversible and it corresponds to a 1D Random Sequential Adsorption (RSA) for which the kinetics and microscopic quantities are also known exactly.[5] Several properties can be calculated analytically for the limit of infinitely slow desorption, i.e. $K \to \infty$ (which is not the same as for the RSA process with no desorption at all[1]). As shown below, some exact results can also be derived for the kinetics of this model when the desorption rate is finite.[6]

Knight et al. performed experiments[7] in which a column containing monodisperse spherical beads was subject to a long sequence of taps with an intensity characterized by $\Gamma = A/g$ where $A$ is the maximum acceleration and $g$ is the gravitational constant. These experiments showed that the density of the beads increases monotonically and surprisingly slowly with the number of taps, $n_t$, for various intensities of tapping. More specifically, the density approaches the steady state value as $1/\ln(n_t)$.

In two other studies, Nowak et al.[3,4] investigated the effect of cycling $\Gamma$, i.e. vibrating the column for a fixed number of taps with a sequence of $\Gamma$ values. They found that as this parameter increases, the density of the granular material increases. When $\Gamma$ is subsequently reduced by reversing the initial sequence, the density continues to increase. This second branch is reversible in that it is retraced if the increasing sequence of $\Gamma$ is repeated. In the same experiments, Nowak et al. monitored the power spectrum of the density fluctuations near the steady state for different values of $\Gamma$. The spectrum is distinctly non-Lorentzian, with the highest and lowest characteristic frequencies being separated by a non-trivial power-law-like regime. More recently, Josserand et al.[8] measured the response of a vibrated column of beads to an abrupt change in shaking intensity and showed the presence of "memory effects".

These experiments have stimulated an intense theoretical activity. Besides some phenomenological approaches,[9,10] several microscopic models have been proposed, whose basic ingredients are geometrical frustration or/and quenched disorder. A representative example in which quenched disorder is present is the frustrated lattice gas model developed by Nicodemi, Coniglio and Hermann.[11,12] Models with purely geometrical frustration and with kinetic constraints, such as the Tetris model,[13] kinetic Ising models[14,15] or atomistic models,[16] have also been studied. The PLM is more akin to this latter class of models in that it is free of disorder: the geometrical constraint, namely that particles cannot overlap, induces the sluggish kinetics and the associated cooperativity as the system becomes more dense.

The connection of the PLM to the granular compaction experiment can be made by regarding the particles on the line as an average layer in the vibrating column. The effect of a tap is to eject particles from the layer (desorption); this is followed by the replacement of the particles, generally in different positions, and possibly by the incorporation of additional particles. $1/K$ plays a role similar to that of $\Gamma$. In this approach, mechanical stability is implicitly included by the absence of motion of the particles when the desorption is switched off. This microscopic and off-lattice description of granular compaction is suitable for describing all previous experiments and, as shown below and elsewhere,[17,6,18] is in fair agreement with the experimental results.

The paper is organized as follows: in Sec. 2, the densification kinetics at constant $K$ is studied with an empty line as the initial state. In Sec. 3, we investigate how the density evolves when the control parameter $K$ is abruptly changed at a fixed time and when $K$ is cycled up and down. Section 4 focuses on the time-dependent correlation function of the density-density fluctuations when the system is close to equilibrium. In the last section we consider aging properties: the out-of-equilibrium, two-time response and correlation functions are investigated as well as the violation of the fluctuation-dissipation theorem.

## 2. KINETICS AT CONSTANT $K$

Formally, the densification kinetics of the system obeys

$$\frac{d\rho(t)}{dt} = \Phi(t) - \frac{\rho(t)}{K}, \tag{1}$$

where $\Phi(t)$, the insertion probability, is the fraction of the substrate that is available for the insertion of a new particle. The presence of a relaxation mechanism, i.e. competing desorption and adsorption with an equilibrium constant $K$, implies that the system eventually reaches a steady state that corresponds to an equilibrium configuration of hard particles with $\rho_{\text{eq}} = K\Phi_{\text{eq}}(\rho_{\text{eq}})$, where $\rho_{\text{eq}}$ denotes the equilibrium density. At equilibrium, the insertion probability is given exactly by

$$\Phi_{\text{eq}}(\rho) = (1-\rho)\exp(-\rho/(1-\rho)), \tag{2}$$

with the rods taken as being of length one. Inserting Eq. (2) in Eq. (1) and setting $d\rho/dt = 0$ leads to the following expression for the equilibrium density:

$$\rho_{\text{eq}} = \frac{L_w(K)}{1+L_w(K)}, \tag{3}$$

where $L_w(x)$ (the Lambert-W function) is the solution of $x = ye^y$. In the limit of small $K$, the isotherm takes the Langmuir form, $\rho_{eq} \sim K/(1+K)$, while for large $K$, $\rho_{eq} \sim 1 - 1/\ln(K)$. At small values of $K$, equilibrium is rapidly obtained, but at large values the densification process is dramatically slowed.

The time evolution of the PLM can be described by using the gap distribution function, $G(h,\, t)$, that represents the density of voids of length $h$. This quantity evolves according to

$$\begin{aligned}\frac{\partial G(h,\, t)}{\partial t} &= -H(h-1)(h-1)G(h,\, t) + 2\int_{h+1}^{\infty} dh' G(h',\, t) - \frac{2}{K}G(h,\, t)\\ &\quad + \frac{H(h-1)}{K\rho(t)}\int_0^{h-1} dh' G(h',\, h-1-h',\, t)\,,\end{aligned} \tag{4}$$

where $H(x)$ is the Heaviside function. This equation is not closed in that it involves the gap-gap distribution function $G(h,\, h',\, t)$, a quantity that represents the probability of finding two consecutive gaps of length $h$ and $h'$. An infinite hierarchy of equations can be written with each successive equation involving a higher-order gap distribution function. At equilibrium, i.e. when the time derivatives are equal to zero, the higher-order gap distribution functions obey a factorization property which allows an exact solution for all microscopic quantities. This is no longer true at finite times, and the hierarchical structure prevents the derivation of a full analytic solution of the model. Many partial results can, however, be obtained.

Once the $G(h,\, t)$ is known, the properties of interest may be calculated from the following sum rules:

$$\rho(t) = \int_0^\infty dhG(h,\, t)\,, \tag{5}$$

$$1 - \rho(t) = \int_0^\infty dhhG(h,\, t)\,, \tag{6}$$

and the insertion probability can be expressed as

$$\Phi(t) = \int_1^\infty dh(h-1)G(h,\, t)\,. \tag{7}$$

The kinetic and equilibrium properties of the PLM have been previously investigated.[4,6,17,19,20] For sufficiently large values of $K$, it is possible to identify three distinct regimes in the densification process. These are illustrated in Fig. 1 that displays a linear-logarithm plot of the density versus time.

In the first stage, the process mimics the irreversible Random Sequential Adsorption (RSA) in which the asymptotic kinetics are described by a power law, $1/t$. In the second regime, the density increases very slowly as $1/\ln(t)$ until it nears the steady-state value, at which point the kinetics cross over to an exponential form (see Fig. 1).

For the inverse power law behavior, desorption events are negligible. The kinetics can be understood in terms of the irreversible filling of isolated "targets" or regions of the line that can accept at most one additional rod. In this regime the gap distribution function evolves essentially according to $\partial G(h,\, t)/\partial t = -(h-1)G(h,\, t)$ with $h > 1$, which yields $G(h,\, t) = G(h,\, t_0)\exp(-(h-1)(t-t_0))$ where $t_0$ is a reference time. Since each target will eventually be occupied by one rod, the difference between the density at time $t$ and the jamming limit density, $\rho^{RSA}(\infty)$ (=0.747...), is given by $\rho^{RSA}(\infty) - \rho(t) \simeq \int_1^{h_0} G(h,\, t)dh \propto 1/t$, where $h_0$

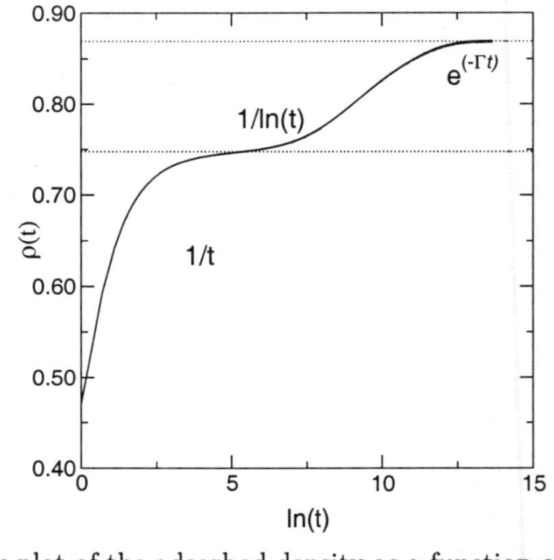


**Fig. 1** Linear-logarithmic plot of the adsorbed density as a function of time for a large value of $K$ (K = 5000). The process is characterized by three slow kinetic regimes: (i) RSA-like regime whose final stage is described by a $1/t$ behavior, (ii) $1/\ln(t)$ regime, and (iii) exponential approach towards equilibrium.

is an irrelevant upper cut-off. When the desorption rate is finite, in practice one switches to the second regime just before the density $\rho^{RSA}(\infty)$ is obtained.

The intermediate regime in which the density varies as $1/\ln(t)$ can be explained by free volume arguments. A simple heuristic one is as follows.[7] The free volume (here the free length) available to each rod is $(1/\rho) - 1$. In order to add one additional rod to the system, a void of length at least one must appear, which implies that a number $N = \rho/(1-\rho)$ of rods must contribute their free length. The rate of increase of density is proportional to the probability of this event which is $\exp(-N) = \exp(-\rho/(1-\rho))$, i.e. essentially the same as that in Eq. (2). Integration of the rate equation then gives the inverse logarithmic time dependence.[a] It should be stressed that this is a remarkably robust result that is observed in all of the models for granular compaction. It is independent of the system dimensionality, it is the same for continuous and lattice systems and it is unchanged by any (reasonable) polynomial function of $\rho$ multiplying the exponential term.

A naive approach to estimate the characteristic relaxation time $\tau_{eq}$ of the final exponential regime is the following adiabatic or mean-field approximation. Denoting $\delta\rho(t) = \rho(t) - \rho_\infty$, with $\rho_\infty = \rho_{eq}(K)$, and inserting in Eq. (1) the equilibrium form of $\Phi$ given in Eq. (2), one obtains at first order in density difference $\delta\rho$

$$\frac{d}{dt}\delta\rho = \delta\rho/\tau_{eq}^{MF} + O(\delta\rho^2) \tag{8}$$

with

$$\tau_{eq}^{MF}(K) = \frac{K}{(1+L_w(K))^2}, \quad \simeq K/\ln(K)^2 \text{ when K is large.} \tag{9}$$

This mean-field assumption leads thus to the paradoxical result that the characteristic time for the rearrangement of the line is smaller than $K$, the time for a desorption event. Since surface diffusion is absent in the PLM, significant rearrangement can only occur on a timescale longer than $K$, and no significant densification can occur on the timescale $\tau_{eq}^{MF}$.

To obtain a correct estimate of the relaxation time of the process, we now turn to a description in terms of the gap distribution approach. We assume that, as for the steady state (or equilibrium), $G(h,\, t) \sim \exp(-\Pi h)$, with $\Pi \sim \ln K \sim (1-\rho)^{-1}$ when $K$ is very large. As a consequence, if one defines $\rho_n(t) = \int_n^{n+1} G(h,\, t)dh$ and $\Phi_n(t) = \int_n^{n+1}(h-1)G(h,\, t)dh$, then $\rho_n \sim \Phi_n \sim K^{-n}$, so that if one looks for the dominant behavior in $1/K$, it is sufficient to consider the first intervals in $h$. As in the mean-field approximation, one can expand the gap densities in power of $\delta\rho(t)$ and keep only the linear term that gives rise to the exponential decay.

By using the various sum rules and assuming that the two-gap distribution function can be written as a product of 1-gap functions to $O(1/K)$, we have obtained after a tedious calculation[6] the relaxation time $\tau_{eq}$ whose expansion for large $K$ is

$$\frac{1}{\tau_{eq}} \simeq 2\frac{(\ln K)^3}{K^2} - 4\frac{(\ln K)^2}{K^2} + 2\frac{(\ln K)}{K^2} + O(\frac{1}{K^2})\,. \tag{10}$$

This time is indeed much slower than that predicted by the mean-field approximation. The reason for this dramatic difference, which is of several orders of magnitude for large $K$, is due to the absence in the PLM of a fast process that would reshuffle the particles on the line at constant density, so that the structure of the system would be that of a typical

[a]A more rigorous derivation can be done in the limit $K \to \infty$: cf Ref. 1.

equilibrium configuration. In the PLM, rearrangement of the particles and densification occur on the same timescale.

## 3. KINETICS WITH VARYING $K$: DENSIFICATION AND MEMORY EFFECTS

In this section, we focus on the dynamical properties of the system when the control parameter $K$ is varied along the process.

The irreversible/reversible cycles[3,4] and the recently demonstrated presence of "memory effects"[8] in the compaction of glass beads can be reproduced by the PLM.

Following the experimental procedure of compaction cycles,[4,3] we performed simulations in which the relative rate of desorption, $1/K$, is first increased at a given rate and then cycled down and up, the simulation being stopped after the same time $t = 40000$. (Recall that $1/K$ plays the same role as the tapping intensity and that, when expressed in units of $k_+$, time is assumed to measure the number of taps; this leads to the reasonable behavior that the number of desorption events increases as the tapping strength increases.) Figure 2(a) displays the density as $1/K$ varies between $10^{-4}$ and $10^{-3}$ and it is similar to the behavior observed experimentally as well as numerically in the FLG models.[21,22]

Along the first (irreversible) branch the density increases rapidly and then passes through a maximum as $1/K$ increases. When the initial sequence of $1/K$ is reversed (decreasing $1/K$), the density increases monotonically. When the initial (increasing) sequence of $1/K$ is repeated, the density now decreases monotonically, nearly retracing the second branch. The residual hysteresis observed between the second and third branches is also present in other models,[10] and it diminishes as one considers larger values of $K$ and larger time intervals between changes of desorption rate. Note that the densities attained in the present

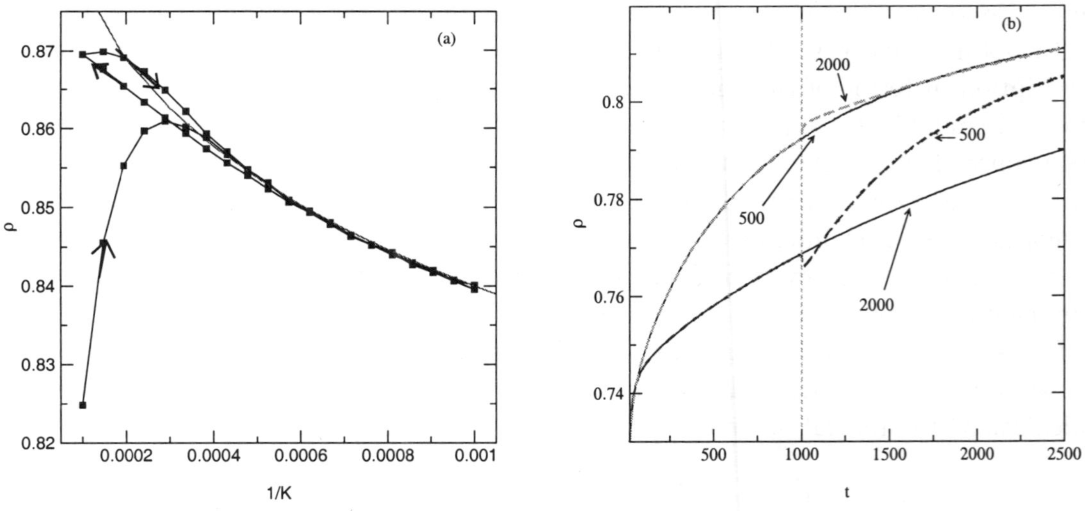


(a) (b)

**Fig. 2** (a) Irreversible and reversible densification branches. Starting from a "loose" packed state, the process consists of a sequence of varying values of $K$ between 1000 and 10000. For each $K$, the duration is 40000. The arrows show the sense of the cycle. The dotted line is the equilibrium curve. (b) "Memory" effect at short times after an abrupt change of the desorption rate: the full curves correspond to a process with a constant $K$, whereas the dashed curves show the kinetics of a process where $K$ is switched from 500 to 2000 and vice-versa at $t = 1000$.

adsorption-desorption model are typical of a one-dimensional system and that more realistic values would be obtained by employing the two-dimensional version of the model.

To further compare the model predictions to the available experimental data on vibrated granular materials, we have studied the effect of an abrupt change in $K$ on the densification kinetics. This is illustrated in Fig. 2(b) in which $K$ is switched from 500 to 2000 and vice-versa. An increase in $K$ corresponds to reducing the vibration intensity and leads, contrary to simple intuition, to an initial *increase* in the density. The outcome is reversed if $K$ is decreased. The same "anomalous" behavior was observed experimentally by Josserand et al.[8]

## 4. BEHAVIOR NEAR EQUILIBRIUM

As mentioned in Sec. 2, when the time is large enough compared to the relaxation time $\tau_{eq}$, the system is at equilibrium. We denote time by $\tau$ in equilibrium regime. In this case, the density of adsorbed particles fluctuates around the average equilibrium value. The time-dependent correlation function of the density fluctuations (at equilibrium) is defined as

$$C(\tau) = \frac{\langle \rho(\tau + t')\rho(t')\rangle - \langle \rho(\tau + t')\rangle\langle \rho(t')\rangle}{\langle \rho(t')^2\rangle - \langle \rho(t')\rangle^2}, \tag{11}$$

where $\langle \rho(\tau + t')\rangle = \langle \rho(t')\rangle = \rho_{eq}$ and the angular brackets denote an average at equilibrium. Since the system is at equilibrium, the correlation function only depends on the time difference $\tau$ and not on $t'$. Around equilibrium, the Onsager principle of regression of fluctuations implies that for a large enough time, $C(\tau)$ approaches zero in the same (exponential) way as $\rho(\tau)$ approaches $\rho(\infty) = \rho_{eq}$, with the same relaxation time. Figure 3 displays a plot of $C(\tau)$ for $K = 500$. The lower right inset shows a linear-logarithmic plot of $C(\tau)$ which confirms the exponential approach at long times.

Figure 3 shows that the correlation function $C(\tau)$ first decays to about 0.8 on a timescale much shorter than that of the relaxation time. To understand this behavior, highlighted in the upper right inset, we note that for this regime, $\tau \ll K$, desorptions are rare and the

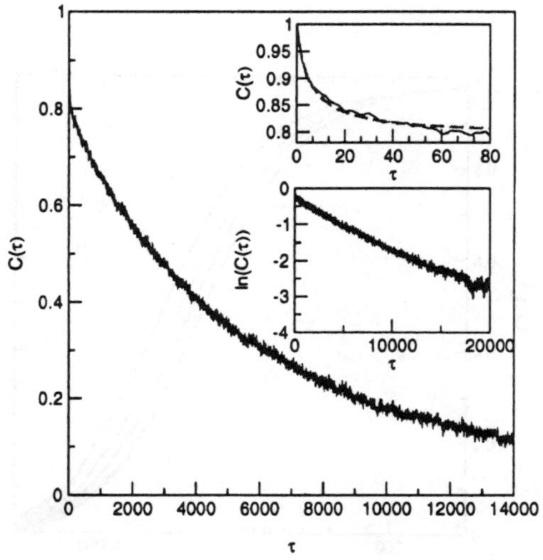


**Fig. 3** Equilibrium density-density correlation function $C(\tau)$ versus time for $K = 1000$. The inset in the upper right corner displays the first step in the decay of the correlation function (full curve) as well as the prediction, Eq. (12) (dashed curve). The other inset shows the exponential decay of $C(\tau)$ at long times on a logarithmic-linear plot.

regression of the density fluctuations is dominated by the filling of isolated targets (without large scale rearrangement of the configuration). On the other hand, the overall amplitude of this initial decay is a function of the desorption rate and it goes to zero as $K$ goes to infinity. The target filling being similar to that in a simple RSA process, one predicts a $1/\tau$ kinetics. At times $\tau \sim K$ desorption introduces a cut-off to this algebraic behavior, leading to a $(1/\tau)\exp(-\tau/K)$-law. A more complete calculation gives[6]

$$C_{short}(\tau) = \frac{P_\infty^2}{K(1-\rho_\infty)^2}\frac{\exp(-\tau/K)}{(\tau + P_\infty)} + O(1/K^2). \tag{12}$$

The upper right inset of Fig. 3 displays a comparison between the simulation result and Eq. (12) at short times. We note that the above heuristic argument allows to predict the short-time behavior of the correlation function for the parking-lot model in higher dimensions. For RSA, the asymptotic kinetics, resulting from the filling of isolated targets, is given by $\tau^{-1/D}$. Correspondingly, the power spectrum should behave as $\omega^{-1+1/D}$ for $\omega > 1/K$. In particular, this predicts a power law $\omega^{-1/2}$ for $D = 2$, which is compatible with the experimental data in vibrated granular media.[4]

## 5. TWO-TIME FUNCTIONS, VIOLATION OF THE FDT AND INTERRUPTED AGING

In this section, we consider the two-time correlation and response functions in the out-of-equilibrium regime, as well as their relation. The two-time correlation function of the density fluctuations, after normalization, is defined as follows:

$$C(t, t_w) = \frac{\langle\rho(t)\rho(t_w)\rangle - \langle\rho(t)\rangle\langle\rho(t_w)\rangle}{\langle\rho(t_w)^2\rangle - \langle\rho(t_w)\rangle^2}, \tag{13}$$

where the brackets denote an average over independent runs and $t \geq t_w$. Figure 4 shows $C(t, t_w)$ as a function of $t - t_w$ for different values of $t_w$, often referred to as the waiting time. Contrary to the situation described in the previous section, the system exhibits aging: the larger $t_w$, the longer the memory of the initial configuration persists. This aging regime is thus characterized by a breakdown of the time translation invariance.[23]

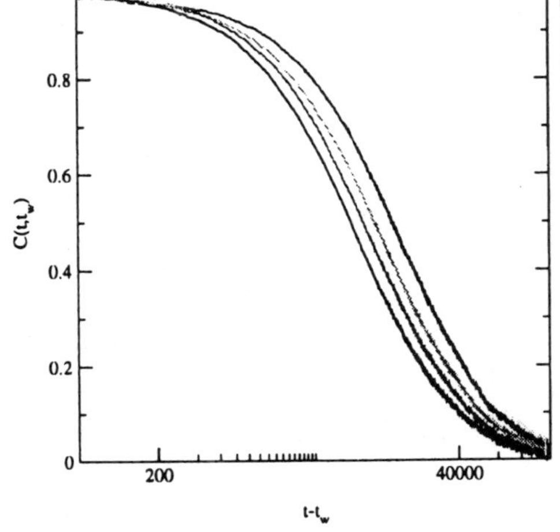


**Fig. 4** Two-time correlation function $C(t, t_w)$ as a function of elapsed time $t - t_w$ for different waiting times ($t_w$ = 2000, 4000, 6000, 10000, 16000) (from left to right) and with $K = 5000$.

When $t$ and $t_w$ are large enough (but still smaller than the equilibration time $\tau_{eq}$), the aging behavior is usually described by a non-trivial form involving a single scaling variable that is the ratio of a function of $t$ (or of the elapsed time $t-t_w$) divided by the same function of $t_w$. In the FLG models, it was found numerically that the appropriate scaling variable is $\ln(t-t_w)/\ln(t_w)$.[21] On the other hand, for the PLM, we showed numerically[18] that the aging regime is characterized by a simple $(t-t_w)/t_w$ scaling, a behavior similar to that found in a large variety of systems.[23] This result suggests that a promising way to establish whether quenched disorder, as in the FLG models, or simple geometrical frustration, as in the PLM, better describe the physics of vibrated granular materials is to study experimentally the two-time correlation function of the density fluctuations in the course of the densification process. (Recall that all other phenomena displayed by vibrated granular materials are, at least qualitatively, reproduced by both the FLG and the PLM.)

We have also calculated the two-time response function as follows. Starting from an empty line, the system evolves at a fixed $K$. After a given time, $t_w$, two clones of the system are made. The original system continues to evolve with the same $K$, and the two clones now evolve with $K+\delta K$ and $K-\delta K$. The response at time $t$ then corresponds to the difference in density between one clone and the original system. In order to obtain reasonable statistics, it is necessary to average over many independent runs (typically $10^4$). Since the PLM in the $t,\, t_w \to \infty$ limit reduces to an equilibrium system in the grand canonical ensemble with a chemical potential given by $\beta\mu = \ln(K)$ with $\beta = 1/k_BT$, $\ln(K)$ is the external "field" conjugate to the density $\rho$. We thus consider the response to a change in $\ln(K)$,

$$R(t,\, t_w) = \frac{\partial \rho(t)}{\partial \ln(K(t_w))}\,. \tag{14}$$

In the simulation, it is approximated by

$$R^{\pm}(t,\, t_w) = K\frac{\rho^*(t,\, t_w,\, K\pm\delta K) - \rho(t,\, K)}{\pm\delta K}\,, \tag{15}$$

where $\rho^*(t,\, t_w,\, K\pm\delta K)$ denote the density of the clones at a time $t$ larger than $t_w$. $R^{\pm}(t,\, t_w)$ are integrated response functions since the perturbation is applied over an extended period time. In the linear-response regime, which is the one of most interest, the response function must be independent of the amplitude and sign of the perturbation. One should have in particular $R^+(t,\, t_w) = R^-(t,\, t_w)$. We have used this requirement as a test to check that our simulation data were indeed, to a very good approximation, in the linear response regime. In practice, we have used $\delta K/K = 0.1$ or $0.07$ and in order to minimize the non-linear contribution of the response, we have taken the average of $R^+(t,\, t_w)$ and $R^-(t,\, t_w)$ to approximate Eq. (14). Choosing smaller values of $\delta K/K$ would have drastically increased the number of independent runs necessary to obtain a well-defined curve for the response function.

When $t_w$ and $t-t_w$ become much larger than the relaxation (or "ergodic") time $\tau_{eq}$, the response function reaches the equilibrium value $R_{eq} = \rho_{eq}(1-\rho_{eq})^2$ and the fluctuation-dissipation theorem (FDT) applies when $t_w > \tau_{eq}$. One then recovers the time translation invariance, i.e. $R(t,\, t_w) = R(\tau = t-t_w)$; in this case, the response function associated with changes in the external field $\beta\mu = \ln(K)$ (coupled to the density) and the density-density correlation function are related by

$$R(\tau) = \tilde{C}(0) - \tilde{C}(\tau)\,, \tag{16}$$

where $\tilde{C}(\tau) = \langle \delta\rho(\tau)\delta\rho(0)\rangle$. (Note that, as usual for hard objects, the temperature is irrelevant and does not explicitly enter in Eq. (16) because it is included in $\ln(K) = \beta\mu$).

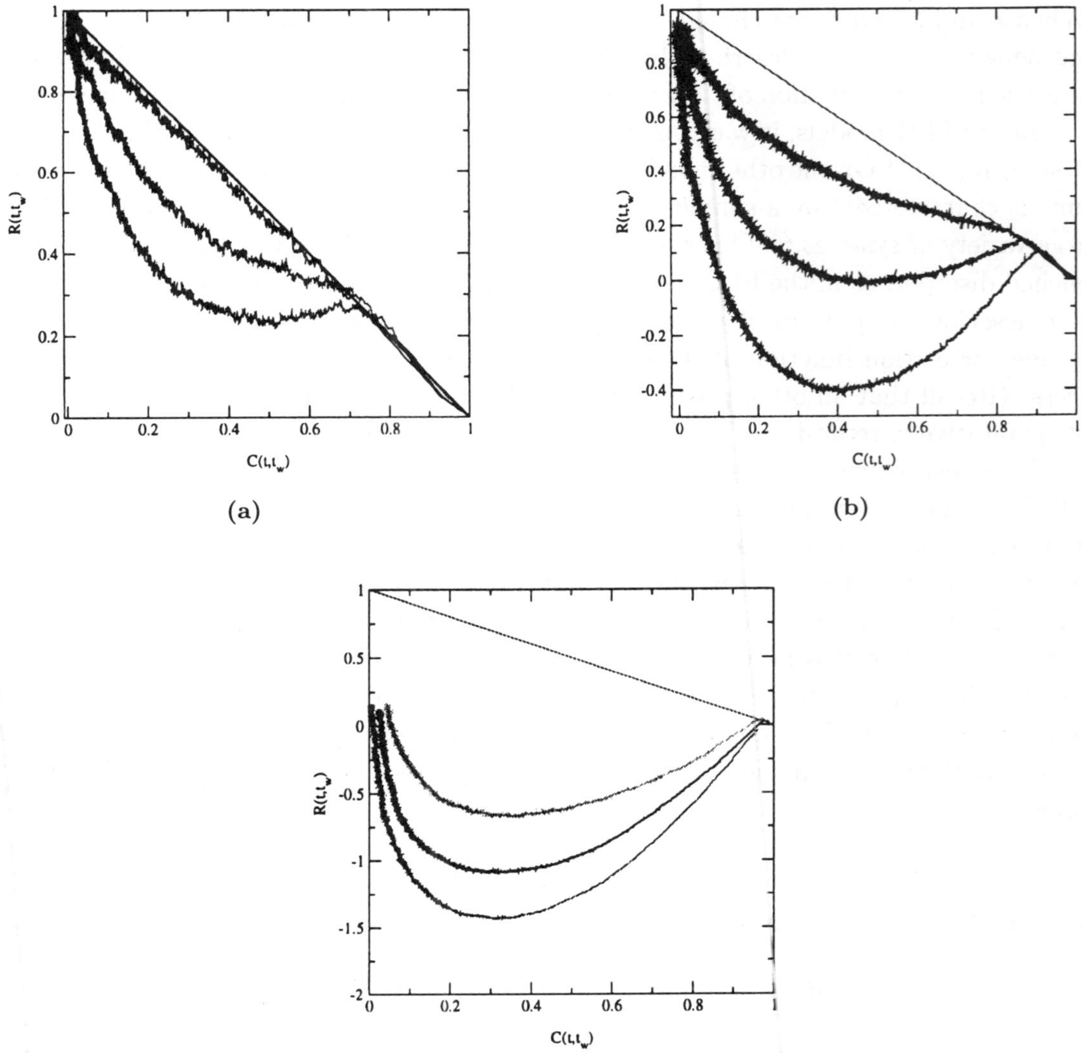


**Fig. 5** Response function $R(t,\, t_w)$, normalized by its equilibrium value, versus normalized correlation function $C(t,\, t_w)$ **(a)** for $K = 300$ and for three waiting times $t_w = 500,\, 1000,\, 3000$ (from bottom to top), **(b)** for $K = 1000$ and for three waiting times $t_w = 1000,\, 3000,\, 8000$ (from bottom to top), and **(c)** for $K = 5000$ and for three waiting times $t_w = 2000,\, 6000,\, 16000$ (from bottom to top).

It is worth stressing that with the definition adopted above of a response to $\ln(K)$, a positive response means that an increase of $K$ (i.e. a decrease in tapping strength) increases the density, whereas a negative response corresponds to a density decrease. In the PLM, the initial response is always positive, although it is very small for large values of $t_w$.

When $t_w$ is less than the relaxation time $\tau_{eq}$, the FDT no longer holds. We illustrate the "violation" of the FDT in Figs 5(a)–(c) where the response function $R(t,\, t_w)$, normalized by its equilibrium value, is plotted versus the normalized correlation function $C(t,\, t_w)$. On such a plot, the FDT, i.e. Eq. (16), implies that all curves fall on the same straight line. One notices that this is the case when the correlation function $C$ is close to one, although the range over which the FDT is obeyed diminishes when $K$ increases and when $t_w$ decreases. The fact that Eq. (16) is satisfied initially can be explained by noting that at short times ($t - t_w$ small), both the response of the system and the decorrelation of the density fluctuations involve only the arrival and departure of particles in small isolated segments of the line. No large-scale rearrangements of the system are possible, and the kinetics are simply governed by the superposition of adsorption-desorption events in small independent subsystems. This is analogous to the short-time situation described in Sec. 4, and the equilibrium relation, Eq (16) thus applies. At very long times $t$, i.e. when $\tau = t - t_w > \tau_{eq}$ and $C$ is very close to zero, the FDT also applies because the system relaxes to equilibrium in the usual Onsager-like regime: this is a manifestation of "interrupted aging".[23] This return to standard FDT behavior is visible in Fig. 5(a) for a moderate value of $K$ ($K = 300$); for larger values of $K$, the relaxation time to equilibrium, $\tau_{eq}$, is prohibitively long for computer simulations (for instance, the time required to obtain the data in Fig. 5(c) is $t = 200\,000$).

For intermediate times $t - t_w$, and intermediate values of the correlation function $C$, the out-of-equilibrium kinetics are characterized by strong violations of the FDT. As shown in Figs. 5(a)–(c), the departure from FDT behavior, when considering decreasing values of $C$, is strongly dependent on the waiting time $t_w$: the larger $t_w$, the smaller the violation of the FDT (for $K = 300$, the system is completely at equilibrium for $t_w = 3000$).

The slope of $R$ versus $C$ when violation first appears (i.e. for values of $C$ near 1, corresponding to times at which aging is just starting) has been used as a central quantity for characterizing aging systems. For out-of-equilibrium systems in which the entropy production is small,[24] this slope defines an effective temperature that is larger than the temperature of the bath. As is obvious from Figs. 5(a)–(c), the concept of effective temperature does not apply, at least not in a straightforward way, to the aging behavior of the PLM: indeed, not only is the slope (for $C > 0.5$) strongly dependent on the waiting time, but it can also reach positive values, which would imply negative effective temperatures (recall that the sign of the slope is the opposite of the ratio of the effective temperature to that of the bath and is thus equal to $-1$ when the FDT holds). A similar feature has been observed in the FLG model[12,25] and it suggests that such models, and possibly real vibrated granular materials, do not fall in the category of systems with small entropy production. We are presently studying the time-dependent entropy of the PLM to further elaborate on this explanation and to make the connection with the formalism developed by Edwards,[26] as has been done recently with the Kob-Andersen[27] and Tetris models.[28]

## ACKNOWLEDGMENTS

We are glad to dedicate this article to Antonio Coniglio, a good friend who has made many innovative contributions to statistical mechanics and the study of disordered and frustrated systems.

J. T. thanks the National Science Foundation (CHE-9814236) and the CNRS for financial support.

## REFERENCES

1. X. Jin, G. Tarjus and J. Talbot, *J. Phys. A. Math. Gen.* **27**, L195 (1994).
2. P. L. Krapivsky and E. Ben-Naim, *J. Chem. Phys.* **100**, 6778 (1994).
3. E. R. Nowak, J. B. Knight, M. Povinelli, H. M. Jaeger and S. R. Nagel, *Powder Technol.* **94**, 79 (1997).
4. E. R. Nowak, J. B. Knight, E. Ben-Naim, H. M. Jaeger and S. R. Nagel, *Phys. Rev.* **E57**, 1971 (1998).
5. J. Talbot, G. Tarjus, P. R. van Tassel and P. Viot, *Colloids and Surf.* **A165**, 287 (2000).
6. J. Talbot, G. Tarjus and P. Viot, *Phys. Rev.* **E61**, 5429 (2000).
7. J. B. Knight, C. G. Fandrich, C. N. Lau, H. M. Jaeger and S. R. Nagel, *Phys. Rev.* **E51**, 3957 (1995).
8. C. Josserand, A. Tkachenko, D. M. Mueth and H. M. Jaeger, *Phys. Rev. Lett.* **85**, 3632 (2000).

9. T. Boutreux and P. G. de Gennes, *Physica* **A244**, 59 (1997).
10. D. Head, *Phys. Rev.* **E62**, 2439 (2000).
11. M. Nicodemi, A. Coniglio and H. J. Hermann, *Phys. Rev.* **E59**, 6830 (1999).
12. A. Coniglio and M. Nicodemi, *J. Phys: Cond. Mat.* **12**, 6601 (2000).
13. E. Caglioti, V. Loreto, H. J. Herrmann and M. Nicodemi, *Phys. Rev. Lett.* **79**, 1575 (1997).
14. J. J. Brey, A. Prados and B. Sánchez-Rey, *Phys. Rev.* **E60**, 5685 (1999).
15. A. Prados, J. J. Brey and B. Sánchez-Rey, *Physica* **A284**, 277 (2000).
16. A. Mehta and G. C. Barker, *J. Phys: Cond. Mat.* **12**, 6619 (2000).
17. J. Talbot, G. Tarjus and P. Viot, *J. Phys. A. Math. Gen.* **32**, 2997 (1999).
18. J. Talbot, G. Tarjus and P. Viot, *cond-mat/0008183.*
19. E. Ben-Naim, J. B. Knight, E. R. Nowak, H. M. Jaeger and S. R. Nagel, *Physica* **D123**, 380 (1998).
20. A. J. Kolan, E. R. Nowak and A. V. Tkachenko, *Phys. Rev.* **E59**, 3094 (1999).
21. M. Nicodemi and A. Coniglio, *Phys. Rev. Lett.* **82**, 916 (1999).
22. A. Barrat and V. Loreto, *J. Phys. A. Math. Gen.* **33**, 4401 (2000).
23. J.-P. Bouchaud, L. F. Cugliandolo, J. Kurchan and M. Mezard, in *Spin Glasses and Random Fields* (World Scientific, Singapore, 1998), p. 161.
24. L. Cugliandolo, D. S. Dean and J. Kurchan, *Phys. Rev. Lett.* **79**, 2168 (1997).
25. M. Nicodemi, *Phys. Rev. Lett.* **82**, 3734 (1999).
26. S. F. Edwards, in *Granular Matter: An Interdisciplinary Approach*, ed. A. Mehta (Springer-Verlag, New York, 1994).
27. A. Barrat, J. Kurchan, V. Loreto and M. Sellitto, *Phys. Rev. Lett.* **85**, 503 (2000).
28. A. Barrat, J. Kurchan, V. Loreto and M. Sellitto, *Phys. Rev.* **E63**, 051301 (2001).

# STANDARD SCALING AND MULTISCALING IN PHASE ORDERING DYNAMICS

MARCO ZANNETTI
*Istituto Nazionale per la Fisica della Materia, Unità di Salerno and Dipartimento di Fisica, Università di Salerno, 84081 Baronissi (Salerno), Italy*

CLAUDIO CASTELLANO
*Fachbereich Physik, Universität GH Essen, 45117 Essen, Germany and Istituto Nazionale per la Fisica della Materia, Unità di Roma "La Sapienza", P.le Aldo Moro 2, 00185 Roma, Italy*

**Abstract**

The phase-ordering dynamics with conserved scalar order parameter is studied simulating the Cahn-Hilliard equation and the kinetic Ising model. In both cases a preasymptotic multiscaling regime is found revealing that the late stage of phase-ordering is always approached through a crossover from multiscaling to standard scaling, independently from the nature of the microscopic dynamics.

The material covered in this talk contains the answers to some questions concerning the scaling behavior in phase ordering processes which were initially raised in joint work with Antonio over ten years ago.[1] The celebration of his 60th birthday seems to be a most nice occasion for the presentation of a brief overview of the development of the problem.

It had been known for quite some time[2] that in the late stage of the phase ordering dynamics of a system quenched below the critical point[3] the structure factor (Fourier transform of the equal time order parameter correlation function $G(\mathbf{r}, t) = \langle \phi(\mathbf{x}, t)\phi(\mathbf{x}+\mathbf{r}, t)\rangle$) obeys the scaling form

$$C(\mathbf{k}, t) \sim L^d(t)F(kL) \tag{1}$$

where the characteristic length grows with a power law $L(t) \sim t^{1/z}$, $d$ is the space dimensionality of the system and $F(x)$ is a scaling function. This is the growth of the Bragg peak due to ordering, namely in the limit $t \to \infty$ the form (1) goes to $M^2\delta(\mathbf{k})$, where $M$ is the equilibrium value of the order parameter in the ordered phase. Although this type of behavior, which we will refer to as standard scaling, is well supported by a large experimental and numerical evidence, its derivation from first principles remains an unsolved problem.

The Cahn-Hilliard (CH) equation

$$\frac{\partial \phi}{\partial t} = \nabla^2 \left[ \frac{\partial V(\phi)}{\partial \phi} - \nabla^2 \phi \right] \tag{2}$$

with $V(\phi) = -\phi^2/2 + \phi^4/4$, describes the dynamics of a system with a scalar, continuous, conserved order parameter and is known to obey standard scaling. Assuming that the system is prepared in a high temperature disordered state the structure factor is constant

$$C(\mathbf{k},\, t = 0) = \Delta \tag{3}$$

due to the lack of correlations. The simulation of the CH equation then yields a structure factor, which, starting from the initial condition (3), after a sufficiently large time evolves into the scaling form (1) with the dynamical exponent $z = 3$.

On the other hand, it is possible to consider the dynamics of a continuous $N$-component conserved order parameter; solving exactly such a model for large-$N$ Antonio and one of us (MZ) found[1] in place of (1) the qualitatively different multiscaling behavior[4]

$$C_0(\mathbf{k}, t) \sim L(t)^{\alpha_0(x)} \tag{4}$$

with $z = 4$ and $x = k/k_m(t)$, where $k_m(t)$ is the peak wave vector related to $L(t)$ by $(k_m L)^4 = d \ln L + (2-d)\ln(k_m L)$. The novelty with respect to the standard scaling form (1) is that now there is a spectrum of wave vector dependent exponents (Fig. 1) given by

$$\alpha_0(x) = d[1 - (1 - x^2)^2]\,. \tag{5}$$

It can be shown that as $t \to \infty$ also the form (4) generates the Bragg peak. However, in this case, the physical meaning of the delta function is different, since it is not due to an ordering process, rather to the macroscopic growth of fluctuations in the $k = 0$ mode[5] with a Gaussian probability distribution. We refer to this process as condensation,[6] due to the close analogy with the ideal Bose gas condensation.

Shortly afterwards Bray and Humayun (BH)[7] through an approximate model did show that if $1/N$ corrections are taken into account the standard scaling form (1) is asymptotically restored. The main open question then was whether multiscaling is strictly a property only of the large-$N$ model or does it enter also in the time development of systems with finite $N$. In order to address this point we decided to investigate systematically how the asymptotic form (1) is reached following the overall time evolution of the system. To gain some insight on what goes on in the preasymptotic region it is quite useful to start with the BH model.

Considering the general case of a vector order parameter, BH have derived a closed equation of motion for the structure factor whose solution can be written as the sum of two contributions

$$C(\mathbf{k},\, t) = \Delta C_0(\mathbf{k},\, t) + \frac{1}{N} C_{nl}(\mathbf{k},\, t)\,. \tag{6}$$

If one takes the limit $N \to \infty$ the second term in the right-hand-side (rhs) disappears and $C_0(\mathbf{k},\, t)$ coincides with the exact solution[1] of the large-$N$ model. In order to simplify the discussion let us rewrite (6) as the sum of the two terms, each in the scaling form they would have if they were present separately[8]

$$C(\mathbf{k},\, t) = \Delta L^{\alpha_0(x)} + \frac{1}{N} L^d. \tag{7}$$

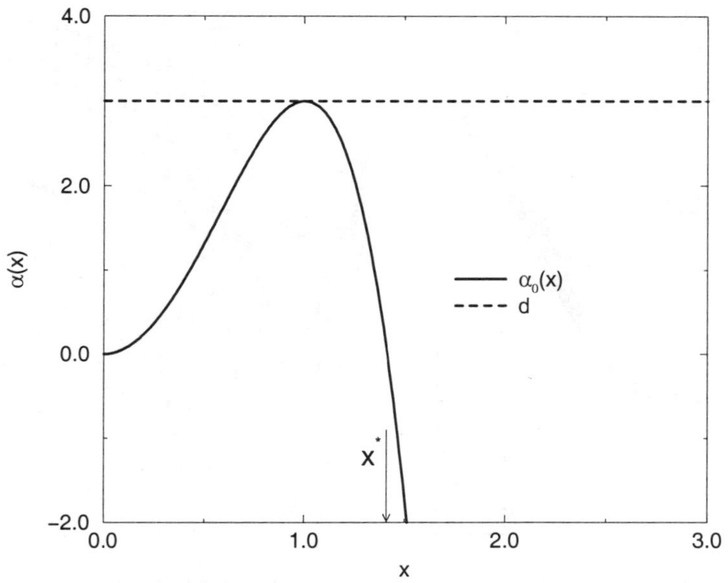


**Fig. 1** Spectrum of multiscaling exponents $\alpha(x)$ in the large-$N$ model with $d = 3$.

In this form the behavior of $C(\mathbf{k}, t)$ is the outcome of the competition between two different scaling forms, which depends both on the magnitude of the prefactors and on the difference of the exponents. Furthermore, the latter one depends on the wave vector giving rise to a wave vector dependent crossover. As it can be seen from Fig. 1 the difference $\delta\alpha(x) = d - \alpha_0(x)$ vanishes for $x = 1$, namely at the peak of the structure factor, while it is positive for $x \neq 1$ and in particular becomes very large for $x > x^* = \sqrt{2}$, where $\alpha_0(x)$ is negative.

Therefore, even if the second term in the rhs of (7) is bound to dominate, by modulating the magnitude of $\Delta$ the relative weight of the two terms may be adjusted so that multiscaling can be observed over a sizable preasymptotic time interval, with a spread in $x$ about the peak which depends on the actual value of $\Delta$. With a continuous order parameter the value of $\Delta$ can be varied at will and in Ref. 8 it was shown that when $\Delta$ is very small ($\Delta \ll M^2$) there exists an observable mean field preasymptotic behavior practically only at the peak, while for large values of $\Delta$ ($\Delta \simeq M^2$) preasymptotic multiscaling is clearly observed over the range $x \leq x^*$.

The interest of this result is that it is not just a property of the BH model, but the same pattern of behavior is shown here to hold also in the numerical solution of the nonlinear scalar Cahn-Hilliard (CH) equation and in the simulation of the Ising model with Kawasaki dynamics. Moreover a preasymptotic multiscaling regime is observed also for the Kawasaki dynamics in $d = 1$ at $T_F > 0$, where ordering occurs only up to the finite equilibrium correlation length $\xi(T_F)$.

The analysis of the crossover between multiscaling and standard scaling is carried out by computing the effective spectrum of exponents $\alpha(x, t)$ that one obtains when fitting the structure factor to the form

$$C(\mathbf{k}, t) = \mathcal{L}_1^{\alpha(x,t)} F(x) \tag{8}$$

with $x = k\mathcal{L}_2$ and $\mathcal{L}_2(t) = k_m^{-1}(t)$, the inverse of the peak position. A delicate point here is the choice of $\mathcal{L}_1$. Adopting straightforwardly the definition of the large-$N$ model $\mathcal{L}_1(t) = L(t)[k_m(t)L(t)]^{2/d-1}$. This amounts for $d = 2$ to put in by hand the growth law $L(t) = t^{1/4}$. On the other hand, in the large-$N$ model $\mathcal{L}_1$ is also given by $[C_0(k_m, t)]^{1/d}$.

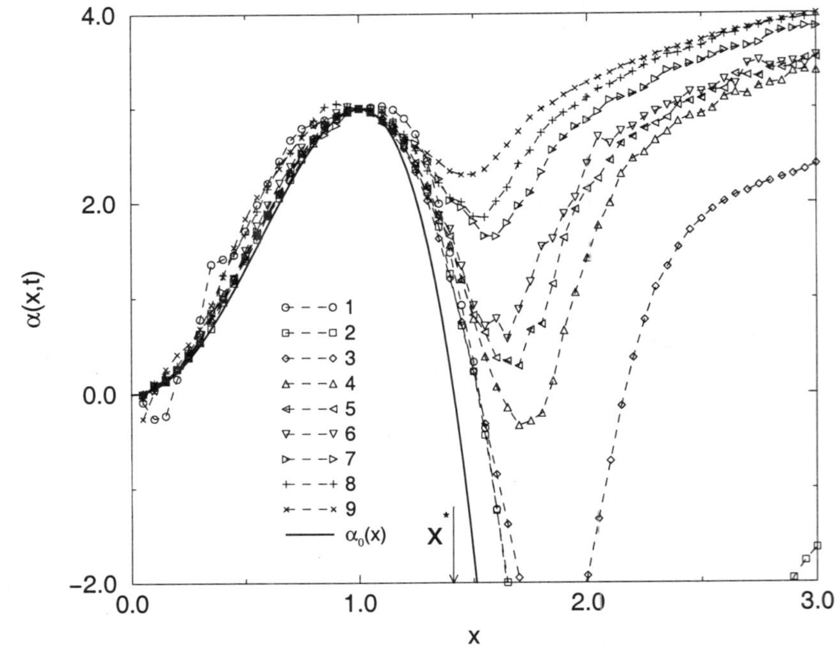


**Fig. 2** Effective exponent for the CH equation.

Therefore we take as an unbiased choice $\mathcal{L}_1 = [C(k_m,\, t)]^{1/d}$, dropping an inessential costant factor $F(1)$. In practice $\alpha(x,\, t)$ is obtained as the slope of the plot of $\ln C(x/\mathcal{L}_2(t),\, t)$ versus $\ln \mathcal{L}_1(t)$ for fixed $x$. When the time dependence of $\alpha(x,\, t)$ disappears there is scaling, which is of the standard type if $\alpha(x)$ is independent of $x$ and it is of the multiscaling type if $\alpha(x)$ actually displays a dependence on $x$.

We have solved numerically the CH equation and divided the entire duration of the simulation into nine subsequent and non-overlapping time intervals of exponentially growing duration. We have computed the effective exponent in the different time intervals. In Fig. 2 we have plotted the effective exponent as a function of $x$ in the subsequent time intervals for the quench to $T_F = 0$ with $\Delta = 2.083$ and $d = 3$.

Next we have considered an Ising system[9] on a two-dimensional lattice of size $512 \times 512$ with periodic boundary conditions and initially prepared in an infinite temperature configuration. We have then let the system evolve with Kawasaki spin-exchange dynamics after a sudden quench to the temperature $T_F = T_c/2$. We have computed the structure factor by averaging over $10^2$ realizations of the time histories and we have then extracted the effective exponent $\alpha(x,\, t)$ (Fig. 3), following the same procedure illustrated for the CH equation, during five non-overlapping time intervals.

By inspection it is clear that Figs. 2 and 3 are quite similar, as it should be expected since in the Ising case $\Delta = 1$. The gross feature is that there are two markedly distinct behaviors for $x < x^*$ and for $x > x^*$. To the left of $x^*$ the effective exponent is essentially time independent and follows clearly $\alpha_0(x)$. To the right of $x^*$, instead, $\alpha(x,\, t)$ appears to be essentially flat with respect to $x$ and time dependent over all the intervals. This pattern fits quite well with the previous analysis of the competition between the first and the second term on the rhs of (7). The multiscaling behavior for $x < x^*$ shows that the relative size of the prefactors is such that, for these values of $x$, there exists a long time interval during which the structure factor is dominated by $\Delta L^{\alpha_0(x)}$. This is a preasymptotic mean field scaling which precedes the crossover towards the eventual standard scaling. The duration

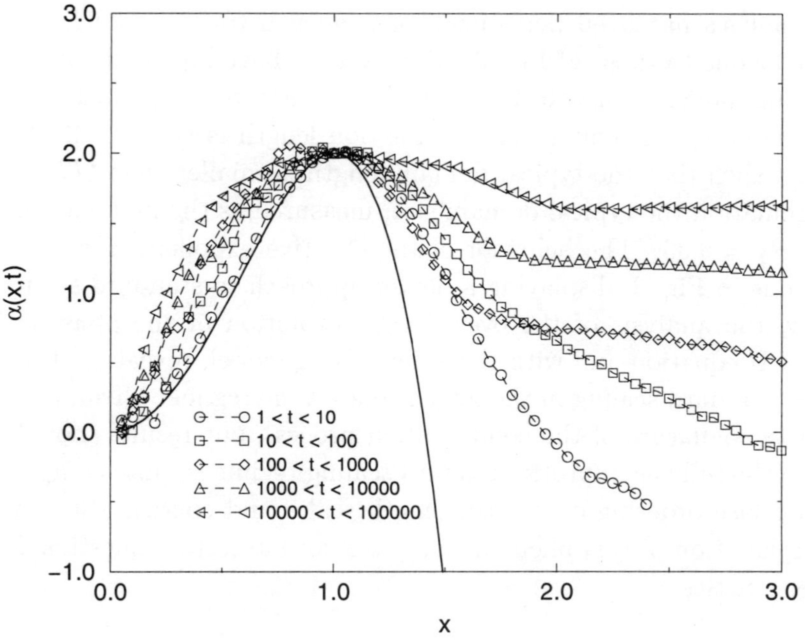


**Fig. 3** Effective exponent for the Kinetic Ising model in $d = 2$.

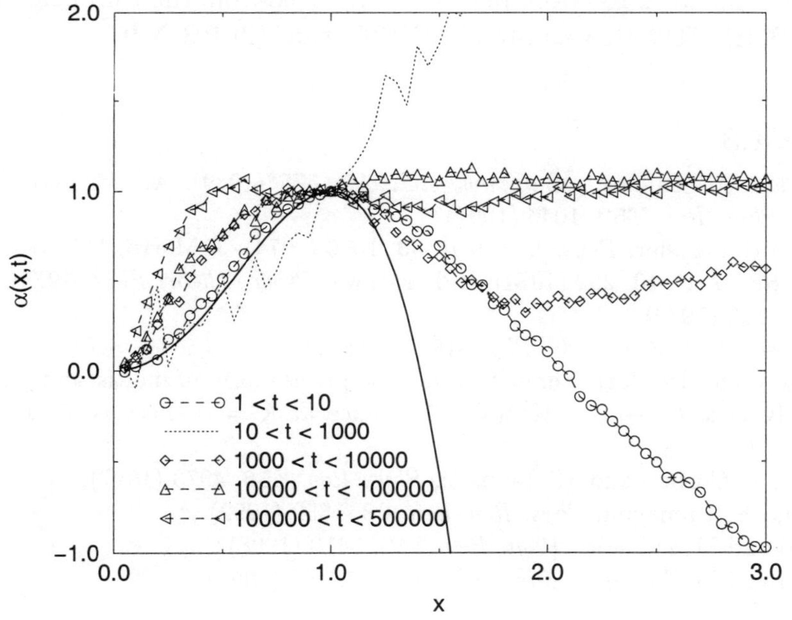


**Fig. 4** Effective exponent for the Kinetic Ising model in $d = 1$.

of our simulations allows to detect just the onset of the crossover and it is not long enough to see the definitive establishment of standard scaling.

For $x > x^*$, instead, $\delta\alpha(x) = d - \alpha_0(x)$ is much too big for observing any multiscaling and the time dependence of the effective exponent, together with the flat behavior as a function of $x$, shows that the crossover towards standard scaling is fully under way. Implicit in this analysis is that the finite value of the temperature $T_F = T_c/2$ is too small to play any role in the preasymptotic regime.[8]

The same analysis has been performed for a quench to $T_F = 0.9T_c$, obtaining results completely analogous to those of Fig. 3. Moreover we have repeated the analysis for a one dimensional Ising model of size 1024 quenched from infinite temperature to $T_F = 0.5J/k_B$. For such a value of $T_F$ the equilibrium correlation length is $\xi(T_F) \simeq 27.3$ and the system orders for times such that the typical domain length is smaller than $\xi(T_F)$. At the longest time in our simulation the typical domain size, measured as the first zero of the correlation function, was $R_g = 9.33$. The behavior of the effective exponent $\alpha(x, t)$ in Fig. 4 is very similar to the one in Fig. 3, displaying a neater approach to standard scaling for $x > 1$.

In summary, the analysis of the overall time evolution of the phase-ordering process, both with the CH equation and with the kinetic Ising model, reveals a clear crossover from multiscaling to standard scaling in the intermediate time regime. Given that multiscaling is the characteristic signature of the condensation process, our results show that the mechanism leading to the fully asymptotic standard scaling regime is quite complex, involving the coexistence of phase ordering over short length scales and condensation over large length scales. The explanation of this phenomenon poses an interesting question for the theory of phase-ordering kinetics.

## ACKNOWLEDGMENTS

We thank the Istituto di Cibernetica of CNR (Arco Felice, Napoli) for generous grant of computer time. This work has been partially supported from the European TMR Network-Fractals c.n. FMRXCT980183 and from MURST through PRIN-97.

## REFERENCES

1. A. Coniglio and M. Zannetti, *Europhys. Lett.* **10**, 575 (1989); A. Coniglio, P. Ruggiero and M. Zannetti, *Phys. Rev.* **E50**, 1046 (1994).
2. K. Binder and D. Stauffer, *Phys. Rev. Lett.* **33**, 1006 (1974); J. Marro, J. L. Lebowitz and M. H. Kalos, *Phys. Rev. Lett.* **43**, 282 (1974); H. Furukawa, *Progr. Theor. Phys.* **59**, 1072 (1978); *Phys. Rev. Lett.* **43**, 135 (1979).
3. For a review, see A. J. Bray, *Adv. Phys.* **43**, 357 (1994).
4. For simplicity we neglect logarithmic factors. The precise form of multiscaling is given in Ref. 1.
5. More precisely $C(\mathbf{k}, t) \to_{t\to\infty} M^2\delta(\mathbf{k} - \mathbf{0}^+)$ since at $\mathbf{k} = 0$ there is no growth due to the conservation law.
6. C. Castellano, F. Corberi and M. Zannetti, *Phys. Rev.* **E56**, 4973 (1997).
7. A. J. Bray and K. Humayun, *Phys. Rev. Lett.* **68**, 1559 (1992).
8. C. Castellano and M. Zannetti, *Phys. Rev.* **E58**, 5410 (1998).
9. C. Castellano and M. Zannetti, *Europhys. Lett.* **47**, 158 (1999).

# Part III

# Scaling

# SCALING IN THE ATMOSPHERE: ON GLOBAL LAWS OF PERSISTENCE AND TESTS OF CLIMATE MODELS

ARMIN BUNDE
*Institute of Theoretical Physics, Justus-Liebig-University Giessen*
*Heinrich-Buff-Ring 16, 35392 Giessen, Germany*

SHLOMO HAVLIN
*Department of Physics, Bar Ilan University, Israel*

**Abstract**

Characterizing the complex atmospheric variability at all pertinent temporal and spatial scales remains one of the most important challenges to scientific research today.[1–5] The main issues are to quantify, within reasonably narrow limits, the potential extent of global warming, and to downscale the global results in order to describe and quantify the regional implications of global change.

## 1. INTRODUCTION

To face these challenges, atmospheric and oceanographic research usually proceeds along two main paths that we are going to decribe next: (i) Analysis of available meteorological records by appropriate time series analysis techniques, and (ii) generation of and analysis of observation-based, interpolated or model simulated weather and climate records, respectively, through the use of a hierarchy of simulation models.

### 1.1 Analysis of Meteorological Methods

Among the standard mathematical techniques that have been used are calculations of means, variations and power spectra, decomposition into empirical orthogonal functions and/or principle components. Very recently, more advanced techniques such as detrended fluctuation analysis and wavelet analysis have been used which are able to systematically separate trends from fluctuations at different time scales.

A considerable amount of effort has been devoted to analyzing temporal correlations that characterize the persistence of weather and climate regimes. The persistence of short term weather states is a well-known phenomenon: there is a strong tendency for subsequent days to remain similar, a warm day is more likely to be followed by a warm day than by a cold day and vice versa. The typical time scale for weather changes is about one week, a time period which corresponds to the average duration of so-called "general weather regimes" or "Grosswetterlagen". This property of persistence is often used as a "minimum skill" forecast for assessing the usefulness of short to medium range numerical weather forecasts. Longer term persistence of synoptic regimes up to time scales of several weeks is often related to circulation patterns associated with blocking.[6] A blocking situation occurs when a very stable high pressure system is established over a particular region and remains in place for several weeks, as opposed to the usual time scale of 3–5 days for synoptic systems. As a result the weather in the region of the high remains fairly persistent throughout the period. Furthermore, transient low pressure systems are deflected around the blocking high so that the region downstream of the high experiences a larger than usual number of storms.

There have been also indications that weather persistence exists over many months or seasons,[7] between successive years, and even over several decades.[8,9] Such persistence is usually associated with slowly varying external (boundary) forcing such as sea surface temperatures and anomaly patterns. On the scale of months to seasons, one of the most pronounced phenomenon is the El Nino Southern Oscillation (ENSO) event which occurs every 3–5 years and which strongly affects the weather over the tropical Pacific as well as over North America.[10] It has also been recently suggested that El Nino years are associated with increased rainfall over Eastern Mediterenean.[11] Although the link between extratropical weather/climate and sea surface temperature has been more difficult to establish, several recent studies have successfully proven that a connection exists on multiyear to decadal time scales between (i) the climate of North America and the North Pacific Ocean,[12] and (ii) the climate over Europe and the North Atlantic Ocean as expressed by the North Atlantic Oscillation (NAO) index.[8,9,13] On the even longer multidecadal to century time scales, external forcing associated with anthropogenic effects (e.g. increasing greenhouse gases and changing land use) also appear to play an important role in addition to the natural variability of the climate system.[14] Clearly separating the anthropogenic forcing from the natural variability of the atmosphere may prove to be a major challenge since the anthropogenic signal may project onto and therefore be hidden in the modes of natural climate variability.[15]

To avoid detection of spurious correlations arising from nonstationarities, new statistical-physics tools (wavelet techniques (WT) and detrended fluctuation analysis (DFA)) have been developed recently (see, e.g. Refs. 16–19). DFA and WT can systematically eliminate trends in the data and thus reveal intrinsic dynamical properties such as distributions, scaling and long-range correlations very often masked by nonstationarities. In a recent series of studies[20] we have used DFA and WT to study temperature correlations in different climatic zones on the globe. The results indicate that a universal long range power law correlation may exist which governs atmospheric variability at all spatiotemporal scales: The persistence, characterized by the auto-correlation $C(s)$ of temperature variations separated by $s$ days, approximately decays as

$$C(s) \sim s^{-\gamma}, \tag{1}$$

with roughly the same exponent $\gamma \simeq 0.7$ for all stations considered. The range of this universal persistence law exceeds one decade, and is possibly even longer than the range of the temperature series considered. There are two major consequences:

- Conventional methods based on moving averages can no longer be used to separate trends from fluctuations.
- Conventional methods for the evaluation of the frequency of extreme low or extreme high temperatures are based on the hypotheris that the temperature fluctuations are essentially uncorrelated. The appearance of long range correlations sheds doubts on these methods.

## 1.2 The Modeling Approach

Regarding the modeling approach towards simulating and explaining atmospheric variability on various time scales, major progress has been made during the last two decades. Today, the research community routinely and extensively makes use of atmospheric, oceanic and coupled ocean-atmosphere circulation models where the major physical processes are included. Moreover, these models include representations of land-surface processes, sea-ice related processes and many other complex forcing mechanisms within the climate system. All processes are represented by mathematical equations which are solved numerically using a three-dimensional grid with a domain, resolution, and complexity determined by the topic of interest. For very long time integrations of thousands of years it is impossible to apply full general circulation models (GCM) due to their extremely high computational demand. Consequently, intermediate complexity models are used in these cases.

For global climate simulations on time scales ranging from months to decades or centuries, GCMs are used with typical resolutions of 200–300 km in the horizontal and 1 km in the vertical.[21] For regional climate simulations, similar models are used but the domain is limited to a few thousand km and the horizontal resolution is typically increased to 50 km or less.[22,23] In any case the grid resolution can never explicitly simulate all relevant scales of motion. This necessitates the parameterization of the smaller, subgrid scale processes such as cloud physics, radiative transfer, and macroscale turbulent mixing. Recently, it has also been found that previously neglected processes such as dust induced heating may also be important.[24]

The state-of-the-art coupled ocean-atmosphere general circulation models are able to simulate many of the important large-scale features of the climate system rather well. This includes seasonal, horizontal and vertical variations. They also explain the response to greenhouse gases and aerosols in terms of physical processes. In addition to this, other less pronounced variations in climate are reproduced with reasonable accuracy (e.g. the relationship between El Nino and rainfall in Central America and the northern part of South America).

The systematic evaluation and intercomparison of climate model results has proven to be a useful and effective mechanism for identifying common model weaknesses. In general, evaluations have been conducted for the atmospheric, oceanic, land-surface and sea-ice components of the models, and for the sensitivity of the links among these components. Until now these validations have not addressed the question of whether such models can reproduce the long-term climate memory in an appropriate way. If the simulations of the model are valid, then the patterns and relationships discovered by analyzing real observations and data must also be identifiable in the virtual world as represented by the model outputs.

This article is organized as follows: In Secs. 2 and 3 we describe the detrending methods and its application to temperature records of several meteorological stations around the globe. In Sec. 4 we show how the current GCMs can be tested by applying the detrending techniques of Sec. 2 to the model data. We show that we can judge the models by their ability of reproducing the proper type of trends and long-range correlations inherent in the real data.

## 2. RECORD ANALYSIS

Consider, e.g. a record $T_i$ of maximum daily temperatures measured at a certain meteorological station. The index $i$ counts the days in the record, $i = 1, 2, \ldots, N$. For eliminating the periodic seasonal trends, we concentrate on the departures of the $T_i$, $\Delta T_i = T_i - \overline{T}_i$, from the mean maximum daily temperature $\overline{T}_i$ for each calendar date $i$, say 1st of April, which has been obtained by averaging over all years in the temperature series.

There exist two powerful detrending analysis methods: (a) the detrended fluctuation analysis (DFA) and (b) the wavelets methods (WT). The DFA was originally developed by Peng et al.[18] to investigate long-range correlations in DNA sequences and heart beat intervals, where nonstationarities similar to the nonstationarities in the temperature records[20] can occur. The wavelet methods in general are very convenient techniques to investigate fluctuating signals.[25] In this article we shall focus on the DFA. A very useful introduction to the wavelet technique with several applications is given in.[26]

Both DFA and wavelets techniques have been used to analyze the correlation function $C(s)$ of temperature records. The correlation function describes, how the persistence decays in time. $C(s)$ is defined by $C(s) = \langle \Delta T_i \Delta T_{i+s} \rangle$. The average $\langle \ldots \rangle$ is over all pairs with same time lag $s$. For reducing the level of noise present in the finite temperature series, we consider the "temperature profile"

$$Y_n = \sum_{i=1}^{n} \Delta T_i, \quad n = 1, 2 \ldots N\,. \tag{2}$$

We can consider the profile $Y_n$ as the position of a random walker on a linear chain after $n$ steps. The random walker starts at the origin and performs, in the $i$th step, a jump of length $\Delta T_i$ to the right, if $\Delta T_i$ is positive, and to the left, if $\Delta T_i$ is negative. According to random walk theory, the fluctuations $F^2(s)$ of the profile, in a given time window of size $s$, are related to the correlation function $C(s)$. For the relevant case (1) of long-range power-law correlations, $C(s) \sim s^{-\gamma}, 0 < \gamma < 1$, the mean-square fluctuations $\overline{F^2(s)}$, obtained by averaging over many time windows of size $s$ (see below) increase by a power law,[27]

$$\overline{F^2(s)} \sim s^{2\alpha}, \quad \alpha = 1 - \gamma/2\,. \tag{3}$$

For uncorrelated data (as well as for correlations decaying faster than $1/s$), we have $\alpha = 1/2$.

To find how the square-fluctuations of the profile scale with $s$, we first divide each record of $N$ elements into $K_s = [N/s]$ nonoverlapping subsequences of size $s$ starting from the beginning and $K_s$ nonoverlapping subsequences of size $s$ starting from the end of the considered temperature series. We determine the square-fluctuations $F_\nu^2(s)$ in each segment $\nu$ and obtain $\overline{F^2(s)}$ by averaging over all segments. When plotted in a double logarithmic way, the fluctuation function

$$F(s) \equiv [\overline{F^2(s)}]^{\frac{1}{2}} \sim s^{\alpha} \tag{4}$$

is a straight line at large $s$ values, with a slope $\alpha > \frac{1}{2}$ in the case of long range correlations.

The various methods differ in the way, the fluctuation function is calculated.

## 2.1 Fluctuation Analysis (FA)

In the simplest type of analysis (where trends are not going to be eliminated), we obtain the fluctuation functions just from the values of the profile at both endpoints of the $\nu$th segment,

$$F_\nu^2(s) = [Y_{\nu s} - Y_{(\nu-1)s}]^2, \tag{5}$$

and average $F_\nu^2(s)$ over the $2K_s$ subsequences

$$\overline{F^2(s)} = (1/K_s) \sum_{\nu=1}^{K_s} F_\nu^2(s). \tag{6}$$

Here, $\overline{F^2(s)}$ can be viewed as mean square displacement of the random walker on the chain, after $s$ steps. We obtain Ficks diffusion law $\overline{F^2(s)} \sim s$ for uncorrelated $\Delta T_i$ values.

We like to note that this fluctuation analysis corresponds to the R/S method introduced by Hurst (for a review, see e.g. Ref. 28). Since both methods do not eliminate trends, they do not give a clear picture when used alone. In many cases they can not distinguish between trends and long-range correlations when applied to a time records without supplementary calculations.

## 2.2 Detrending Fluctuation Analysis (DFA)

There are different orders of DFA that are distinguished by the way the trends in the data are eliminated. In lowest order (DFA1) we determine, for each subsequence $\nu$, the best *linear* fit of the profile, and identify the fluctuations by the standard deviation $F_\nu^2(s)$ of the profile from this straight line. This way, we eliminate the influence of possible linear trends on scales larger than the segment sizes. Note that linear trends in the profile correspond to patch-like trends in the original record. DFA1 has been proposed originally by Peng et al.[18] when analyzing correlations in DNA.

DFA1 can be generalized straightforwardly to eliminate higher order trends: In second order DFA (DFA2) one calculates the standard deviations $F_\nu^2(s)$ of the profile from best *quadratic* fits of the profile, this way eliminating the influence of possible linear and parabolic trends on scales larger than the segment considered. In general, in the $n$th-order DFA technique, we calculate the deviations of the profile from the best $n$th-order polynomial fit and can eliminate this way the influence of possible $(n-1)$th-order trends on scales larger than the segment size.

It is essential in the DFA-analysis that the results of several orders of DFA (e.g. DFA1–DFA5) are compared with each other. The results are only reliable when above a certain order of DFA they yield the same type of behavior.[29] When compared with FA one can get additional insight into possible nonstationarities in the data.

# 3. ANALYSIS OF TEMPERATURE RECORDS

Figures 1 and 2 show the results of the FA and DFA analysis of the maximum daily temperatures $T_i$ of the following weather stations (the length of the records is written within the parentheses): Luling (USA, 90 y), and Kasan (Russia, 96 y) [Figs. 1(a) and (c)], Tuscon (USA, 97 y), Melbourne (136 y), Seoul (86 y), Prague (218 y) Figs. 2(a)–(d)]. The results are typical for a large number of records that we have analyzed so far (see Ref. 20).

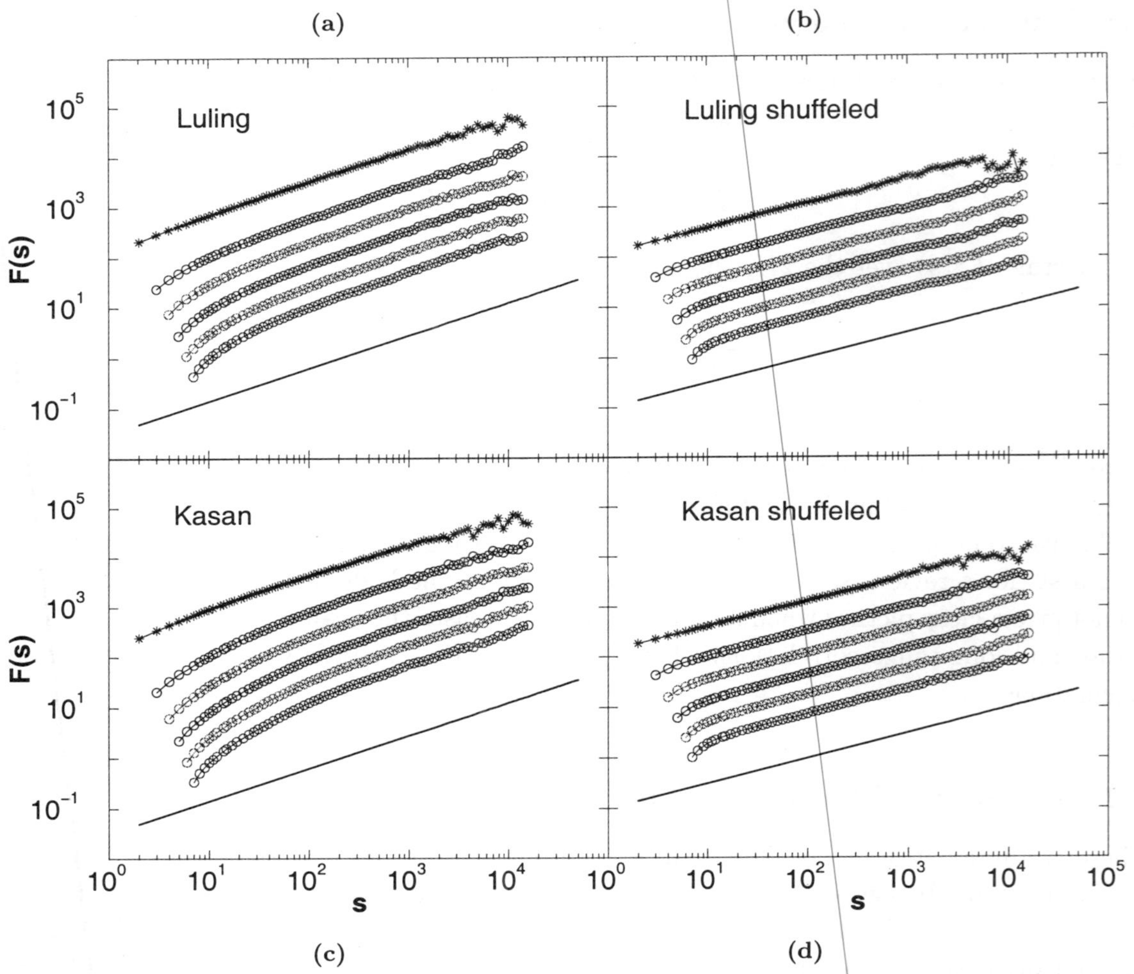


**Fig. 1** Analysis of daily maximum temperature records of Luling and Kasan. The analysis of the real data shown in **(a)** and **(c)** is compared with the analysis of the corresponding shuffled data shown in **(b)** and **(d)**. The four figures show the fluctuation functions obtained by FA, DFA1, DFA2, DFA3, DFA4, and DFA5 (from top to bottom) for the four sets of data. The scale of the fluctuation functions is arbitrary.

In the log-log plots, all curves are (except at small $s$-values) approximately straight lines, with a slope $\alpha \cong 0.65$. There exists a natural crossover (above the DFA-crossover) that can be best estimated from FA and DFA1. As can be verified easily, the crossover occurs roughly at $t_c = 10d$, which is the order of magnitude for a typical Grosswetterlage. Above $t_c$, there exists long-range persistence expressed by the power-law decay of the correlation function with an exponent $\gamma = 2 - 2\alpha \cong 0.7$. The results seem to indicate that the exponent is universal, i.e. does not depend on the location and the climatic zone of the weather station. Below $t_c$, the fluctuation functions do not show universal behavior and reflect the different climatic zones.

To test our claim that the slope $\alpha \cong 0.65$ is due to long-range correlations, and does not result from a singular behavior of the probalitity distribution function of the $\Delta T_i$ we have eliminated the correlations by randomly shuffling the $\Delta T_i$. By definition this shuffling has no effect on the probability distribution function of the $\Delta T_i$, which we found to be approximately Gaussian at large temperature variations. The right hand side of Fig. 1 [Figs. 1(b) and (d)] show the effect of shuffling on the fluctuation functions. By comparing the left hand sides of the figures with the right hand sides we see the effect of correlations.

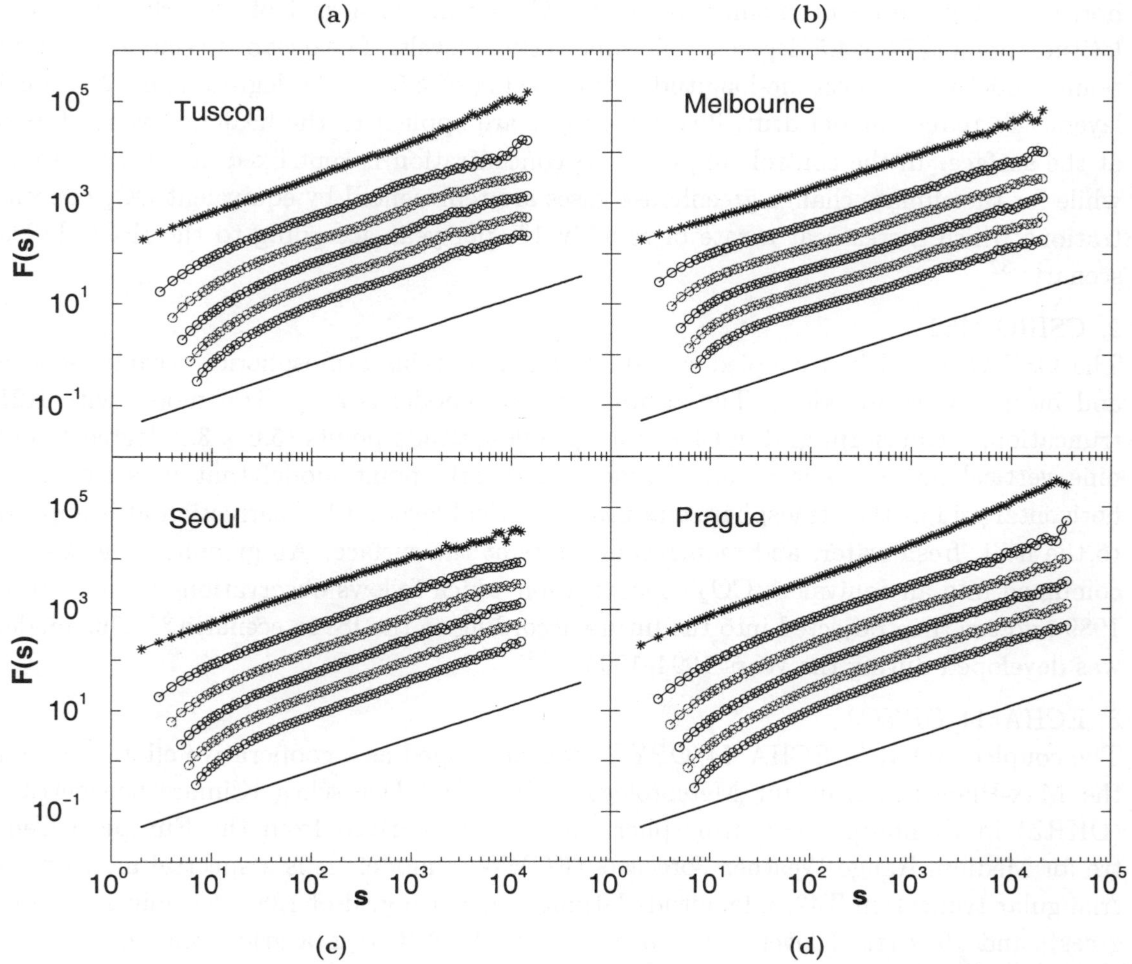


**Fig. 2** Analysis of the daily maximum temperature records of Tucson, Melbourne, Seoul and Prague (as Fig. 1).

The exponent $\alpha$ characterizing the fluctuations in the shuffled uncorrelated sequence is $1/2$, as expected.

Since the exponent does not depend on the location of the meteorological station and its local environment, the power law behavior can serve as an ideal test for climate models where regional details cannot be incorporated and therefore regional phenomena like urban warming cannot be accounted for. The power law behavior seems to be a global phenomenon and therefore should also show up in the simulated data of the GMCS. As mentioned earlier, the presence of long range correlations has far-reaching consequences on the possible detection of trends directly from the record and on the evaluation of extreme events.

## 4. ANALYSIS OF SIMULATED TEMPERATURE RECORDS

Next we turn to the analysis of simulated data that were obtained by four general circulation models around Praque. We have chosen Prague, since the Prague record is the longest record we could get. The models are:

- 1. GFDL-R15-a

  This is the latest version of a coupled atmosphere-ocean model (AOGCM) that has been developed over many years.[30,31] The atmosperic sub-model is a spectral model with a horizontal truncation of rhomboidal 15 (R15), a transform grid of 48 × 40 longitude-latitude points (7.5 × 4.5 degrees), and nine vertical levels. The ocean sub-model is a grid point model with a latitude-longitude grid spacing of 4.5 × 3.75 degrees and 12 vertical layers. To reduce model drift, flux corrections are applied to the heat and water fluxes at the surface. In the control run, the $CO_2$ concentration is kept fixed at the 1958 value while for the climate change greenhouse gases are represented by equivalent $CO_2$ concentrations which increase at a rate of roughly 1% per year according to the IPCC IS92a scenario.[32]

- 2. CSIRO-Mk2
  The CSIRO model is a coupled AOGCM which contains atmospheric, oceanic, sea-ice and biospheric sub-models. The atmospheric sub-model is a spectral model with R21 truncation, a transform grid of 64 × 56 longitude-latitude points (5.6 × 3.2 degrees), and nine vertical layers. The ocean sub-model is a grid point model that uses the same horizontal grid as the atmosphere and has 21 vertical levels. Flux correction at is applied to the heat, fresh water, and momentum fluxes at the surface. All greenhouse gases are combined into an equivalent $CO_2$ concentration which follows observations from 1880 to 1989 and are then pojected into the future according to the IS92a scenario.[32] This model was developed during the years 1994–1995.[33,34]

- 3. ECHAM4/OPYC3
  The coupled AOGCM ECHAM4/OPYC3 was developed as a cooperative effort between the Max-Planck-Institut für Meteorologie (MPI) and Deutsches Klimarechenzentrum (DKRZ) in Hamburg. The atmospheric model was derived from the European Centre for Medium Range Weather Forecasts (ECMWF) model. It is a spectral model with triangular truncation T42, a longitude-latidue transform grid of 128 × 64 points (2.8 degrees), and 19 vertical levels. The ocean model (OPYC3) is a grid point model with 11 isopycnal layers and it is run on the same grid as the atmosphere. Flux correction is applied to the heat, fresh water and momentum fluxes at the surface.[35–37] Historic greenhouse gas concentrations are used from 1860–1989 and from 1990 onward they are projected according to the IS92a scenario.

- 4. HADCM3
  HADCM3 is the latest version of the coupled AOGCM developed at the Hadley Centre.[38] Unlike the other models described above, here the atmospheric model is a grid point model with a longitude-latitude grid of 96 × 73 points (3.75 × 2.5 degree spacing) and it has 19 vertical levels. The ocean model has a horizontal resolution of 1.25 degrees in both latitude and longitude and 20 vertical levels. No flux correction is applied at the surface. Historic greenhouse gas concentrations are used during the period 1860–1989. From 1990 and onward they are increased according to the IS95a scenario (a slightly modified version of IS92a).

For each model, we obtained the temperature records (mean monthly data) of the four grid points closest to Prague from the internet.[39] We interpolated the data at the location of Prague.

Figure 3(a) shows the results of the FA- and DFA-Analysis for the real temperature record of Prague that starts in 1775 and ends in 1992. Figures 3(b)–(d) show the results obtained from ECHAM4, CSIRO and HADCM, that end up at the same year as the real record. The available data of GFDL cover only 40y, so we do not present them here.

We are interested in the way the models can reproduce the actual data regarding (a) trends and (b) long-range correlations. Of course, we cannot expect the models to reproduce local

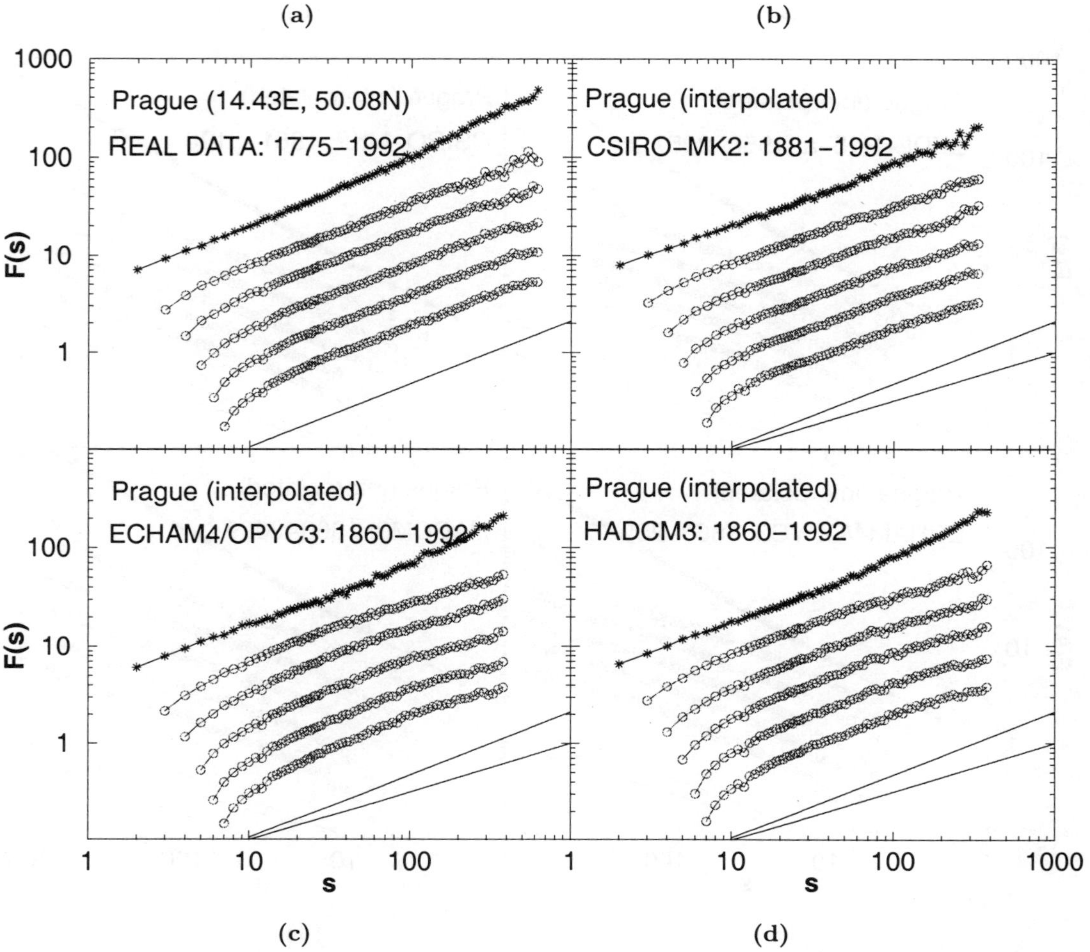


**Fig. 3** FA- and DFA-Analysis of **(a)** the monthly mean temperature record of Prague and **(b)** simulated interpolated monthly mean temperature records at the geographical position of Prague, for three general circulation models: **(b)** CSIRO, **(c)** ECHAM4 and **(d)** HADCM3. The four figures show the fluctuations functions obtained by FA, DFA1, DFA2, DFA3, DFA4 and DFA5 (from top to bottom) for the four sets of data. The scale of the fluctuations is arbitrary (after Govindan et al.).[40]

trends like urban warming or short-term correlation structures. But the long-range correlations we discussed in the previous section show characteristic universal features that are actually independent of the local environment around a station. So we can expect that successful models with good prognostic features will be able to reproduce them.

As discussed already above (for the daily data), the FA- and DFA fluctuation functions for the real temperature record of Prague have approximately the same slope of $\alpha \simeq 0.65$ in the double-logarithmic plot (shown as straight line in the figure). At large time scales there is a slight increase of the FA-function (which clearly indicates a weak trend). In contrast, the FA-results for the HADLEY and ECHAM4 data show a pronounced trend above 100 months represented by a large slope. For CSIRO, the FA-result is not so conclusive since they scatter considerably at large scales. It seems that two of the three models overestimate the trend in the past. Regarding scaling, the DFA curves for CSIRO show good straight lines in a double logarithmic presentation. The exponents are close to the exponents from the Prague record. In contrast, ECHAM4 and HADLEY show a crossover to an exponent $\alpha = 0.5$ after about 3 years. The exponent $\alpha = 0.5$ indicates loss of persistence. Hence ECHAM4 and HADLEY reproduce data sets that show a linear trend at large time scales and simultaneously the lack of correlations exceeding 3y, in contrast to the reality.

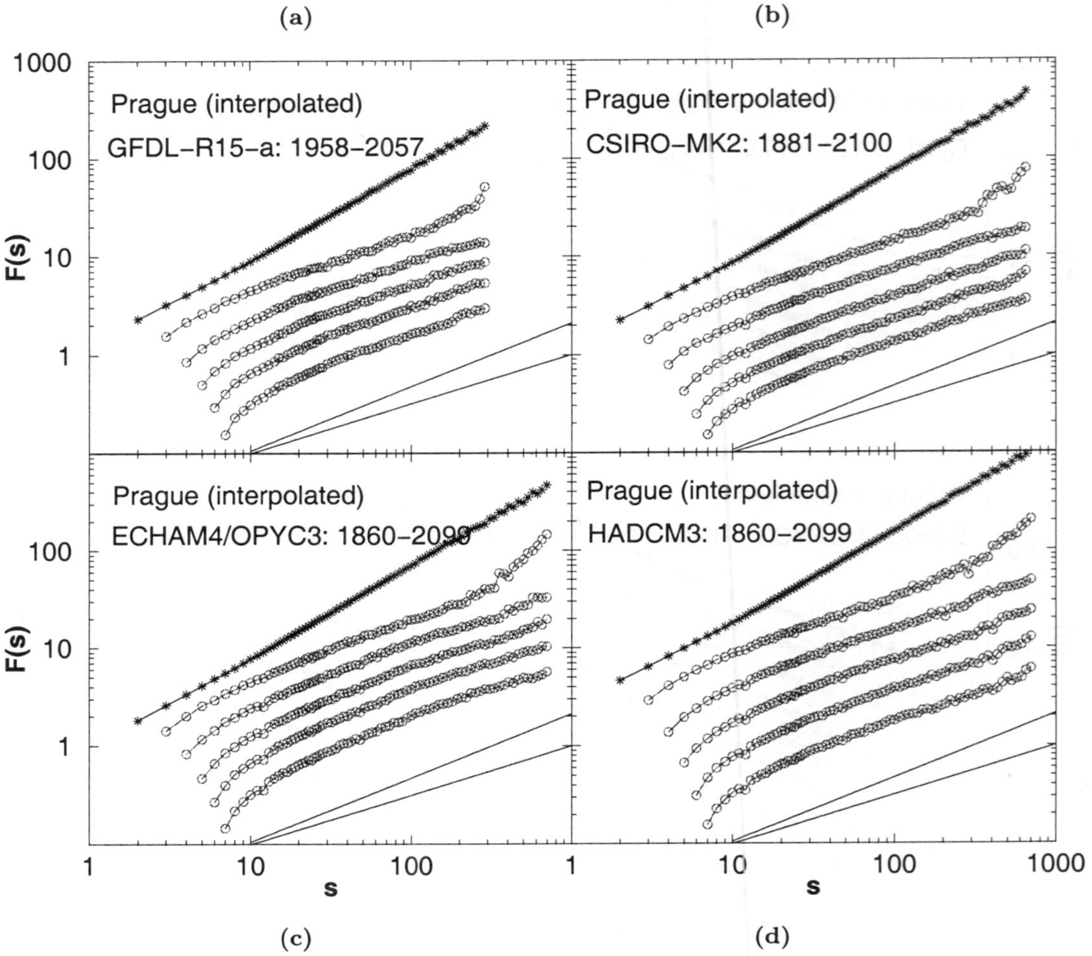


**Fig. 4** FA- and DFA-Analysis of the simulated interpolated monthly mean temperature records of the geographical position of Prague, for four general circulation models: **(a)** GFDL, **(b)** CSIRO, **(c)** ECHAM4 and **(d)** HADCM3. While Fig. 3 considered only data in the past, Fig. 4 considers the whole set of data (past and future). The data is available in the internet[39] and, the analysis is after Govindan et al.[40]

The scaling features remain very similar when the data sets from the models are extended into the 21st century (see Fig. 4).[40] We consider this an important issue, since it shows an internal consistency of the models. However, the trends, which show up in FA and DFA1, are much more pronounced compared to the data from Fig. 3.

We have obtained similar qualitative behavior also for other simulated temperature record. From the trends, one estimates the warming of the atmosphere in the future. Since the trends are almost not visible in the real data and overestimated by the models in the past, it seems possible that the trends are also overestimated by the models in the future. From this point of view it cannot be excluded that the global warming in the next 100y will be less pronounced than predicted by the models.

## REFERENCES


1. IPCC, *The Regional Impacts of Climate Change. An Assessment of Vulnerability* (University Press, Cambridge, 1998).
2. P. M. Blaikie, T. Cannon, I. Davi and B. Wisner, *At Risk: Natural Hazards, Vulnerability and Disasters* (Routledge, London, 1994).
3. H. Grassl, "Global Climate Change," *Interdisc. Sci. Rev.* **24**(3), 185–194 (1999).

4. H. J. Schellnhuber, *Nature* **402**(6761), C19–C23 Suppl. (1999).
5. K. Hasselmann, *Clim. Dynam.* **13**, 601 (1997), and the references therein.
6. J. G. Charney and J. G. Devore, *J. Atmos. Sci.* **36**, 1205 (1979).
7. J. Shukla, *Science* **282**, 728–731 (1998).
8. R. L. Molinari, D. A. Mayer, J. F. Festa and H. F. Bezdek, *J. Geophys. Res.* **102**, 3267 (1997).
9. R. T. Sutton and M. R. Allen, *Nature* **388**, 563 (1997).
10. S. G. Philander, *El Nino, La Nina and the Southern Oscillation.* International Geophysics Series, Vol. 46, 1990.
11. C. Price, L. Stone, A. Huppert, B. Rajagopalan and P. Alpert, *Geophys. Res. Lett.* **25**, 3963 (1998).
12. M. Latif and T. P. Barnett, *Science* **266**, 634–637 (1994).
13. M. J. Rodwell, D. P. Rodwell and C. K. Folland, *Nature* **398**, 320–323 (1999).
14. K. Hasselmann, *Science* **276**, 914–915 (1997).
15. S. Corti, F. Molteni and T. N. Palmer, *Nature* **398**, 799–802 (1999).
16. P. Ch. Ivanov, A. Bunde, L. A. N. Amaral, S. Havlin, J. Fritsch-Yelle, R. M. Baevsky, H. E. Stanley and A. L. Goldberger, *Europhys. Lett.* **48**(59), 4 (1999).
17. A. Bunde, S. Havlin, J. W. Kantelhardt, T. Penzel, J. H. Peter and K. Voigt, *Phys. Rev. Lett.* **85**, 3736 (2000).
18. C.-K. Peng, S. V. Buldyrev, A. L. Goldberger, S. Havlin, F. Sciortino, M. Simons and H. E. Stanley, *Nature* **356**, 168 (1992); S. V. Buldyrev, A. L. Goldberger, S. Havlin, C. K. Peng, M. Simons and F. Sciortino, *Phys. Rev. Lett.* **71**, 1776 (1993); S. V. Buldyrev, A. L. Goldberger, S. Havlin, R. N. Mantegna, M. E. Matsa, C.-K. Peng, M. Simons and H. E. Stanley, *Phys. Rev.* **E51**, 5084 (1995).
19. A. Arneodo, E. Bacry, P. V. Graves and J. F. Muzy, *Phys. Rev. Lett.* **74**, 3293 (1995); A. Arneodo et al., *Physica* **D96**, 291 (1996).
20. E. Koscielny-Bunde, A. Bunde, S. Havlin, H. E. Roman, Y. Goldreich and H.-J. Schellnhuber, *Phys. Rev. Lett.* **81**, 729 (1998); E. Koscielny-Bunde, H. E. Roman, A. Bunde, S. Havlin and H. J. Schellnhuber, *Phil. Mag.* **B77**, 1331 (1998); E. Koscielny-Bunde, A. Bunde, S. Havlin and Y. Goldreich, *Physica* **A231**, 393 (1996).
21. R. Voss and U. Mikolajewicz, Max-Planck-Institut fur Meteorologie, Report No. 298 (1999); S. Brenner, *J. Climate* **9**, 3337 (1996).
22. F. Giorgi, *J. Climate* **3**, 941–963 (1990).
23. E. S. Takle, W. J. Gutowski Jr., R. W. Arritt, Z. Pan, C. Anderson, R. Silva, D. Caya, S. Chen, J. H. Christensen, S-Y. Hong, H.-M. H. Juang, H. Katzfey, W. Lapenta, R. Laprise, P. Lopez, J. McGregor and J. R. Roads, "Project to Intercompare Regional Climate Simulations (PIRCS)," *J. Geophys. Res.*, in press (1999).
24. P. Alpert, Y. J. Kaufman and Y. Shay-El, *Nature* **395**(6700), 367–370 (1998).
25. M. Vetterli and J. Kovacevic, *Wavelets and Subbane Coding* (Prentice Hall, Englewood Cliffs, New Jersey, 1995).
26. A. Arneodo, E. Bacry, P. V. Graves and J. F. Muzy, in *The Science of Disasters*, eds. A. Bunde, J. Kropp and H. J. Schellnhuber (Springer, Berlin, 2001).
27. A. Bunde and S. Havlin (eds.), *Fractals in Science* (Springer, New York, 1995).
28. J. Feder, *Fractals* (Plenum, New York, 1989).
29. J. W. Kantelhardt, E. Koscielny-Bunde, H. A. Rego, A. Bunde and S. Havlin, *Physica A*, in press (2001).
30. S. Manabe, R. J. Stouffer, M. J. Spelman and K. Bryan, *J. Climate* **4**, 785 (1991).
31. S. Manabe, M. J. Spelman and R. J. Stouffer, *J. Climate* **5**, 105 (1992).
32. J. Leggett, W. J. Pepper and R. J. Swart, in *Climate Change 1992: The Supplementary Report to the IPCC Scientific Assessment*, eds. J. T. Houghton, B. A. Callander and S. K. Varney (Cambridge University Press, Cambridge, 1992).
33. H. B. Gordon and S. P. O'Farrell, *Mon. Wea. Rev.* **125**, 875 (1997).
34. A. C. Hirst, H. B. Gordon and S. P. O'Farrell, *Geophys. Res. Lett.* **23**, 3361 (1996).
35. L. Bengtsson, K. Arpe and E. Roeckner et al., *Clim. Dynam.* **12**(4), 261 (1996).
36. ECHAM3, *Atmospheric General Circulation Model*, ed. Dkrz-Model User Support Group (October 1996).

37. *The OPYC Ocean General Circulation Model*, ed. Joseph M. Oberhuber (October 1992). Deutsches Klimarechenzentrum (DKRZ), Report No. 7, (1997).
38. C. Gordon, C. Cooper and C.A. Senior et al., *Clim. Dynam.* **16**(2–3), 147 (2000).
39. Internet address: http://*ipcc – ddc.cru.uca.ac.uk/dkrz/dkrz – index.html*
40. Govindan et al., *Physica A*, in press.

# NONLINEAR RELAXATION IN POPULATION DYNAMICS

MARKUS A. CIRONE*
*INFM, Unitá di Palermo, and Dipartimento di Fisica e Tecnologie Relative Universitá di Palermo, Viale delle Scienze, Palermo, I-90128, Italy*

FERDINANDO DE PASQUALE
*INFM, Unitá di Roma, and Dipartimento di Fisica, Universitá di Roma "La Sapienza", P.le A. Moro 2, Roma, I-00185, Italy*

BERNARDO SPAGNOLO
*INFM, Unitá di Palermo, and Dipartimento di Fisica e Tecnologie Relative Universitá di Palermo, Viale delle Scienze, Palermo, I-90128, Italy*

**Abstract**

We analyze the nonlinear relaxation of a complex ecosystem composed of many interacting species. The ecological system is described by generalized Lotka-Volterra equations with a multiplicative noise. The transient dynamics is studied in the framework of the mean field theory and with random interaction between the species. We focus on the statistical properties of the asymptotic behaviour of the time integral of the $i$th population and on the distribution of the population and of the local field.

## 1. INTRODUCTION

Systems of interacting biological species evolve through a dynamical complex process that can be conveniently described, within relatively short time scales, by generalized Lotka-Volterra equations.[1] The nonlinearity of these equations complicates their analytical investigation, especially in the case of a great number of interacting species. Nevertheless some analytical approximation for the mean field interaction between the species as well

*Present address: Abteilung für Quantenphysik, Universität Ulm, D-89069 Ulm, Germany.

as numerical simulations give some insight into the behaviour of complex ecosystems.[2–4] Basic elements of a Lotka-Volterra model are the growth parameter and the interaction parameter. For a large number of interacting species, it is reasonable, as a phenomenological approach, to choose these parameters at random from given probability distributions. Within this type of representation, the dynamics of coevolving species can be characterized by statistical properties over different realizations of parameter sets. Though the generalized Lotka-Volterra model has been explored in detail,[1] it seems that a full characterization, either deterministic or statistical, of the conditions under which a population extinguishes or survives in the competition process has not been achieved.[5,6]

In this paper we analyze the role of the noise on the transient dynamics of the ecosystem of many interacting species in the presence of an absorbing barrier, i.e. extinction of the species. Two type of interaction between the species are considered: (a) mean field interaction, and (b) random interaction. We focus on the asymptotic behaviour of the time integral of the $i$th population and on the distributions of the population and of the local field, which is the total interaction of all species on the $i$th population.

By introducing an approximation for the time integral of the average species concentration $M(t)$ we obtain analytical results for the transient behaviour and the asymptotic statistical properties of the time average of the $i$th population. We find that for a very large number of interacting species the statistical properties of the time average of the $i$th population process are determined asymptotically from the statistical properties of the Wiener process. At the critical point and around the stability-instability transition, for mean field interaction, the system goes from a purely long time tail behaviour (namely $M(t) \sim \sqrt{t}$ to a new long time modified regime $M(t) \sim \sqrt{t}e^{\sqrt{t}}$. Specifically for random interaction we find that the local field and the cavity field,[7] which is the total interaction of all species on the $i$th population when this population is absent, are different in absence of external noise while overlap quite well in the presence of the noise. This behaviour, which is very different from the analogue spin glasses problem in statistical mechanics, is reminiscent of a phase transition phenomenon. It suggest that, becuase all population are positive and can grow during the dynamical process of the ecosystem, each population play an important role on the total interaction between the species.

The paper is organized as follows. In the next section we describe the model. The mean field and the random interactions are considered in Sec. 3 and 4. Our results are discussed in Sec. 5.

## 2. THE MODEL

We consider an N-species generalization of the usual Lotka-Volterra system with a Malthus-Verhulst modelization of the self regulation processes for a fully connected ecological network. Therefore the Ito stochastic differential equation describing the dynamical evolution of the ecosystem is

$$dn_i(t) = \left[\left(\gamma + \frac{\varepsilon}{2}\right) - n_i(t) + \sum_{j \neq i} J_{ij} n_j(t)\right] n_i(t)dt + \sqrt{\varepsilon} n_i(t) dw_i\,, \quad i = 1, \ldots, N \tag{1}$$

where $n_i(t) \geq 0$ is the number of elements of the $i$th species. In Eq. (1) $\gamma$ is the growth parameter, the interaction matrix $J_{ij}$ modelizes the interaction between different species $(i \neq j)$ and $w_i$ is the Wiener process whose increment $dw_i$ satisfies the usual statistical properties

$$\langle dw_i(t)\rangle = 0; \quad \langle dw_i(t) dw_j(t')\rangle = \delta_{ij}\delta(t - t')dt\,. \tag{2}$$

Our ecosystem is composed of $N = 1000$ species. We consider all species equivalent so that the characteristic parameters of the ecosystem are independent of the species. The random interaction with the environment (climate, disease, etc.) is taken into account by introducing a multiplicative noise in the Eq. (1). The solution of the dynamical equation Eq. (1) is given by

$$n_i(t) = \frac{n_i(0) \exp\left[\delta t + \sqrt{\varepsilon} w_i(t) + \int_0^t dt' \sum_{j \neq i} J_{ij} n_j(t')\right]}{1 + \gamma n_i(0) \int_0^t dt' \exp\left[\delta t' + \sqrt{\varepsilon} w_i(t') + \int_0^{t'} dt'' \sum_{j \neq i} J_{ij} n_j(t'')\right]} . \tag{3}$$

We consider two different types of interaction between the species: (a) a mean field approximation with a symbiotic interaction between the species; (b) a random interaction between the species with different types of mutual interactions: competitive, symbiotic and prey-predator relationship.

## 3. MEAN FIELD APPROXIMATION

We consider a mean field symbiotic interaction between the species. As a consequence the growth parameter is proportional to the average species concentration

$$\sum_{j \neq i} J_{ij} n_j(t) = \frac{J}{N} \sum_j n_j(t) = Jm(t) , \tag{4}$$

and the stochastic differential equation Eq. (1) becomes

$$dn_i = \left[\left(Jm + \gamma + \frac{\varepsilon}{2}\right) n_i - n_i^2\right] dt + \sqrt{\varepsilon} n_i dw_i . \tag{5}$$

In the limit of a large number of interacting species the stochastic evolution of the system is given by the following integral equation

$$M(t) = \frac{1}{N} \sum_i \ln\left(1 + n_i(0) \int_0^t dt' e^{JM(t') + \gamma t' + \sqrt{\varepsilon} w_i(t')}\right) , \tag{6}$$

where

$$M(t) = \frac{1}{N} \sum_i \int_0^t dt' n_i(t') = \int_0^t dt' m(t') . \tag{7}$$

is the time integral of the site population concentration average. We introduce an approximation of Eq. (6) which greatly simplifies the noise affected evolution of the system and allows us to obtain analytical results for the population dynamics. We note that in this approximation the noise influence is taken into account in a nonperturbative way, and that the statistical properties of the time average process $M(t)$ are determined asymptotically from the statistical properties of the process $w_{\max}(t) = \sup_{0<t'<t} w(t')$, where $w$ is the Wiener process. Starting from the following approximated integral equation for $M(t)$

$$M(t) \simeq \frac{1}{N} \sum_i \ln\left(1 + n_i(0) e^{\sqrt{\varepsilon} w_{\max_i}} \int_0^t dt' e^{JM(t') + \gamma t'}\right) \tag{8}$$

it is possible to analyze the role of the noise on the stability-instability transition in three different regimes of the nonlinear relaxation of the system: (i) towards the equilibrium population ($\gamma > 0$), (ii) towards the absorbing barrier ($\gamma < 0$), (iii) at the critical point ($\gamma = 0$). Specifically at the critical point we obtain for the time average process $M(t)/t$ as a dominant asymptotic behaviour in the stability region (namely when $J < 1$)

$$\frac{M(t)}{t} \simeq \left(\frac{1}{1-J}\right)\sqrt{\frac{2\varepsilon}{\pi}}\frac{1}{\sqrt{t}}, \tag{9}$$

and in the instability region (namely when $J > 1$)

$$\frac{M(t)}{t} \simeq e^{\langle \ln(n_i(0))\rangle}\sqrt{\frac{2\pi}{\varepsilon}}\,\frac{\exp\left[\sqrt{\frac{2\varepsilon}{\pi}}\sqrt{t}\right]}{\sqrt{t}}. \tag{10}$$

We obtain for the case (i) an explicit expression of the transition time $t_c$ as a function of the noise intensity ($\varepsilon$), the initial population distribution ($n_i(0)$) and the parameters of the system ($\gamma$, $J$)

$$t_c \simeq \frac{1}{\gamma}\left\{\left[\frac{\varepsilon}{2\pi\gamma} + \ln\left(1 + \left(\frac{\gamma}{(J-1)}\right)e^{-\langle \ln(n_i(0))\rangle}\right)\right]^{1/2} - \sqrt{\frac{\varepsilon}{2\pi\gamma}}\right\}^2. \tag{11}$$

For the cases (ii) and (iii) we obtain two implicit expressions in terms of exponential and error functions of the same quantities: $t_c, \varepsilon$ and the system paramters.[3] The transition time increases from $\gamma > 0$ to $\gamma < 0$ according to the following inequality

$$(t_c)_{\gamma<0} > (t_c)_{\gamma=0} > (t_c)_{\gamma>0}\,. \tag{12}$$

This means that when the interaction between the species prevails over the resources, the presence of a hostile environment ($\gamma < 0$) causes a late start of the divergence of some population (i.e. the instability). The noise forces the system to sample more of the available range in the parameter space and therefore moves the system towards the instability. The effect of the noise is to make unstable the system earlier than in the deterministic case ($\varepsilon = 0$). If we raise the intensity of the noise and keep fixed the initial distribution, we obtain the same effect of the enhancement of the variance of the Gaussian initial distribution of the population for moderate values of noise intensity (namely $\varepsilon = 0.1$). For high values of noise intensity (namely $\varepsilon = 1$) we strongly perturb the population dynamics and because of the presence of an absorbing barrier we obtain quickly the extinction of the populations.

## 4. RANDOM INTERACTION

The interaction between the species is assumed to be random and it is described by a random interaction matrix $J_{ij}$, whose elements are independently distributed according to a Gaussian distribution

$$P(J_{ij}) = \frac{1}{\sqrt{2\pi\sigma_J^2}}\exp\left[-\frac{J_{ij}^2}{2\sigma_J^2}\right], \quad \sigma_J^2 = \frac{J^2}{N}\,. \tag{13}$$

where $J$ is the interaction strength and

$$\langle J_{ij}\rangle = 0, \quad \langle J_{ij}J_{ji}\rangle = 0\,. \tag{14}$$

With this asymmetric interaction matrix our ecosystem contains 50% of prey-predator interactions (namely $J_{ij} < 0$ and $J_{ji} > 0$), 25% competitive interactions ($J_{ij} < 0$ and $J_{ji} < 0$)

and 25% symbiotic interactions ($J_{ij} > 0$ and $J_{ji} > 0$). The initial values of the populations $n_i(0)$ have also Gaussian distribution

$$P(n) = \frac{1}{\sqrt{2\pi\sigma_n^2}} \exp\left[-\frac{(n - \langle n \rangle)^2}{2\sigma_n^2}\right], \; \sigma_n^2 = 0.01, \text{ and } \langle n \rangle = 1\,. \tag{15}$$

The strength of interaction between the species $J$ determines two differnt dynamical behaviours of the ecosystem. Above a critical value $J_c$ the system is unstable, this means that at least one of the populations diverges. Below the critical interaction strength, the system is stable and reaches asymptotically an equilibrium state. For our ecosystem this critical value is approximately $J = 1.1$. The equilibrium values of the populations depend both on their initial values and on the interaction matrix. If we consider a quenched random interaction matrix, the ecosystem has a great number of equilibrium configurations, each one with its attraction basin. For vanishing noise ($\varepsilon = 0$), the steady state solutions of Eq. (1) are obtained by the fixed-point equation

$$(\gamma - n_i + h_i)n_i = 0 \tag{16}$$

where

$$h_i = \sum_j J_{ij} n_j(t) \tag{17}$$

is the local field. For a great number of interacting species we can assume that the local field $h_i$ is Gaussian with zero mean and variance $\sigma_{h_i}^2 = \langle h_i^2 \rangle = J^2 \langle n_i^2 \rangle$

$$P(h_i) = \frac{1}{\sqrt{2\pi\sigma_{h_i}^2}} \exp\left[-\frac{h_i^2}{2\sigma_{h_i}^2}\right]. \tag{18}$$

The solutions of Eq. (16) are

$$n_i = 0, \text{ i.e. extinction} \tag{19}$$

and

$$n_i = (\gamma + h_i)\Theta(\gamma + h_i), \; n_i > 0\,, \tag{20}$$

where $\Theta$ is the Heaviside unit step function. From this equation and applying the self-consistent condition we can calculate the steady state average population and its variance. Specifically we have

$$\begin{aligned} \langle n_i \rangle &= \langle (\gamma + h_i)\Theta(\gamma + h_i) \rangle \\ &= \frac{1}{\sqrt{2\pi\sigma_{h_i}^2}} \left[ \sigma_{h_i}^2 \exp\left[\frac{\gamma^2}{2\sigma_{h_i}^2}\right] + \frac{\gamma\sqrt{2\sigma_{h_i}^2 \pi}}{2} \left(1 + erf\left(\frac{\gamma}{\sqrt{2\sigma_{h_i}^2}}\right)\right)\right], \end{aligned} \tag{21}$$

and

$$\begin{aligned} \langle n_i^2 \rangle &= \left\langle (\gamma + h_i)^2 \Theta^2(\gamma + h_i) \right\rangle \\ &= \left[\left(\frac{\gamma^2 + \sigma_{h_i}^2}{2}\right)\left(1 + erf\left(\frac{\gamma}{\sqrt{2\sigma_{h_i}^2}}\right)\right) + \frac{\gamma}{2}\sqrt{\frac{2\sigma_{h_i}^2}{\pi}} \exp\left[\frac{\gamma^2}{2\sigma_{h_i}^2}\right]\right]. \end{aligned} \tag{22}$$

For an interaction strength $J = 1$ and an intrinsec growth parameter $\gamma = 1$ we obtain: $\langle n_i \rangle = 1.4387, < n_i^2 = 4.514$, and $\sigma_{n_i}^2 = 2.44$. These values are in good agreement with that

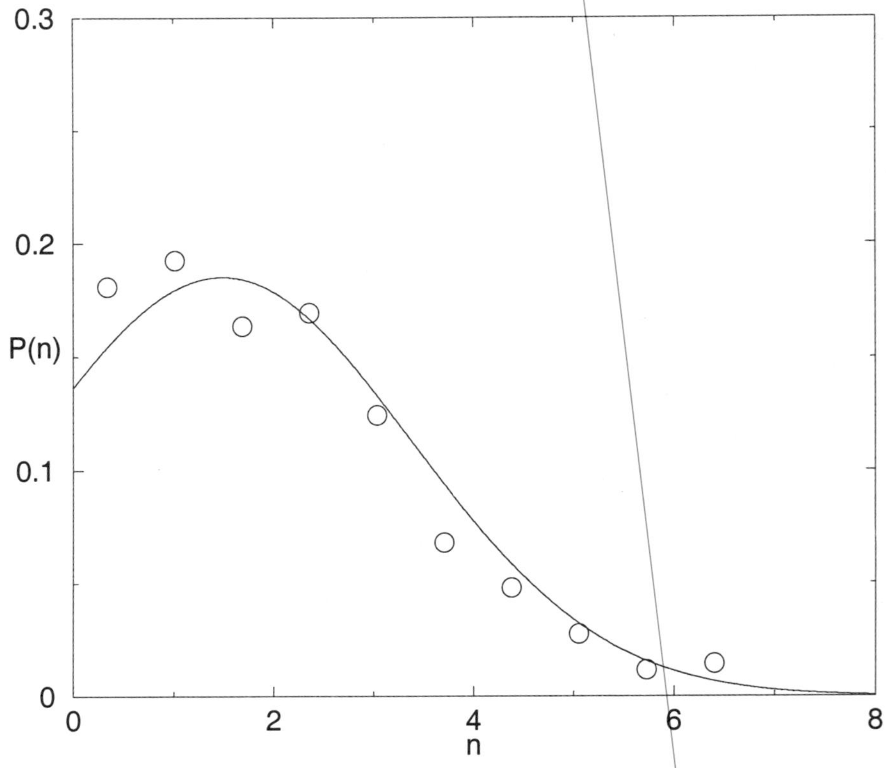


**Fig. 1** The stationary probability distribution [Eq. (23)], without the extinct species, in comparison with the histograms arising from numerical simulations (open circles). The system parameters are: $N = 1000$ species, $J = 1$ and $\gamma = 1$.

obtained from numerical simulation of Eq. (1). The choice of this particular value for the interaction strength, based on a preliminary investigation on the stability-instability transition of the ecosystem, ensures us that the ecosystem is stable. The stationary probability distribution of the populations is the sum of a delta function and a truncated Gaussian

$$P(n_i) = n_{e_i}\delta(n_i) + \Theta(n_i)\frac{\exp\left[-\dfrac{(n_i - n_{io})^2}{2J^2\sigma_{n_i}^2}\right]}{\sqrt{2\pi J^2 \sigma_{n_i}^2}} . \tag{23}$$

In Fig. 1 we report the stationary probability distribution of the population densities, without the extinct species, in comparison with the computer simulations for systems with $N = 1000$ species and for an interaction strength $J = 1$, and $\gamma = 1$.

In the case of non-vanishing noise ($\varepsilon \neq 0$) the population densities fluctuate around their mean values, which for low enough noise are not very different from the steady state solutions of Eq. (20). From our simulations we confirm this behaviour and found also that the external multiplicative noise cause very large fluctuations of the population densities $n(t)$. To study the transient dynamics therefore we focus on the statistical properties of the time integral of the $i$th population $N_i(t)$

$$N_i(t) = \int_0^t dt' n_i(t') , \tag{24}$$

in the asymptotic regime, which we expect to have a smooth behaviour. From Eq. (3) we have

$$N_i(t) = \ln\left[1 + n_i(0)\int_0^t dt' \exp\left[\gamma t' + \sqrt{\varepsilon} w_i(t') + \sum_{J\neq i} J_{ij} N_j(t')\right]\right] , \tag{25}$$

In Eq. (25) the term $\sum_j J_{ij}N_j$ gives the influence of other species on the differential growth rate of the time integral of the $i$th population and represents a local field acting on the $i$th population[3,7]

$$h_i = \sum_j J_{ij}N_j(t) = J\eta_i \,. \tag{26}$$

We use the same approximation of the Eq. (8) and, after differentiating, we get the asymptotic solution of Eq. (25)

$$N_i(t) \simeq \ln\left[n_i(o)e^{\sqrt{\varepsilon}w_{\max_i}(t)+J\eta_{\max_i}(t)}\int_0^t dt' e^{\gamma t'}\right] \tag{27}$$

where $w_{\max_i}(t) = \sup_{0<t'<t} w(t')$ and $\eta_{\max_i}(t) = \sup_{0<t'<t} \eta(t')$. Equation (27) is valid for $\gamma \geq 0$, that is when the system relaxes towards an equilibrium population and at the critical point. Evaluating Eq. (27) for $\gamma \geq 0$, after making the ensemble average, we obtain for the time average of the $i$th population $\bar{N}_i$

$$\langle \bar{N}_i \rangle \simeq \frac{1}{t}[N_w\sqrt{\varepsilon t} + \ln t + \langle \ln[n_i(o)]\rangle], \; \gamma = 0\,, \tag{28}$$

and

$$\langle \bar{N}_i \rangle \simeq \frac{1}{t}\left[N_w\sqrt{\varepsilon t} + \gamma + N_\eta + \left\langle \ln\left[\frac{n_i(o)}{\gamma}\right]\right\rangle\right], \; \gamma > 0\,, \tag{29}$$

where $N_w$ and $N_\eta$ are variables with a semi-Gaussian distribution[2] and $N_\eta$ must be determined self-consistently from the Eq. (26). These asymptotic behaviours are consistent with those obtained using a mean field approximation. We obtain in fact the typical long time tail behaviour ($t^{-1/2}$) dependence, which characterize nonlinear relaxation regimes when $\gamma \geq 0$. Besides the numerical results confirm these analytical asymptotic behaviours of $\bar{N}_i$.[4] When the system relaxes towards the absorbing barrier ($\gamma < 0$) we get from Eq. (25) in the long time regime

$$\langle \bar{N}_i \rangle \simeq \frac{1}{t}\left[\ln(n_i(0)) + \ln\left[\int_0^t dt' e^{\gamma t' + \sqrt{t'}w_i(t') + j\eta_i(t')}\right]\right]. \tag{30}$$

In this case the time average of the $i$th population $\langle \bar{N}_i \rangle$ is a functional of the local field and the Wiener process, and it depends on the history of these two stochastic processes.

We have also analyzed the dynamics of the ecosystem when one species is absent. Specifically we considered the cavity field, which is the field acting on the $i$th population when this population is absent.[7] In Fig. 2 we report the probability distributions of the local and of the cavity fields obtained by our simulations after a time $t = 100$ (expressed in arbitrary units) in absence of external noise and for two species (namely species 1 and 33). We note that the probability distributions of the cavity fields differ substantially from that of the local fields for the same species unlike the spin glasses dynamics, where the two fields coincide. We calculate also the same quantities in the presence of the external noise. The results of our simulations are reported in Figs. 3 and 4. The effect of the external noise is to overlap the two fields in such a way that for some particular species they coincide. Specifically this happens for the species 1 (see Fig. 3). For the species 33, we obtain a partial overlap (see Fig. 4). We found this interesting phenomenon, which is reminiscent of the phase transition phenomenon, for some populations. The main reasons for this behaviour are: (i) all the populations are positive; (ii) the particular structure of the attraction basins of our ecosystem; (iii) the initial conditions, which differ for the value of one population, belong to different attraction basins. Some populations have a dynamical behaviour such that after

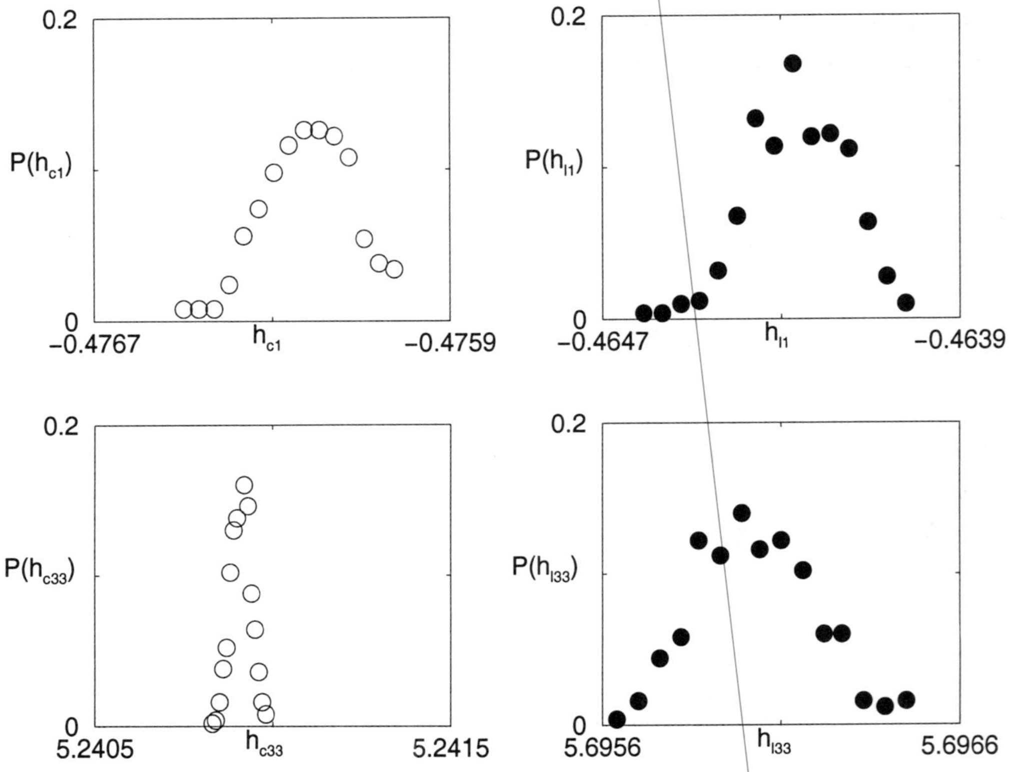


**Fig. 2** The probability distribution of the cavity fields $P(h_c)$ (open circles) and of the local fields $P(h_l)$ (black circles) for the species 1 and 33 after time $t = 100$. The system parameters are the same of Fig. 1.

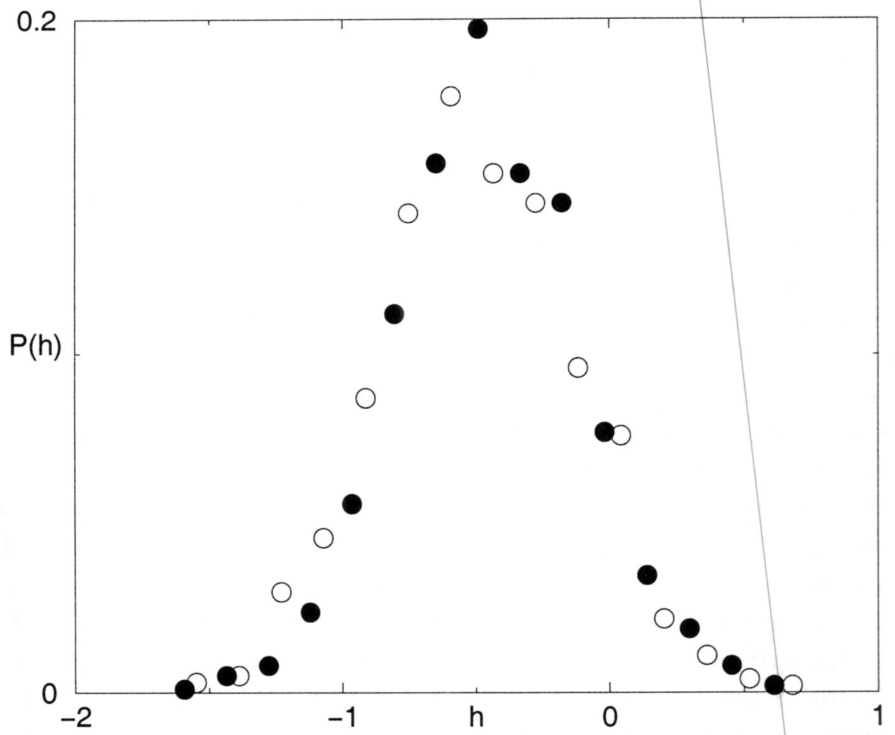


**Fig. 3** The probability distribution $P(h_1)$ of the local (black circles) and of the cavity (open circles) fields for the species 1 after time $t = 100$, in the presence of external noise. The noise intensity is $\varepsilon = 0.1$. The other system parameters are the same of Fig. 1.

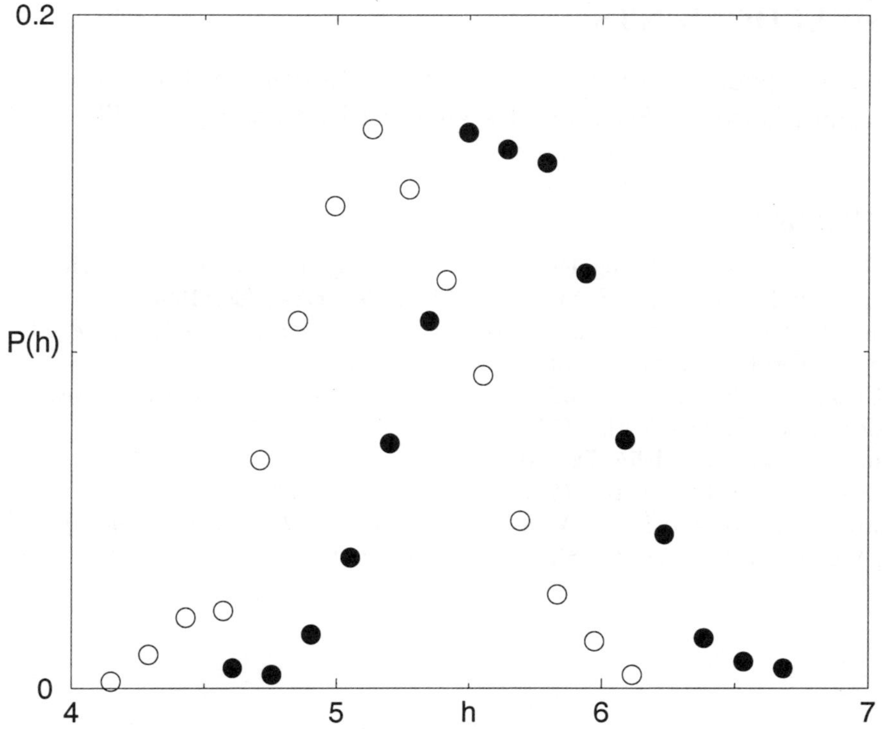


**Fig. 4** The probability distribution $P(h_{33})$ of the local (black circles) and of the cavity (open circles) fields for the species 33 after time $t = 100$ in the presence of external noise. The noise intensity is $\varepsilon = 0.1$. The other system parameters are the same of Fig. 1.

a long time they influence in a significant way the dynamics of other species. While in the presence of noise all the populations seem to be equivalent from the dynamical point of view. We found also that for strong noise intensity (namely $\varepsilon = 1$) all species extinguish on a long time scale ($t \approx 10^6$ a.u.). Whether extinction occurs for any value of noise intensity or not is still an open question, because of time-consuming numerical calculations.

## 5. CONCLUSIONS

We studied a stochastic model of an ecosystem of $N$ interacting species. By means of an approximation of the integral equation, which gives the stochastic evolution of the system, we obtain analytical results reproducing very well almost all the transient. We investigate the role of the noise on the stability-instability transition and on the transient dynamics. For random interaction we obtain asymptotic behaviour for three different nonlinear relaxation regimes. We obtain the stationary probability distribution of the population, which is the sum of two contributions: (i) a delta function around $n = 0$ for the extinct species and (ii) a truncated Gaussian for the alive species. When we switch on the external noise an interesting phenomenon is observed: the local and the cavity fields, whose probability distributions are different in the absence of noise, coincide for some populations. This phenomenon can be ascribed to the peculiarity of the attraction basins of our ecosystem. We have also investigated the overlap between the asymptotic values of the populations and the eigenvector of the interaction matrix with the maximum eigenvalue and we have not found any ordering regime phenomenon like in the spin glasses system. A more detailed investigation concerning the probability distribution of the populations and the local fields in the presence of noise is the subject of work in progress.

## ACKNOWLEDGMENTS

This work was supported in part by the National Institute of Physics of Matter (INFM) and the Italian Ministry of Scientific Research and University (MURST).

## REFERENCES

1. Yu. M. Svirezhzv and D. O. Logofet, *Stability of Biological Communities* (Mir, Moscow, 1983).
2. S. Ciuchi, F. de Pasquale and B. Spagnolo, *Phys. Rev.* **E54**, 706 (1996).
3. F. de Pasquale and B. Spagnolo, in *Chaos and Noise in Biology and Medicine*, eds. M. Barbi and S. Chillemi (World Scientific, Singapore, 1998), p. 305.
4. M. A. Cirone, F. de Pasquale and B. Spagnolo, in *Nuclear and Condensed Matter Physics*, ed. A. Messina (AIP, New York, 2000), p. 365.
5. G. Abramson, *Phys. Rev.* **E55**, 785 (1997).
6. G. Abramson and D. H. Zanette, *Phys. Rev.* **E57**, 4572 (1998).
7. M. Mezard, G. Parisi and M. A. Virasoro, *Spin Glasses Theory and Beyond*, Lecture Notes in Physics, Vol. 9 (World Scientific, Singapore, 1987), p. 65.

# SCALING BEHAVIOUR OF A MULTIPLY CONNECTED FLUCTUATING INTERFACE IN TWO DIMENSIONS

AYŞE ERZAN
*Department of Physics, Istanbul Technical University, Maslak, Istanbul, Turkey*

HÜSEY.IN KAYA
*Department of Biochemistry, Faculty of Medicine, University of Toronto, Toronto, Ontario M5S 1A8, Canada*

ALKAN KABAKÇIOĞLU
*Department of Physics of Complex Systems, Weizmann Institute of Science, Rehovot 76 100, Israel*

**Abstract**

We consider a one-parameter kinetic model for a fluctuating interface which can be thought of as an infinite string decorated with infinitely many closed strings. Numerical simulations show that a number of scaling exponents describing this string system may be related to the Kardar-Parisi-Zhang exponents. However, as the average velocity of the infinite string is taken to zero, and the string system becomes an isotropic fractal set, we also find new exponents which cannot be reduced to previously known ones.

## 1. INTRODUCTION

Interfaces, or reaction fronts formed in heterogenous systems, where the reaction takes place on a two dimensional substrate, model pattern formation in many different contexts besides catalytic phenomena.[1–6] The model we will describe here was first conceived[7] to represent an adsorption-reaction system without diffusion, which goes over to the Eden[4] model in one limit, and exhibits a transition to a completely delocalized fluctuating interfacial region, in another. The model can equally well describe the evolution of a shoreline or the fluctuations of a string. The only constraint is that the pieces of string should either extend to infinity or have closed ends.

Our model is defined on a square lattice, on a strip of width $L$. We associate a "site" with each plaquette of the square lattice. As an initial state we take a flat interface, just consisting of a line at $y = 0$, separating two absorbing states, the $A$ phase ($y < 0$) from the $\emptyset$ ($y > 0$) phase. Periodic boundary conditions are applied in the transverse ($x$) direction. In the longitudinal ($y$) direction, the strip extends, in principle, out to infinity in both directions. Due to the periodic boundary conditions, the interface is isomorphic to a circle, and will remain so, in whichever way we may deform it. In this system there will always be at least one such line going around the cylinder, and we will call it the "infinite string," (IS) in reference to the limit $L \to \infty$ which one may eventually take.

The kinetics of the strings is defined stochastically. At each time step we randomly chose an interface site, namely a site in the $\emptyset$ phase, separated from the $A$ phase by a segment of the interface. Then, with probability $p$, the interface is deformed so as to include this site in the $A$ phase, and with probability $1-p$, it is deformed so that an $A$ site which is nearest neighbor to the chosen $\emptyset$ site, now becomes included in the $\emptyset$ phase. In the latter case, if there are more than one nearest neighbor $A$ sites to the chosen interface site, one of them is chosen at random.

As the system evolves, the interface roughens, and becomes multiply connected. The IS advances or recedes in the $y$ direction with an average velocity proportional to $\varepsilon \equiv p - 1/2$. Meanwhile, as loops get pinched off the IS, rings, or closed strings (CS) keep getting formed. Note that no CS can be spontaneously created in either of the phases $A$ or $\emptyset$. It is easy to see that as each CS, in its turn, gets deformed in the stochastic manner described above, its center of mass performs a random walk. Since the IS has a finite drift velocity proportional to $\varepsilon$, on the average it catches up with those CS lying ahead of it and annihilates them by merging with them. On the other hand, for $\varepsilon < 0$ ($\varepsilon > 0$) those CS left behind in the $\emptyset$ ($A$) phase shrink, until they eventually disappear. Thus, for $\varepsilon \neq 0$, the system of strings reaches a steady state of finite width. In the rest frame of the IS, one has a wiggling string which keeps boiling off ringlets which eventually evaporate. In the rest of the paper, we will use "interface" and "string system" interchangably.

Clearly the limit of $|\varepsilon| \to 0.5$ ( $p \to 1$ or $p \to 0$) corresponds to Eden[4] growth. Overhangs are not prohibited but they have a characteristic length scale and therefore they can be eliminated under a one-step coarse graining of the IS. Also the CS are few and have a small characteristic size, therefore their inclusion does not affect the scaling behaviour, which is described by Kardar, Parisi, Zhang (KPZ) exponents.[5] Indeed, for $|\varepsilon|$ sufficiently far from 0, we find that any finite section of the IS is self-affine, and can be described by a scaling relation of the type[10]

$$w \sim t^{\beta} g(\ell/t^{1/z}), \tag{1}$$

where $g(u) \sim$ const. for $u < 1$ and $\sim u^{\chi}$ for $u \gg 1$, with a roughness exponent of $\chi = 1/2$, a growth exponent $\beta = 1/3$, and a dynamical exponent $z = \beta/\chi = 2/3$, namely KPZ[5] scaling.

As $|\varepsilon| \to 0$, and the drift velocity of the IS vanishes, the string system is no longer confined to a region of finite width. The CS multiply and grow in such a way that, for finite $L$, they may give rise to more than one IS at any given time. (This phenomenon is similar to the formation of Liesegang bands[8,9] in the language of chemical reactions.) Note that, strictly speaking, for finite $L$ our spanning string is also closed, due to the periodic boundary conditions, and this is the reason it can proliferate. On the other hand, in the limit of $L \to \infty$ there is a unique IS at all times.

For finite $\varepsilon$, the set of possible configurations of the strings is similar to that of an interface between two pure phases of the two-dimensional Ising model at low temperatures and has already been studied by Galavotti.[11] As $\varepsilon \to 0$, the system resembles more and more the critical Ising interface. On the other hand, in the Ising system at the critical temperature one could flip a spin within one of the pure phases, spontaneously creating a CS, which is not allowed in the present model. In any case, in the present non-equilibrium system no Gibbs distribution is available and the statistics of the fluctuations have to be determined purely from the kinetics.

In the next section we summarize the results of computer simulations and scaling arguments[7] which describe the scaling behaviour of this string system.

## 2. SCALING BEHAVIOUR

We characterise[7] the width of the string system (the interfacial region), with the function

$$w(\ell)^2 = \left\langle \frac{1}{N(\ell)} \sum_{i}^{N(\ell)} (h_i - \langle h(\ell) \rangle)^2 \right\rangle \tag{2}$$

where $N(\ell)$ is the total number of interfacial sites $h_i$ the vertical position of the $i$th interfacial site and $\langle h(\ell) \rangle$ the mean position of the interface within an interval of size $\ell$. Note that up to fluctuations of the size of the lattice spacing, the positions of interface sites are related to the positions of the strings in a one to one manner.

For sufficiently long times, $w(\ell) \sim \ell^{\chi_{\text{eff}}}$ for $\ell \ll L$, where $L$ is the transverse system size. The effective roughness exponent $\chi_{\text{eff}}$ goes continuously to zero as $|\varepsilon| \to 0$. We would like to understand the nature of this delocalization transition, to describe the crossover behaviour and to characterise the self-similar string system formed as $|\varepsilon| \to 0$.

As we decrease $|\varepsilon|$, the IS becomes increasingly convoluted, and decorated with CS of all sizes. As a measure of the "thickness" of the string system, let us define the variance of the height,

$$y(i) = \left( \frac{1}{n_i} \sum_{j=1,n_i} (h_{ij} - \langle h_i \rangle)^2 \right)^{1/2} \tag{3}$$

with $n_i$ being the number of interfacial sites above the point $i$ along the $x$ axis, and $\langle h_i \rangle$ being the average height at the point $i$. The thickness obeys a skewed–Gaussian distribution; the average $y_L$ and the variance of this distribution diverge as $|\varepsilon| \to 0$ with the same exponent $\nu = 0.55 \pm 0.05$, as $\sim |\varepsilon|^{-\nu}$. The scaling form for $y_L$ is

$$y_L \sim t^{\tilde{\beta}} G(|\varepsilon|^{-\nu}/t^{1/\zeta}) \tag{4}$$

where $G(v) \sim v$ for $v < 1$ and does not depend on $v$ for $v > 1$. The new growth exponent $\tilde{\beta}$ we find to be equal to $1/2$. In Eq. (4) we have also introduced the (longitudinal) dynamical critical exponent $\zeta$, which must satisfy $\zeta = 1/\tilde{\beta}$ as can be seen from the behaviour of the crossover function $G$.

Normalizing $w(\ell)$ by $y_L$ yields a collapse of the data for all different $\varepsilon$, and one obtains, $w(\ell) \sim \ell^{1/2}$ for $\ell^{1/2} \gg y_L$, whereas, for $\ell^{1/2} \ll y_L$, one has $w(\ell) \sim y_L$. Thus, we see that $\chi_{\text{eff}}$ is depressed to zero as the self-affine excursions of the interfacial region are blurred by the thickness of the interface or as $y_L$ becomes greater than $\ell^{1/2}$. It is interesting to note that the width of the IS does not diverge as $|\varepsilon| \to 0$, so that the disconnected parts, i.e. the CS, make up almost all of the interfacial region at the delocalization transition.

It has previously been pointed out[12,13] that the presence of overhangs may cause the small-scale structure of an interface to crossover from being self-affine to self-similar, while the large scale behaviour is self-affine. To test this hypothesis we investigated two different ways of defining single-valued profiles of the interface. The first consists of simply taking the average height at any given point, $\langle h_i \rangle$, while the second,[11] the extremal height (maximum for $\varepsilon > 0$, minimum for $\varepsilon < 0$). We find that the width functions

$$w_s(\ell)^2 = \left\langle \frac{1}{\ell} \sum_i^{\ell} (\langle h_i \rangle - \langle h(\ell) \rangle)^2 \right\rangle \tag{5}$$

and

$$w_{\rm ext}(\ell)^2 = \left\langle \frac{1}{\ell} \sum_i^{\ell} (h_i^{\rm ext} - \langle h^{\rm ext}(\ell) \rangle)^2 \right\rangle \tag{6}$$

exhibit KPZ like scaling with $\ell$, both for the case where we include the CS into our definition of the interface, and where we take only the IS. We find $w_{\rm ext} \sim w_s(\ell) \sim \ell^{\chi}$, with $\chi = 0.45 \pm 0.05$, recovering the KPZ scaling behaviour for $\ell^{1/2} > y_L$.

We would now like to focus on the *length* of a particular section of the string system, and to examine its scaling behaviour in the transverse and longitudinal directions. We define the correlation functions

$$C_x(l) = (\langle (x(r+l) - x(r))^2 \rangle)^{1/2} \tag{7}$$

$$C_y(l) = (\langle (h(r+l) - h(r))^2 \rangle)^{1/2} \tag{8}$$

where both $r$ and $l$ are the ("chemical") length measured along a single spanning string (IS) in the interface, and $x$ and $h$ are the horizontal and vertical coordinates of the interface site.

We find that for small times, $l \gg t^{1/2}$,

$$C_x \sim \begin{cases} l & y_L \ll l^{1/2} \\ l/t^{\psi} & y_L \gg l^{1/2} \end{cases} \tag{9}$$

the horizontal projection decreasing with $t$ as the string system evolves in the critical ($|\varepsilon| \to 0$) region. The value of the exponent $\psi$ is 1/6. On the other hand, for $l \ll t^{1/2}$, i.e., in the steady state,

$$C_x \sim \begin{cases} l & y_L \ll l^{1/2} \\ l^{\chi^{\rm isot}} & y_L \gg l^{1/2} \end{cases} \tag{10}$$

where $\chi^{\rm isot} = 2/3$.

The longitudinal correlation function

$$C_y \sim t^{1/3} \tag{11}$$

for early times in both regions, while in the steady state ($l \ll t^{1/2}$),

$$C_y \sim \begin{cases} l^{1/2} & y_L \ll l^{1/2} \\ l^{\chi^{\rm isot}} & y_L \gg l^{1/2}\,. \end{cases} \tag{12}$$

One sees that for $y_L \gg l^{1/2}$, the string system becomes isotropic. This region is characterised by $\chi^{\rm isot} = 2/3$, with $C_x \sim C_h \sim l^{\chi^{\rm isot}}$. In the opposite limit, $C_y \sim C_x^{1/2}$, as expected for the self-affine Eden surface.

These results are consistent with the scaling functions

$$C_y \sim \alpha_y(\varepsilon) t^{\beta} f(u,\, v) \tag{13}$$

and

$$C_x \sim \alpha_x(\varepsilon) t^{1/z} g(u,\, v) \tag{14}$$

over the whole range of $\varepsilon$. They are given in terms of the scaling variables $u = l/t^{1/z}$, $s = y_L/l^{\chi}$ with

$$f(u,\, s) \sim \begin{cases} \text{const.} & u \gg 1 \\ u^{\chi} & u \ll 1,\ s \ll 1 \\ u s^{1/z} & u \ll 1,\ s \gg 1 \end{cases} \tag{15}$$

and

$$g(u,\, v) \sim \begin{cases} u & s \ll 1 \\ u^{\chi}/s & u \gg 1,\ s \gg 1 \\ s^{-1/\tilde{\beta} z} & u \ll 1,\ s \gg 1\,, \end{cases} \tag{16}$$

where $\chi$, $z$ and $\beta$ have the KPZ values. The amplitudes $\alpha_y$ and $\alpha_x$ are given by $\alpha_y = 1/(a^{-1} + |\varepsilon|^{\psi})$ and $\alpha_x = a + |\varepsilon|^{\psi}$, with $a$ being some constant.

These scaling forms and Eq. (4) imply that the scaling exponents obey the relationships

$$\psi = \tilde{\beta} - \frac{1-\chi}{z}\,, \tag{17}$$

and

$$\chi^{\text{isot}} = \zeta \chi / z\,, \tag{18}$$

which yields,

$$\tilde{\beta} = \beta / \chi^{\text{isot}}\,. \tag{19}$$

Let us define the graph dimension $D_g$ of the IS in the isotropic regime, by $l \sim C_x^{D_g}$. We see that $D_g = 1/\chi^{\text{isot}}$. Since in two dimensions the graph dimension is related to the roughness exponent by $D_g = 2 - \chi$, we find indeed that $\chi^{\text{isot}} = 2/3$. This gives, $\tilde{\beta} = 1/2$, as we have found. From Eqs. (17) and (18) we also confirm that $\psi = 1/6$ and $\zeta = 2$.

Since we have found that the string system crosses over from being self-affine to self-similar for length scales smaller than $y_L \sim |\varepsilon|^{-\nu}$, one may compute its fractal dimension.[14] We find, for length scales $\ell < y_L^2$ that $D_I = 1.85 \pm 0.05$. The distribution of $s$, the area enclosed in a CS, obeys[14]

$$P_s(s) \sim \exp(-s/s_c)/s^{\tau}\,, \tag{20}$$

with $s_c \sim |\varepsilon|^{-\sigma}$, where we find $\sigma \simeq 0.9$ and $\tau = 1.6 \pm 0.1$. At the critical point, the distribution of the perimeters $\pi$ of the CS, is likewise a power law[14] $P_\pi(\pi) \sim \pi^{-\omega}$, with the exponent $\omega = 1.8 \pm 0.1$, from which we deduce that the perimeter to area scaling exponent $\phi$, defined by $p \sim s^{\phi}$ is $\phi = (\tau - 1)/(\omega - 1) \simeq 3/4$. From this we may conclude that the enclosed area is compact, i.e., $s \sim C_x^2$. Note that $\pi \sim l$ and $C_x \sim C_h \sim l^{\chi^{\text{isot}}}$, giving $\phi^{-1} = 2\chi^{\text{isot}}$.

## 3. DISCUSSION

Many interface problems with a well-defined growth direction, such as the Eden model,[4] or the Edwards–Wilkinson model[15] (EW), where the interface is a single valued, self-affine curve, can be thought of as describing a fluctuating (infinite) string. The Eden model is in the KPZ universality class, with $z = 2/3$, $\chi = 1/2$, and $\beta = 1/3$. The EW interface can also be described by such a scaling form as in Eq. (1), with $z = 2$, $\chi = 1/2$ and $\beta = 1/4$. It has been argued that in the limit where the average velocity of the string goes to zero, or is independent of the slope, the scaling behaviour of the infinite string should go over to that of the EW model.[10,16]

In this paper, we have recast a novel interface model[7] in the language of a fluctuating string system consisting of an infinite string decorated with many closed strings. Although in a certain limit our string is identical to the Eden interface and obeys KPZ scaling, it does not go over to the EW model as its velocity is taken to zero. On the contrary, in this limit the string becomes highly convoluted, closed strings proliferate, and the string systems becomes an isotropic fractal. We have performed numerical simulations and found scaling arguments that relate certain exponents of the model in this isotropic limit to those of the KPZ model. Nevertheless, there are exponents, like $\nu$, describing how the "thickness" of the string system blows up as one nears the critical point, or the fractal dimension and the size distribution of closed loops at the isotropic critical point, which cannot be deduced from such scaling relations. It remains as a challenge to build a theory which would allow the computation of these quantities from the stochastic kinetics of this system.

## ACKNOWLEDGMENTS

A. E. would like extend her best wishes to Antonio Coniglio on the occasion of his 60th birthday, and to thank the organizers of "Scaling and Disordered Systems" for a stimulating meeting. She would also like to thank the Turkish Academy of Sciences for partial support.

## REFERENCES

1. R. M. Ziff, E. Gulari and Y. Barshad, *Phys. Rev. Lett.* **56**, 2553 (1986).
2. H. Kaya, F. Kadirgan and A. Erzan, *J. Chem. Phys.* **98**, 9030 (1993).
3. H. Kaya and A. Erzan, *Surface Sci.* **320**, 185 (1994).
4. M. Eden, *Proceedings of the Fourth Symposium on Mathematical Statistics and Probability*, ed. F. Neymar (University of California, Berkeley, 1961), Vol. 4, p. 273.
5. M. Kardar, G. Parisi and Y. C. Zhang, *Phys. Rev. Lett.* **56**, 889 (1986).
6. P. Meakin, *Fractals, Scaling and Growth Far From Equilibrium* (Cambridge University, Cambridge Press, 1998).
7. H. Kaya, A. Kabakçıoğlu and A. Erzan, *Phys. Rev.* **B61**, 1102 (2000).
8. B. Chopard, M. Droz, J. Magnin and Z. Rácz, *Phys. Rev.* **E56**, 5343 (1997).
9. R. E. Lisegang, *Naturwiss. Wochenschr.* **11**, 353 (1896).
10. T. Halpin-Healy and Y.-C. Zhang, *Phys. Rep.* **254**, 215 (1995).
11. G. Galavotti, *Commun. Math. Phys.* **27**, 103 (1972).
12. C. S. Nolle, B. Koiller, N. Martyrs and M. O. Robbins, *Phys. Rev. Lett.* **71**, 2074 (1993).
13. M. Cieplak, A. Maritan and J. R. Banavar, *J. Phys.* **A27**, L765 (1994).
14. H. Kaya, Ph.D. Thesis, Istanbul Technical University (1999), unpublished.
15. S. F. Edwards and D. R. Wilkinson, *Proc. R. Soc. London, Ser. A* **381**, 17 (1982).
16. J. Krug and S. Spohn, in *Solids Far From Equilibrium*, ed. C. Godrèche (Cambridge University Press, New York, 1992), p. 479.

# MULTILINEAL RANDOM PATTERNS EVOLVING SUBDIFFUSIVELY IN SQUARE LATTICE

ADAM GADOMSKI
*Institute of Mathematics and Physics, University of Technology and Agriculture, PL-85796 Bydgoszcz, Poland*

**Abstract**

Stochastic multiline evolution in square lattice is studied. It turns out that the emerging patterns evolve subdiffusively, which is characterized by the $\frac{1}{4}$ exponent. Possible origin of such a slow behavior is discussed, and some elucidation, supporting the small fractional value is given. A notion of (dynamic) phase transition concept may sometimes help in understanding the presented random kinetic behavior.

## 1. INTRODUCTION

Retrospective view towards history of statistical physics shows that simple models are really worth developing. As a certain quite "out-of-date" general example, traced back to the twenties, the Ising model should be mentioned. A slightly younger "toy" model, known to almost everyone, appears to be the percolation model. After the era of applying the computer and its very capacities had emerged, they both have been visibly put forward, and stand nowadays for basis of many applications in microelectronics, materials engineering and technology, chemical processing, etc. Other examples of simple statistical-physical models can undoubtedly be the growth models, intensively explored since the early eighties, or even earlier. On the list of those models one can mostly find the following simple "computer-aided" phenomena, like e.g. diffusion-limited aggregation (first invented by Witten and Sander in 1981), and its variations; ballistic growth; Eden cluster formation or chemical reaction-limited growth. One has also to mention a large class of models, that may be named deposition models. They led to formation of random deposits on a plane (line), e.g. ballistic and/or diffusional deposits, just mentioning a few. They gave for sure rise to some intense theoretical understandings, leading to formulation of the conditions of surface dynamics, known as the Kardar-Parisi-Zhang (KPZ) system. This, in turn, initiated an eruption of theoretical approaches in surface science (thin film formation), crystal growth theory, polymer dynamics or turbulence studies, etc.[1–3]

It seems to be well-known that all the above mentioned dynamic processes deserved fruitful interpretation and description in terms of the random walk and/or phase transition concepts. As for the former, one may notice either the standard (normal) random walk or the anomalous (sub- or super-normal) one, realized possibly under some constrained conditions.[4] As for the latter, one easily checks the seminal literature and concludes thereafter, that either the continuous phase transition or that of order-disorder type should be considered while examining the above listed complex systems.[5] These notion will be used throughout this study, sometimes supported by certain kinetic arguments, *cf.* a review paper.[6]

The paper is structured as follows. In Sec. 2, a sketchy presentation of the algorithm, involving random evolution rules as well as initial conditions (IBCs) is given. In Sec. 3, some versatile physically-motivated explanation of the subdiffusive random motion of the evolving multiline is provided. In Sec. 4, one has to expect final remarks.

## 2. ALGORITHM, RULES AS WELL AS INITIAL AND BOUNDARY CONDITIONS

Because in Refs. 7 and 8, the algorithm of the random multiline evolution has been presented in sufficient detail, let us refer to its main points. Thus, let us start from a sketch of the algorithm. It is as follows:

- Initialize a starting configuration (a straight-line,[8] or a rectangular seed,[9] for example);
- Apply a set of random rules, i.e. choose a suitable one taken from 8 available rules (3 of reversible type, which means six forth and back rules in total, and some 2 remaining, being of irreversible nature), causing some evolution, or structural rearrangement, of the multilineal pattern(s) under investigation (the rules are specified below);
- Go permanently back to the above point unless desired statistics and/or other estimated quantities have been obtained; do not forget to refresh accordingly all the possible counters involved in the simulational process.

For introducing the rules in a possibly concise way, let us label four edges of the unit square on which the multiline may land by $d$, $u$, $l$ and $r$, respectively, and assume for simplicity, that $d$-edge (lying on $X$-direction), being just occupied, will be shifted upwards along $Y$-direction, see Fig. 1 for some exemplified realization within a small system.

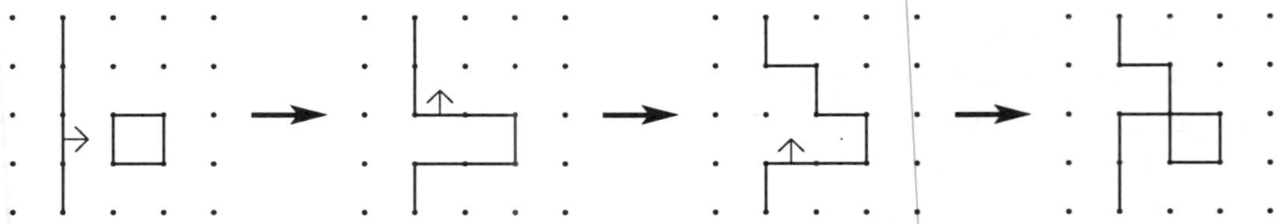

**Fig. 1** A few snapshots taken from an evolution of the so-called small system, assumed that the first two steps are given as in Fig. 2 in Ref. 8, but the next few steps are accidentally realized in a different manner, just presented here.

Now, one can draw 8 basic annihilation-creation (or, neutral) rules, driving the system, schematically by means of a single formula:

$$X_i^d + X_j^u + Y_k^l + Y_l^r \rightleftharpoons Y_{1-l}^r + Y_{1-k}^l + X_{1-j}^u + X_{1-i}^d \,, \tag{1}$$

where: $i$, $j$, $k$, $l$ can be equal to 0 (means: empty edge) or 1 (means: non-empty edge), and where the following elementary local rearrangements of the multilineal (discrete) patterns are possible to occur:

- Rule No. 1: Eq. (1) for $i = 1$, $j = k = l = 0$, i.e. with $\pm 2$ gain/loss of elementary units (creative/annihilative); note, that the multiline is composed of elementary units, and that the length of the elementary unit is always equal to the square lattice constant
- Rule No. 2: Eq. (1) for $i = l = 1$, $j = k = 0$, i.e. with 0 gain/loss of elementary units (neutral)
- Rule No. 3: Eq. (1) for $i = j = 1$, $k = l = 0$, i.e. with 0 gain/loss of elementary units (neutral); see remarks on indirect reversibility in[7]
- Rule No. 4: Eq. (1) for $i = j = k = 1$, $l = 0$, i.e. with +2 loss of elementary units (moderately annihilative)
- Rule No. 5: Eq. (1) for $i = j = k = l = 1$, i.e. with +4 loss of elementary units (strongly annihilative),

where the last two cases listed above have to be taken irreversibly, so that one has to put "$\rightharpoonup$" into Eq. (1) instead of the double arrows indicating reversibility, valid exclusively for the three first points mentioned above. To get wider outlook about the (ir)reversibility in question, *cf.* Refs. 7–9. (For another realization, but presenting a large system, see Fig. 2.)

As to the IBCs applied, let as ascertain the following. In a study,[7] some nearly periodic boundary conditions (BCs), i.e. vertically periodic, but horizontally free, have been applied. The IC was a vertical straight-line, usually placed in the middle of the lattice. In another work,[8] the IC remained the same, but the BCs have been changed subject to physical reality (constraints, see Ref. 4), which implies, that some reflecting BCs have been assumed. In Ref. 9, the authors decided to use fully periodic (on-torus-concept based) BCs. Quite versatile ICs have also been probed, like the rectangular seed(s) or two parallel lines, constituting an interfacial stripe, as a certain extension of the single-line IC concept.

The performed simulations, *cf.* Refs. 7–9, enable to arrive at some elucidations, mostly in terms of the subdiffusive (random walk) behavior, shown in the simulation time scale, measured in Monte Carlo steps. (Note, however, that another time scale applied, mostly

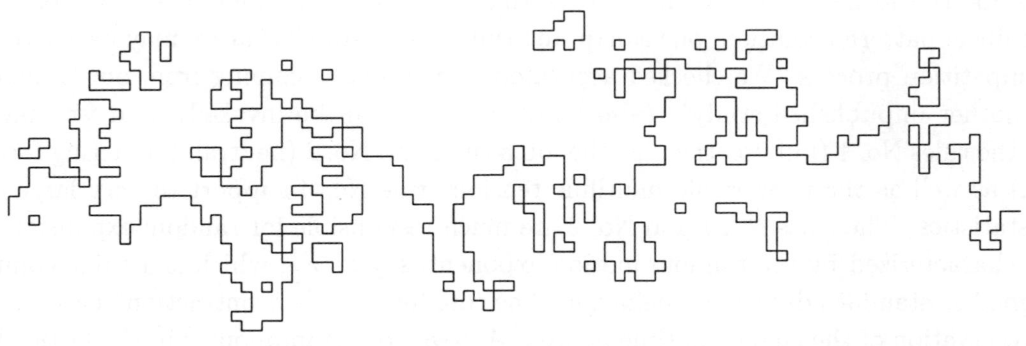

**Fig. 2** A large configuration, resulting from simulation of the random multiline (*Courtesy of Martin Schönhof and Lutz Schimansky-Geier*).

motivated by some real or Gillespie time scale concept,[9b] may result in slightly different time behavior.) It will be explained in the next section.

## 3. $\frac{1}{4}$-EXPONENT DRIVEN SUBDIFFUSIVE BEHAVIOR

Since a subdiffusive time-behavior has been checked quite thoroughly for many different realizations,[7–9] even quite irrespective of the BCs applied (mentioned in the preceding section), but preferentially for the single-line IC condition, it is certainly worth elucidating why and when such a behavior survives. It is interesting to note that such $\frac{1}{4}$ exponent driven characteristics are observed even if we use slightly different measures,[7,8] though they are chosen to describe some linear characteristics of the multilineal evolution in the course of time. To be more specific, let us state explicitly, that in Ref. 7 we studied the overall averaged length of that interline (termed also a diffusion-reaction front), whereas in Ref. 8 one has to notice, that we considered the problem of survival of the multiline, since boundary conditions under use (reflecting BCs) helped the multiline to disappear. We simply wanted to know how those disappearence events are related to a linear measure of the available evolution space, which in this particular case means the overall lattice (square or triangular)[8] size. All the characteristics obtained, but measured for long enough simulation times as well as averaged over a set of single simulations for each lattice size, $L$, end up with the following asymptotic formula, namely

$$\mathrm{L} \sim t^{\alpha}, \tag{2}$$

where $\alpha \approx \frac{1}{4}$; here L means a characteristic length, while t stands for the simulation time, see above. Such a dependency has also been revealed for the fully periodic BCs.[9] Let us then try to explain it briefly in the subsequent subsections.

### 3.1 Line-Tension Controlled Expansion Behavior

In the following, we wish to rationalize, what we have done in computer simulations, that we performed. First, we are willing to rely on some simple, or even verbal argumentation, supporting the evolution in question. Second, we are trying to utilize a proportionality law, quite well-known while offering a stochastic description of grain growth, e.g. in metals or ceramics. A common denominator of both of them appears to be everypresence of the line tension, which is a factor controlling (damping) multiline expansion process.

As for the former, we have to observe that based on our numerical study of random multilineal pattern evolution, one can quite firmly risk to state[9] that we may have to do with a competition process. We checked accurately the statistics coming from our simulations, *cf.* another unpublished study[10] (also another study[9] can be invoked), and we may state that the rules No. 1 (but creative, i.e. the forth rule) and No. 2 (neutral, but acting forth and back) as well as the irreversible multiline tension[7] rule No. 4 support to very large extent the statistics. The rules No. 1 and No. 2 are much responsible for random expansion of the line, characterized by the random motion exponent, say $\nu \approx \frac{1}{2}$, which is a value commonly assigned to standard diffusional behavior. The countereffect (a "contraction" case, or a case of preservation of the change in time of area, $A$, covered instantanously in the lattice by the line, must be given by roughly the same exponent, denoted however by $\mu$, where $\mu \approx \frac{1}{2}$, just for underlying that two different mechanisms are involved in the evolution. This is because a change in time of the area $A$, scavenged (swept) by the evolving multiline, is given by

$$\frac{dA}{dt} \simeq const., \tag{3}$$

which is fulfilled if we realize that the overall multilineal expansion-contraction process is readily going to proceed purely at random (this is just the case!), so that none of the major competition forces can be *a priori* priviledged, and the competition, because of every present random circumstances ("random mixing"), can be thought of to be a kind of liazon, which must then be expressed by the fact, that $\nu \times \mu$ is the overall characteristic exponent of the evolution, so that any L has to follow

$$\mathrm{L} \sim t^{\nu\times\mu}, \tag{4}$$

where $\nu \times \mu \approx \frac{1}{2} \times \frac{1}{2} = \frac{1}{4}$, and where L has to be measured for sufficiently large times.

As for the latter (read: more sophisticated elucidation will appear), let us present the following reasoning. To do so, however, let us start with the following question. What is possible to ascertain while trying to get a relative change of that variable $A \equiv A(t)$ in two consecutive time instants $t$ as well as $t + \Delta t$? Our observations, much confirmed by systematic numerical investigations[9] reveal, that

$$\frac{A(t+\Delta t) - A(t)}{A(t)} \propto \varepsilon\,, \tag{5}$$

where $\varepsilon$ is a random fraction, *cf.* a study on polycrystals and their boundaries,[11] being independent of time. (By the way, Eq. (5) is called the law of proportionate effects, or simply a proportionality law. It appears often to be a signature of the log-normal behavior in the system.[11]) But we do not have to forget that from Eq. (3) a constancy of changes of the area $A$ in time follows, and that this way a line tension effect is manifested.

Following Eq. (3) we have crudely to assume that $A \sim N\Delta t$ holds, where $\Delta t \ll 1$ (but, unfortunately $\Delta t \to 0$ is not guaranteed) and $N \gg 1$ is expected, where $N$ (a positive constant) can be the number of the simulation steps, or some quantity very related to it.

If the above is accepted, we may, loosing a bid on accuracy, write down a Langevin-like equation of the form

$$\frac{dA}{dt} \propto \chi\,, \tag{6}$$

where $\chi$ equals $N\varepsilon$, and stands for a random but uncorrelated "force", having a constant average value, $\chi_o$. Note that $A$ scales with some $t = N \times \Delta t$, so that a characteristic line tension influenced scaling behavior can be detected this way, presumed that $N$ would remain constant or insensitive to such a time scale; notice that if $A \propto \mathrm{L}^2$ one automatically provides the known standard relation, namely $\mathrm{L}^2 \propto t$ (in fact, characteristic for both typical line (surface) tension-driven as well as standard diffusional behaviors, but argued just for the latter).

The above physical picture notwithstanding, and knowing that such a Langevin-like equation has its counterpart in the so-called Fokker-Planck-Kolmogorov representation,[12a] we conclude that this is exactly the standard diffusion equation

$$P_t(\mathrm{L}^2, t) = D\Delta P(\mathrm{L}^2, t)\,, \tag{7}$$

[a]One may certainly argue whether such a continuous representation may stand for some argumentation for supporting discretized pictures, that usually come from the computer experiment. Nevertheless, we are of the opinion, that it can provide at least another direction of how to deal with problems of such types in approximate way.

where $P_t$ and $\Delta$ (Laplacian) have their usual meanings; here $D$ is the strength of the noise (or, the diffusion coefficient, but assumed that the diffusion process in the space of the quantity $L^2$, proportional to the instantaneous area swept by the multiline, but not in the position space is provided). The mean squared displacement, $\langle (L^2)^2(t) \rangle$, evaluated as the second (statistical) moment of the stochastic process, namely

$$\langle (L^2)^2(t) \rangle = \int_0^\infty (L^2)^2 P(L^2, t) dL^2, \tag{8}$$

reads

$$\langle (L^2)^2(t) \rangle \propto t, \tag{9}$$

so that the $\frac{1}{4}$ exponent behavior has this way been recovered, based on the above continuum approximation. Summing up, one has to consider the evolution in question as a nonstandard random walk behavior, where such a nonstandard (decelerated) kinetic character is anticipated because the line tension effect somehow balances random (standard diffusional) expansion of the multilineal pattern (it is clearly not in contradiction to define that random walk as self-avoiding, which comes directly from computer simulation[7,9]).

### 3.2 Back to Smoluchowski: Diffusion-Controlled Annihilation Problem Revisited

In Ref. 7 it was mentioned that some most elementary "chemical reactions" in our model phenomenon are the following

$$\begin{aligned} X^* + X &\rightharpoonup 0 \\ Y^* + Y &\rightharpoonup 0 \\ X^* + 0 &\rightleftharpoons X \\ Y^* + 0 &\rightleftharpoons Y, \end{aligned}$$

where $X$ means a horizontal "particle", whereas $Y$ represents a vertical one; notice that the superscript "$*$" means that a corresponding "particle" will be chosen (shifted; movable). It was stated therein,[7] that the two first reactions are far from being trivial. Even more, following seminal literature one has to argue that they constitute a theoretical framework for the diffusion-controlled annihilation process, e.g. characteristic of self-segregation phenomena, which in one dimension obeys the $\frac{1}{4}$ temporal law,[4] when presumed that $X^*$ (shifted along $Y$-direction), or $Y^*$ (shifted along $X$-direction, accordingly), do realize a one-dimensional random walk (time $t$ stands here for a survival time) until they are trapped by unmovable (not selected at random for being shifted!) traps $X$ and/or $Y$, respectively. While looking at the formula 5.57 in Ref. 4a (Chap. 5), one sees, that for sufficiently long times, namely $t \gg 1$, one gets similarly to Eq. (9)

$$\langle L(t) \rangle \propto t^{1/4}, \tag{10}$$

where here $\langle L(t) \rangle$ stands for the average nearest distance between $X$ (or, $Y$) "particle" and another of the same type, just before meeting it at random (in the above equation a certain prefactor may appear, which depends upon details of the model[4a]); notice, that according to the rule No. 3 both of them are annihilated, but some two new counterparticles $Y$ (or, $X$), respectively, emerge. This scenario agrees well with our simulational observations.

It touches also visibly the problem of determining the mean first passage time for the multiline,[9b] since the above distance quantity is defined as

$$\langle \mathrm{L}(t) \rangle = \int_0^\infty Q(\mathrm{L},\, t) dt\,, \tag{11}$$

where $Q(\mathrm{L},\, t)$ stands for a survival probability.[4] It is, by the way, noteworthy that such a kinetic anomaly, concerning the exploration of $X^* + X \rightharpoonup 0$ (or, $Y^* + Y \rightharpoonup 0$) reactions, has first been pointed out by Marian Smoluchowski (also, later touched upon by Ovchinnikov and Zeldovich[3]), who showed, that it can be fully manifested in a restricted (e.g. two-dimensional) geometry, where effective stirring, controlled by diffusion, seems to fail.[4] Finishing this subsection, let us underscore that the annihilation process in a $1D$ system described is not only thought of when a single unmovable trap stands for sink for randomly walking individuals, but also under a set of more relaxed conditions. We believe, that our conditions belong also to such a group. (Notice that in such problems all the realizations have to be averaged over the available set of ICs, and a thorough confirmation of the above reasonable suppositions needs just a repetition of the experiment towards satisfying this ascertion.)[4]

### 3.3 Naive Renormalization Group Considerations

The most naive (probably, oversimplified) explanation for the $\frac{1}{4}$ exponent driven temporal multiline behavior may come from real space renormalization[5] considerations presented below, and valid preferentially for certain $2D$ systems, like $XY$ model. We may also refer to Fig. 3, where the most primitive as well as self-consistent configurations (say directly, most essential) both for the square lattice[7,8] but also for a triangular one[8] are depicted.

The number of them, $M_n$, on each level of the renormalization (designated by $n$, being in general not mismatched with $N$) equals

$$M_n = 4^n = 2^{2n}, \tag{12}$$

whereas the (square root) "renormalization length" or factor, $l_n$, typically reads[13]

$$l_n = 2^{n/2}. \tag{13}$$

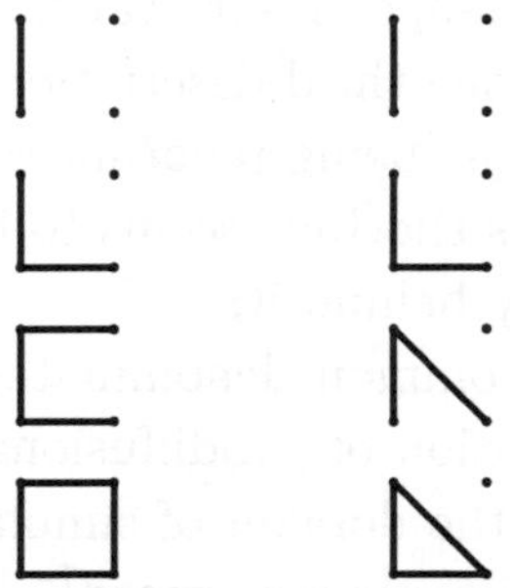

**Fig. 3** Most primitive configurations, constituting the random multiline for square as well as triangular lattices, respectively (*cf.* Figs. 3 and 4 in Ref. 8).

Defining the critical exponent for the proposed renormalization procedure, namely

$$\bar{\nu} = \frac{\ln(l_n)}{\ln(M_n)} \tag{14}$$

taken for $n \to \infty$, one gets immediately the desired value of $\bar{\nu} = \frac{1}{4}$. Such a value can mostly be met while inspecting certain numerical investigations on the $2D$ Ising system dynamics near equilibrium.[14,15] This will be the subject of subsequent section. (We may also find interesting to perform quite simple calculations, according to what is done in Ref. 13, Chap. 5 for the $2D$ bond percolation problem, i.e. to derive the so-called renormalization group equation. The result can be that we get, as in the percolation problem two seemingly different limits, namely the low- as well as high-temperature ones at the rearrangement probability equal to 0 and 1. Moreover, instead of one intermediate probability point somewhere in between, as in the bond percolation problem, one provides now two such points: A certain one closer to 0, and another one closer to 1. This result may indicate that in our system, instead of the continuous order first transition, or that of the second type according to Landau,[5] one would have to do with the order-disorder phase change of Kosterlitz-Thouless-like type, with an intermediate, or "hexatic" (tetratic) phase, placed just in between, *cf.* Refs. 5, 13 and 16. This attractive supposition must however be checked, perhaps in another study.)

## 4. CONCLUSIONS

The following final remarks, based on what has been presented before, can be juxtaposed. These are:

- The multilineal evolution under study can be analyzed on three levels of description, and by applying three seemingly different approaches, *cf.* Sec. 3.
- The first description presented (Subsec. 3.1) seems to be most physically motivated, and is readily realized in two-dimensional space. It takes into account that the evolution studied can be understood in terms of cooperation of two randomly acting processes. The first process is due to random expansion (random walk in the area space) whereas its "*alter ego*" is the surface tension (random contraction), which in the case of line tension effect, present in the system, causes mostly a continuous (dynamic) phase transition,[17] quite differently than in a $3D$ case, where some first order phase transition effects are present. (Phases can be differentiated in our problem as sets of points inside and outside the area covered by the line(s), so that a problem of intermingled phases, like in biomembranes, may likely arise.)
- The second description (Subsec. 3.2) is realized in a $1D$ space, and borrows something from the concept of mean first passage or survival time.[4]
- Last but hopefully not the least, the third description (Subsec. 3.3) is most general, and refers to the problem in question as being renormarizable,[5,15] or statistically self-similar (or, even more, *e.g.* self-affine); its shortage seems to be that no straightforward physical argumentation is observed to stay behind it.
- It is interesting to note, that a common denominator of all of the above presented descriptions appears to be confirmation of subdiffusional random motion of the multiline, characterized by $\frac{1}{4}$ exponent (in the domain of simulation times, and surely, within the accuracy level of computer experiments performed[7–10]).
- Finally, let us state clearly, that the problem studied has been treated on kinetic level,[6–10] and that no dynamic assignements[14] have so far been proposed.

## ACKNOWLEDGMENTS

A. G. wishes to acknowledge a partial financial support of the University of Technology and Agriculture, Bydgoszcz, Poland, for realizing the opportunity of presenting the invited lecture during Antognio Coniglio's International Workshop on Scaling and Disordered Systems, Paris, 13–14 April, 2000. He also thanks Lutz Schimansky-Geier for critical remarks, and Martin Schönhof for drawing the figures as well as for providing some of his results prior publication. Stimulating discussions with Mr. Michał Cyrankiewicz are appreciated.

## REFERENCES

1. F. Family and T. Vicsek (eds), *Dynamics of Fractal Surface* (World Scientific, Singapore, 1991).
2. T. M. Liggett, *Interacting Particle Systems* (Springer-Verlag, New York, 1985), Chaps. 3 and 4.
3. J. Schmelzer, G. Röpke and R. Mahnke, *Aggregation Phenomena in Complex Systems* (Wiley-VCH, Weinheim, 1998).
4. G. H. Weiss, *Aspects and Applications of the Random Walk* (North-Holland, Amsterdam, 1994); R. Kopelman and Y.-E. Koo, *Israel J. Chem.* **31**, 147 (1991); R. Kopelman, *Science* **241**, 1620 (1988).
5. Z. Gulasi and M. Gulasi, *Adv. Phys.* **47**, 1 (1998); K. Huang, *Statistical Mechanics* (Wiley and Sons, Inc., New York, 1963), Chaps. 2 and 15.
6. A. Gadomski, *Nonlinear Phenomena in Complex Systems* **3/4**, 321 (2000).
7. A. Gadomski, L. Schimansky-Geier and H. Rosé, *Acta Phys. Pol.* **B29**, 1647 (1998).
8. A. Gadomski, M. Schönhof and K. Bończak, in *Stochastic Processes in Physics, Chemistry and Biology*, eds. J. A. Freund and T. Pöschel, Lecture Notes in Physics No. 557 (Springer-Verlag, Berlin, 2000), pp. 496–506.
9. M. Cyrankiewicz, *Diploma Thesis* (*in Polish*) (University of Technology and Agriculture, Bydgoszcz, 2001); M. Schönhof, *M.Sc. Thesis* (*in German*) (Humboldt University, Berlin, 2001).
10. K. Bończak, *Diploma Thesis* (*in English*) (University of Technology and Agriculture, Bydgoszcz, 2000).
11. C. S. Pandè, *Acta Metall.* **36**, 2161 (1988); S. K. Kurtz and F. M. A. Carpay, *J. Appl. Phys.* **51**, 5725 (1981).
12. H. Malchow and L. Schimansky-Geier, *Noise and Diffusion in Bistable Nonequilibrium Systems* (Teubner Verlagsgesellschaft, Berlin, 1985); G. Czajkowski, *Z. Phys.* **B43**, 87 (1981).
13. J. J. Binney, N. J. Dowrick, A. J. Fisher and M. E. J. Newman, *The Theory of Critical Phenomena. An Introduction to the Renormalization Group*, (Clarendon Press, Oxford, 1992).
14. D. Stauffer and D. P. Landau, *Phys. Rev.* **B39**, 9650 (1989); D. Stauffer and N. Jan, *Canad. J. Phys.* **66**, 187 (1988).
15. Cieplak and M. Suan Li, *Fractals* **2**, 481 (1994).
16. A. Gadomski, *Phys. Rev.* **E60**, 1252 (1999).
17. V. Privman, *Int. J. Mod. Phys.* **C3**, 857 (1992).

# HOW TO BECOME A DICTATOR: A SIMPLE MODEL FROM PHYSICS

SERGE GALAM
*Laboratoire des Milieux Désordonnés et Hétérogènes, UMR-CNRS n° 7603*
*Université Paris 6, Tour 13 – Case 86, 4 place Jussieu*
*Paris, 75252 Cedex 05, France*

**Abstract**

The dynamics of majority rule voting in hierarchical structures is studied using concepts from collective phenomena in physics. In the case of a two-party competition a very simple model to a democratic dictatorship is presented. For each running group, a critical threshold (in the overall support) is found to ensure full and total power at the hierarchy top. However, the respective value of this threshold may vary a lot from one party to the other. It is this difference which creates the dictatorian nature of the democratic voting system. While climbing up the hierarchy, the initial majority-minority ratio can be reversed at the profit of actual running party. Such a reversal is shown to be driven by the natural inertia of being in power. The model could shed light on last century Eastern European Communist collapse.

## 1. MODELLING THE PERFECT DICTATORSHIP

A very new trend of statistical physics is its application to a rather wide range of topics outside of physics.[1] A model was even derived to describe former Yugoslavia fragmentation.[2]

In this paper, we study the voting dynamics underlying a two-party competition for power within a hierarchical structure. The analysis is done using real space renormalization group techniques.[3] From the hierarchy bottom, local groups of randomly chosen individuals are formed to elect representatives using a local majority rule. Once elected, these persons aggregate randomly into new groups which again elect representatives using the same local majority rule (see Fig. 1). And so on and so forth up to the last top unique group which elects the president. The hierarchy is built up acccording to a tree-like structure. At each level the number of voting groups is divided by the same fixed number.

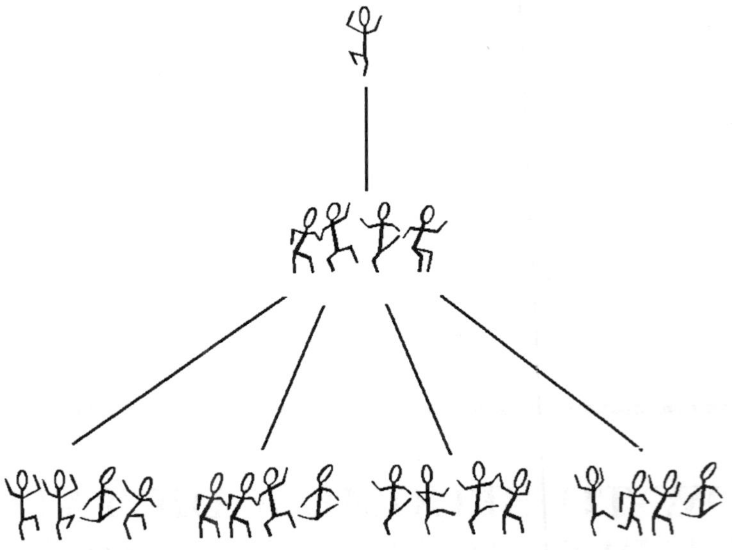

**Fig. 1** Schematic of a 2-level hierarchy with local groups of 4.

Within this hierarchical voting scheme, critical thresholds to power are obtained. An initial support above the critical threshold guarantees to win the presidency. At contrast, starting below the threshold leads to a democratic disappearance at the top hierarchy. However, the natural inertia of being in power is shown to produce a drastic breaking in the two-party running symmetry. The critical threshold is split into one below 50% and one above. For instance, the politically correct rule: "a majority is needed to change a leadership" puts the threshold to power simultaneously at 77% for the opposition, and at 23% for actual ruling party. The result is a democratic dictatorship.

The following paper is organized as follows. Next section presents the hierarchical voting model for competition between two tendencies A and B with local groups of size 3. For an overall initial support lower than 50%, the A party self-eliminates within a few voting levels. At the opposite, starting above 50% will result in total control of the hierarchy. Section 3 considers 4-person groups. It is a simple way to account for the very complex process by which the ruling party takes benefit from its actual dominant position. Accordingly, a 2A-2B tie elects a ruling party member. Such a natural bias of "you need a majority to turn down a leadership" is found to shift the value of the critical threshold to power from 50% to 77% for the opposition. Simultaneously it reduces the threshold to stay in power to 23% for the ruling party. The system has turned to a democratic dictatorship. Results of a numerical simulation are presented.[4] Last century Eastern European Communist party collapse is briefly discussed in the last section.

## 2. THE SIZE 3 MODEL

We consider a population with individuals of either A or B opinion. Respective overall proportions are $p_0$ and $1 - p_0$. Each member does have an opinion. The frame can be a political group, a firm or a society,

We start with the simplest case. At the hierarchy bottom level, people are aggregated randomly by 3. It can correspond to home localization or working place. Each group then elects representatives using a local majority rule. Groups with either 3 A or 2 A elect an A and otherwise a B. Therefore the probability to have an A elected from the bottom level is,

$$p_1 \equiv P_3(p_0) = p_0^3 + 3p_0^2(1 - p_0)\,, \tag{1}$$

where $P_3(p_n)$ denotes the voting function.

Above process of group forming and voting is repeated within level-1. The elected persons (from level-0) form groups which in turn elect new higher representatives. The new elected persons constitute level-2. The process can be repeated again and again. At each vote, a new additional higher hierarchical level is created. At the last level is the president. The probability to have an A elected at level $(n + 1)$ from level-n is then,

$$p_{n+1} \equiv P_3(p_n) = p_n^3 + 3p_n^2(1 - p_n)\,, \tag{2}$$

where $p_n$ is the proportion of A elected persons at level-n.

Analyzing the voting function $P_3(p_n)$ singles out the existence of 3 fixed points $p_l = 0$, $p_{c,3} = \frac{1}{2}$ and $p_L = 1$. First one corresponds to no A elected. On the opposite, last one $p_L = 1$ represents a totalitarian state with only A elected. Both $p_l$ and $p_L$ are stable fixed points. At contrast, in between, $p_{c,3}$ is unstable. It determines indeed the threshold to democratic flowing towards either full power (with $p_L$) or to total disappearance (with $p_l$). Starting from $p_0 < \frac{1}{2}$ leads to $p_l$ while $p_0 > \frac{1}{2}$ drives to $p_L$.

Majority rule voting thus produces the self-elimination of any A proportion as long as its initial support is less than 50%. However, for completion, this self-elimination dynamics requires a sufficient number of voting levels.

Therefore, the instrumental question is to determine the number of levels required to ensure full leadership of the initial larger tendency. To make any sense, this level number must be small enough since most organizations have only a few number of levels (less than 10).

To get a feeling about above voting dynamics, it is enough to take any initial value for $p_0 < 50\%$ and to plug into Eq. (1). Then take the result and put it back into the same equation and so forth up to getting zero. For instance from $p_0 = 0.45$, we get successively $p_1 = 0.42$, $p_2 = 0.39$, $p_3 = 0.34$, $p_4 = 0.26$, $p_5 = 0.17$, $p_6 = 0.08$ down to $p_7 = 0.02$ and $p_8 = 0.00$. Within 8 levels, 45% of the population has been self-eliminated.

The overall process preserves however the democratic character of majority rule voting. It is the bottom leading tendency (more than 50%) which eventually gets the full leadership at the organization top level. The situation is fully symmetry with respect to A and B tendencies with the same threshold to full power (50%).

## 3. THE INERTIA OF BEING IN POWER

To go over 50% of full support and to turn down a top leadership seem to be a fair constraint. On this basis, one would then expect regular power changes. However from real life, power changes are indeed rather rare. A simple majority appears often not to be enough to take over power. In other words, it seems more easy to stick to power than to get to power.

While institutions are built in to ensure stability over some time intervals, they often turn to oppose any change, even if it is requested by a huge majority of people. One major reason is that once in power, a given party automatically gets a lot of political and financial advantages. There exists a real tip to the ruling party. We call it the "inertia" to stay in power. It thus breaks quite naturally the symmetry between the two competing tendencies. This tip is sometimes materialized, for instance, giving one additional vote to the committee president.

To account in a simple schematic manner for such a very complex process, by which the ruling party takes benefit from its actual dominant position, we consider groups of 4 persons. There, a majority does not always exists. Accordingly, we will use the associated tie configuration 2A-2B to favor the ruling party. Along the politically correct criterium "you need a majority to turn down a leadership", a tie configuration will result in the election of a ruling party member.

Such a natural bias of *a priori* no dramatic consequences, is found to shift the value of the critical threshold to power from 50% to 77% for the opposition. Simultaneously it reduces the same threshold to power down to 23% for the ruling party. The system has turned to a democratic dictatorship.

Consider even 4-size groups with B being the ruling party. Then, to be locally elected an A needs either 4 or 3 A in a given group. The tie case 2A-2B votes for a B together with 3 and 2B. Therefore, the probability to get an A elected from level-n, at level $n+1$ is,

$$p_{n+1} \equiv P_4(p_n) = p_n^4 + 4p_n^3(1-p_n)\,, \tag{3}$$

where $p_n$ is the proportion of elected A at level-n. At constrast, at the same $n+1$ level, for a B to be elected the probability is,

$$1 - P_4(p_n) = p_n^4 + 4p_n^3(1-p_n) + 2p_n^2(1-p_n)^2\,, \tag{4}$$

where last term embodies the bias in favor of B.

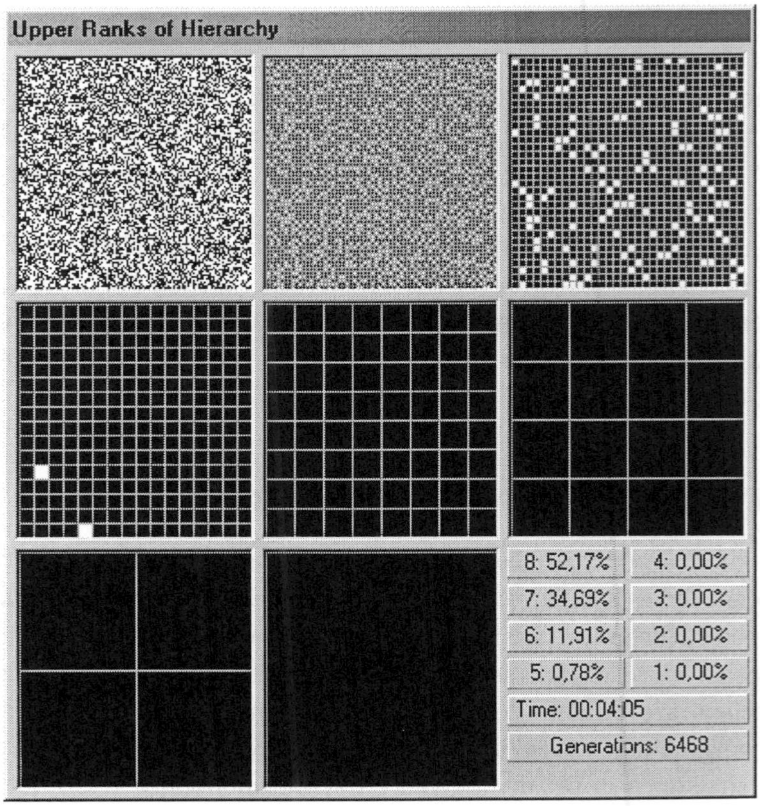


**Fig. 2** A 8-level hierarchy for even groups of 4 persons. The two A and B tendencies are represented respectively in white and black with the bias in favor of the black squares, i.e. a tie 2-2 votes for a black square. Written percentages are for the white representation at each level. The "Time" and "Generations" indicators should be discarded. The initial white support is 52.17%.

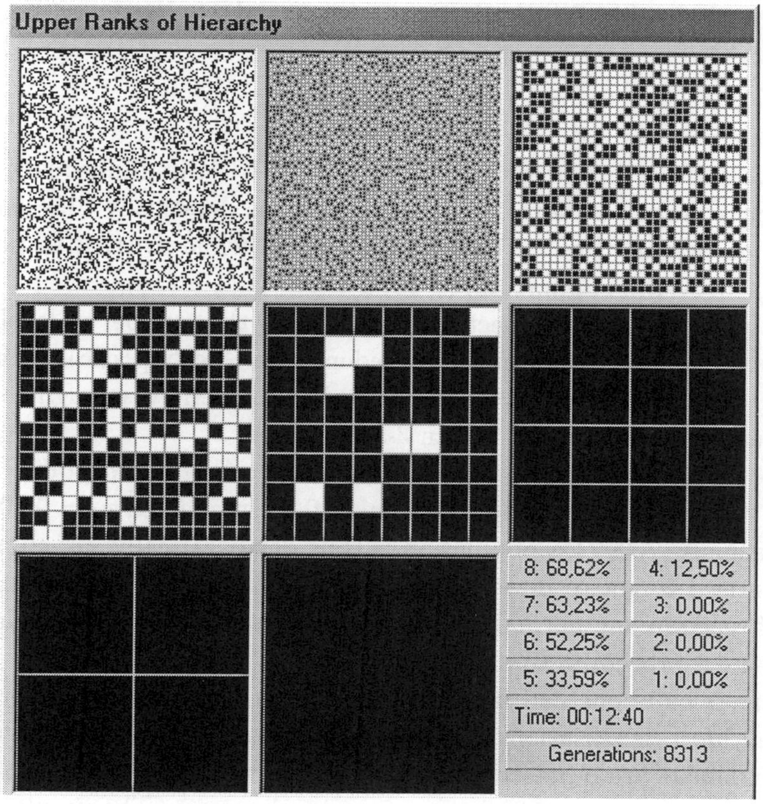


**Fig. 3** The same as Fig. 1 with an initial white support of 76.07%

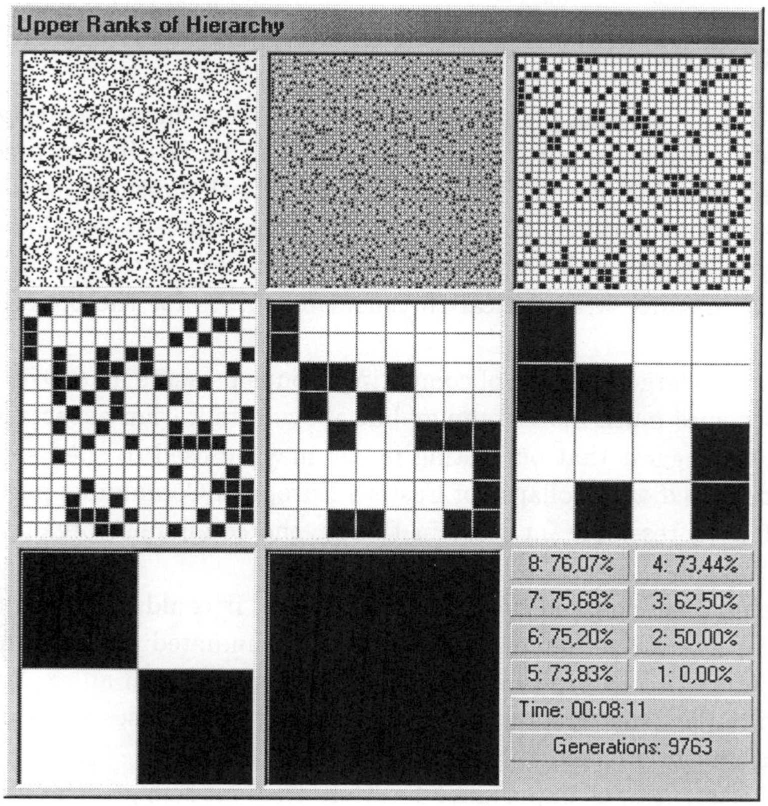


**Fig. 4** The same as Fig. 1 with an initial white support of 77.05.

From Eq. (3), the stable fixed points are still 0 and 1. However the unstable one is drastically shifted to,

$$p_{c,4} = \frac{1+\sqrt{13}}{6}, \tag{5}$$

for the A. It makes the A threshold to power at about 77%. Simultaneously, the B threshold to stay in power is around 23% making both situations drastically different. To take over the A needs to go over 77% of initial bottom support while to stick to power the B only needs to keep their overall support above 23%.

In addition to the asymmetry effect, the bias makes the number of levels to democratic self-elimination even smaller than in the precedent case (3-group size). From $p_0 = 0.45$, we have $p_1 = 0.24$, $p_2 = 0.05$ and $p_3 = 0.00$. Three levels are now enough to make the A to disappear instead of the 8 required within 3-size groups.

To emphazise above effect, let us start for instance from $p_0 = 0.70$, quite away from 50%. The associated voting dynamics is $p_1 = 0.66$, $p_2 = 0.57$, $p_3 = 0.42$, $p_4 = 0.20$, $p_5 = 0.03$, and $p_6 = 0.00$. Amazingly, 70% of a population is thus democratically self-eliminated within 6 voting levels. Indeed to get to power, the A must pass over 77% of overall support which is almost out of reach in any normal and democratic two-tendency situation. Majority rule voting has become a procedure to dictatorship.

Some analytic calculations can indeed be performed[4] as well as numerical simulations together with a "life" vizualisation.[5] Some snapshots[5] are shown in Figs. 2, 3 and 4. The two A and B tendencies are represented respectively in white and black squares. The bias is in favor of the black ones, i.e. a tie 2-2 votes for a black square. The simulation is done for a 8-level hierarchy. We can see how a huge white square majority can become self-eliminated. Written percentages are for the white representation at each level.

## 4. CONCLUSION

We have treated only very simple cases to single out some main trends produced by repeated democratic voting. In particular, it was found to be possible to predict an election outcome knowing initial supports. The existence of critical thresholds was shown to be an instrumental feature of majority rule voting. Moreover, these thresholds were found not to be always symmetrical with respect to both tendencies. Dictatorship situations were demonstrated for instance with critical thresholds of 0.77% for the opposition and 0.23% for the rulers.

The extension to a larger number of competing groups presents no *a priori* difficulty. The situation just becomes much more complex but above totalitarian aspects are reinforced.

To conclude, we suggest that our voting model may be used to shed some new light on last century generalized auto-collapse of Eastern European Communist parties. Our voting hierarchy can indeed resemble, at least in theory, the communist organization known as democratic centralism.

Instead of being a sudden and opportunistic change, it could have resulted from a very long and massive internal movement which was self-eliminated from top hierarchies. But once it reaches there, it has been full and then irreversible. Such an explanation does not oppose any additional constraints and mechanisms but emphazises some possible trend in the internal mechanism of communist organizations.

Last but not least, it is of importance to stress that modeling social and political phenomena is not aimed at stating an absolute truth but instead to single out some basic trends within a very complex situation.

## REFERENCES


1. S. M. de Oliveira, P. M. C. de Oliveira and D. Stauffer, *Non-Traditional Applications of Computational Statistical Physics: Sex, Money, War and Computers* (Springer, 2000).
2. R. Florian and S. Galam, *Eur. Phys. J.* **B16**, 189 (2000).
3. Sh-k Ma, *Modern Theory of Critical Phenomena* (Reading, MA, 1976).
4. S. Galam, *J. Stat. Phys.* **61**, 943 (1990).
5. S. Galam and S. Wonczak, *Eur. Phys. J.* **B18**, 183 (2000).

# ON FRACTIONAL RELAXATION

R. HILFER
*ICA-1, Universität Stuttgart*
*Pfaffenwaldring 27, 70569 Stuttgart, Germany*
*Institut f. Physik, Universität Mainz, 55099 Mainz, Germany*

**Abstract**

Generalized fractional relaxation equations based on generalized Riemann-Liouville derivatives are combined with a simple short time regularization and solved exactly. The solution involves generalized Mittag-Leffler functions. The associated frequency dependent susceptibilities are related to symmetrically broadened Cole-Cole susceptibilities occurring as Johari Goldstein $\beta$-relaxation in many glass formers. The generalized susceptibilities exhibit a high frequency wing and strong minimum enhancement.

## 1. INTRODUCTION

An ubiquitous feature of the dynamics of supercooled liquids and amorphous polymers is the nonexponential relaxation exhibited in numerous experiments such as dielectric spectroscopy, viscoelastic modulus measurements, quasielastic light scattering, shear modulus and shear compliance as well as specific heat measurements.[1] Derivatives and integrals of noninteger order (fractional calculus[2]) with respect to time are proposed in this paper as a mathematical framework for the slow relaxation near the glass transition.

Most glass forming liquids and amorphous polymers exhibit strong deviations from the Debye relaxation function $f(t) = \exp(-t/\tau)$ where $\tau$ is the relaxation time. All relaxation functions in this article will be normalized such that $f(0) = 1$ unless stated otherwise. In dielectric spectroscopy, the dominant $\alpha$-relaxation peak is broadened and asymmetric when plotted against the logarithm of frequency.[3] One popular phenomenological method to generalize the universal exponential relaxation behaviour is to introduce a fractional stretching exponent thereby arriving at the "stretched exponential" or Kohlrausch relaxation function given as

$$f(t) = \exp[-(t/\tau)^{\beta_K}] \tag{1}$$

with fractional exponent $\beta_K$. Relaxation in the frequency domain is usually described in terms of the normalized complex susceptibility

$$\hat{\chi}(\omega) = \frac{\chi(\omega) - \chi_\infty}{\chi_0 - \chi_\infty} = 1 - i\omega\mathcal{L}\{f\}(i\omega) \tag{2}$$

where $\chi_0 = \mathrm{Re}\chi(0)$, $\chi_\infty = \mathrm{Re}\chi(\infty)$ and $\mathcal{L}\{f\}(i\omega)$ is the Laplace transform of the relaxation function $f(t)$ evaluated at purely imaginary argument $i\omega$. Extending the method of stretching exponents to the frequency domain one obtains the Cole-Cole susceptibility[4]

$$\frac{\chi(\omega) - \chi_\infty}{\chi_0 - \chi_\infty} = \frac{1}{1 + (i\omega\tau)^{\alpha_C}} \tag{3}$$

or the Davidson-Cole expression[5]

$$\frac{\chi(\omega) - \chi_\infty}{\chi_0 - \chi_\infty} = \frac{1}{(1 + i\omega\tau)^{\gamma_D}} \tag{4}$$

as empirical expressions for the broadened relaxation peaks. My objective in this paper is to generalize the Cole-Cole expression using fractional calculus.[2]

Differentiation and integration of noninteger order are defined and discussed in the theory of fractional calculus as a natural generalization of conventional calculus.[2] Exponential relaxation functions are of fundamental importance because the exponential function is the eigenfunction of the time derivative operator that in turn is the infinitesimal generator of the time evolution defined as a simple translation

$$\mathcal{T}(t)f(s) = f(s - t) \tag{5}$$

acting on observables or states $f(t)$. It is therefore of interest to ask whether fractional time derivatives can arise as infinitesimal generators of time evolutions and, if yes, to find their eigenfunctions and the corresponding susceptibilities.

## 2. FRACTIONAL TIME DERIVATIVES

Given that time evolutions in physics are generally translations in time it is a well known consequence that the infinitesimal generator of the time evolution (5)

$$\lim_{t\to 0} \frac{\mathcal{T}(t)f(s) - f(s)}{t} = -\frac{\mathrm{d}}{\mathrm{d}s}f(s) \tag{6}$$

is generally given as the first order time derivative. Long time scales and time scale separation require coarse graining of time, i.e. an averaging procedure combined with a suitably defined long time limit in which $t \to \infty$ and $s \to \infty$.[6] On long time scales the rescaled macroscopic time evolution can differ from a translation and instead become a fractional convolution semigroup in the rescaled macroscopic time $t$ of the form

$$\mathrm{T}_\alpha(t)f(t_0) = \int_0^\infty f(t_0 - s)h_\alpha\left(\frac{s}{t}\right)\frac{ds}{t} \tag{7}$$

as first conjectured in Ref. 7 and later shown in Refs. 6–9. References 6–9 have shown that the parameters obey $t \geq 0$ and $0 < \alpha \leq 1$, and the kernel function ($b \geq 0$, $c \in \mathbb{R}$)

$$h_\alpha(x;\, b,\, c) = \frac{1}{b^{1/\alpha}} h_\alpha\left(\frac{x-c}{b^{1/\alpha}}\right) = \frac{1}{\alpha(x-c)} H^{10}_{11}\left(\frac{b^{1/\alpha}}{x-c}\middle|\begin{matrix}(0,1)\\(0,1/\alpha)\end{matrix}\right) \tag{8}$$

is defined through its Mellin transform[10]

$$\int_0^\infty H^{10}_{11}\left(x\middle|\begin{matrix}(0,1)\\(0,1/\alpha)\end{matrix}\right) x^{s-1}\, dx = \frac{\Gamma(s/\alpha)}{\Gamma(s)}\,. \tag{9}$$

It follows that the infinitesimal generator of $\mathrm{T}_\alpha(t)$ given as

$$\mathrm{D}^\alpha f(s) = \lim_{t\to 0} \frac{\mathrm{T}_\alpha(t) f(s) - f(s)}{t}\,. \tag{10}$$

is a fractional time derivative[2] of order $\alpha$. My objective in the rest of this paper is to discuss various fractional relaxation equations with time derivatives of Riemann-Liouville type as infinitesimal generators of time evolutions giving rise to nonexponential eigenfunctions. Before entering into the discussion of fractional differential equations the basic definitions of fractional derivatives are recalled for the convenience of the reader.

A fractional (Riemann-Liouville) derivative of order $0 < \alpha < 1$

$$\mathrm{D}^\alpha_{a+} f(t) = \frac{\mathrm{d}}{\mathrm{d}t} \mathrm{I}^{1-\alpha}_{a+} f(t) \tag{11}$$

is defined via the fractional (Riemann-Liouville) integral

$$(\mathrm{I}^\alpha_{a+} f)(x) = \frac{1}{\Gamma(\alpha)} \int_a^x (x-y)^{\alpha-1} f(y)\mathrm{d}y \quad (x > a) \tag{12}$$

with lower limit $a$. The following generalized definition was introduced by the present author.[6] The (right-/left-sided) fractional derivative of order $0 < \alpha < 1$ and type $0 \leq \beta \leq 1$ with respect to $x$ is defined by

$$\mathrm{D}^{\alpha,\beta}_{a\pm} f(x) = \left(\pm \mathrm{I}^{\beta(1-\alpha)}_{a\pm} \frac{\mathrm{d}}{\mathrm{d}x} (\mathrm{I}^{(1-\beta)(1-\alpha)}_{a\pm} f)\right)(x) \tag{13}$$

for functions for which the expression on the right hand side exists. The Riemann-Liouville fractional derivative $\mathrm{D}^\alpha_{a\pm} := \mathrm{D}^{\alpha,0}_{a\pm}$ corresponds to $a > -\infty$ and type $\beta = 0$. For subsequent calculations it is useful to record the Laplace transformation

$$\mathcal{L}\left\{\mathrm{D}^{\alpha,\beta}_{a+} f(x)\right\}(u) = u^\alpha \mathcal{L}\{f(x)\}(u) - u^{\beta(\alpha-1)} (\mathrm{D}^{(1-\beta)(\alpha-1),0}_{a+} f)(0+) \tag{14}$$

where $u$ denotes the dual variable and the initial value $(\mathrm{D}^{(1-\beta)(\alpha-1),0} a{+}f)(0+)$ is the Riemann-Liouville derivative for $t \to 0+$. Note that fractional derivatives of type 1 involve nonfractional initial values.

## 3. FRACTIONAL RELAXATION EQUATIONS

Consider now the fractional relaxation or eigenvalue equation for the generalized Riemann-Liouville operators in Eq. (13)

$$\tau^\alpha_\alpha \mathrm{D}\alpha, \beta_{0+} f(t) = -f(t) \tag{15}$$

for $f$ with initial condition

$$\mathrm{I}_{0+}^{(1-\beta)(1-\alpha)} f(0+) = \tau_\beta^{(1-\beta)(1-\alpha)} f_0 \tag{16}$$

where $\tau_\alpha$ is the relaxation time and the second time scale of the initial condition, $\tau_\beta$, becomes important only if $\beta \neq 1$. It is clear from the discussion of fractional stationarity in Ref. 6 that for $\beta \neq 1$ this initial condition conflicts with the requirement $f(0) = 1$ because it implies $f(0) = \infty$. Hence $f(t)$ cannot itself represent a relaxation function and a regularizing short time dynamics is required. The function $f(t)$ represents a metastable level to which the relaxation function $g(t)$ decays. An example is the mestastable equilibrium position of an atom in a glass. It evolves slowly as the structural relaxation proceeds. The metastable level is itself time dependent and relaxes according to Eq. (15). The short time dynamics is assumed to be exponential for simplicity and hence $g(t)$ obeys

$$\tau_{\mathrm{D}} \mathrm{D}^1 g(t) + g(t) = f(t) \tag{17}$$

with initial conditions $g(0+) = 1$.

Without short time regularization, i.e. for $\tau_{\mathrm{D}} = 0$, Laplace Transformation of Eq. (15) gives

$$f(u) = \frac{\tau_\alpha^\alpha \tau_\beta^{(1-\beta)(1-\alpha)} u^{\beta(\alpha-1)} f_0}{1 + (\tau_\alpha u)^\alpha} . \tag{18}$$

To invert the Laplace transform rewrite this equation as

$$f(u) = \frac{\tau_\beta^{1-\gamma} u^{\alpha-\gamma}}{\tau_\alpha^{-\alpha} + u^\alpha} = \tau_\beta^{1-\gamma} u^{-\gamma} \frac{1}{(\tau_\alpha u)^{-\alpha} + 1} = \tau_\beta^{1-\gamma} \sum_{k=0}^{\infty} (-\tau_\alpha^{-\alpha})^k u^{-\alpha k - \gamma} \tag{19}$$

with

$$\gamma = \alpha + \beta(1 - \alpha) . \tag{20}$$

Inverting the series term by term using $\mathcal{L}\{x^{\alpha-1}/\Gamma(\alpha)\} = u^{-\alpha}$ yields the result

$$f(t) = (t/\tau_\beta)^{\gamma-1} \sum_{k=0}^{\infty} \frac{(-(t/\tau_\alpha)^\alpha)^k}{\Gamma(\alpha k + \gamma)} . \tag{21}$$

The solution may be written as

$$f(t) = f_0 \, (t/\tau_\beta)^{(1-\beta)(\alpha-1)} E_{\alpha,\alpha+\beta(1-\alpha)}(-(t/\tau_\alpha)^\alpha) \tag{22}$$

using the generalized Mittag-Leffler function defined by

$$E_{a,b}(x) = \sum_{k=0}^{\infty} \frac{x^k}{\Gamma(ak + b)} \tag{23}$$

for all $a > 0$, $b \in \mathbb{C}$. This function is an entire function of order $1/a$.[11] Moreover it is completely monotone if and only if $0 < a \leq 1$ and $b \geq a$.[12] As a consequence the generalized relaxation function in (22) is smooth and decays monotonically. It resembles closely the exponential function. Using the asymptotic expansion of $E_{a,b}(x)$ for $x \to -\infty$ shows that for $t \to \infty$ the relaxation function decays as

$$f(t) \sim \frac{f_0}{\Gamma(\beta(1-\alpha))} \left(\frac{t}{\tau_\beta}\right)^{(1-\beta)(\alpha-1)} \left(\frac{t}{\tau_\alpha}\right)^{-\alpha} . \tag{24}$$

For $\tau_\alpha = \infty$ the result reduces to

$$f(t) = \frac{f_0\,(t/\tau_\beta)^{(1-\beta)(\alpha-1)}}{\Gamma((1-\beta)(\alpha-1)+1)}\,. \tag{25}$$

because $E_{a,b}(0) = 1/\Gamma(b)$. The same result is obtained asymptotically in the limit $t \to 0$. Hence $f(t)$ diverges at short time and needs to be regularized by Eq. (17). Of special interest is again the case $\beta = 1$ because in this special case there is no divergence and no regularization is necessary. It has the well known solution

$$f(t) = f_0 E_\alpha(-(t/\tau_\alpha)^\alpha) \tag{26}$$

where $E_\alpha(x) = E_{\alpha,1}(x)$ denotes the ordinary Mittag-Leffler function.

## 4. FRACTIONAL SUSCEPTIBILITIES

To calculate fractional susceptibility based on Eq. (15) it is necessary to include the regularization defined by Eq. (17). This regularization controls the divergence of the solution in Eq. (22) for $t \to 0$. Laplace transforming the system (15)–(17) and solving for $g$ yields with the help of Eq. (2) the susceptibility

$$\frac{\chi(\omega)-\chi_\infty}{\chi_0-\chi_\infty} = 1 - \left[\frac{(\tau_\alpha/\tau_\beta)^\alpha (i\omega\tau_\beta)^{\beta(\alpha-1)+1} f_0}{(1+(i\omega\tau_\alpha)^\alpha)(1+i\omega\tau_D)} + \frac{i\omega\tau_D}{1+i\omega\tau_D}\right] \tag{27}$$

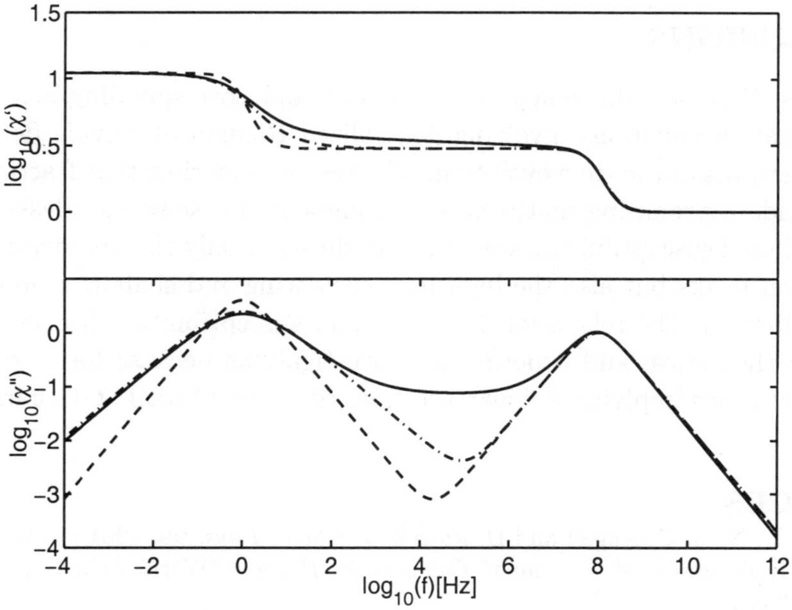


**Fig. 1** Real part (upper figure) $\chi'(\omega)$ and imaginary part $\chi''(\omega)$ (lower figure) of the complex frequency dependent susceptibility given in Eq. (27) versus frequency $\omega$ in a doubly logarithmic plot. All curves have $\tau_\alpha = 1$ s, $\tau_D = \tau_\beta = 10^{-8}$ s, $f_0 = 0.8$, $\chi_\infty = 1$ and $\chi_0 = 0$ and their two main peaks are located at relaxation frequencies $1/\tau_\alpha = 1$ Hz and $1/\tau_D = 10^8$ Hz. The dashed line shows two Debye-peaks with $\alpha = \beta = 1$. The dash-dotted curve corresponds to a pure Cole-Cole peak at $1/\tau_\alpha = 1$ Hz with $\beta = 0.7$, $\beta = 1$ and a Debye peak at $1/\tau_D = 10^8$ Hz. The solid curve corresponds to a generalized Cole-Cole peak with $\alpha = 0.7$ and $\beta = 0.98$ and a regularizing Debye peak at $1/\tau_D = 10^8$ Hz.

where the restrictions $0 < \alpha \leq 1$, $0 \leq \beta \leq 1$, $\tau_D > \tau_\beta$, $\tau_D > \tau_\beta$ and $f_0 \leq 1$ must be fulfilled. For $\alpha = 1$ this reduces to two Debye peaks, for $\alpha = 1$ and $\tau_D = 0$ it reduces to a simple Debye peak. For $\beta = 1$ and $\tau_D = 0$ the formula reduces the Cole-Cole Eq. (3).

In Fig. 1 the real and imaginary part of $\chi(\omega)$ are depicted for three special cases. In all cases the relaxation strengths were fixed by setting $\chi_\infty = 1$ and $\chi_0 = 0$ combined with $f_0 = 0.8$. The time scale of the inital condition was fixed in all cases at $\tau_\beta = 10^{-8}$. Large values of this constant destroy the regularization for $t \to 0$ by increasing the influence of the divergence in Eq. (22). Small values have little to no influence on the results. In all cases two main relaxation peaks can be observed at relaxation frequencies given roughly by $1/\tau_\alpha = 1$ Hz and $1/\tau_D = 10^8$ Hz.

The first case is obtained by setting $\alpha = 1$. It gives rise to two Debye peaks and is shown as the dashed curve in Fig. 1. In this case $\beta$ drops out from the right hand side of Eq. (27). This results in two Debye peaks with a deep minimum between them. The second case, shown as the dash-dotted line in Fig. 1, is obtained by setting $\beta = 0.7$ and $\beta = 1$. This results in a symmetrically broadened low frequency peak of Cole-Cole type. Such peaks are often observed as $\beta$-relaxation peaks in polymers.[13] Again the two peaks are separated by a deep minimum. The experimental results never show such a deep minimum. The relaxation function corresponding to the low frequency peak is the ordinary Mittag Leffler function as given in Eq. (26). The third case, shown as the solid line, corresponds to the parameters to $\beta = 0.7$ and $\beta = 0.98$. As a consequence of the minute change in $\beta$ from $\beta = 1$ to $\beta = 0.98$ there appears a high frequency wing and the minimum becomes much more shallow. The shallow minimum is a consequence of the divergence of the generalized relaxation function in Eq. (22) for $t \to 0$. The relaxation function corresponding to the low frequency peak is now the generalized Mittag-Leffler function as given in Eq. (22).

## 5. CONCLUSIONS

The paper has discussed the relaxation functions and corresponding susceptibilities for fractional relaxation equations involving generalized Riemann-Liouville derivatives as infinitesimal generators of the time evolution. The results underline that fractional relaxation equations provide a promising mathematical framework for slow and glassy dynamics. In particular fractional susceptibilities seem to reproduce not only the broadening or stretching of the relaxation peaks but also the high frequency wing and shallow minima observed in experiment. However, the relaxation functions and susceptibilities discussed in this work require further theoretical and experimental investigations because for $\beta \neq 1$ they are divergent at short times implying a departure from the conventional stationarity concept.

## REFERENCES


1. R. Böhmer, K. Ngai, C. Angell and D. Plazek, *J. Chem. Phys.* **99**, 4201 (1993).
2. R. Hilfer, *Applications of Fractional Calculus in Physics* (World Scientific Publishing Co., Singapore, 2000).
3. P. Lunkenheimer, U. Schneider, R. Brand and A. Loidl, *Contemp. Phys.* **41**, 15 (2000).
4. K. Cole and R. Cole, *J. Chem. Phys.* **9**, 341 (1941).
5. D. Davidson and R. Cole, *J. Chem. Phys.* **19**, 1484 (1951).
6. R. Hilfer, in *Applications of Fractional Calculus in Physics*, ed. R. Hilfer (World Scientific, Singapore, 2000), p. 87.
7. R. Hilfer, *Phys. Rev.* **E48**, 2466 (1993).
8. R. Hilfer, *Fractals* **3**, 549 (1995).
9. R. Hilfer, *Physica* **A221**, 89 (1995).

10. C. Fox, *Trans. Am. Math. Soc.* **98**, 395 (1961).
11. L. Bieberbach, *Lehrbuch der Funktionentheorie* (Teubner, Leipzig, 1931), Vol. 2.
12. W. Schneider, *Expo. Math.* **14**, 3 (1996).
13. F. K. A. Hofmann, E. Fischer and A. Schönhals, in *Disorder Effects on Relaxational Processes*, eds. R. Richert and A. Blumen (Springer, Berlin, 1994), p. 309.

# STRIPE PATTERNS IN SELF-STRATIFICATION EXPERIMENTS

NOÉ LECOCQ and NICOLAS VANDEWALLE
*GRASP, Institut de Physique B5, Université de Liège, B-4000 Liège, Belgium*

**Abstract**

We report pattern formation of complex stripes with binary and ternary granular mixtures. Ternary mixtures lead to a particular ordering of the strates which was not accounted for in former explanations. Bouncing grains are found to have an important effect on strate formation. A complementary mechanism for self-stratification of binary and ternary granular mixtures is proposed. This mechanism leads to a simple model. Eventually we report the observation of self-stratification for a binary mixture of grains of the same size but with different shapes.

## 1. INTRODUCTION

When a mixture of grains differing in size is poured between two vertical slabs, there is a global tendency for large and small grains to segregate in different regions of the pile.[1,2] Additionally, a self-stratification of the mixture in alternating layers of small and large grains is observed if large grains have a larger angle of repose than the small ones. It has been also shown[3] that the phase segregation takes also place when grains of same size but with different shapes are mixed together. We suggest that strates might also appear in those conditions as will be shown below.

For describing the self-stratification phenomenon, both continuous[4,5] and discrete models[5–7] have been proposed up to now for binary mixtures. These models are able to reproduce the alternating layers of the different granular species. The more elaborated models consider phenomenological continuous equations which account mainly for the angles of repose of the various species, the percolation of grains in the rolling phase, and the kink formation on the pile surface. A challenge for physicists is the generalization of both experiments and models to the case of a continuous distribution of grain sizes.

Some experimental work concerning ternary mixtures were first reported in Ref. 8. We expose them here with some additionnal perspectives and report new observations for mixtures of grains of the same size but with different shapes.

## 2. EXPERIMENTAL SETUP

A vertical Hele-Shaw (HS) cell was specially built for our purpose. The distance $e$ separating the vertical planes of the HS could be continuously adjusted between 0 and 50 mm. The cell was inclined in such a way that its slope was equal to the slope of the heap in formation (see Fig. 1). This allow us to have a constant lengh for the heap in the direction of the strata. This lenght was typically around 20 cm. Various granular species have been used: sand, wheat semolina, poppy seeds, Fe filings. We controlled the granulometry of each species by a preliminary sifting. Important properties of each type of grains are given in Table 1. Granular mixtures contained an equal volume of each species. They were poured in the HS cell with a funnel.

A CCD camera with a lens 12/120 mm F5.6 was placed perpendicular to the HS cell for examining the physical processes involved at the grain scale. Top views have also been taken. They allowed us to observe the dynamics and composition of the upper surface during the flow. We also used a color camera when needed for contrast (7/42 mm, F1.4). Precise video imaging is necessary for studying the evolution of granular piles. Our results are quite reproducible.

## 3. OBSERVATIONS

### 3.1 Mixtures of Grains Differing in Size

First, let us consider the case of ternary mixtures being composed of small ($S$), medium ($M$) and large ($L$) grains. Figure 2 presents three pictures of ternary piles in the HS cell.

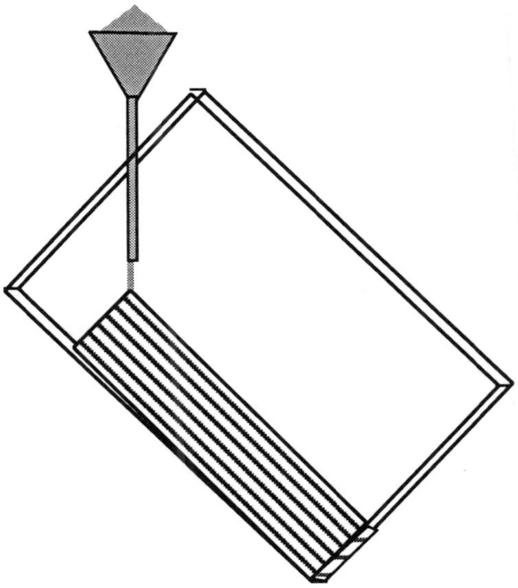

**Fig. 1** The Hele-Shaw cell was inclined in such a way that the length of the slope was approximately constant during heap formation.

**Table 1 Properties of the grains we used**

| Material | Size | Repose Angle | Density |
|---|---|---|---|
| Sand 1 ($e = 6$ mm) | 0.25 – 0.315 mm | $37° \pm 2$ | 2.6 g/cm$^3$ |
| Sand 1 ($e = 2$ mm) | 0.25 – 0.315 mm | $41° \pm 1$ | 2.6 g/cm$^3$ |
| Sand 2 | 0.2 – 0.25 mm | $42° \pm 1$ | 2.5 g/cm$^3$ |
| Sand 3 | 0.25 mm | $35° \pm 1$ | 2.8 g/cm$^3$ |
| Sand 4 | 0.25 – 0.315 mm | $40° \pm 1$ | 2.5 g/cm$^3$ |
| Fe filings ($e = 6$ mm) | $< 0.1$ mm | $35° \pm 2$ | 5.2 g/cm$^3$ |
| Fe filings ($e = 2$ mm) | $< 0.1$ mm | $42° \pm 1$ | 5.2 g/cm$^3$ |
| Wheat semolina | 0.8 mm | $41° \pm 1$ | 1.3 g/cm$^3$ |
| Poppy seeds | 1 mm | $39° \pm 1$ | 1.1 g/cm$^3$ |

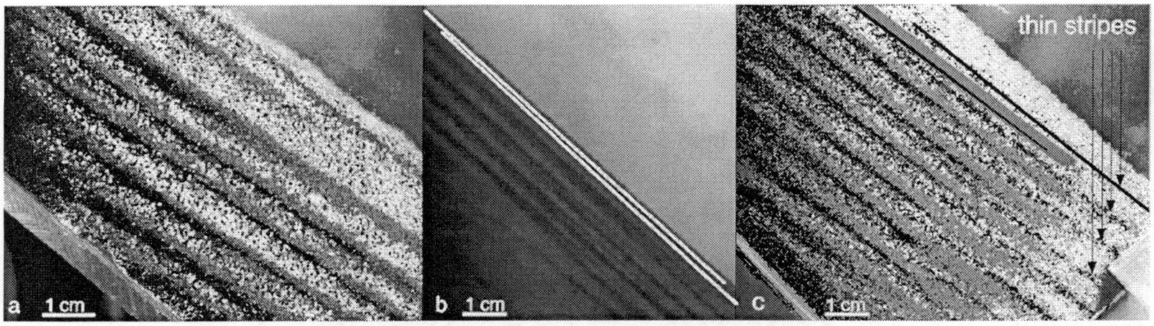


**Fig. 2** Three different patterns observed in ternary granular mixtures. The mixtures are composed of small ($S$), medium ($M$) and large ($L$) grains. **(a)** A pile exhibiting a $\ldots SMLSMLSML\ldots$ type of stripes ordering ($S$ in black, $M$ in grey and $L$ in white). Close look at the tails of the stripes reveals a $\ldots SMLMSMLM\ldots$ ordering. **(b)** $\ldots SMLMSMLM\ldots$ ordering with clear $M$ grains and dark $S$ and $L$ grains. A difference in angle between $L$ stripes and $S$ stripes ($S$ in the upper part of the heap) is emphasised. **(c)** Additional thin stripes are visible mainly in the tail of the pile. The same mixture was used in $a$ and $c$ (Fe filings ($S$), sand 1 ($M$) and wheat semolina ($L$)) with the same plate separation (6 mm). Only the funnel used was different (2 cm$^3$/s in $a$ and 2.5 cm$^3$/s in $c$, with faster grains). The mixture used in $b$ was composed of Fe filings ($S$), sand 2 ($M$) and sand 1 ($L$) with a plate separation of $e = 2$ mm. Images $a$ and $c$ have been processed by computer for better contrast between the three different granular species.

Two different mixtures were used. In each case, size segregation is clearly observed. Indeed, small grains have a global tendency to locate in the center of the pile while large grains are located in the tail. The medium-sized grains are dispersed in between the center and the tail. In addition, the self-stratification phenomenon is present. Three different patterns are obtained. In the first pile, a $\ldots SMLSMLSML\ldots$ ordering is displayed as found in.[1] The second pile is characterised by a $\ldots SMLMSMLM\ldots$ ordering. The third pile is quite similar to the first, as far as wide stripes are concerned. However, the ordering seems to be more complex, especially in the tail: a thin stripe of small grains is visible in the center of wide stripes made of large grains. These different orderings require special attention and the presence of the thin stripe cannot be explained with conventional models for self-stratification. Another important observation is that for some mixtures, a difference in inclinaison angle appears between the different stripes [Fig. 2(b)]. This difference of stripe angles, the particular ordering and the occurrence of a thin additional stripe indicates that quite different mechanisms are present during the heap formation.

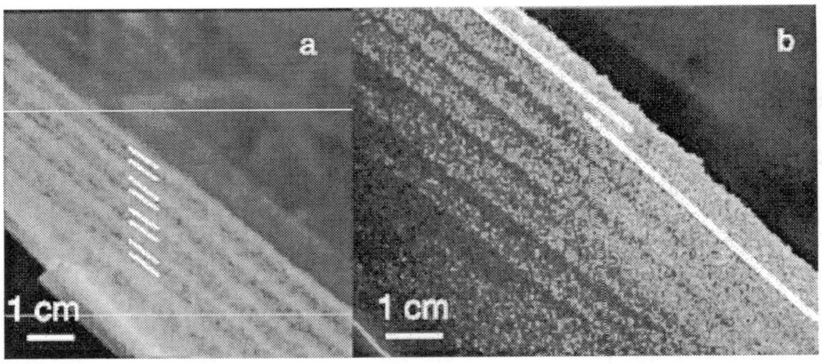


**Fig. 3** Additional stripe of small grains in binary mixtures with $e = 6$ mm plate separation. **(a)** A binary granular mixtures of sand 3 ($S$) and poppy seeds ($L$) gives rise to pairs of $L$ strates. Another way to describe the pattern is to say that a thin $S$ stripe is in the center of each $L$ stripe. **(b)** Sand 1 ($S$) and wheat semolina ($L$). Notice the difference in angle between the wide $S$ layers in the upper part of the pile and the thin $S$ layers of the lower part.

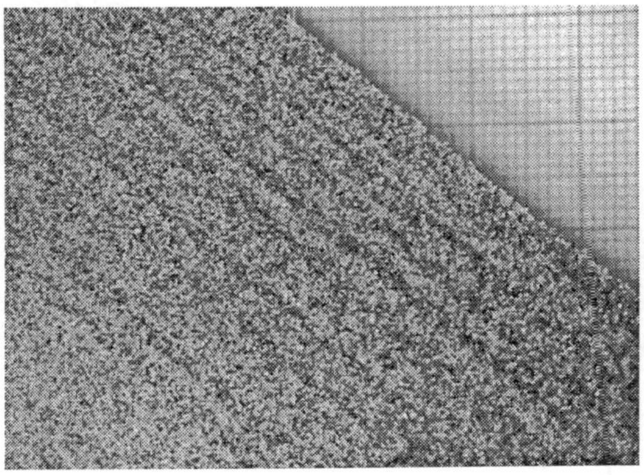

**Fig. 4** Stratification obtained with grains of the same size but with different shapes and coefficients of friction. The walls of the HS cell are separated by $e = 2$ mm. Dark grains (sand 1) are more spherical and have a bigger angle of repose than light grains (sand 4).

In the case of binary mixtures, Koeppe et al.[9] have reported some anomalous stripe pattern that they called "pairing". Indeed, stripes of the same grains seemed to be formed by pairs in particular experimental conditions. We made careful experiments with binary mixtures and we observed again the above-mentioned thin stripes (see Fig. 3). We suggest that thin stripes can be at the origin of the previously reported "pairing". Similarily to ternary mixtures, the successive stripes may have different angles.

## 3.2 Mixtures of Grains with the Same Size

We mentionned that Grasselli et al.[3] have observed that phase segregation can occur with grains of the same size but with different shape. They found that stratification can only occur for size ratio greater than 1.5. We made observations that for mixtures of grains of the same size we could get some stratification patterns (see Fig. 4). Those grains where selected with the same sieve and differ only in shape and surface properties (coefficient of friction). It is worth noticing that rough grains can have smaller angle of repose than spherical ones if the surface friction coefficients are different. This is indeed the case here. In order to get the stripes, extremely low flux of grain must be maintained during heap formation. The impact of the incoming flux of grains on pattern formation was underlined by Baxter et al.[10]

It must be said that not everything is understood on the processes coming into play for demixtion with grains of the same size. On the other hand, demixtion for grains differing in size is better understood, as the reader can see in the next section.

## 4. PHYSICAL PROCESSES

### 4.1 Fundamental Mechanism

#### 4.1.1 *For Binary Mixtures*

Self-stratification for binary mixtures of grains differing in size has been explained[1,11] by the combination of two main processes. They are illustrated in Fig. 5. The first process is the size segregation mechanism. There is a global tendency of large grains to roll farther downhill than the small ones due to their larger mass/inertia. This effect is strongly increased by the percolation of small grains through the gap between large grains in the flow. Large grains will be on the top of small grains, in the upper part of the moving layer. Small grains form a relatively smooth surface, on which large grains roll easily. Thus the moving layer is characterized by a strong velocity gradient such that the upper large grains are the fastest. As a result, large grains also locate in the front of the avalanche while small grains are confined in the lower part of the tail of the moving layer. Large grains will travel farther than small grains. The second process coming into play is the so-called kink mechanism. When the head of an avalanche reaches the end of the pile, grains will stop moving due to the horizontal slope of the basis. If the flux is not too important down the slope, grains in the tail of the avalanche will come to rest on the wall just formed by preceding grains. This wall is the kink and it moves up the slope as more grains arrive on it. This process is similar to the formation and backing of traffic jams. A pair of layers is formed through the kink mechanism with the small grains in the lower layer and the large grains in the top layer, starting at the basis of the pile. Since small grains were located in the tail of the avalanche, the wide stripe of small grains does not extend to the basis of the pile. The resulting surface involves an efficient capture for small grains during the next avalanche.

#### 4.1.2 *For Ternary Mixtures*

This explanation can be extended to ternary mixtures. It accounts for patterns in Figs. 2(a) and (b). The $\ldots SMLMSMLM \ldots$ ordering reflects the dynamics and composition of the lower layer of the avalanche: large grains in the head, followed by medium sized grains, themselves followed by small grains. So when an avalanche flow down on top of the surface formed by the preceeding kink, after the head of large grains has passed, medium grains will pass and some will be captured before small grains pass. A very close examination of the heap displayed in Fig. 2(a) reveals that there are indeed some medium sized grains caught below the stripes of small grains in the lower part of the heap. On the other hand, in the upper part of the heap, the $\ldots SMLSMLSML \ldots$ ordering is visible because there was not enough time for medium grains to move in front of small grains in the lower part of the moving layer. Thus, small grains were the first to be captured during the avalanche in the upper part of the pile. These dynamical processes depend strongly on the incoming flux of grains and the relative velocities of different granular species. However, the above mechanism for self-stratification cannot explain the thin layers of small grains observed in Fig. 2(c) and in Fig. 3.

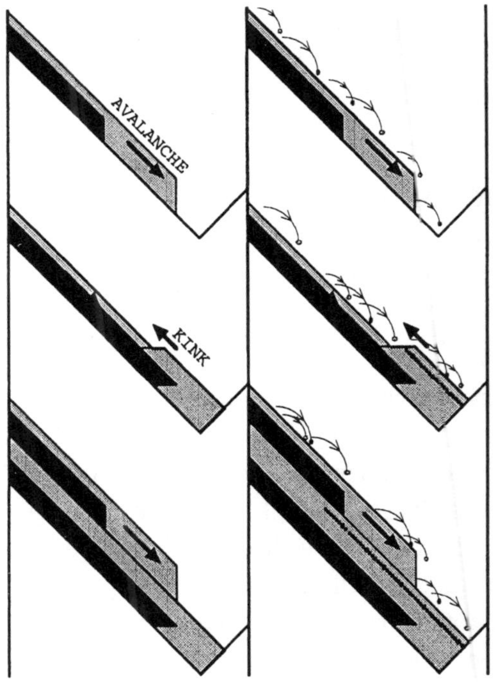


**Fig. 5** Schematic representation of self-stratification mechanisms described in the text for a flow of binary mixtures in a HS cell. Small grains are in dark grey and large grains are in light grey (left column). Three stages of the formation of pairs of strates in earlier explanations (right column). Three stages of the formation of strates in our model.

## 4.2 Complementary Mechanism

### 4.2.1 *Presence of Bouncing Grains*

The complementary mechanism we propose is illustrated in Fig. 5. Considering again a binary mixture, the first steps are identical to previous explanation. The kink, however, does not catch all moving grains. Indeed, in any granular flow, various transport mechanisms take place. From bottom to top in the moving layer, one encounters: combined movement of grains, motion of individual grains with friction and collisions, and above the rolling-slipping phase, a layer with boucing grains. Typically, the thickness of the rolling-slipping phase is about five to ten large grain diameters in our mixtures. The bouncing grains layer is characterised by a higher velocity and a much smaller density (number of grains/unit of volume). Indeed grains does not touch each other permanently ike in the rolling-slipping phase. We found that bouncing grains can jump typically 2 cm and then rebounce or be captured. We stress that most bouncing grains found at the bottom of the pile have bounce all their way down. Both kinds of grains can bounce. The efficiency of this mechanism depends on resilience of the grains. We observed that a non-negligible amount of those bouncing grains can flow down the slope over the kink (see Fig. 6). We stress that bouncing grains are those involved in the formation of the additional thin stripe observed in the center of the wide stripe made of large grains.

### 4.2.2 *How Bouncing Grains Build the Thin Stripes*

Figure 7 shows also the evolution of a binary pile. The first picture was taken as the kink was moving up. The thickness of the kink is about the same that the thickness of the rolling

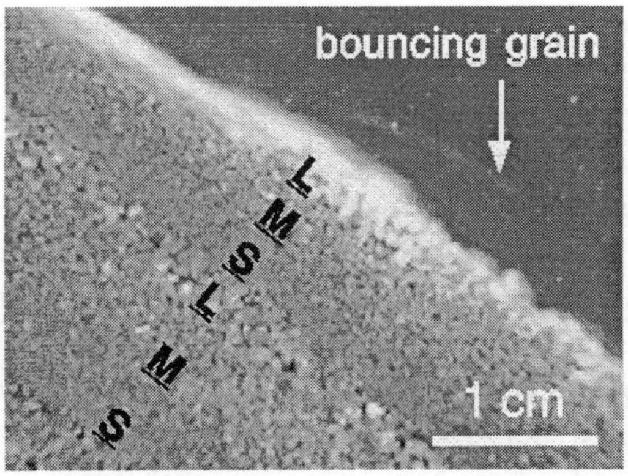


**Fig. 6** Upper part of pile 1c during formation. Bouncing grains are flowing over the kink.

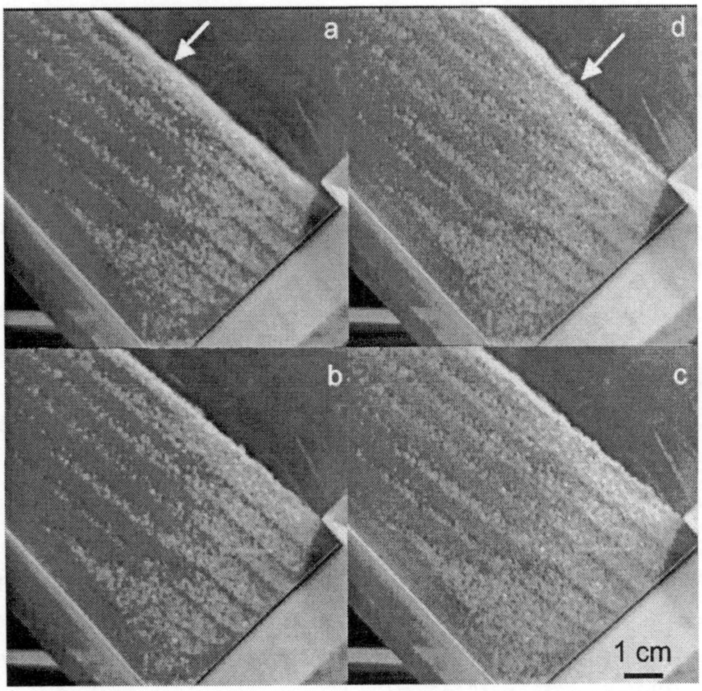


**Fig. 7** Apparition of a thin stripe of small grains in a binary mixture (Fe filings ($S$) and wheat semolina ($L$), plate separation $e = 6$ mm). **(a)** The kink is pointed by an arrow. **(b)** Just after the kink has passed. **(c)** Just before the next avalanche. The thin $S$ stripe has appeared. **(d)** The head of the next avalanche is marked by an arrow.

layer. As a consequence, the kink moves fast upward. The second picture shows the pile just after the kink has passed. The third picture was taken just before the next avalanche comes down, and the last picture shows the pile as the avalanche is flowing.

It is clearly visible that the thin stripe of small grains appears mainly during the time interval separating pictures (b) and (d), that is between the passage of the kink and the next avalanche. In addition, a close examination shows that this thin stripe is composed of both kinds of grains, in opposition to wider stripes where the granular species are quasi-pure. In the case of pile 1(c), close look reveals that small, medium and large sized grains are involved in the thin stripes.

The reason these strates are thinner is that there are much less grains involved — only those that are not immediately stopped by the kink. When a small bouncing grain is captured by the pile, it will fall through the gaps left by the large grains forming the static surface. This is why the stripes are formed of both kinds of grains. This also explains the fact that the thin stripe is observed to form below the static surface, about one large grain deep. Thin stripes have the same inclinaison angle as the surface left by the kink. The difference in angle for thin stripes and wide stripes is observed in Fig. 3(b) ($\Delta\theta \approx 3°$). It indicates that the kink builds at a greater angle than the angle of the avalanche (avalanches erode mainly the upper part of the pile). This can be related to another observation we made when pouring granular materials through the funnel. We noticed that the jamming limit diameter was lower in the case of mixture made of large and small grains than in the case of large grains only. Thus the presence of small grains seems to reduce the arching properties of large grains. We do not know if this is a general feature.

### 4.2.3 *Cases Where No Thin Stripes Are Observed*

Several causes can inhibe or hide the phenomena of thin stripe formation. In the case of Figs. 2(a) and (c), the only difference is the funnel used for pouring the grains. As a consequence, not only the flux of grains was lower in the case of pile *a* but also, and this is related, the initial velocity of grains seemed to be lower in *a*. The main effect is a lowering of the number of bouncing grains. The kink may become perfectly efficient, i.e. it may completely stop all moving grains it encounters. Then the thin layer will not form. Indeed we observed no bouncing grains flowing over the kink during heap formation of (a). Next, if the avalanches are important, they can erase by erosion the very thin layer that forms the top of the pile.

### 4.2.4 *Pairing*

All the mixtures we used are characterised by a large aspect ratio (> 2, up to 10) and by a larger density for small grains. Koeppe et al.[9] used a mixture of sand and sugar that corresponds to these characteristics when they obtained their "pairing". The difference in size causes percolation to be efficient and might favorise an important velocity gradient in the moving layer. These make self-stratification particularly pronounced. High velocity in the top of the rolling-slipping layer favorise the transport of bouncing grains. The large density of small grains allows them to bounce as far as large grains do. Indeed, in the case of grains all of the same density, small grains are much lighter than large grains. They would have much less kinetic energy and so they would travel less far than large grains.

### 4.2.5 *Model*

The number of grains involved in the thin stripe, depending on the position along the slope, can be computed from the time interval between the passage of the kink and next avalanche, assuming that one knows the probability of capture by unit of lenght for bouncing grains. Let $x$ be the position along the slope, starting at the top. If we assume constant velocity for the avalanche ($v_a$) and for the kink ($v_k$), the time interval $\Delta t(x)$ between the kink and

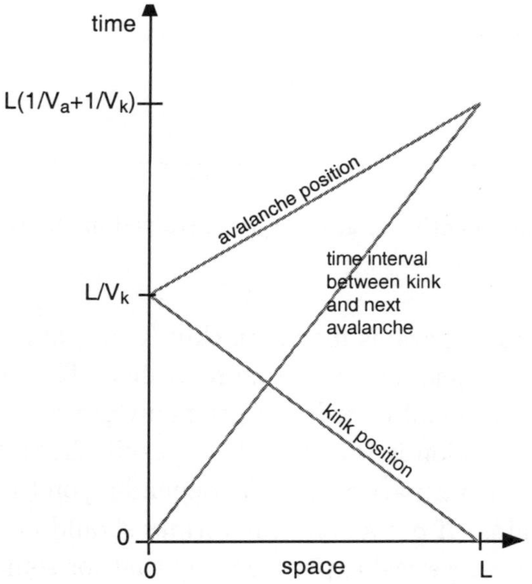


**Fig. 8** Depending on the position along the slope, the time interval between one kink and next avalanche varies.

the next avalanche occuring for a position $x$ is given by (see Fig. 8)

$$\Delta t(x) = x\left(\frac{1}{v_a} + \frac{1}{v_k}\right). \tag{1}$$

We will assume that bouncing grains have a constant probability $pdx$ to be captured along the slope. This means that bouncing grains are supposed to have a constant kinetic energy along the slope. Their losses of kinetic energy by collisions must be equal to their gains in free fall. Then, the quantity of bouncing grains $q(x)$ per unit length along the slope will be

$$q(x) = q_0 e^{-px}, \tag{2}$$

where $q_0$ is the initial quantity of bouncing grains. (We consider that no new grains are set to bounce in the slope.) If $Q(x)$ represents the quantity of grains by unit length involved in the thin stripe, we have

$$Q(x) = q(x) P \Delta t(x) \tag{3}$$

where $P$ is the probability of capture by unit of time. Let's write $v_b$ the velocity of bouncing grains (which has to be constant since we said that kinetic energy of bouncing grains is constant in this model), then

$$dx = v_b dt\,, \tag{4}$$

$$Pdt = pdx\,, \tag{5}$$

$$P = pv_b\,. \tag{6}$$

And we can write

$$Q(x) = pv_b x\left(\frac{1}{v_a} + \frac{1}{v_k}\right) q_0 e^{-px}. \tag{7}$$

This result is displayed in Fig. 9.

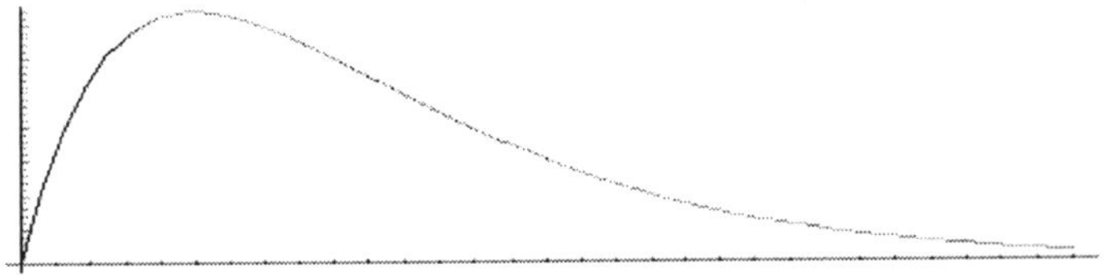

**Fig. 9** Graph of the quantity of grains $Q(x)$ involved in the thin stripe ($\sim x\,e^{-x}$).

We conclude that if bouncing grains moves quickly ($v_b \gg$) and/or are initially numerous ($q_0 \gg$), then the thin stripes should involve more grains. The velocities of the kink ($v_k$) and avalanche ($v_a$) should be small in order to get many grains in the thin stripes.

Another interesting observation is that the above result allows for the thin stripes to be more visible in the upper or lower part of the pile, depending on the density of probability of capture $p$ for bouncing grains. If $p$ is small, thin stripes should be located farther down the slope. On the other hand we observed experimentally that for some mixtures with narrower density difference between small and large grains, thin stripes do not extend to the basis of the pile. We thus conclude that bouncing transport was strongly damped by capture. Experimental study of thin stripes could provide an efficient way to study the transport properties of bouncing mechanism on a granular surface.

## 5. CONCLUSION

In summary, we have investigated self-stratification phenomenon in vertical HS cells with both ternary and binary granular mixtures. We have found some particular patterns. Firstly, stratification was obtained with grains of the same size but different shape. Secondly, for mixtures of grains differing in size, we observed the formation of an additional thin stripe which we relate to the previously reported "pairing".[9] This cannot be fully explained by former self-stratification models. We have proposed a complementary mechanism which is compatible with observations of binary/ternary patterns. The mechanism is based on the presence of bouncing grains on top of the rolling-slipping layer. Size ratio and denstity difference between small and large grains seems to be decisive properties of mixtures for thin stripe forming. In particular, denser small grains can bounce as far as large grains. Two important experimental parameters were incoming flux and initial velocity of grains. A simple model was proposed to account for the transport and capture of bouncing grains.

## ACKNOWLEDGMENTS

NV thanks the FNRS (Brussels, Belgium) for financial support. This work is also supported by the Belgian Royal Academy of Sciences through the Ochs-Lefebvre prize. Valuable discussion with M. Ausloos, J. Rajchenbach, E. Clément and S. Galam are acknowledged.

## REFERENCES

1. H. A. Makse, S. Havlin, P. R. King and H. E. Stanley, *Nature* **386**, 379 (1997).
2. P. Meakin and R. Jullien, *J. Physique* **48**, 1651 (1987).
3. Y. Grasselli and H. J. Herrmann, *Granular Matter* **1**, 43 (1998).
4. T. Boutreux, *Euro. Phys. J.* **B6**, 419 (1998).
5. P. Cizeau, H. A. Makse and H. E. Stanley, *Phys. Rev.* **E59**, 4408 (1999).

6. D. A. Head and G. J. Rodgers, *Phys. Rev.* **E56**, 1976 (1997).
7. N. Vandewalle, *Physica* **A272**, 450 (1999).
8. N. Lecocq and N. Vandewalle, *Phys. Rev.* **E62**, 8241 (2000).
9. J. P. Koeppe, M. Enz and J. Kakalios, *Phys. Rev.* **E58**, 4104 (1998).
10. J. Baxter, U. Tüzün, D. Heyes, I. Hayati and P. Fredlund, *Nature* **391**, 136 (1998).
11. H. A. Makse, R. C. Ball, H. E. Stanley and S. Warr, *Phys. Rev.* **E58**, 3357 (1998).

# SCALING IN COSMIC STRUCTURES

LUCIANO PIETRONERO, MAURIZIO BOTTACCIO and MARCO MONTUORI
*Physics Department & INFM Sezione di Roma 1,*
*Università "La Sapienza", P.le A. Moro 2, 00185 Roma, Italy*
FRANCESCO SYLOS LABINI
*Département de Physique Théorique, Université de Genève,*
*24, quai E. Ansermet, 1211 Genève*

**Abstract**

The study of the properties of cosmic structures in the universe is one of the most fascinating subject of the modern cosmology research. Far from being predicted, the large scale structure of the matter distribution is a very recent discovery, which continuosly exhibits new features and issues. We have faced such topic along two directions; from one side we have studied the correlation properties of the cosmic structures, that we have found substantially different from the commonly accepted ones. From the other side, we have studied the statistical properties of the very simplified system, in the attempt to capture the essential ingredients of the formation of the observed strucures.

## 1. INTRODUCTION

The existence of a cosmic structures in the universe is one of the most fascinating findings of the two last decades in observational cosmology. Galaxy distribution is far from homogeneous on small scale and large scale structures (filaments and walls) appear to be limited only by the boundary of the sample in which they are detected. There is currently an acute debate on the result of the statistical analysis of large scale features. In the past years we propose a new statistical approach, which has shown surprisingly a fractal structure, extending from small scales to distance beyond to $40h^{-1}$ $Mpc$ and with larger statistical uncertainity, even up to $100h^{-1}$ $Mpc$. In the following, we report our findings along two directions of investigations: in the first part we describe the standard analysis and its limits of applicability. We describe our novel analysis and the results we got in the characterization of large scale structures. Contrary to standard claims, this analysis is completely consistent with the results of the standard analysis: it's the interpretation of the latter which is radically different. In the second part, we refer on a study of the dynamics of a very simple model of gravitational formation of structures. We have analysed the evolution and the statistical spatial properties of a *N-body* system of point masses, interacting through gravity. The system is arranged as to simulate an infinite system of particles and with very simple initial conditions.

## 2. COSMIC STRUCTURES

The usual way to investigate the properties of the galaxy spatial clustering is to measure the two point autocorrelation function $\xi(r)$.[8,9] This is the most used statistical tool, since it can be measured quite accurately with current redshift surveys. $\xi(r)$ is the spatial average of the fluctuations in the galaxy number density at distance $r$, with respect to an homogeneous distribution with the same number of galaxies.

Consider a little volume $\delta V$ at position $\mathbf{r}_i$; let $n(\mathbf{r}_i)$ be the density of galaxies in $\delta V$ and $\langle n \rangle = N/V$ the density of galaxy in the whole sample.

The galaxy density fluctuations in $\delta V$ with respect to the average galaxy density $\langle n \rangle$, i.e. the galaxy relative density fluctuations, is:

$$\frac{n(\mathbf{r}_i) - \langle n \rangle}{\langle n \rangle} = \frac{\delta n(r_i)}{\langle n \rangle}. \tag{1}$$

The two-point correlation function $\xi(r)$ at the scale $r$ is the spatial average of the product of the relative density fluctuations in two little volumes at distance $r$:

$$\xi(r) = \left\langle \frac{\delta n(\mathbf{r}_i + r)}{\langle n \rangle} \frac{\delta n(\mathbf{r}_i)}{\langle n \rangle} \right\rangle_i = \frac{\langle n(\mathbf{r}_i) n(\mathbf{r}_i + \mathbf{r}) \rangle_i}{\langle n \rangle^2} - 1 \tag{2}$$

where the average is performed over the sample. Roughly speaking, a set of points is correlated on scale $r$ if $\xi(r) > 0$; it is uncorrelated over a certain scale $R$ if $\xi(r) = 0$ for $r > R$. In the latter case, the points are evenly distributed at scale $R > r$ or, in another words, they have an homogeneous distribution at scale $R > r$. In the definition of $\xi(r)$, the use of the sample density $\langle n \rangle$ as reference value for the fluctuations of galaxies is the *conceptual assumption* that the galaxy distribution is *homogeneous at the scale of the sample.*

Clearly such an approach is valid if the average density $\langle n \rangle$ of the sample *is* the average density of the distribution, or, in other words, if the distribution is homogeneous on the scale of the sample. For this reason, $\xi(r)$ analysis *assumes the homogeneity* and it is unreliable for *testing* it.

In order to use $\xi(r)$ analysis, the density of galaxies in the sample must be a good estimation of the density of the whole distribution of galaxies. This may either be true or not; in any case, it should be checked *before* applying $\xi(r)$ analysis.[7]

In addition to such criticisms, the usual interpretation of $\xi(r)$ measure is uncorrect for an another aspect; it is customary to define a characteristic scale for the correlations in any spatial distribution of points with respect the amplitude of the $\xi(r)$. The *correlation length of the distribution* $r_0$ is indeed defined as the scale such that $\xi(r_0) = 1$.[5]

Such a definition is incorrect since, in statistical mechanics, the *correlation length* of the distribution is defined by how fast the correlations vanish as a function of the scale, i.e. by the functional form of the $\xi(r)$ and not by its amplitude.

The quantity $r_0$, then, does not concern the *scale* of fluctuations and it is not correct referring to this as a measure of the characteristic size of correlations.[10,11] According to the $\xi(r)$ definition, $r_0$ simply separates a regime of large fluctuations $\delta n/\langle n\rangle \gg 1$ from a regime of small fluctuations $\delta n/\langle n\rangle \ll 1$: this is correct *if the average density of the sample* $\langle n\rangle$ *is the average density of galaxy distribution.*

Such problems of $\xi(r)$ approach can be avoided analysing the spatial correlations of the data set without any *a priori* assumptions on the homogeneity scale of the data itself.[7]

The way to perform such an unbiased analysis is to study the behaviour of the *conditional average number of galaxies* $\langle N(<r)\rangle$ or the *average conditional galaxy density* $\Gamma^*(r)$ versus the scale $r$. The two quantities are respectively:

$$\langle N(<r)\rangle = B \cdot r^D \tag{3}$$

where $N(<r)$ is the number of galaxies contained in a sphere of radius $r$ centered on a galaxy of the sample. $\langle N(r)\rangle$ is the average of $N(<r)$ computed in all the spheres centered on every galaxy of the sample

$$\Gamma^*(r) = \frac{\langle N(<r)\rangle}{4/3\pi r^3} = \frac{3B}{4\pi} \cdot r^{D-3}. \tag{4}$$

$\Gamma^*(r)$ is then the corresponding average density of galaxies in spheres of radius $r$:[7,12] The exponent $D$ is called *the fractal dimension* and characterises in a quantitative way how the system fills the space, while the prefactor $B$ depends on the lower cut-off of the distribution. $\langle N(<r)\rangle$ and $\Gamma^*(r)$ are the suitable statistical tools to detect the two-point correlation properties of a spatial distribution of objects and the possible crossover scale between different statistical distributions.

If the point distribution has a *crossover* to an homogeneity distribution at scale $R$, $\Gamma^*(r)$ shows a flattening toward a constant value at such scale. In this case, the fractal dimension of Eqs. (3) and (4) has the same value of the dimension of embedding space $d$, $D = d$ (in three-dimensional space $D = 3$).[7,12,13]

If this does not happen, the density sample will not correspond to the density of the distribution and it will show correlations up to the sample size. The simplest distribution with such properties is a fractal structure.[13] A fractal consists of a system in which more and more structures appear at smaller and smaller scales and the structures at small scales are similar to the ones at large scales. The distribution is then self-similar. It has a value of $D$ smaller than $d$, $D < d$. In 3-dimensional space, $d = 3$, a fractal has $D < 3$ and $\Gamma^*(r)$ is a *power law.* The value of $N(<r)$ largely fluctuates by both *changing the starting point*, from which we compute $N(<r)$, and *the scale* $r$. This is due to the scale invariant feature of a fractal structure, which does not have any *characteristic length.*[13,14] It is simple to show that if we analyse a fractal structure with $\xi(r)$, we can obtain a value for the *correlation length* $r_0$, which evidently does not have any relation with the correlation properties of the system. In fact such a value is simply a fraction to the size of the sample under analysis. Larger is the sample size, larger is the corresponding $r_0$.

According to our criticism to the standard analysis, we have performed the measure of galaxy conditional average density $\Gamma^*(r)$ in all the three-dimensional catalogs available. Our analysis is carried out on several 3D galaxy samples. The results are are collected in Fig. 1.[12]

$\Gamma^*(r)$, measured in different catalogues, is a *power law* as a function of the scale $r$, extending from $\approx 0.5$ to 30-40$h^{-1}$ *Mpc*, without any tendency towards homogenization

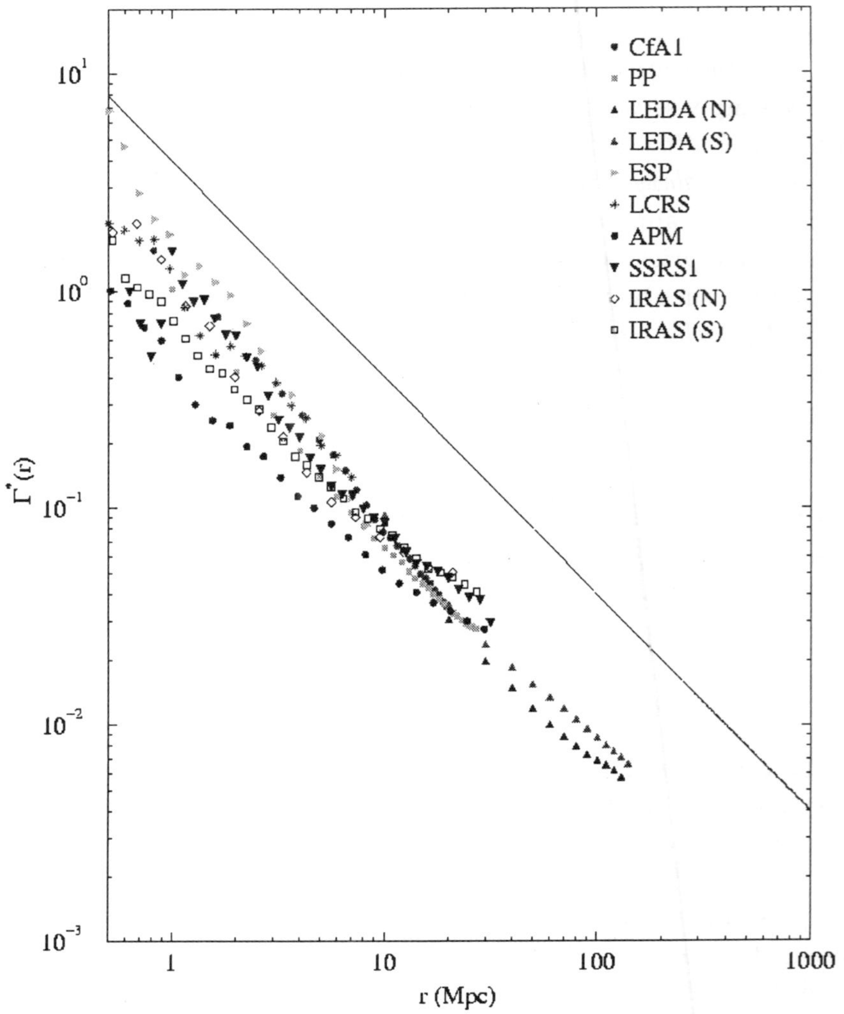


**Fig. 1** $\Gamma^*(r)$ in the range of scales $0.5 \div 100h^{-1}$ $Mpc$ for all the available 3D galaxy data. A reference line with a slope $-1$ is also shown (i.e. fractal dimension $D = 2$).

(flattening).[12] In a single case, the LEDA sample,[16] it is possible to reach larger scales, $\sim 100h^{-1}$ $Mpc$. The scaling $\Gamma^*(r)$ appears to continue with the same properties observed at smaller scales. This data sample has been largely criticised, but to our knowledge, never in a quantitative way. The statistical tests we performed show clearly that up to $50h^{-1}$ $Mpc$ the results are consistent with all other data.[12]

Such results imply that the $\xi(r)$ analysis is inappropriate as it describes correlations as deviations from an assumed underlaying homogeneity. The galaxy distribution shows instead fractal properties at least in the range $r \approx 0.5 - 30 - 40h^{-1}$ $Mpc$, which seem to extend in a single sample up $r \approx 100h^{-1}$ $Mpc$.

By consequence, the *correlation length* $r_0$, i.e. the amplitude of the $\xi(r)$, for samples with such a linear extention, should be a fraction of the sample size: $r_0$ should be larger for samples whose size is larger.

This is evident in Fig. 2, where we plot the results of the standard analysis $\xi(r)$ performed on the same data sets analysed with $\Gamma^*(r)$ (Fig. 1).

In this case, then, $r_0$ has no relation with the correlation properties of the system; its variation in different samples is not related to any variation of the clustering of the corresponding data set.

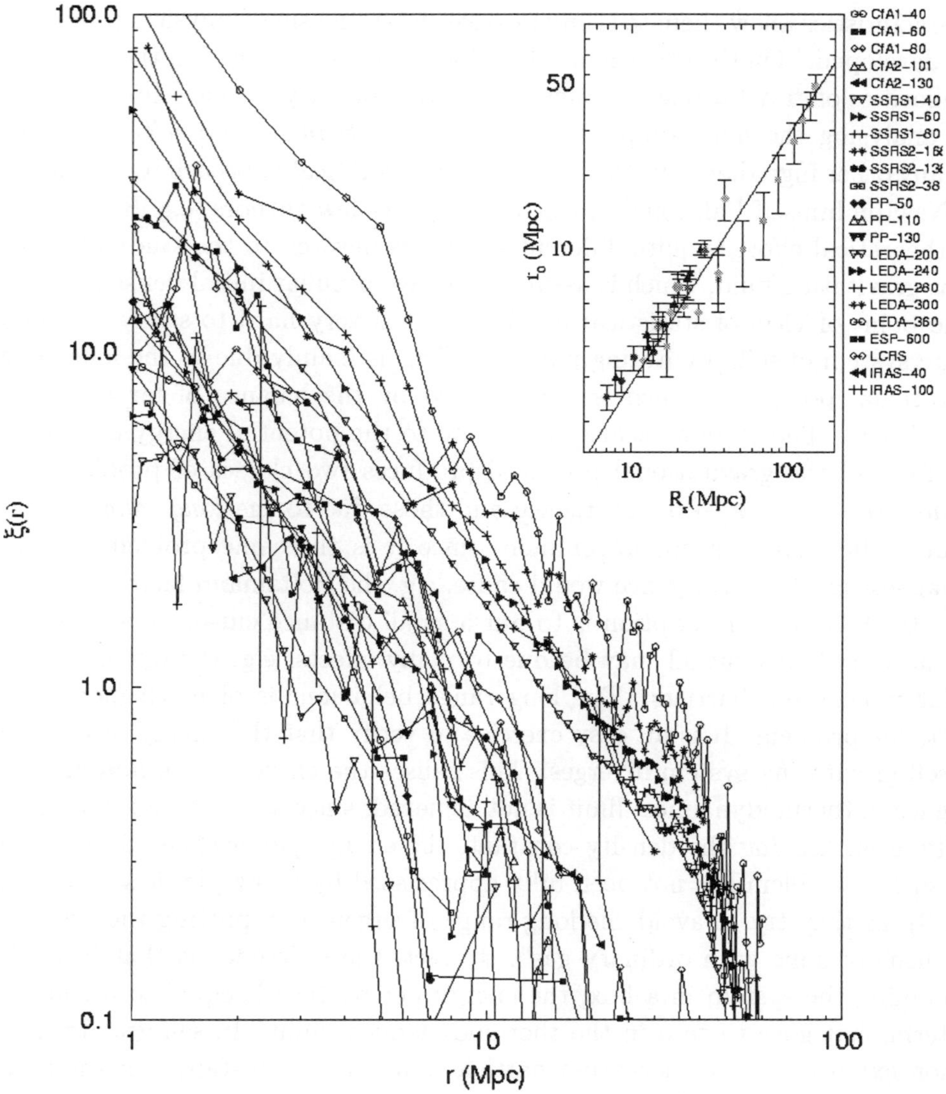


**Fig. 2** $\xi(r)$ measure in various VL galaxy samples. The general trend is an increase of the $\xi(r)$ amplitude for brighter and deeper samples. In the *insert panel* we show the dependence of *correlation length* $r_0$ on *sample size* $R_s$ for all samples. The linear behaviour is a consequence of the fractal nature of galaxy distribution in these samples.

## 3. SIMULATIONS OF GRAVITATIONAL CLUSTERING

The study of the formation of the cosmic strucures we have analysed in the above section is one of the most challenging problems in astrophysics. Gravity is the most natural candidate for the explanation of the variety of structures we observe. Indeed, the range of scales on which the gravitational clustering takes place is really impressive: from $10^{-1}$ $pc$ to $10^{8}$ $pc$ ($1\ pc = 3.2615\ light - yr = 3.0856\ x10^{18}\ cm$). This implies interactions of gravity with other physical processes depending on the scale: from turbulence in cold molecular clouds to cosmological expansion above galaxy cluster scale. Because such a richness of physical processes can be involved in modelling the various structures we observe in cosmos, it is actually very difficult to retrieve a clear picture of the statistical properties of self-gravitating system. Current astrophysical simulations have reached a high level of refinement, both in resolution and in the number of different physical processes which they take into account.

Such characteristics allow them to study in great detail the single physical problem for which they are developed.[4] On the other hand they don't allow a clarification the common role and the peculiarities of gravitational interaction. On the contrary, we have tried to focus on such features analysing the most simplest case of *a many-body infinite self-gravitating system*, without any other ingredient but the gravity. The theoretical approach to such a system goes back to Newton himself,[1] although it has faced by very few authors (e.g. Ref. 2). Indeed, the current theoretical effort is quite different since it is devoted to the study of evolution of a continuos gravitating fluid, which is assumed to have peculiar initial density fluctuations.[5,6] From the point of view of statistical mechanics, it is very hard to study the properties of an infinite system of self-gravitating particles. This is mainly due to the long range nature of gravitational potential, which is not shielded by the balance of far away charges, as e.g. in a plasma. Therefore all scales contribute to the potential energy of a particle. The peculiar form of the gravitational potential produces two classes of problems: those due to the *short range* (i.e. $r \to 0$) divergence and those due to the *long range* (i.e. $r \to +\infty$) behaviour. The former is not uncommon, since it is the same problem which arises in electromagnetism. The divergence would cause, e.g. the Boltzmann factor to diverge in the limit $r \to 0$. A typical prescription is to put a small distance cut-off in the potential. The physical nature of this cut-off may be due to many effects, e.g. the dynamical emergence of angular momentum barriers. The long range behaviour is of much more concern and is, in fact, *the* problem. It is an easy exercise to verify that the energy of a particle in an infinite self-gravitating system diverges. This causes the energy to be *non-extensive.* As a consequence, a thermodynamical limit is not achieved, since as the number of particles goes to infinity, even keeping the density constant, the energy *per particle* diverges. Strangely enough, such a problem has not been fully appreciated by many physicists in the field (see e.g. Ref. 3), as they try to avoid the long range divergence by putting the system in a box "as it is usually done with ordinary gas". In fact, the difference is that in ordinary gas, when confining the system in a box, the energy per particle is equal to a constant plus a surface term that goes to zero in the thermodynamical limit. In self-gravitating systems, due to non extensivity, the energy per particle is neither a constant, nor the surface term goes to zero (in fact, it is of the same order of magnitude as the potential energy due to particles belonging to the system).

Another very interesting consequence, which is often not appreciated, is that the thermodynamical definition of temperature, as the parameter which controls the equilibrium of the system, doesn't hold for a self-gravitating system, since one cannot divide a system into smaller subsystems with the same thermodynamic properties of the larger system.

As a consequence of such difficulties, a satisfying thermodynamic equilibrium treatment of such systems is still lacking.

However we are much more interested in what happens *out of equilibrium*, during the evolution of a system.

The system we intend to simulate is a infinite many body system. At this aim, the $N$ particles we effectively consider are confined in a cube of size $L$ submitted to periodic boundary conditions. Every particle in the simulation box interacts with all other particles and with the periodic replicas of the whole system.

The initial conditions we consider are:

(1) *random (white noise) initial positions of particles*;
(2) *no cosmological expansion*;
(3) *zero initial velocities*;
(4) *equal mass particles*;

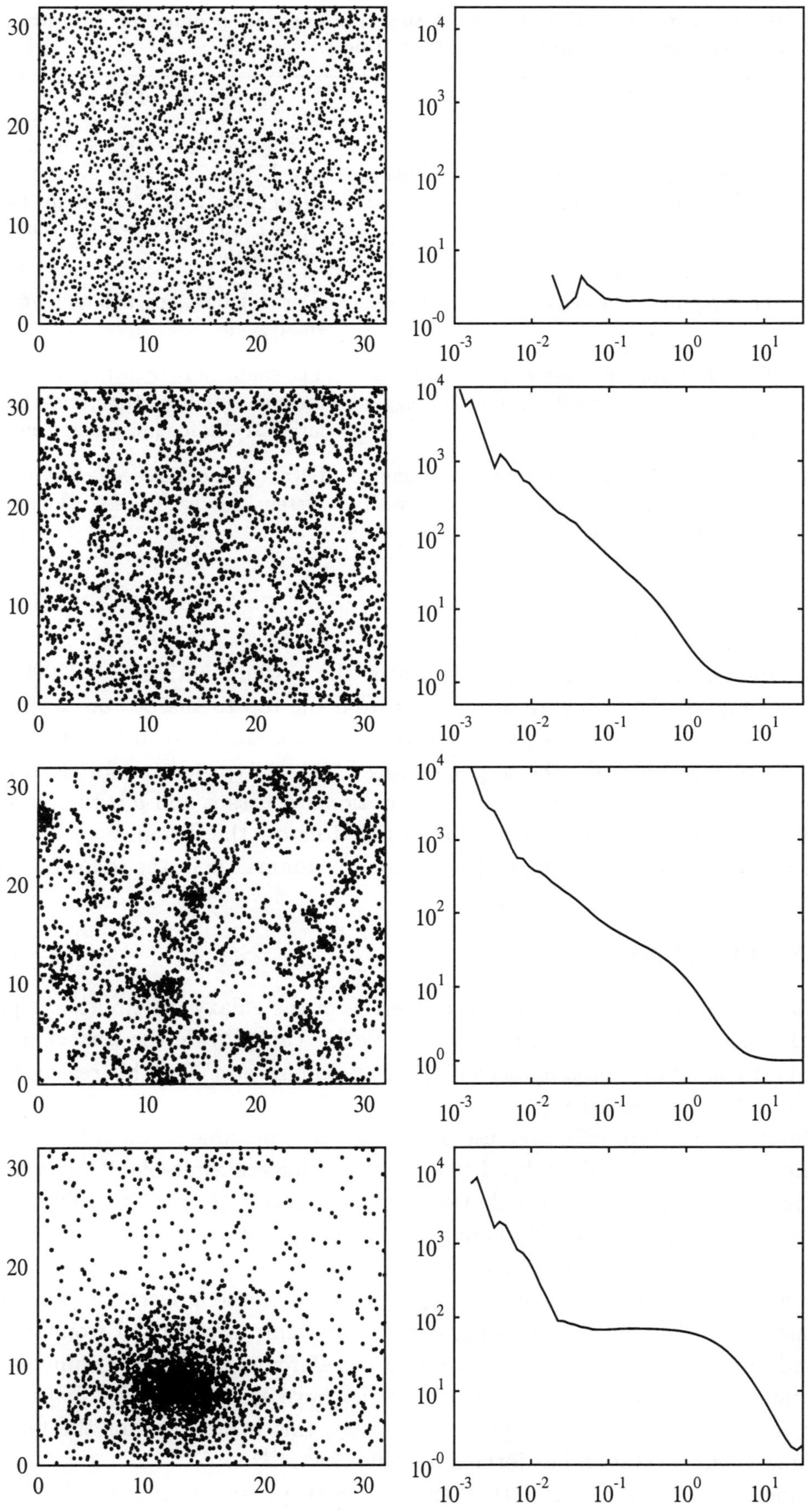


**Fig. 3** *On the left*: snapshots of system evolution. The corresponding time is: $t = 0$, $t = 2/3\tau$, $t = \tau$ and $t = 4 \cdot \tau$. *On the right*: the corresponding $\Gamma^*(r)$ for the snapshots on the left.

**Table 1** ***On the left*: project of the simulation box onto $x - y$ plane at different time. *On the right*: corresponding measure of $\Gamma^*(r, t)$ at time $t$.**

| $t$ time | Description | $\Gamma^*(r)$ |
|---|---|---|
| $t_0 = 0$ | The system is composed by $N$ *particles* at rest with spatial poissonian distribution in the simulation box. | The system has a constant number density $\Gamma^*(r, t_0) = 1$ at all the scales. At small scale, $\Gamma^*(r)$ is more fluctuating, because of the larger *poisson noise* $1/\sqrt{N}$ at small scale. |
| $t_1 \approx 2/3\tau$ | The system starts to cluster at small scale. | $\Gamma^*(r, t_1)$ develops a larger amplitude at small scale. |
| $t_2 \approx \tau$ | The clustering process evolves with the merging of the small clusters in bigger ones. | The shape of $\Gamma^*(r, t_2)$ appears to be quite independent from time, but shifts toward larger scales for increasing time. |
| $t_f \approx 4\tau$ | All the clusters have merged in a single big one, with dimension comparable with the simulation box size. | $\Gamma^*(r, t_f)$ does not evolve anymore and the system has reached a stationary state. |

Some snapshots of the temporal evolution of the system, with $N = 32000$ particles, are shown in Fig. 3. The time evolution goes from the top to the bottom and the initial unclustered distribution of mass points evolves toward a clustered distribution. For each snapshot, we plot on the *right*, the corresponding $\Gamma^*(r)$. A typical time for the evolution of the system is $\tau = 1/\sqrt{G\rho}$, where $G$ is the gravitational constant and $\rho$ is the density of the system. It is roughly the time needed to a particle to cross the system.

Figure 3 shows some interesting features, that we summarize in the following table.

## 4. CONCLUSIONS

The standard analysis of the correlation properties of the galaxy distribution is performed through the measure of the $\xi(r)$ function. The latter can provide the correct information *if* the set under analysis is homogeneous inside the sample size. For this reason $\xi(r)$ is not reliable for *testing* homogeneity.

This should be checked before the use of $\xi(r)$ and it is possible through $\Gamma^*(r)$ analysis of the sample set. Such an analysis has been performed for the available 3D galaxy samples, with the result that the galaxy distribution appear fractal from $\approx 0.5$ to $100h^{-1}$ $Mpc$.

For such a range of scales, the $\xi(r)$ analysis does not give the correct informations on the statistical properties of the galaxy distribution.

To investigate the formation of such a fractal structure in the universe, we are performing simulations of an $N$-body infinite self-gravitating system. We started from the simplest initial conditions and analysed the system with the aforementioned $\Gamma^*(r)$ function.

Simulations with a different number of particles have shown that the shape of $\Gamma^*(r, t)$ for $t \leq \tau$ is independent from $N$. On the contrary, this is not true for the final state of equilibrium, when the system has formed a single cluster.[15]

These measures seem to show that the transient phase, during which the collapse occurs, possess a well-defined thermodynamical limit, which we are currently analysing.[15] The discrete nature of the N-body system seems to be a fundamental ingredient in the development of the spatial correlations. The latter, indeed, grow at the small scale, where the discreteness of the point distribution has to be taken into account. At the moment is not clear if such a system can develop fractal correlations as seen in the galaxy distribution.

## ACKNOWLEDGMENTS

We would like to thank A. Gabrielli and M. Joyce for enlighting discussions. We acknowledges financial support from EC TMR Research Network under contract ERBFMRXCT960062 and the project *Clustering* by INFM.

## REFERENCES

1. I. Newton, quoted in *Theories of the Universe*, ed. M. K. Munitz (Toronto, 1957), p. 211.
2. W. C. Saslaw, *The Distribution of the Galaxies* (Cambridge University Press, 2000).
3. T. Padmanabhan, *Phys. Rep.* **188**, 285 (1990).
4. For references to astrophysical $N$-body simulations see:
   *http://star-www.dur.ac.uk/frazerp/virgo/virgo.html*
   *http://star-www.dur.ac.uk/cole/mocks/main.html*
   *http://lca.ncsa.uiuc.edu:8080/GC3_Home_Page.html*
5. P. J. E. Peebles *Principles of Physical Cosmology* (Princeton University Press, 1993).
6. T. Padmanabhan *Structure Formation in the Universe* (Cambridge University Press, 1993).
7. P. Coleman and L. Pietronero *Phys. Rep.* **213**, 311 (1992).
8. P. J. E. Peebles, *Appl. J.* **185**, 413 (1973).
9. H. Totsuji and T.Kihara, *Pub. Astron. Soc. Jap.* **21**, 221 (1969).
10. J. Gaite, A. Domnguez and J. Pérez-Mercader, *Appl. J. Lett.* **522**, L5 (1999).
11. A. Gabrielli, F. Sylos Labini and R. Durrer, *Appl. J. Lett.* **531**, L1 (2000).
12. F. Sylos Labini, M. Montuori and L. Pietronero, *Phys. Rep.* **293**, 66 (1998).
13. B. Mandelbrot, *The Fractal Geometry of Nature* (Freeman, San Francisco, 1983).
14. M. Montuori, F. Sylos Labini, A. Gabrielli, A. Amici and L. Pietronero, *Europhys. Lett.* **39**, 103 (1997).
15. M. Bottaccio, A. Amici, P. Miocchi, R. Capuzzo Dolcetta, M. Montuori and L. Pietronero, *Phys. Rev. Lett.*, submitted (2001).
16. Di Nella H. et al., *A&A Lett.* **308**, L33 (1996).

# REVISITING THE DERIVATION OF THE FRACTIONAL DIFFUSION EQUATION

ENRICO SCALAS
*Dipartimento di Scienze e Tecnologie Avanzate,*
*Università del Piemonte Orientale,*
*Corso Borsalino 54, Alessandria, I–15100, Italy*

RUDOLF GORENFLO
*Erstes Mathematisches Institut, Freie Universität Berlin,*
*Arnimallee 3, Berlin, D–14195, Germany*

FRANCESCO MAINARDI
*Dipartimento di Fisica, Università di Bologna e INFN Sezione di Bologna,*
*Via Irnerio 46, Bologna, I–40126, Italy*

MARCO RABERTO
*Dipartimento di Ingegneria Biofisica ed Elettronica, Università di Genova,*
*Via all'Opera Pia 11a, Genova, I–16145, Italy*

**Abstract**

The fractional diffusion equation is derived from the master equation of continuous time random walks (CTRWs) via a straightforward application of the Gnedenko-Kolmogorov limit theorem. The Cauchy problem for the fractional diffusion equation is solved in various important and general cases. The meaning of the proper diffusion limit for CTRWs is discussed.

## 1. INTRODUCTION

This paper provides a short, but self-contained, introduction to fractional diffusion. The readers will find the basic ideas behind the derivation of the fractional diffusion equation starting from continuous-time random walks. We have included formulae for the solution of the Cauchy problem which can be numerically implemented and used for applications. Special care has been used to avoid unessential mathematical technicalities. Even if far from exhaustive, the bibliography should give a sufficient number of entry points for further reading. The following sections are based on a series of papers about the application of fractional calculus to finance.[1–4]

It is our hope that theoretical and experimental condensed matter physicists will find this work useful.

The paper is divided as follows. In Sec. 2, we outline the theory leading to the time-fractional master equation. In Sec. 3, the transition to the space-time fractional diffusion equation is discussed. Section 4 is devoted to the solutions of the Cauchy problem for the fractional diffusion equation. The main results are briefly summarized and discussed in Sec. 5. In Appendices A and B, we introduce the definitions of fractional derivatives in time and space, respectively, entering the fractional diffusion equation.

## 2. STEP ONE: TRANSITION TO THE TIME-FRACTIONAL MASTER EQUATION

Let $x$ be the position of a diffusing particle in one dimension. Let us assume that both jumps $\xi_i = x(t_i) - x(t_{i-1})$ and waiting times between two consecutive jumps $\tau_i = t_i - t_{i-1}$ are i.i.d. random variables described by two probability density functions: $w(\xi)$ and $\psi(\tau)$. According to the model of continuous-time random-walk (CTRW), introduced by Montroll and Weiss,[5,6] the evolution equation for $p(x, t)$, the probability of finding the random walker at position $x$ at time instant $t$, can be written as follows, assuming the initial condition $p(x, 0) = \delta(x)$ (i.e. the walker is initially at the origin $x = 0$),[2]

$$p(x, t) = \delta(x)\Psi(t) + \int_0^t \psi(t - t') \left[\int_{-\infty}^{+\infty} w(x - x')p(x', t')dx'\right] dt', \tag{2.1}$$

where

$$\Psi(t) = \int_t^{\infty} \psi(t')dt' = 1 - \int_0^t \psi(t')dt', \quad \psi(t) = -\frac{d}{dt}\Psi(t). \tag{2.2}$$

The *master equation* of the CTRW can be also derived in the Fourier-Laplace domain.

The integral $\int_0^{\tau} \psi(t')dt'$ represents the probability that at least one step is taken at some instant in the interval $[0, \tau)$. Thus, $\Psi(\tau)$ is the probability that the diffusing quantity $x$ does not change value during the time interval of duration $\tau$ after a jump.

In a paper by Mainardi et al.,[2] an alternative form of Eq. (2.2) was presented in terms of a convolution between the first time derivative of $p(x, t)$ and a suitable kernel. The resulting equation can be interpreted as an *evolution* equation of generalized *Fokker-Planck-Kolmogorov* kind. It reads:

$$\int_0^t \Phi(t - t')\frac{\partial}{\partial t'} p(x, t')dt' = -p(x, t) + \int_{-\infty}^{+\infty} w(x - x')p(x', t)dx', \tag{2.3}$$

where the "auxiliary" function $\Phi(t)$ is such that $\Psi(t) = \int_0^t \Phi(t - t')\psi(t')dt'$. Equation (2.3) can be obtained by Fourier-Laplace transforming Eq. (2.1) and by suitable assumptions on the Laplace transform of the function $\Phi(t)$.

In general, a CTRW is a non-Markovian process. A CTRW becomes Markovian if (and only if) the above memory function is proportional to a delta function so that $\Psi(t)$ and $\psi(t)$ differ only by a multiplying positive constant. By an appropriate choice of the unit of time, we can write $\Phi(t) = \delta(t)$, $t \geq 0$. In this case, Eq. (2.3) becomes:

$$\frac{\partial}{\partial t} p(x, t) = -p(x, t) + \int_{-\infty}^{+\infty} w(x - x')p(x', t)dx', \quad p(x, 0) = \delta(x). \tag{2.4}$$

Up to a change of the unit of time, this is the most general *master equation* for a *Markovian CTRW*; Saichev and Zaslavsky call it the *Kolmogorov-Feller equation.*[7]

Equation (2.3) allows a natural characterization of a peculiar class of non-Markovian processes, where the memory function, $\Phi(t)$ has power-law time decay. Within this class, an interesting choice is the following:

$$\Phi(t) = \frac{t^{-\beta}}{\Gamma(1-\beta)}, \quad t \geq 0, \quad 0 < \beta < 1. \tag{2.5}$$

In this case, $\Phi(t)$ is a weakly singular function that, in the limit $\beta \to 1$, reduces to $\Phi(t) = \delta(t)$, according to the formal representation of the Dirac generalized function, $\delta(t) = t^{-1}/\Gamma(0)$, $t \geq 0$.[8] As a consequence of the choice (2.5),[2] Eq. (2.3) can be written as:

$$\frac{\partial^\beta}{\partial t^\beta} p(x,\, t) = -p(x,\, t) + \int_{-\infty}^{+\infty} w(x-x')p(x',\, t)dx', \quad p(x,\, 0) = \delta(x)\,, \tag{2.6}$$

where $\partial^\beta/\partial t^\beta$ is the pseudo-differential operator explicitly defined in the Appendix A, that we usually call the *Caputo* fractional derivative of order $\beta$. Equation (2.6) is a time-fractional generalization of Eq. (2.4) and can be called *time-fractional Kolmogorov-Feller equation.*

Our choice for $\Phi(t)$ implies peculiar forms for the functions $\Psi(t)$ and $\psi(t)$ generalizing the exponential behaviour of the waiting time density in the Markovian case. In fact, we have for $t \geq 0$:[2]

$$\Psi(t) = E_\beta(-t^\beta), \quad \psi(t) = -\frac{d}{dt}E_\beta(-t^\beta), \quad 0 < \beta < 1\,, \tag{2.7}$$

where $E_\beta$ denotes an entire transcendental function, known as the Mittag-Leffler function of order $\beta$, defined in the complex plane by the power series

$$E_\beta(z) := \sum_{n=0}^{\infty} \frac{z^n}{\Gamma(\beta n + 1)}, \quad \beta > 0, \quad z \in \mathbf{C}\,. \tag{2.8}$$

Detailed information on the Mittag-Leffler type functions is available in the literature.[9–11]

From the properties of the Mittag-Leffler function, it can be shown that the corresponding survival probability and waiting-time *pdf* interpolate between a stretched exponential, for small waiting times, and a power-law decay, for large waiting times. Such a behaviour has been observed in Mainardi et al.[2] and in Raberto et al.,[3] where the function $\Psi(\tau)$ has been estimated from empirical financial data.

As a final remark, it is important to notice that different choices of the kernel $\Phi(t)$ in Eq. (2.3) are possible, leading to different properties of the waiting-time pdf and different generalized Kolmogorov-Feller evolution equations for $p(x,\, t)$.

## 3. STEP TWO: THE DIFFUSION LIMIT

In the physical literature, many authors have discussed the connection between continuous–time random walks and diffusion equations of fractional order, with different degrees of detail.[12–23] A sound proof of the equivalence between fractional diffusion and CTRW has been given by Hilfer and Anton.[16] However, in order to perform the transition to the diffusion limit, we shall use a different approach. We shall start from Eqs. (2.4) and (2.6), and pass through their Fourier-Laplace counterparts. The stochastic process whose probability density evolves according to those equations is a random walk originating from a sequence of jumps, each jump being a sample of a real random variable $Y$. During the time interval $t_n \le t < t_{n+1}$, the particle position is $Y_1 + Y_2 + \dots Y_n$. The $Y_k$ are i.i.d. random variables all described, as $Y$, by the *pdf* $w(x)$. Let us denote by $\hat{w}(\kappa)$ the characteristic function corresponding to the probability density $w(x)$.

Let us specify some conditions on the *pdf* $w(x)$. The requirement is that, if $\alpha = 2$:

$$\sigma^2 = \int_{-\infty}^{+\infty} x^2 w(x) dx < \infty \,, \tag{3.1}$$

whereas, if $0 < \alpha < 2$:

$$w(x) = (b + \varepsilon(|x|))|x|^{-(\alpha+1)}, \quad b > 0\,, \quad \varepsilon(|x|) \to 0 \text{ as } |x| \to \infty\,. \tag{3.2}$$

In Eq. (3.2), $b > 0$ and $\varepsilon(|x|)$ is bounded and $O(|x|^{-\eta})$ with $\eta > 0$ as $|x| \to \infty$. Let us furthermore recall the necessary requirements $w(x) \ge 0$, and the normalization condition $\int_{-\infty}^{+\infty} w(x) dx = 1$.

Let us now consider a sequence of random process *pdf*'s $p_h(x, t)$ describing scaled jumps of size $hY_k$ instead of $Y_k$, with a speed increase of the process by a factor (*the scaling factor*) $\mu^{-1/\beta} h^{-\alpha/\beta}$, where $\mu$ must satisfy some conditions which will be specified later. The *pdf* of the jump size is $w_h(x) = w(x/h)/h$, so that its characteristic function is $\hat{w}_h(\kappa) = \hat{w}(\kappa h)$. For $0 < \alpha \le 2$ and $0 < \beta \le 1$, Eq. (2.6) (including (2.4) in the special case $\beta = 1$) is replaced by the sequence of equations

$$\mu h^\alpha \frac{\partial^\beta}{\partial t^\beta} p_h(x, t) = -p_h(x, t) + \int_{-\infty}^{+\infty} w_h(x - x') p_h(x', t) dx'. \tag{3.3}$$

By Fourier-Laplace transforming and by recalling the Laplace transform of the Caputo time-fractional derivative, defined by Eq. (A.2), we have

$$\mu h^\alpha \{ s^\beta \hat{\tilde{p}}_h(\kappa, s) - s^{\beta-1} \} = [\hat{w}_h(\kappa) - 1] \hat{\tilde{p}}_h(\kappa, s)\,. \tag{3.4}$$

We shall now present arguments based on the *classical central limit theorem* or on the *Gnedenko limit theorem*, (see the book by Gnedenko and Kolmogorov[24]) both expressed in terms of the characteristic functions. The Gnedenko limit theorem is a suitable generalization of the classical central limit theorem for space *pdf*'s with infinite variance, decaying according to condition (3.2).

The transition to the *diffusion limit* is based on the following Lemma introduced by Gorenflo:[25]

*With the scaling parameter*

$$\mu = \begin{cases} \dfrac{\sigma^2}{2}\,, & \text{if} \quad \alpha = 2\,, \\[2ex] \dfrac{b\pi}{\Gamma(\alpha+1)\sin(\alpha\pi/2)}\,, & \text{if} \quad 0 < \alpha < 2\,, \end{cases} \tag{3.5}$$

*we have the relation*

$$\lim_{h \to 0} \frac{\hat{w}(\kappa h) - 1}{\mu h^\alpha} = -|\kappa|^\alpha, \quad 0 < \alpha \le 2\,, \quad \kappa \in \mathbf{R}\,. \tag{3.6}$$

Now, it is possible to set

$$\rho_h(\kappa) = \frac{\hat{w}(\kappa h) - 1}{\mu h^\alpha}\,, \tag{3.7}$$

and the sequence of Eq. (3.4) reads

$$s^{\beta}\hat{\tilde{p}}_h(\kappa,\, s) - s^{\beta-1} = \rho_h(\kappa)\hat{\tilde{p}}_h(\kappa,\, s). \tag{3.8}$$

Then, passing to the limit $h \to 0$, thanks to (3.6), we get:

$$s^{\beta}\hat{\tilde{p}}_0(\kappa,\, s) - s^{\beta-1} = -|\kappa|^{\alpha}\hat{\tilde{p}}_0(\kappa,\, s)\,, \quad 0 < \alpha \le 2\,, \quad 0 < \beta \le 1\,. \tag{3.9}$$

By inversion and using the Fourier transform of the Riesz space-fractional derivative, defined in Eq. (B.2), we finally obtain the equation:

$$\frac{\partial^{\beta}}{\partial t^{\beta}}\, p_0(x,\, t) = \frac{\partial^{\alpha}}{\partial |x|^{\alpha}}\, p_0(x,\, t)\,, \quad p_0(x,\, 0) = \delta(x)\,, \tag{3.10}$$

which is a space-time fractional diffusion equation. In the limiting cases $\beta = 1$ and $\alpha = 2$, Eq. (3.10) reduces to the standard diffusion equation.

We have presented a formally correct transition to the diffusion limit starting from the general master equation of the CTRW, namely Eq. (2.1) or Eq. (2.3). By invoking the continuity theorem of probability theory, see e.g. the book by Lukacs,[26] we can see that the random variable whose density is $p_h(x,t)$ converges in distribution ("weakly" or "in law") to the random variable with density $p_0(x,\, t)$.

Solving (3.8) for $\hat{\tilde{p}}_h(\kappa,\, s)$, and (3.9) for $\hat{\tilde{p}}_0(\kappa,\, s)$, gives:

$$\hat{\tilde{p}}_h(\kappa,\, s) = \frac{s^{\beta-1}}{s^{\beta} - \rho_h(\kappa)}\,, \quad \hat{\tilde{p}}_0(\kappa,\, s) = \frac{s^{\beta-1}}{s^{\beta} + |\kappa|^{\alpha}}\,, \tag{3.11}$$

which yields:

$$\hat{p}_h(\kappa,\, t) = E_{\beta}(\rho_h(\kappa)t^{\beta})\,, \quad \hat{p}_0(\kappa,\, t) = E_{\beta}(-|\kappa|^{\alpha}t^{\beta})\,. \tag{3.12}$$

By (3.6) $\rho_h(\kappa) \to -|\kappa|^{\alpha}$ as $h \to 0$, hence

$$p_h(x,\, t) \to p_0(x,\, t)\,, \qquad \text{for} \quad t > 0\,, \quad h \to 0\,. \tag{3.13}$$

## 4. SOLUTIONS AND THEIR SCALING PROPERTIES

For the determination of the fundamental solutions of Eq. (3.12) in the general case $\{0 < \alpha \le 2,\, 0 < \beta \le 1\}$ the reader can consult Gorenflo et al.[27] and Mainardi et al.[29] We also refer to the above references for the particular cases $\{0 < \alpha \le 2,\, \beta = 1\}$ and $\{\alpha = 2,\, 0 < \beta \le 1\}$, already dealt with in the literature.

For parameters in the interval $0 < \alpha \le 2$, and $0 < \beta \le 1$, the Cauchy problem in Eq. (3.10) can be solved by means of the Fourier-Laplace transform method.

The solution (Green function) turns out to be:

$$p_0(x,\, t) = \frac{1}{t^{\beta/\alpha}} W_{\alpha,\beta}\left(\frac{x}{t^{\beta/\alpha}}\right). \tag{4.1}$$

The function $W_{\alpha,\beta}(u)$ is the Fourier transform of a Mittag-Leffler function:

$$W_{\alpha,\beta}(u) = \frac{1}{2\pi}\int_{-\infty}^{+\infty} e^{-iqu} E_{\beta}(-|q|^{\alpha})dq\,. \tag{4.2}$$

Indeed, $E_\beta$ is the Mittag-Leffler function of order $\beta$ and argument $z = -|q|^\alpha$

In the limiting case $0 < \alpha < 2$ and $\beta = 1$, the solution is:

$$p_0(x,\, t) = \frac{1}{t^{1/\alpha}} L_\alpha \left( \frac{x}{t^{1/\alpha}} \right), \tag{4.3}$$

where $L_\alpha(u)$ is the Lévy standardized probability density function:

$$L_\alpha(u) = \frac{1}{2\pi} \int_{-\infty}^{+\infty} e^{-iqu-|q|^\alpha} dq\,, \tag{4.4}$$

whereas, in the case $\alpha = 2$, $0 < \beta < 1$,[28] we get

$$p_0(x,\, t) = \frac{1}{2} t^{-\beta/2} M_{\beta/2} \left( \frac{x}{t^{\beta/2}} \right), \tag{4.5}$$

where $M_{\beta/2}$ denotes the $M$ function of Wright type of order $\beta/2$.

Remarkably, a composition rule holds true, and it can be shown that the Green function for the space-time fractional diffusion equation of order $\alpha$ and $\beta$ can be written in terms of the Green function for the space-fractional diffusion equation of order $\alpha$ and the Green function for the time-fractional diffusion equation of order $2\beta$:[29]

$$p_0(x,\, t) = t^{-\beta} \int_0^\infty r^{-1/\alpha} L_\alpha(x/r^{1/\alpha}) M_\beta(r/t^\beta) dr\,. \tag{4.6}$$

Finally, as written before, in the case $\alpha = 2$, $\beta = 1$, Eq. (3.10) reduces to the standard diffusion equation, and the Cauchy problem is solved by:

$$p_0(x,\, t) = t^{-1/2} \frac{1}{2\sqrt{\pi}} \exp(-x^2/(4t)) = t^{-1/2} G \left( \frac{x}{t^{1/2}} \right), \tag{4.7}$$

where $G(x)$ denotes the Gaussian *pdf*

$$G(x) = \frac{1}{2\sqrt{\pi}} \exp(-x^2/4)\,. \tag{4.8}$$

## 5. SUMMARY AND DISCUSSION

Applications of fractional diffusion equations have been recently reviewed by Uchaikin and Zolotarev[30] and by Metzler and Klafter.[22] After that, other contributions appeared on this issue, among which we quote the paper of Zaslavsky in the book edited by Hilfer,[31] the papers by Meerschaert et al.[32] and by Paradisi et al.[33] and the letter by West and Nonnenmacher.[34]

In this paper, a scaling method has been discussed to get the transition to the diffusion limit in a correct way, starting from the CTRW master equation describing the time evolution of a stochastic process. Moreover, the solutions of the Cauchy problem for the fractional diffusion equation have been listed for the various relevant values of the fractional derivative orders $\alpha$ and $\beta$.

Various formulae which can be useful for applications have been presented. In principle, given a diffusing quantity, the waiting–time density, the jump density, and the probability of finding the random walker in position $x$ at time $t$ are all quantities which can be empirically determined. Therefore, many relationships presented above can be corroborated or falsified in specific contexts.

As a further remark, it may be useful to add some comments on the meaning of the diffusion limit taken in Sec. 3.

The factor $\mu h^\alpha$ can be viewed as causing the jump process to run faster and faster (the waiting times becoming shorter and shorter) as $h$ becomes smaller and smaller. Replacing the density $w(x)$ by the density $w_h(x) = w(x/h)/h$, and, accordingly, the jumps $Y$ by $hY$, means that the jump size becomes smaller and smaller as the scaling length $h$ tends to zero.

An alternative interpretation is that we look at the same process with a discrete number of jumps occurring after finite times, from far away and after long time, so that spatial distances and time intervals of normal size appear very small, being $x$ replaced by $x/h$, $t$ replaced by $t/(\mu^{1/\beta} h^{\alpha/\beta})$.

## REFERENCES


1. E. Scalas, R. Gorenflo and F. Mainardi, *Physica* **A284**, 376 (2000).
2. F. Mainardi, M. Raberto, R. Gorenflo and E. Scalas, *Physica* **A287**, 468 (2000).
3. M. Raberto, E. Scalas, R. Gorenflo and F. Mainardi, *J. Quant. Fin.*, submitted (preprint in `http://xxx.lanl.gov/abs/cond-mat/0001253`).
4. R. Gorenflo, F. Mainardi, E. Scalas and M. Raberto, in *Proceedings of the Workshop on Mathematical Finance, University of Konstanz, Germany, October 5-7, 2000*, eds. M. Kohlmann and S. Tang (Birkhäuser, Boston, 2001), to appear.
5. E. W. Montroll and G. H. Weiss, *J. Math. Phys.* **6**, 167 (1965).
6. M. F. Shlesinger, in *Encyclopedia of Applied Physics*, ed. G. L. Trigg (VCH Publishers, Inc., New York, 1996), p. 45.
7. A. I. Saichev and G. M. Zaslavsky, *Chaos* **7**, 753 (1997).
8. I. M. Gel'fand and G. E. Shilov, *Generalized Functions*, Vol. 1 (Academic Press, New York, 1964).
9. A. Erdélyi, W. Magnus, F. Oberhettinger and F. G. Tricomi, *Higher Transcendental Functions*, Bateman Project, Vol. 3 (McGraw-Hill, New York, 1955), Chap. 18, pp. 206–227.
10. R. Gorenflo and F. Mainardi, in *Fractals and Fractional Calculus in Continuum Mechanics*, eds. A. Carpinteri and F. Mainardi (Springer-Verlag, Wien and New York, 1997), p. 223.
11. F. Mainardi and R. Gorenflo, *J. Comput. Appl. Math.* **118**, 283 (2000).
12. H. E. Roman and P. A. Alemany, *J. Phys. A*: Math. Gen. **27**, 3407 (1994).
13. H. C. Fogedby, *Phys. Rev.* **E50** 1657 (1994).
14. B. D. Hughes, *Random Walks and Random Environments, Vol. 1. Random Walks* (Oxford Science Publication Clarendon Press, Oxford, 1995), p. 631.
15. M. Kotulski, *J. Stat. Phys.* **81**, 777 (1995).
16. R. Hilfer and L. Anton, *Phys. Rev.* **E51**, R848 (1995).
17. A. Compte, *Phys. Rev.* **E53**, 4191 (1996).
18. B. J. West, P. Grigolini, R. Metzler and T. F. Nonnenmacher, *Phys. Rev.* **E55**, 9 (1997).
19. R. Metzler and T. F. Nonnenmacher, *Phys. Rev.* **E57**, 6409 (1998).
20. R. Hilfer, in *Applications of Fractional Calculus in Physics*, ed. R. Hilfer (World Scientific, Singapore, 2000), p. 87.
21. E. Barkai, R. Metzler and J. Klafter, *Phys. Rev.* **E61**, 132 (2000).
22. R. Metzler and J. Klafter, *Phys. Rep.* **339**, 1 (2000).
23. V. V. Uchaikin *Int. J. Theor. Phys.* **39**, 2087 (2000).
24. B. V. Gnedenko and A. N. Kolmogorov, *Limit Distributions for Sums of Independent Random Variables* (Addison-Wesley, Cambridge, Massachusetts, 1954).
25. R. Gorenflo and F. Mainardi, in *Problems in Mathematical Physics*, eds. J. Elschner, I. Gohberg and B. Silbermann (Birkhäuser Verlag, Basel, 2001), p. 120.
26. E. Lukacs, *Characteristic Functions* (Griffin, London, 1960).
27. R. Gorenflo, A. Iskenderov and Yu. Luchko, *Fractional Calculus Appl. Anal.* **3**, 75 (2000).
28. F. Mainardi, *Chaos, Solitons & Fractals* **7**, 1461 (1996).
29. F. Mainardi, Yu. Luchko and G. Pagnini, *Fractional Calculus Appl. Anal.* **4**, in press (2001).
30. V. V. Uchaikin and V. M. Zolotarev, *Change and Stability. Stable Distributions and Their Applications* (VSP, Utrecht, 1999).

31. G. M. Zaslavsky, in *Applications of Fractional Calculus in Physics*, ed. R. Hilfer (World Scientific, Singapore, 2000), p. 377.
32. M. M. Meerschaert, D. A. Benson and B. Baeumer, *Phys. Rev.* **E63**, in press (2001).
33. P. Paradisi, R. Cesari, F. Mainardi and F. Tampieri, *Physica* **A293**, in press (2001).
34. B. J. West and T. Nonnenmacher, *Phys. Lett.* **A278**, 253 (2001).
35. M. Caputo, *Geophys. J. R. Astr. Soc.* **13**, 529 (1967).
36. M. Caputo, *Elasticità e Dissipazione* (Zanichelli, Bologna, 1969).
37. M. Caputo and F. Mainardi, *Riv. Nuovo Cimento* (Ser. II) **1**, 161 (1971).
38. S. G. Samko, A. A. Kilbas and O. I. Marichev, *Fractional Integrals and Derivatives, Theory and Applications* (Gordon and Breach, Amsterdam, 1993).
39. I. Podlubny, *Fractional Differential Equations* (Academic Press, San Diego, 1999).
40. R. Gorenflo and F. Mainardi, *Fractional Calculus Appl. Anal.* **1**, 167 (1998).
41. R. Gorenflo and F. Mainardi, *J. Anal. Appl.* (*ZAA*) **18**, 231 (1999).

## APPENDIX A: THE CAPUTO TIME-FRACTIONAL DERIVATIVE

For readers' convenience, here, we present an introduction to the *Caputo* fractional derivative starting from its representation in the Laplace domain and pointing out its difference from the standard *Riemann-Liouville* fractional derivative. In so doing we avoid the subtleties lying in the inversion of fractional integrals.

If $f(t)$ is a (sufficiently well-behaved) function with Laplace transform $\mathcal{L}\{f(t);s\} = \tilde{f}(s) = \int_0^\infty e^{-st} f(t)dt$, we have

$$\mathcal{L}\left\{\frac{d^\beta}{dt^\beta} f(t); s\right\} = s^\beta \tilde{f}(s) - s^{\beta-1} f(0^+), \quad 0 < \beta < 1\,, \tag{A.1}$$

if we define

$$\frac{d^\beta}{dt^\beta} f(t) := \frac{1}{\Gamma(1-\beta)} \int_0^t \frac{df(\tau)}{d\tau} \frac{d\tau}{(t-\tau)^\beta}\,. \tag{A.2}$$

We can also write

$$\frac{d^\beta}{dt^\beta} f(t) = \frac{1}{\Gamma(1-\beta)} \frac{d}{dt}\left\{\int_0^t [f(\tau) - f(0^+)] \frac{d\tau}{(t-\tau)^\beta}\right\}, \tag{A.3}$$

$$\frac{d^\beta}{dt^\beta} f(t) = \frac{1}{\Gamma(1-\beta)} \frac{d}{dt}\left\{\int_0^t \frac{f(\tau)}{(t-\tau)^\beta} d\tau\right\} - \frac{t^{-\beta}}{\Gamma(1-\beta)} f(0^+)\,. \tag{A.4}$$

Equations (A.1) to (A.4) can be extended to any non integer $\beta > 1$ (see e.g. the survey by Gorenflo and Mainardi[10]). We refer to the fractional derivative defined by (A.2) as the *Caputo* fractional derivative, as it was used by Caputo for modelling dissipation effects in *linear viscoelasticity* in the late sixties.[35–37]

This definition differs from the usual one named after Riemann and Liouville, given by the first term in the R.H.S. of (A.4), and defined e.g. in the treatise on Fractional Calculus by Samko, Kilbas and Marichev.[38]

Gorenflo and Mainardi[10] and Podlubny[39] have pointed out the usefulness of the Caputo fractional derivative in the treatment of differential equations of fractional order for *physical applications*. In fact, in physical problems, the initial conditions are usually expressed in terms of a given number of boundary values assumed by the field variable and its derivatives of integer order, despite the fact that the governing evolution equation may be a generic integro-differential equation and therefore, in particular, a fractional differential equation.

## APPENDIX B: THE RIESZ SPACE-FRACTIONAL DERIVATIVE

If $f(x)$ is a (sufficiently well-behaved) function with Fourier transform

$$\mathcal{F}\{f(x);\,\kappa\} = \hat{f}(\kappa) = \int_{-\infty}^{+\infty} e^{i\kappa x} f(x)\,dx\,, \quad \kappa \in \mathbf{R}\,,$$

we have

$$\mathcal{F}\left\{\frac{d^\alpha}{d|x|^\alpha}\,f(x);\,\kappa\right\} = -|\kappa|^\alpha \hat{f}(\kappa)\,, \quad 0 < \alpha < 2\,, \tag{B.1}$$

if we define

$$\frac{d^\alpha}{d|x|^\alpha}\,f(x) = \Gamma(1+\alpha)\,\frac{\sin(\alpha\pi/2)}{\pi}\int_0^\infty \frac{f(x+\xi) - 2f(x) + f(x-\xi)}{\xi^{1+\alpha}}\,d\xi\,. \tag{B.2}$$

The fractional derivative defined by (B.2) can be called *Riesz fractional derivative*, as it is obtained from the inversion of the fractional integral originally introduced by Marcel Riesz, known as the *Riesz potential.*[38] The representation (B.2),[25] is more explicit and convenient than others found in the literature.[7,38] It is based on a suitable regularization of a hyper-singular integral.

For $\alpha = 2$, the Riesz derivative reduces to the standard derivative of order 2, as $-|\kappa|^2 = -\kappa^2$.

For $\alpha = 1$, the Riesz derivative is related to the Hilbert transform, resulting in the formula

$$\frac{d}{d|x|}\,f(x) = -\frac{1}{\pi}\,\frac{d}{dx}\int_{-\infty}^{+\infty}\frac{f(\xi)}{x-\xi}\,d\xi\,. \tag{B.3}$$

We note, by writing $-|\kappa|^\alpha = -(\kappa^2)^{\alpha/2}$, that the Riesz derivative of order $\alpha$ can be interpreted as the opposite of the $\alpha/2$ power of the (positive definite) operator $-D^2 = -\frac{d^2}{dx^2}$, namely

$$\frac{d^\alpha}{d|x|^\alpha} = -\left(-\frac{d^2}{dx^2}\right)^{\alpha/2} \tag{B.4}$$

The notation used above is due to Saichev and Zaslavsky.[7] A different notation which takes into account asymmetries was used by Gorenflo and Mainardi.[40,41]



# LEARNING IN AN OSCILLATORY CORTICAL MODEL

SILVIA SCARPETTA
*Department of Physics "E. R. Caianiello" Salerno University,*
*I-84081 Baronissi (SA) IT,*
*INFM, Sezione di Salerno, (SA), IT*

ZHAOPING LI
*Gatsby Computational Neuroscience Unit, UCL, London, UK*

JOHN HERTZ
*Nordita, DK-2100 Copenhagen, Denmark*

**Abstract**

We study a model of generalized-Hebbian learning in asymmetric oscillatory neural networks modeling cortical areas such as hippocampus and olfactory cortex. The learning rule is based on the synaptic plasticity observed experimentally, in particular long-term potentiation and long-term depression of the synaptic efficacies depending on the relative timing of the pre- and postsynaptic activities during learning. The learned memory or representational states can be encoded by both the amplitude and the phase patterns of the oscillating neural populations, enabling more efficient and robust information coding than in conventional models of associative memory or input representation. Depending on the class of nonlinearity of the activation function, the model can function as an associative memory for oscillatory patterns (nonlinearity of class II) or can generalize from or interpolate between the learned states, appropriate for the function of input representation (nonlinearity of class I). In the former case, simulations of the model exhibits a first order transition between the "disordered state" and the "ordered" memory state.

## 1. INTRODUCTION

Synaptic connection in the brain are almost never symmetrical; indeed, coupled excitatory-inhibitory systems are intrinsically asymmetric. Unfortunately, our generic mathematical understanding of asymmetric networks is far less extensive than that of symmetric ones. In the latter case, the system settles to static states which are local minima of a Lyapunov function,[1] while real brain states are seldom static and often exhibit oscillatory behavior. Some recent work (see Ref. 9 and references therein) has begun to investigate generic computational differences between symmetrical networks and asymmetrical excitatory-inhibitory ones. However, whereas we know something about appropriate learning algorithms for associative memories and input representations in symmetric networks,[1] and there is some mathematical literature on oscillatory networks, there is no previous work on learnings in oscillatory networks.

Recent studies on synapses between pyramidal neocortical and hippocampal neurons[2–5] have revealed that changes in synaptic efficacy can depend on the relative timing of pre- and postsynaptic spikes. Typically, a presynaptic spike followed by a postsynaptic one leads to an increase in efficacy (long term potentiation or LTP), while the reverse temporal order leads to a decrease (long term depression or LTD). The dependence of the change in synaptic efficacy on the difference $\tau$ between the two spike times may be characterized by a kernel which we denote $A(\tau)$.[5] For hippocampal pyramidal neurons, the half-width of this kernel is around 20 ms. Many important neural structures, notably hippocampus and olfactory cortex, exhibit oscillatory activity in the 20–50 Hz range. Here the temporal variation of the neuronal firing can clearly affect the synaptic dynamics, and vice versa.

In this paper, we study learning in a model of asymmetric neural networks. The model structure and elements are based on the physiological and anatomical findings in the CA3 hippocampal region and the olfactory cortex. These regions contain the principal excitatory pyramidal cells and the inhibitory interneurons. The pyramidal cells project long range axons to other pyramidal cells and the interneurons, whereas the interneurons only project locally. The neural population shows oscillatory activities in response to sensory and environmental inputs such as odor objects or spatial locations.

We introduce a simple model of learning for oscillatory patterns, based on the structure of the kernel $A(\tau)$ and other known physiology of these areas. We will assume that learning is driven by oscillatory, patterned, input to a network that initially has only local synaptic connections, and that the learning results in synaptic changes in the long range lateral connections. The result is an imprinting of the oscillatory patterns in the long range synapses, such that subsequent inputs resembling any stored pattern evoke strong resonant responses. The learning rule in our model can be viewed as a generalization to oscillatory networks with spike-timing-dependent learning of the standard scenario whereby stationary patterns are stored in Hopfield networks using the conventional Hebb rule.

## 2. MODEL

The model neurons represent local populations of biological neurons that share common input. They follow the equations of motion[6]

$$\dot{u}_i = -\alpha u_i - \beta_i^0 g_v(v_i) + \sum_j J_{ij}^0 g_u(u_j) + I_i \,, \tag{1}$$

$$\dot{v}_i = -\alpha v_i + \gamma_i^0 g_u(u_i) + \sum_{j \neq i} W_{ij}^0 g_u(u_j) \,. \tag{2}$$

Here $u_i$ and $v_i$ are membrane potentials for excitatory and inhibitory (formal) neuron $i$, $\alpha^{-1}$ is their membrane time constant, and the sigmoidal functions $g_u(\ )$ and $g_v(\ )$ model

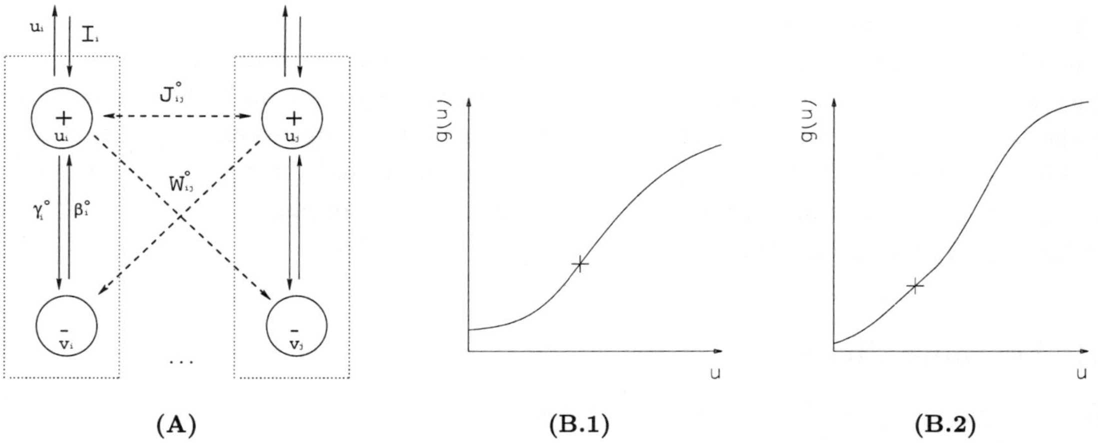


**Fig. 1** **(A)** The model circuit: in addition to the local excitatory-inhibitory connections (vertical solid lines), there are long-range connections (dashed lines) between excitatory units ($J_{ij}$) and from excitatory to inhibitory units ($W_{ij}$). External inputs are fed to the excitatory units, which provide model outputs. Learning affects long range connections only. **(B)** Activation functions used for the excitatory units. **B.1** class I; **B.2** class II. Crosses mark the equilibrium point $(\bar{u}, \bar{v})$ of the system.

the dependence of their outputs (interpreted as instantaneous firing rates) on their membrane potentials. The couplings $\beta_i^0$ and $\gamma_i^0$ are inhibitory-to-excitatory (resp. excitatory-to-inhibitory) connection strengths within local excitatory-inhibitory pairs, and for simplicity we take the external drive $I_i(t)$ to act only on the excitatory units. We include nonlocal excitatory couplings $J_{ij}^0$ between excitatory units and $W_{ij}^0$ from excitatory units to inhibitory ones. In this minimal model, we ignore long-range inhibitory couplings, appealing to the fact that real anatomical inhibitory connections are predominantly short-ranged. (In what follows, we will sometimes use bold and sans serif notation (e.g., $\mathbf{u}$, $\mathsf{J}$) for vectors and matrices, respectively.) The structure of the couplings is shown in Fig. 1(A).

We start by limiting our treatment to an analysis of small oscillations around a stable fixed point $\{\bar{\mathbf{u}}, \bar{\mathbf{v}}\}$ determined by the DC part of the input. Using linear (small amplitude) approximation and eliminating the inhibitory units,[6,7] we obtain

$$\ddot{\mathbf{u}} + [2\alpha - \mathsf{J}]\dot{\mathbf{u}} + [\alpha^2 + \beta(\gamma + \mathsf{W}) - \alpha\mathsf{J}]\mathbf{u} = (\partial_t + \alpha)\delta\mathbf{I}\,. \tag{3}$$

Here $\mathbf{u}$ is now measured from the fixed point $\bar{\mathbf{u}}$, $\delta\mathbf{I}$ is the time-varying part of the input, and the elements of $\mathsf{J}$ and $\mathsf{W}$ are related to those of $\mathsf{J}^0$ and $\mathsf{W}^0$ by $W_{ij} = g_u'(\bar{u}_j)W_{ij}^0$ and $J_{ij} = g_u'(\bar{u}_j)J_{ij}^0$. For simplicity, we have assumed that the effective local couplings $\beta_i = g_v'(\bar{v}_i)\beta_i^0$ and $\gamma_i = g_u'(\bar{u}_i)\gamma_i^0$ are independent of $i$: $\beta_i = \beta$, $\gamma_i = \gamma$. With oscillatory inputs $\delta\mathbf{I} = \boldsymbol{\xi}e^{-i\omega t} + \text{c.c.}$, the oscillatory pattern elements $\xi_i = |\xi_i|e^{-i\phi_i}$ are complex, reflecting possible phase differences across the units. We likewise separate the response $\mathbf{u} = \mathbf{u}^+ + \mathbf{u}^-$ (after the initial transients) into positive- and negative-frequency components $\mathbf{u}^\pm$ (with $\mathbf{u}^- = \mathbf{u}^{+*}$ and $\mathbf{u}^\pm \propto e^{\mp i\omega t}$). Since $\dot{\mathbf{u}}^\pm = \mp i\omega\mathbf{u}^\pm$, Eq. (3) can be written

$$\left[2\alpha \pm \frac{i}{\omega}(\alpha^2 + \beta\gamma - \omega^2)\right]\mathbf{u}^\pm = \mathsf{M}^\pm\mathbf{u}^\pm + \left(1 \pm \frac{i\alpha}{\omega}\right)\delta\mathbf{I}^\pm, \tag{4}$$

a form that shows how the matrix

$$\mathsf{M}^\pm(\omega) \equiv \mathsf{J} \mp \frac{\mathrm{i}}{\omega}(\beta\mathsf{W} - \alpha\mathsf{J})\,. \tag{5}$$

describes the effective coupling between local oscillators. $2\alpha$ is the intrinsic damping and $\sqrt{\alpha^2 + \beta\gamma}$ the frequency of the individual oscillators.

We assume the system has two modes or phases of operation. One is the learning mode in which the oscillating patterns are imprinted in the synaptic connections J and W, the other is the recall mode, when connections do not change any more under any input pattern. In the recall mode the network recognizes an input pattern by resonating in response to it. We use notation $\boldsymbol{\xi}^0 e^{-i\omega_0 t}$ + c.c. for the patterns imprinted during the learning phase, $\delta\mathbf{I} = \boldsymbol{\xi} e^{-i\omega t}$ + c.c. for input patterns presented during the recall phase, and $\mathbf{u} = \mathbf{r}e^{-i\omega t}$ + c.c. for the response of the network.

## 2.1 Learning Phase

We employ a generalized Hebb rule of the form

$$\delta C_{ij}(t) = \eta \int_0^T dt \int_{-\infty}^{\infty} d\tau y_i(t+\tau) A(\tau) x_j(t) \tag{6}$$

for changes in synaptic weight $C_{ij}$, where $x_j$ and $y_i$ are the pre- and postsynaptic activities, measured relative to stationary levels at which no changes in synaptic strength occur. We consider a general kernel $A(\tau)$, although experimentally $A(\tau) > 0$ ($< 0$) for $\tau > 0$ ($< 0$). Applying the rule to both J and W in our linearized network, where the firing rates $g_u(u_i)$ and $g_v(v_i)$ vary linearly with $u_i$ and $v_i$, we will use $x_j = u_j$ and $y_i = u_i$ or $v_i$ (measured from the fixed point $\bar{u}_i$, $\bar{v}_i$), respectively.

In the brain structures we are modeling, cholinergic modulation makes the long-range connections ineffective during learning.[8] Thus we set J = W = 0 in Eq. (3) and find that, to an imprinting input $\delta\mathbf{I} = \boldsymbol{\xi}^0 e^{-i\omega_0 t}$ + c.c., the model response is

$$u_i^+ = \frac{(\omega_0 + i\alpha)\xi_i^0 e^{-i\omega_0 t}}{2\alpha\omega_0 + i(\alpha^2 + \beta\gamma - \omega_0^2)} \equiv U_0 \xi_i^0 e^{-i\omega_0 t} \tag{7}$$

and, from $(\partial_t + \alpha)v_i = \gamma u_i$,

$$v_i^+ = \frac{\gamma}{-i\omega_0 + \alpha} U_0 \xi_i^0 e^{-i\omega_0 t}. \tag{8}$$

Using these in the learning rule (6) leads to

$$J_{ij} = 2J_0 \mathrm{Re}\,[\tilde{A}(\omega_0)\xi_i^0 \xi_j^{0*}], \quad W_{ij} = 2(\eta_W/\eta_J) J_0 \gamma \mathrm{Re}\left[\frac{\tilde{A}(\omega_0)\xi_i^0 \xi_j^{0*}}{\alpha - i\omega_0}\right], \tag{9}$$

where $\tilde{A}(\omega) = \int_{-\infty}^{\infty} d\tau A(\tau) e^{-i\omega\tau}$ is the Fourier transform of $A(\tau)$, $J_0 = 2\pi\eta_J |U_0|^2/\omega_0$, and $\eta_J(W)$ are the respective learning rates. When the rates are tuned such that $\eta_J = \eta_W \gamma\beta/(\alpha^2 + \omega_0^2)$, we have, when $\omega = \omega_0$, $M_{ij}^+ = J_0 \tilde{A}(\omega_0)\xi_i^0 \xi_j^{0*}$, a generalization of the outer-product learning rule to the complex patterns $\boldsymbol{\xi}^0$ from the Hopfield-Hebb form for real-valued patterns. For learning multiple patterns $\boldsymbol{\xi}^\mu e^{i\omega_\mu t}$ + c.c., $\mu = 1, 2, \ldots$, the learned weights are simply sums of contributions like Eqs. (9) from individual patterns with $\xi_i^0$, $\omega_0$ replaced by $\xi_i^\mu$, $\omega_\mu$.

## 2.2 Recall Phase

We return to the single-pattern case and study the simple case when $\eta_J = \eta_W \gamma\beta/(\alpha^2 + \omega_0^2)$. Consider first an input pattern $\delta\mathbf{I} = \boldsymbol{\xi} e^{-i\omega t}$ + c.c. that matches the stored pattern exactly

($\boldsymbol{\xi} = \boldsymbol{\xi}^0$), but possibly oscillating at a different frequency. We then find, using Eqs. (9) in Eq. (3), the (positive-frequency) response

$$\mathbf{u}^+ = \chi(\omega,\, \omega_0)\boldsymbol{\xi}^0 e^{-i\omega t}$$

$$\chi(\omega,\, \omega_0) = \frac{\omega + i\alpha}{2\alpha\omega - \frac{J_0}{2}(\omega+\omega_0)\tilde{A}'(\omega_0) + i[\alpha^2 + \beta\gamma - \frac{J_0}{2}(\omega+\omega_0)\tilde{A}''(\omega_0) - \omega^2]} \tag{10}$$

where $\tilde{A}'(\omega_0) \equiv \mathrm{Re}\,\tilde{A}(\omega_0)$ and $\tilde{A}''(\omega_0) \equiv \mathrm{Im}\,\tilde{A}(\omega_0)$. For strong response at $\omega = \omega_0$, we require $\chi^{-1}(\omega,\, \omega_0) \sim 0$:

$$\omega_0 = \sqrt{\alpha^2 + \beta\gamma - J_0\omega_0\tilde{A}''(\omega_0)}, \quad J_0\tilde{A}'(\omega_0) \approx 2\alpha\,. \tag{11}$$

This means (1) the resonance frequency $\omega_0$ is determined by $\tilde{A}''$, (2) the effective damping $2\alpha - J_0\tilde{A}'$ should be small, and (3) deviation of $\omega$ from $\omega_0$ reduces the responses.

If the drive $\boldsymbol{\xi}$ does not match the stored pattern (in phase and amplitude), the response will consist of two terms. The first has the form of Eq. (10) but reduced in amplitude by an overlap factor $\boldsymbol{\xi}^{0*} \cdot \boldsymbol{\xi} \equiv \langle \boldsymbol{\xi}^0 | \boldsymbol{\xi} \rangle$. (For convenience we use normalized pattern vectors.) The second term is proportional to the part of $\boldsymbol{\xi}$ orthogonal to the stored pattern (let's call it $\boldsymbol{\xi}^\perp$). The J and W matrices do not act in this subspace, so the frequency dependence of this term is just that of uncoupled oscillators, i.e. Eq. (10) with $J_0$ set equal to zero. This response is always highly damped and therefore small. More specifically, for response $\mathbf{u} = \mathbf{r}e^{-i\omega t} + \text{c.c.}$,

$$\mathbf{r} = \chi(\omega,\, \omega_0)\langle \boldsymbol{\xi}^0 | \boldsymbol{\xi} \rangle \boldsymbol{\xi}^0 + \chi_0(\omega)\langle \boldsymbol{\xi}^\perp | \boldsymbol{\xi} \rangle \boldsymbol{\xi}^\perp \tag{12}$$

with $\chi_0(\omega) = \frac{\omega + i\alpha}{2\alpha\omega + i[\alpha^2 + \beta\gamma - \omega^2]} \ll \chi(\omega,\, \omega_0)$.

It is straightforward to extend this analysis to multiple imprinted patterns orthogonal to each other. The response consists of a sum of terms like in Eq. (12), one for each imprinted pattern. Consequently, in a linearized model, an input which overlaps several stored patterns of the same imprinting frequency will, under resonant frequency, evoke a resonant response which is a linear combination of the stored patterns. This property enables the model to interpolate between and generalize from the imprinted patterns, and is useful to function for input representations. Our analysis below shows that this property still holds in a nonlinear model when the nonlinearity around $\bar{\mathbf{u}}$, $\bar{\mathbf{v}}$ is of a particular class termed by us as class I. However, to function for categorical associative memory or classification, when the network should amplify (or respond only to) the dominant pattern component in an input which is a mixture of imprinted patterns, a qualitatively different class (class II) of model nonlinearity is required, as shown below.

## 3. SIMULATIONS

In order to check the validity of our linearized analysis and to study the effect of the nonlinearity in the model, we have performed numerical simulations of the nonlinear equations (1) and (2), using two qualitatively different classes of nonlinearities in $g_u(\,)$. In class I nonlinearity, the resting (fixed) point $\bar{u}_i$ of the potentials $u_i$ lies at a maximum of the slope $g'_u$ of the activation function, and the slope decreases monotonically toward zero as $u_i$ departs from $\bar{u}_i$ [as shown in Figs. 1(B.1)]. This class includes most standard sigmoids used in modeling, including logistic sigmoids and rounded threshold-linear models with a

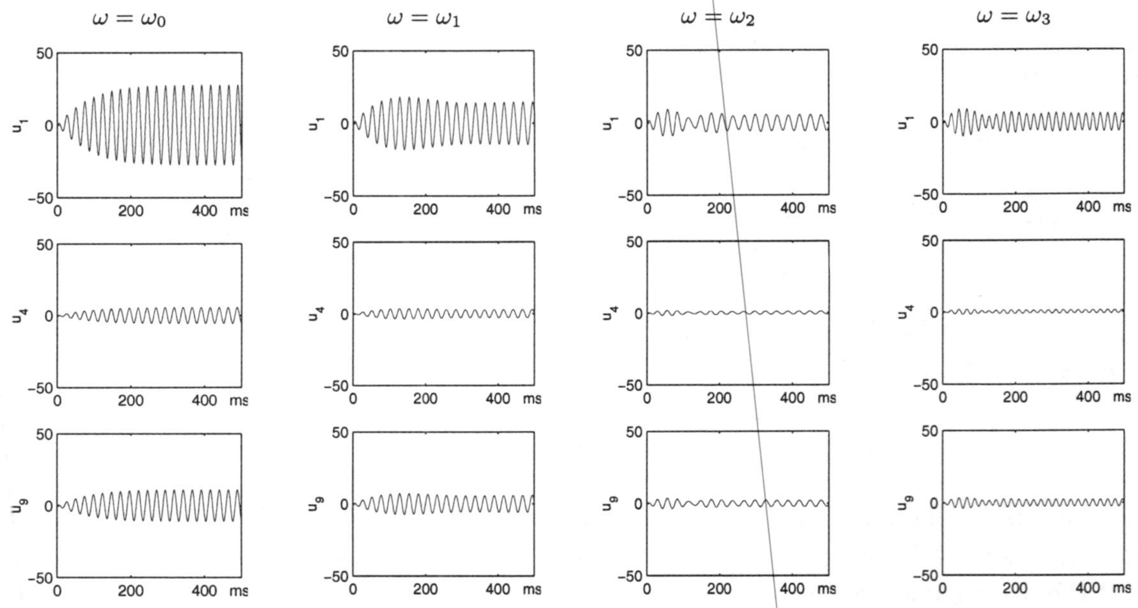


**Fig. 2** Effect of frequency mismatch: Pattern $\boldsymbol{\xi}^0 e^{-i\omega_0 t}$ + c.c. has been imprinted. Shown are responses to inputs at frequencies $\omega_0 = 41$ Hz, $\omega_1 = 38$ Hz, $\omega_2 = 33$ Hz and $\omega_3 = 49$ Hz. The activation function shown in Fig. 1(B.1) (class I) was used for excitatory units.

soft saturation at high input level. In class II nonlinearity, the slope $g'_u(u)$ *increases* for a range of $u$ above the resting point [as shown in Figs. 1(B.2)]. In both cases we have kept the activation function for inhibitory units linear in the neighborhood of the equilibrium point, with saturation for high and low inputs, as for class I excitatory units. Note that the nonlinear effects in the learning phase is ignored, i.e. Eq. (9) is kept unchanged, since the network response during learning is non-resonant or of small amplitudes, making the linear approximation quite accurate and appropriate.

We simulated the recall operation in a network consisting of 10 excitatory and 10 inhibitory cells. Parameters were set in such a way that the selective resonance was around 41 Hz. Our results confirm that when the input pattern matches the imprinted one in frequency, amplitude and phase, the network responds with strong resonant oscillations, and the resonance is absent for non-matching input patterns.

Figure 2 shows the effect of frequency mismatch, using the class I nonlinearity. Here we plot the oscillatory responses evoked on 3 of the 10 excitatory cells by four different inputs patterns $\mathbf{I}_0$, $\mathbf{I}_1$, $\mathbf{I}_2$ and $\mathbf{I}_3$. Only input $\mathbf{I}_0$ has the same frequency as the imprinted oscillatory input. All inputs $\mathbf{I}_0, \ldots, \mathbf{I}_3$ share the same pattern of amplitudes and phases with the imprinted pattern. The model responses become weaker as the input frequency moves away from the resonant frequency. The dependence on the input frequency is shown in Figure 3(A). Both the nonlinear simulation results and the prediction of the linearized analysis are plotted. Away from the resonance, the nonlinear simulation agrees with the linear theory. Close to the resonance, the nonlinear response is suppressed relative to the linear response for class I nonlinearity, while it is amplified to and saturates at a higher value for class II nonlinearity.

In the next set of figures we keep the input frequency on the resonance frequency and vary instead the overlap $|\langle\boldsymbol{\xi}^0|\boldsymbol{\xi}\rangle|$ between the input and a stored pattern. The dependence of the overlap $|\langle\boldsymbol{\xi}^0|\mathbf{r}\rangle|$ between output and stored pattern on $|\langle\boldsymbol{\xi}^0|\boldsymbol{\xi}\rangle|$ is shown in Fig. 3(B). Evidently, the linear approximation is valid for weak input overlap or familiarity (or output response), but the nonlinearity begins to take effect as the component of the input along the

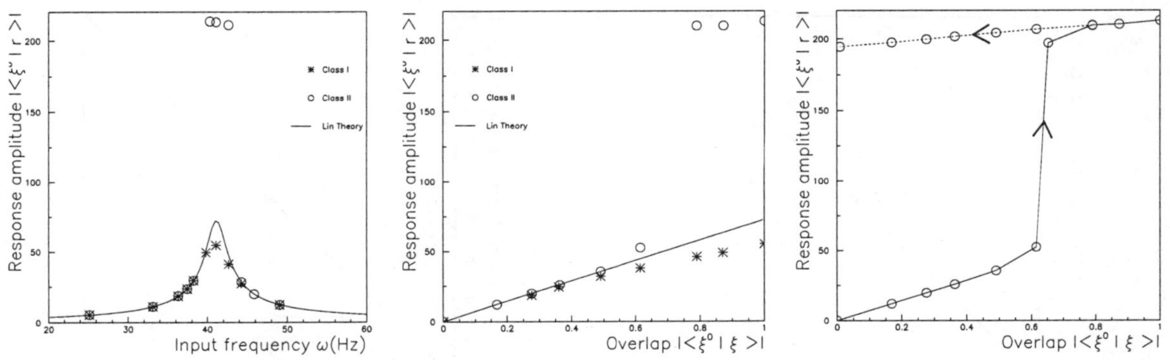


**(A)** **(B)** **(C)**

**Fig. 3** **(A)** Frequency tuning: Response amplitude $|\langle \boldsymbol{\xi}^0|\mathbf{r}\rangle|$ (i.e. overlap between the output and the stored pattern, or simply $|\chi(\omega;\omega_0)|$ in this case) as a function of input frequency $\omega$ when pattern $\boldsymbol{\xi}^0 e^{-i\omega_0 t}$ has been imprinted, and input pattern $\boldsymbol{\xi}^0 e^{-i\omega t}$ is presented. Solid line is the linear prediction, stars show respones from a model of class I nonlinearity while circles show those from class II nonlinearity. **(B)** Pattern tuning: Response amplitude $|\langle \boldsymbol{\xi}^0|\mathbf{r}\rangle|$ as a function of input familiarity or strength $|\langle \boldsymbol{\xi}^0|\boldsymbol{\xi}\rangle|$, under input $\boldsymbol{\xi} e^{-i\omega_0 t}$. Results from linearized theory (solid line) and from models of class I (stars) and II (circles) nonlinearities. **(C)** Hysteresis effects in the class II simulations. Response amplitude depends on the history or initial condition of the system: If the input degrades after the system locks to resonance, the output remains locked (circles connected by dotted line). Circles connected by solid line correspond to the case of random initial conditions or zero overlap initial conditions. The connecting lines are used for clarification only.

stored pattern grows. Whereas class I nonlinearity leads to a smooth reduction of response from that of the linear network, class II nonlinearity amplifies the response to a maximum level determined by the saturation in the nonlinear function $g_u(.)$ beyond its large slope region near $\bar{u}$. Furthermore, the saturating class II response shows hysteresis: the response stays locked to this level even when the external drive is withdrawn, i.e. $|\langle \boldsymbol{\xi}^0|\boldsymbol{\xi}\rangle| \to 0$.

The computational differences, interpolation versus categorization, between the two classes of nonlinear models can be illustrated directly by the model responses to inputs that are linear combinations of two imprinted, orthogonal, patterns of the same imprinting frequency. Let $\boldsymbol{\xi}^1 e^{-i\omega_o t}$ + c.c. and $\boldsymbol{\xi}^2 e^{-i\omega_o t}$ + c.c. denote the imprinted patterns, and $\boldsymbol{\xi} e^{-i\omega_o t}$ + c.c., where $\boldsymbol{\xi} \propto \cos(\psi)\boldsymbol{\xi}^1 + \sin(\psi)\boldsymbol{\xi}^2$, the input to the trained network. Let the response $\mathbf{r}$ be $\mathbf{r} \propto \cos(\phi)\boldsymbol{\xi}^1 + \sin(\phi)\boldsymbol{\xi}^2$. Figure 4 shows $\phi$ plotted as a function of $\psi$. In the linearized theory, $\mathbf{r} \propto \boldsymbol{\xi}$ and so $\phi = \psi$. With class I nonlinearity, $45° \geq \phi > \psi$ when $\psi < 45°$ and $45° < \phi < \psi$ when $\psi > 45°$. Hence, the output $\mathbf{r}$ tends to equalize the response amplitudes to the two input components $\boldsymbol{\xi}^1$ and $\boldsymbol{\xi}^2$ even when they contribute unequally to the input $\boldsymbol{\xi}$. In other words, the system is unable to be attracted to the stronger input component. In contrast, with class II nonlinearity, the response $\mathbf{r}$ gives higher gain to the stronger input component, $\boldsymbol{\xi}^1$ or $\boldsymbol{\xi}^2$, thus amplifying their differential input strengths. Consequently, the system performs a kind of categorization of the input, into three attractors corresponding to the two stored patterns and their symmetric combination.

## 3.1 Nonlinear Analysis

To gain some insight, we analyze nonlinear corrections, due to the nonlinearity in $g_u$, to our linear approximation $\mathbf{u} = \chi(\omega;\omega_0)\delta\mathbf{I}_{\|} + \chi_0(\omega;\omega_0)\delta\mathbf{I}_{\perp}$ during recall. Take for simplicity $g_v$ as linear, as supported by physiology. Note also that nonlinear effects are ignored during learning and thus the expression for J, W and M stay unchanged. During recall, Eq. (3)

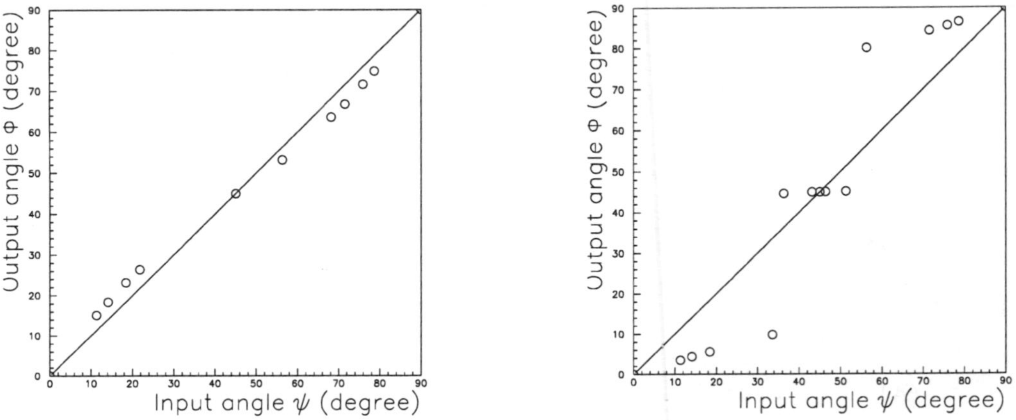


(A) (B)

**Fig. 4** Input-output relationship after two orthogonal patterns, $\boldsymbol{\xi}^1$ and $\boldsymbol{\xi}^2$, have been imprinted at the same frequency $\omega = 41$ Hz. Input $\boldsymbol{\xi} \propto \cos(\psi)\boldsymbol{\xi}^1 + \sin(\psi)\boldsymbol{\xi}^2$ and response $\mathbf{r} \propto \cos(\phi)\boldsymbol{\xi}^1 + \sin(\phi)\boldsymbol{\xi}^2$. Circles show the simulation results, while the dotted line show the analytical prediction for the linearized model. **(A)** Class I nonlinear simulations; and **(B)** Class II nonlinear simulations.

then becomes

$$\begin{aligned}\ddot{\mathbf{u}} + (2\alpha - \mathsf{J})\dot{\mathbf{u}} + [\alpha^2 - \alpha\mathsf{J} + \beta(\gamma + \mathsf{W})]\mathbf{u} - \mathsf{J}(\dot{g}_u(\mathbf{u}) - \dot{\mathbf{u}})\\ +[-\alpha\mathsf{J} + \beta(\gamma + \mathsf{W})](g_u(\mathbf{u}) - \mathbf{u}) = (\partial_t + \alpha)\delta\mathbf{I}\,.\end{aligned} \tag{13}$$

The two classes of nonlinearity can be formulated as

$$\begin{aligned}&\text{class I} \quad g_u(u_i) \sim u_i - au_i^3\\ &\text{class II} \quad g_u(u_i) \sim u_i + au_i^3 - bu_i^5\end{aligned} \tag{14}$$

where $a$, $b > 0$. Neglecting higher-order harmonics, we make the ansatz $\mathbf{u} \approx \mathbf{r}e^{-i\omega t} + \text{c.c.}$. Treating $\gamma$ as a diagonal element of W, Eq. (13) becomes

$$\left[2\alpha + \frac{i}{\omega}(\alpha^2 - \omega^2)\right]\mathbf{u}^+ = \mathsf{M}^+ g(\mathbf{u}) + \left(1 + \frac{i\alpha}{\omega}\right)\delta\mathbf{I}^+. \tag{15}$$

where $g(\mathbf{u})$ is a vector with components $[g(\mathbf{u})]_i = g(u_i)$.

Multiplying both sides of Eq. (15) by $< \boldsymbol{\xi}^0|$ from the left, using the formulae (9), we obtain, for patterns with random phases (i.e. with $N \equiv \sum|\xi_i^0|^2 \gg \sum|\xi_i^0|e^{2i\phi_i}$),

$$\text{class I} \quad \frac{q}{\chi(\omega;\,\omega_0)} + \frac{3aB}{N}\sum_j |\xi_j^0|^4|q|^2 q = h\,, \tag{16}$$

$$\text{class II} \quad \frac{q}{\chi(\omega;\,\omega_0)} - \frac{3aB}{N}\sum_j |\xi_j^0|^4|q|^2 q + \frac{5bB}{N}\sum_j |\xi_j^0|^6|q|^4 q = h\,, \tag{17}$$

where $q = \langle\boldsymbol{\xi}^0|\mathbf{r}\rangle$, $h = \langle\boldsymbol{\xi}^0|\boldsymbol{\xi}\rangle$ and $B = \frac{J_0(\omega+\omega_0)\tilde{A}(\omega_0)}{2(\omega+i\alpha)}$.

Let's consider the case when $\chi^{-1}$ and $B$ have the same phase, i.e. when

$$\chi^{-1} \simeq \varepsilon\frac{\tilde{A}}{\omega + i\alpha}\,, \tag{18}$$

with $\varepsilon \to 0$, which is a special case of the resonance conditions (11). In such a case we have

$$|\chi^{-1}|q + \frac{3a|B|}{N}\sum_j |\xi_j^0|^4|q|^2 q = he^{-i\phi_B} \tag{19}$$

$$|\chi^{-1}|q - \frac{3a|B|}{N}\sum_j |\xi_j^0|^4|q|^2 q + \frac{5b|B|}{N}\sum_j |\xi_j^0|^6|q|^4 q = he^{-i\phi_B}. \tag{20}$$

where $B \equiv |B|e^{i\phi_B}$. We can therefore introduce a sort of Ginzburg-Landau free energy for the order parameter $q$ measuring response amplitude to, or the similarity between the response and, the imprinted pattern:

$$F_I(q) = \frac{|\chi^{-1}|}{2}|q|^2 + \frac{3a|B|}{4N}\sum_j |\xi_j^0|^4|q|^4 - \mathrm{Re}(he^{-i\phi_B}q^*) \tag{21}$$

$$F_{II}(q) = \frac{|\chi^{-1}|}{2}|q|^2 - \frac{3a|B|}{4N}\sum_j |\xi_j^0|^4|q|^4 + \frac{5b|B|}{6N}\sum_j |\xi_j^0|^6|q|^6 - \mathrm{Re}(he^{-i\phi_B}q^*)\,. \tag{22}$$

While $F_I(q)$ has a single minimum, and does not show any phase transition with $h$, the free energy for the class II equations may support a first order transition at a critical value of the external field $h$. This is just the behavior we have observed in our simulations (Fig. 3). The jumps in the response at critical values of the frequency or overlap is the sign of a first order transition, with hysteresis. This difference between the two nonlinear classes essentially gives their different computational properties.

## 4. SUMMARY AND DISCUSSION

We have presented a model of learning for memory or input representations in neural networks with input-driven oscillatory activities. The model structure is an abstraction of the hippocampus or the olfactory cortex. We propose a simple generalized Hebbian rule, using temporal-activity-dependent LTP and LTD, to encode both magnitudes and phases of oscillatory patterns into the long range synapses in the network. After learning, the model responds resonantly to inputs which have been learned (or, for networks which operate essentially linearly, to linear combinations of learned inputs), but negligibly to other input patterns. Encoding both amplitude and phase enhances computational capacity, for which the price is having to learn both the excitatory-to-excitatory and the excitatory-to-inhibitory connections. As for the Hopfield model, we distinguish two functional phases: (1) the learning phase, in which the system is clamped dynamically to the external inputs and (2) the recall phase, in which the system dynamics is determined by both the external inputs and the internal interactions. The computational utilities and limitations of our model are determined mainly by the generalization and categorization capabilities of the model. When the model operates near the linear regime, it has the following generalization capability: at given oscillation frequency, once the system has learned a set of memory or representation states, all other states in the subspace spanned by the learned states can also evoke strong responses. This is a generic property of linear associative memories, and we have seen in our simulations that moderate nonlinearity, of a sort that one expects to be common in cortical networks, does not change things much. Generalization and interpolation properties are desirable for learning input representations such as spatial coding by the place cells in the hippocampus. Of course, this generalization property is not always advantageous. For instance, in categorical memory, one does not want inputs which are linear combinations of stored patterns to elicit responses which are also similar linear combinations. Suitable nonlinearity can (as we saw in the case of class II nonlinear models), enable the system to perform categorization, as implemented in a previous work by two of us for odor recognition.[6] The two classes of nonlinearity for two different computational purposes could be implemented in a single nonlinear cortical model or area, i.e. with a single or fixed activation function $g_u(\ )$. This is because it is possible to change the nature of the nonlinearity near the operating point by shifting the fixed or resting point $\mathbf{u}$, $\mathbf{v}$ of the system through appropriate external DC inputs. It seems likely to us that the brain may employ different kinds and degrees of nonlinearity in different areas or at different times to enhance the versatility of its computations. This work is a preliminary investigation which shows how interesting computational properties can arise from few basic principles such as: excitatory-inhibitory network, asymmetry of plasticity, resonance and degree of non-linearity. More details of our work, including constraints on our learning rule from biology and the vice versa, will be presented in an upcoming longer paper. We hope that our findings can stimulate further study and experimental investigations.

## REFERENCES

1. J. J. Hopfield, "Neurons with Graded Response have Collective Computational Properties like Those of Two-State Neurons," *Proc. Natl. Acad. Sci. USA.* **81**, 3088–3092 (1984).
2. H. Markram, J. Lubke, M. Frotscher and B. Sakmann, "Regulation of Synaptic Efficacy by Coincidence of Postsynaptic APs and EPSPs," *Science* **275**, 213 (1997).
3. J. C. Magee and D. Johnston, "A Synaptically Controlled Associative Signal for Hebbian Plasticity in Hippocampal Neurons," *Science* **275**, 209 (1997).
4. D. Debanne, B. H. Gahwiler and S. M. Thompson "Long-Term Synaptic Plasticity Between Pairs of Individual CA3 Pyramidal Cells in Rat Hippocampal Slice Cultures, *J. Physiol* **507**, 237 (1998).
5. G. Q. Bi and M. M. Poo. "Synaptic Modifications in Cultured Hippocampal Neurons: Dependence on Spike Timing, Synaptic Strength, and Postsynaptic Cell Type," *J. Neurosci.* **18**, 10464 (1998).
6. Z. Li and J. Hertz, "Odor Recognition and Segmentation by a Model Olfactory Bulb and Cortex," *Network Comput. Neural Syst.* **11**, 83–102 (2000).
7. Z. Li and J. J. Hopfield, "A Model of the Olfactory Bulb and Its Oscillatory Processing," *Biol. Cybern.* **61**, 379–392 (1989).
8. M. E. Hasselmo, "Acetylcholine and Learning in a Cortical Associative Memory," *Neural Comp.* **5**, 32–44 (1993).
9. Z. Li and P. Dayan, "Computational Differences Between Asymmetrical and Symmetrical Networks," *Network Comput. Neural Syst.* **10**, 59, 1999.

# FLUID FLOW THROUGH DISORDERED POROUS MEDIA

H. E. STANLEY
*Center for Polymer Studies, Boston University,*
*Boston, MA 02215, USA*

A. D. ARAÚJO, U. M. S. COSTA and J. S. ANDRADE JR.
*Departamento de Física, Universidade Federal do Ceará,*
*60451-970 Fortaleza, Ceará, Brazil*

### Abstract

This talk briefly reviews the subject of fluid flow through disordered media. First, we use two-dimensional percolation networks as a simple model for porous media to investigate the dynamics of viscous penetration when the ratio between the viscosities of displaced and injected fluids is very large. The results indicate the possibility that viscous displacement through critical percolation networks constitutes a single universality class, independent of the viscosity ratio. We also focus on the sorts of considerations that may be necessary to move statistical physics from the description of idealized flows in the limit of zero Reynolds number to more realistic flows of real fluids moving at a nonzero velocity, when inertia effects may become relevant. We discuss several intriguing features, such as the surprisingly change in behavior from a "localized" to a "delocalized" flow structure (distribution of flow velocities) that seems to occur at a critical value of $Re$ which is significantly smaller than the critical value of $Re$ where turbulence sets in.

## 1. INTRODUCTION

### 1.1 What is the Scientific Question Associated with Fluid Flow Through Disordered Structures?

The primary question for physicists to entertain is: what happens to the "laws of fluid flow" when the medium through which the fluid actually flows is disordered? What happens when the particles of the fluid are not passing through a medium that resembles a straight pipe, but one that is full of disorder, e.g. the filter of a cigarette?

## 1.2 Why Do We Care About This Question?

The scientific reason for our interest in this topic is immediately obvious: the laws associated with fluid flow through disordered media are only beginning to be understood.[1–4] There is much interesting work to be done before a complete understanding is obtained. The practical reasons for our interest are numerous:

- The oil in the earth is not contained in some kind of "balloon", released when the balloon is punctured — the time of gushing oil wells in Oklahoma has long since passed. Rather, two-thirds of the world's oil supply is held in porous earth — much the same way as water is held in a sponge. We can squeeze a sponge to release the water, but cannot perform an analogous action on a portion of porous earth to release the oil. One technique that is being used to force oil out of the ground involves drilling a second well some distance away (e.g. one kilometer) from the original well and pumping some other fluid, such as water, into this second well. Because the earth is porous, the fluid from this second well can often "push" the oil through the ground, and out the first well. As we might expect, oil companies are very interested in the numerous laws of physics associated with this flow of oil and other fluid.[5,6] They want to know, among other things, how to calculate the time of breakthrough, i.e. they want to know the amount of time it takes the injected fluid to push the oil out. In order to calculate that time, they need to know the path the oil takes in this very disordered sponge called the earth.
- Fluid flow in branching geometries is related to many phenomena in physics, geology and biology. Examples range from fluid flow through porous media[4,7] to respiration[8] and blood circulation.[9] In particular, the mechanism of flow bifurcation[10] plays a crucial role in the functioning of the respiratory and circulatory systems. Air does not enter a lung the way air fills a balloon, but enters a ramified, tree-like structure with approximately 20 generations or $2^{20}$ branches. As the stream of air enters, it bifurcates at each of these branches. This bifurcation is not symmetric, and different parts of the lung can exhibit very different levels of aeration.[10] The lung is also a disordered medium. Diseases such as emphysema can introduce disorder into this system of asymmetrical branches by partially blocking some of the airways. How this kind of disorder affects lung functioning is another problem for physicists.
- Because cigarette smoke contributes so significantly to lung cancer, cigarette manufactures are extremely interested in the physical laws associated with smoke passing through the disordered medium of the cigarette filter.

There are many, many other examples of the practical interest we might have in the laws associated with fluid flow through disordered media.

## 1.3 What Do We Actually Do in Pursuing This Question?

As physicists, our approach to problems is to discover what we can quantify. But what can we quantify when the subject is, by definition, disordered? It is not perfectly obvious what we are going to be able to quantify in a cigarette filter. If we have no randomness in the flow path of a fluid, we know how to treat the situation. But if we start placing obstacles in the flow path, everything becomes much more complex. Work on percolation models has shown us that the actual positioning of the obstacles is extremely important. Placing them in a completely uncorrelated pattern is a simple first step, and corresponds to an idealized limit. But almost nothing is really uncorrelated. If we have inclusion zones in porous earth, their presence is not uncorrelated. If we have a permeability in zone $k_1$, then the permeability zone $k_2$ has something to do with $k_1$ — its permeability is not simply random. There is a correlation in the spatial disorder.

We should take special note of the fact that almost nothing has been done in the area of correlated percolation with some notable exceptions. For example, Coniglio[11] has noted that correlated disorder is an important form of disorder (e.g., in the growth of cities[12,13]), and random multiplicative correlations have also been studied.[14,15]

So most disorder, most obstacles in the flow path of a fluid, are placed in an uncorrelated fashion — somewhat like the potholes in the streets of Boston. And like Boston potholes, uncorrelated obstacles can often stop the flow completely, and not just impede it.

## 2. IDEALIZED MODEL FOR THE LIMIT OF VISCOUS FLOW

There are many ways of generating percolation. One of the easiest and most useful to us is the algorithm that produces invasion percolation.[1,2,16] This algorithm drives the system

**Fig. 1** The incipient infinite percolation cluster that forms at the percolation threshold, highlighting the path that connects one point to another using the minimum number of bonds. This "minimum path" is a fractal object whose dimension is not known exactly even for the case of a two-dimensional cluster (but is approximately 1.13). The average chemical distance scales with exponent $d_{\min}$, where various estimates of $d_{\min}$ are $d_{\min} \approx 1.130 \pm 0.005$[29] and $d_{\min} \approx 1.1307 \pm 0.0004$[26] (courtesy of S. Schwarzer).

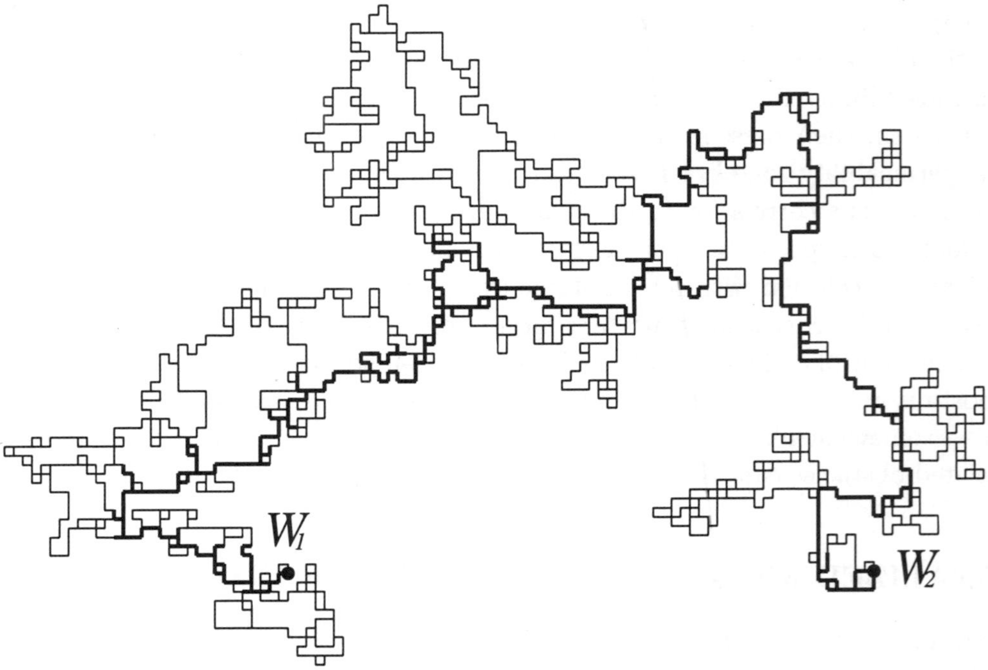


**Fig. 2** Pictorial representation of the viscous penetration process in a typical percolation network of pores. A constant pressure drop $\Delta p$ is applied between the points $W_1$ and $W_2$ separated by the distance $r$. At initial time, the entire network is filled with a fluid of finite viscosity (displaced fluid). The invading fluid of zero viscosity penetrates through $W_1$ and reaches $W_2$ at the breakthrough time $t_b$. The thin lines correspond to pores filled with the displaced fluid, while the thick lines are the pores filled with the invading fluid at $t = t_b$.

until it reaches the critical point. In this algorithm, we "wet" one square pixel near the center of the computer screen, and randomly generate numbers for the four neighboring pixels having replaced the continuum by a lattice.[17,18] The "fluid" then randomly "wets" the pixel with the lowest random number, and a tiny percolation cluster of size 2 is the result. The newly-wetted site is surrounded by three fresh sites. The random numbers are generated for these three sites, the site with the lowest random number is wetted, and the percolation cluster is now size 3. This algorithm is iterated until a large cluster emerges. If we make a histogram of the random numbers associated with the occupied sites, the curve is flat until we reach a critical value, the percolation threshold $p_c$, at which time it plunges almost to zero. The width of that transition region scales as a power law with system size, i.e., the reciprocal of the system size to some critical exponent — the critical exponent of the singly-connected "red" bonds.[19,20]

If we examine our percolation cluster visually, and on several scales (use two vertical and two horizontal lines to divide the screen into nine equal squares and magnify the middle square until it is as big as the screen — repeat as desired), we notice that the texture of the cluster seems to be the same, irrespective of scale. It is difficult to distinguish the original screen-filling cluster from the magnified portion of that original — and a magnified portion of that magnified portion (see Figs. 1 and 2 in Stanley et al.[21]).

This shows that this simple algorithm of wetting neighboring sites on a computer screen generates a fractal structure, i.e., one that is self-similar. Each part of the original is similar to the whole. Indeed, percolation (unlike some of the natural systems discussed in Avnir et al.[22] and Malcai et al.[23]) is a true fractal. The total number of wet sites inside a box as a function of $L$ (the edge of the box) will be some number less than two. This number is called the fractal dimension, which has been solved exactly as 1.89 in 2-dimensional percolation.[1,16] The behavior of this fractal dimension in the non-classical region and below the critical dimension is an unsolved problem. Work can be done using approximations, e.g., brute force computer simulations or position-space renormalization group.

In position-space renormalization group (PSRG), we take a $2 \times 2$ Kadanoff cell in which each site is there with probability $p$. We map this $2 \times 2$ group into a cell that will be wetted with probably $p'$. The goal in this PSRG is to find the expression that relates this wetting probability $p'$ of the cell to the initially wetting probability $p$ of the underlying lattice. In this $2 \times 2$ case, when all four cells are occupied there is a connecting path from bottom to top ("south to north") — remembering that we want to get oil out of the ground — and when there are three cells occupied there is still a connected south-to-north path. If we have only two occupied, two of the six possible configurations still provide a connected south-to-north path. This procedure is not very accurate for $2 \times 2$ cells, but by using a combination of Monte Carlo methods and renormalization group concepts one can consider a sequence of cells, starting with two and going up to one thousand. That sequence of larger and large cells has so many configurations ($2^{L^2}$) that it is impossible to work out, but by using sampling procedures one can get an approximate recursion relation.[24,25]

How does fluid go through a medium when there are obstacles in the water? The structure of the cluster at the exact moment it connects the percolation threshold — the *incipient* infinite cluster — is sometimes called the wetting path or the backbone. Notice that there are dangling ends in this structure. If we inject a current into this idealized structure, then these dead ends are not going to connect to anything. What happens if we throw them away? The first thing to notice is that we no longer have a fractal of dimension 1.89! If what we threw away was a fixed fraction independent of size (e.g., 90% of the total cluster), then the fractal dimension would not be changed. But this is not the case, because these dead ends are such an overwhelmingly large fraction of the total cluster that when we do throw them away, we reduce the fractal dimension of the cluster. The fractal dimension of the cluster without its dead ends — i.e., the fractal dimensions of just the wetting path — is roughly 1.62.[26,27]

In this wetting path we distinguish between two different kinds of bonds: "blobs," which have multiple connections (a variety of routes within the wetting path), and singly-connected bonds ("red bonds"), which must carry 100% of the fluid traversing the wetting path.[19,20] The structure of the wetting path is therefore a kind of necklace of blobs and single-connection bonds. Both the blobs and red bonds are themselves fractals with self-similarity.

## 3. MINIMUM WETTING PATH

The minimum path is a subset of the total wetting path, characterized not by the amount of fluid that flows through it, but by the question of what is the shortest path distance $\ell$ between two points in this model random medium. The length of this minimum path does not increase linearly as the straight-line distance between the beginning and ending of the total wetting path is increased. The longer the total straight-line distance, the more likely significant obstacles will be encountered and the total minimum path lengthened (Fig. 1). The function that describes this increase is once again a fractal object with an exponent $d_{\min}$, an exponent for which an exact result has not yet been found, even for dimension $d = 2$.[28,29]

In real-world problems, such as drilling oil wells, one does not "average over a million realizations". There's a given oil field, and that's it. Simple "appropriate averages" have no bearing on the situation. In a real oil field we care about the entire distribution of shortest paths. The function of interest to us is the conditional probability $P(\ell|r)$ for two sites to be separated by the shortest path $\ell$, given that the geometrical distance between these sites is $r$. At the percolation threshold, it has been shown[30,31] that

$$P(\ell|r) \sim \frac{1}{r^{d_{\min}}} \left(\frac{\ell}{r^{d_{\min}}}\right)^{-g_\ell} \exp\left(-a\left(\frac{\ell}{r^{d_{\min}}}\right)^{-\phi_\ell}\right). \tag{1}$$

The probability distribution of more practical interest is $P'(\ell|r)$, defined in the same way as $P(\ell|r)$ but for any two randomly-chosen points separated by geometrical distance $r$ and on the same cluster, but not necessarily on the incipient infinite cluster. $P'(\ell|r)$ has the same scaling form as in Eq. (1), but with $g_\ell$ replaced by[30,31]

$$g'_\ell = g_\ell + \frac{d - d_f}{d_{\min}}. \tag{2}$$

The complete scaling form of $P'(\ell|r)$, which accounts also for finite size effects and off-critical behavior, has been studied for $d = 2$ and reported in Dokholyan et al.[30,31] Specifically, the following *Ansatz* has been proposed[30,31]

$$P'(\ell|r) \sim \frac{1}{r^{d_{\min}}} \left(\frac{\ell}{r^{d_{\min}}}\right)^{-g'_\ell} f_1\left(\frac{\ell}{r^{d_{\min}}}\right) f_2\left(\frac{\ell}{L^{d_{\min}}}\right) f_3\left(\frac{\ell}{\xi^{d_{\min}}}\right), \tag{3}$$

where $\xi \sim |p - p_c|^{-\nu}$ is the pair connectedness length, and the scaling functions have the form

$$f_1(x) \equiv \exp(-ax^{-\phi}), f_2(x) \equiv \exp(-bx^{\psi}), f_3(x) \equiv \exp(-cx). \tag{4}$$

The function $f_1$ accounts for the lower cut-off due to the constraint $\ell > r$, while $f_2$ and $f_3$ account for the upper cut-offs due to the finite size effect and due to the finite correlation length respectively. Either $f_2$ or $f_3$ becomes irrelevant, depending on the magnitudes of $L$ and $\xi$: for $L < \xi$, $f_2$ dominates the upper cut-off, otherwise $f_3$ dominates. We assume the independence of the finite size effect and the effect of the concentration of the occupied sites, so that Eq. (3) can be represented as a product of the terms which are responsible for the finite size effect ($f_2$) and the effect of the concentration ($f_3$). Simulations performed recently[30,31] support this assumption.

## 4. VISCOUS DISPLACEMENT IN PERCOLATION MEDIA

The displacement phenomenon of a viscous fluid by a less viscous one inside a porous material has been the subject of intensive research in the past and in recent years, in particular due to its close connections with hydrology and oil recovery.[3,32,33] Only a few studies, however, have been devoted to the investigation of the displacement process of a viscous fluid by a less viscous one through percolation porous media. Murat and Aharony[34] showed by numerical simulation with two-dimensional diluted percolation lattices that, although the clusters generated from VF and DLA have the same fractal dimension at the vicinity of the critical point, many other geometrical differences can be observed between these two processes. In two recent studies,[35,36] the dynamics of viscous displacement through percolation porous media has been investigated in the limiting condition of unitary viscosity ratio, $m \equiv \mu_2/\mu_1 = 1$, where $\mu_1$ and $\mu_2$ are the viscosities of the injected and displaced fluids, respectively. In this situation, the displacement front can be approximately modeled by tracer particles that follow the streamlines of the flow. As a result, it was shown that the distributions of shortest path and minimal traveling time of the tracer closely obey the scaling *Ansatz* Eq. (1).

More recently,[37] it was found by numerical simulations on $2d$ percolation networks at criticality that the scaling *Ansatz* (1) also holds for the case of very large viscosity ratio, $m \to \infty$. In this case, the porous media is modeled by bond percolation on a square lattice with bonds that are cylindrical tubes of fixed length $\ell_p$ and radius $r_p$. The percolation backbone is generated at the critical point between two sites ("wells") $W_1$ and $W_2$ separated by a fixed distance $r$ (see Fig. 2). A constant pressure drop $\Delta p = p_{W1} - p_{W2}$ is imposed between the injecting ($W_1$) and extracting wells ($W_2$) during the dynamics. For simplicity, we assume that the flow between nodes $i$ and $j$ follows

$$q_{ij} = g_{ij}(p_i - p_j)\,. \tag{5}$$

Here $p_i$ is the pressure at node $i$, $q_{ij}$ is the volumetric flow rate between nodes $i$ and $j$ and $g_{ij}$ is the hydraulic conductance of the pore. The local variable $x_{ij}$, $0 \leq x_{ij} \leq \ell_p$, is a time-dependent length which corresponds to the part of the pore that is filled with the displacing fluid during the penetration process. Mass conservation at each node of the lattice leads to the following set of linear algebraic equations:

$$\sum_j q_{ij} = \sum_j g_{ij}(p_i - p_j) = 0 \quad i = 1,\, 2, \ldots,\, N\,, \tag{6}$$

where $N$ is the number of sites. Note that because $g_{ij}^{-1} = 0$, the pressure inside the invaded region must be everywhere equal to the pressure $p_{W1}$ applied at the well $W_1$. In order to simulate the dynamics of viscous invasion, the local displacement in each pore is computed at the front as $\Delta x_{ij} = q_{ij}\Delta t_{\min}/\pi r_p^2$, where $\Delta t_{\min}$ is the variable time step of the process, calculated as the minimum value among all the interface pores, necessary for the invading fluid to reach a new node. When the displacing fluid reaches the second well $W_2$, the mass $M_b$ of the invaded cluster and the breakthrough time $t_b$ (i.e., the total time for the invading front to move from $W_1$ to $W_2$) are recorded. For a fixed value of $r$, this operation is repeated for 10000 network realizations of size $L \times L$, where $L = 500 \gg r$. Running these simulations for different values of $r$, we find that there is always a well-defined region where the distributions of $M_b$ and $t_b$ follow the scaling form Eq. (1). For example, in Fig. 3 we show the log-log plot of the distribution $P(M_b)$ for five different values of the well distance, $r = 4$, 8, 16, 32 and 64. The inset of Fig. 3 shows the resulting data collapse obtained by rescaling $M_b$. Surprisingly, the distribution exponents estimated for the limiting case of $m \to \infty$ are statistically identical to the values obtained with $m = 1$.[36] This agreement seems to indicate that the process of viscous penetration in percolation porous media constitutes a single class of universality.

## 5. INERTIAL EFFECTS ON SINGLE-PHASE FLOW

When we're talking about a realistic fluid flow, we're not talking about something that behaves necessarily like an electrical current. Fluid flow can have inertia; if it is too fast, it "plows ahead" like a crowd in a train station. When a slow-moving fluid tries to make a turn in, e.g., a "bent pipe", it is able to make the turn in a more-or-less smooth fashion; since the fluid is laminar, the velocity of the fluid varies smoothly across the cross-section of the bent pipe, being zero on the walls. If the Reynolds number is higher, the fluid does not make the turn in the same way, and a lot of inertial activity results.

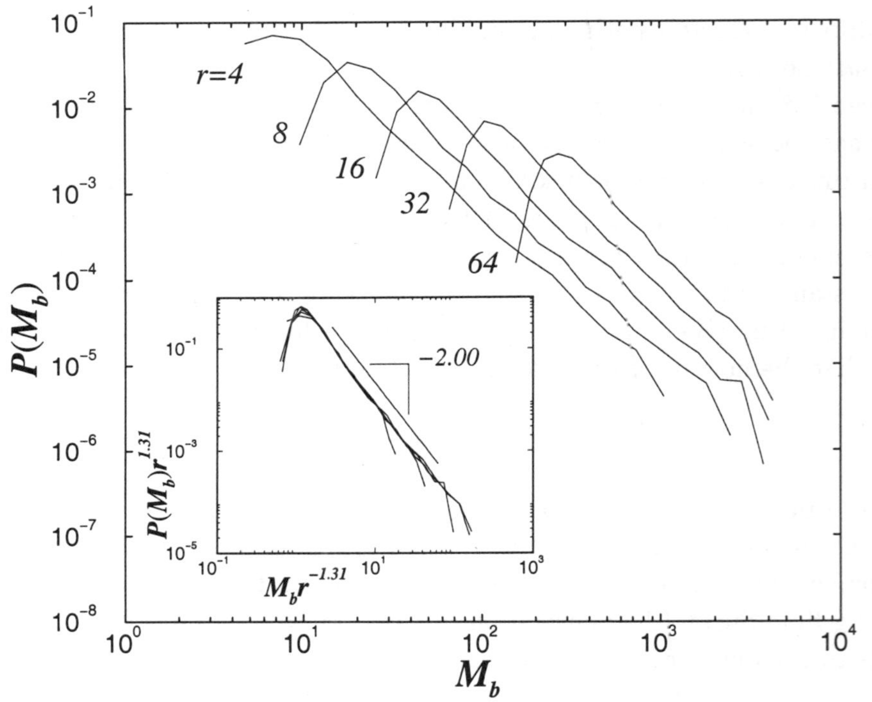


**Fig. 3** Logarithmic plot of the distribution of mass of invaded clusters $P(M_b)$ for different distances $r = 4, 8, 16, 32, 64$ between injection ($W_1$) and extraction ($W_2$) points. For each curve, we determine the characteristic size $M_b^*$ as the peak of the distribution and plot $M_b^*$ versus the distance $r$ in double logarithmic scale. $M_b^*$ follows a power-law behavior on the distance $r$, $M_b^* \sim r^{d_M}$, and the linear fit to the data yields the exponent $d_M = 1.31 \pm 0.02$.[37] The inset shows the collapsed data obtained by rescaling the mass $M_b$ with its corresponding characteristic value $M_b^* \sim r^{1.31}$. As shown, the distribution $P(M_b)$ is consistent with the scaling form $P(M_b) = A_M \left(\frac{M_b}{M_b^*}\right)^{-g_M} f\left(\frac{M_b}{M_b^*}\right)$, where the normalization constant is given by $A_M \sim (M_b^*)^{-1}$ and the scaling function has the form $f(M_b) = \exp(-a_M M_b^{-\phi_M})$. The least-square fit to the data in the scaling region gives $g_M = 2.00 \pm 0.04$. Performing the same statistical analysis with the distribution of breakthrough times $P(t_b)$,[37] we observed a similar behavior, but with exponents $d_t = 2.25 \pm 0.03$ and $g_t = 1.54 \pm 0.03$. Note that the breakthrough time exponents previously reported[36] for the special case $m = 1$ are $d_t \approx 2.3$ and $g_t \approx 1.57$, computed at constant pressure. Furthermore, the scaling exponents of the invaded cluster mass $g_M$ and $d_M$ also coincide with the exponents for breakthrough time reported in Andrade et al.[36] for $m = 1$ at constant flow.

These features of the fluid flow phenomenon in irregular geometries have not been intensively studied before, at least from a microdynamical perspective. In order to investigate them, it is necessary to perform direct simulations of the Navier-Stokes equations at the scale of the pore.[38–40] We adopt the general picture of site percolation disorder as a simplified model for the pore connectivity. Square obstacles are randomly removed from a $64 \times 64$ square lattice until a porous space with a prescribed void fraction $\varepsilon$ is generated. The mathematical description for the detailed fluid mechanics in the interstitial pore space is based on the assumptions that we have steady state flow in isothermal conditions and the fluid is continuum, Newtonian and incompressible. The Reynolds number

$$Re \equiv \frac{\rho d_p \bar{v}}{\mu} \tag{7}$$

characterizes the relative contributions from convective and viscous mechanisms of momentum transfer in fluid flow, where $\rho$ is the density, $d_p$ the particle size, $\mu$ the kinematic viscosity and $\bar{v}$ is the average fluid velocity. For a given realization of the pore geometry

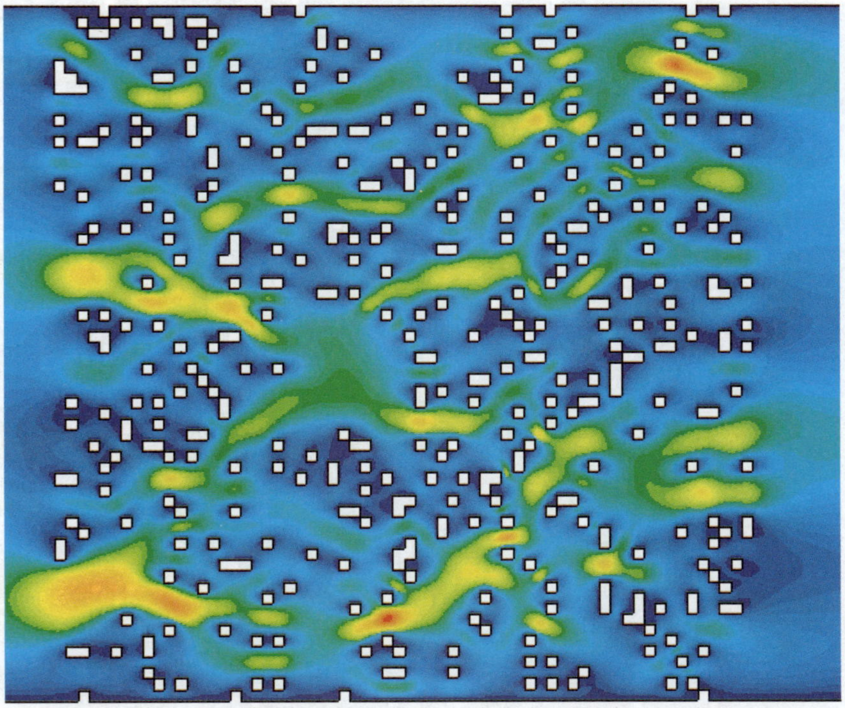

**Fig. 4** Flow in disordered porous media. We show the contour plot of the velocity magnitude for a typical realization of a highly porous void space ($\varepsilon = 0.9$) subjected to low Reynolds conditions, $Re = 0.0156$. Fluid is pushed from left to right. The colors ranging from blue to red correspond to low and high velocity magnitudes, respectively.

and a fixed $Re$, the local velocity and pressure fields in the fluid phase of the void space, are numerically obtained through discretization by means of the control volume finite-difference technique.[41]

First, we study fluid flow under conditions of low Reynolds number to ensure that the contribution from inertial terms (convection) does not prevail over the viscous mechanism of momentum transfer. In spite of the well-connected pathways available for flow at a large porosity value, the predominant viscous forces in the momentum transport through the complex void geometry generates well defined "hot spots" of fluid flow (see Fig. 4). As shown in Fig. 5, the situation is somewhat different at higher Reynolds conditions. Due to the relevant contribution of inertial forces (convection) to the transport at the pore scale, the flow distribution along the direction orthogonal to the main flux becomes more uniform.

The "localization" effect shown in Figs. 4 and 5 can be statistically quantified in terms of the spatial distribution of kinetic energy in the flowing system. In analogy with previous work on localization of vibrational modes in harmonic chains,[42] we define a "participation" number $\pi$,

$$\pi \equiv \left( n \sum_{i=1}^{n} q_i^2 \right)^{-1} \quad \left( \frac{1}{n} \leq \pi \leq 1 \right), \tag{8}$$

where $n$ is the total number of fluid cells in the numerical grid enclosing the physical pore space,

$$q_i \equiv \frac{e_i}{\sum_{j=1}^{n} e_j}, \tag{9}$$

where $e_i \propto (u_i^2 + v_i^2)$ is the kinetic energy associated with each individual fluid cell, and $u_i$ and $v_i$ are the components of the velocity vector at cell $i$ in the $x$ and $y$ directions,

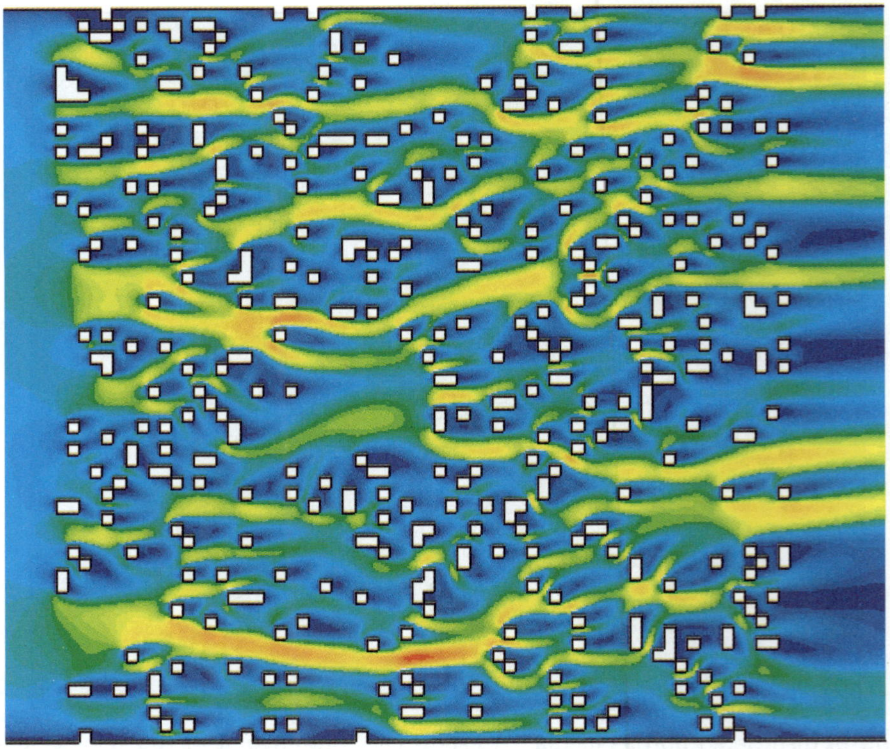

**Fig. 5** Same as Fig. 4, but for a Reynolds number 1000 times larger ($Re = 15.6$).

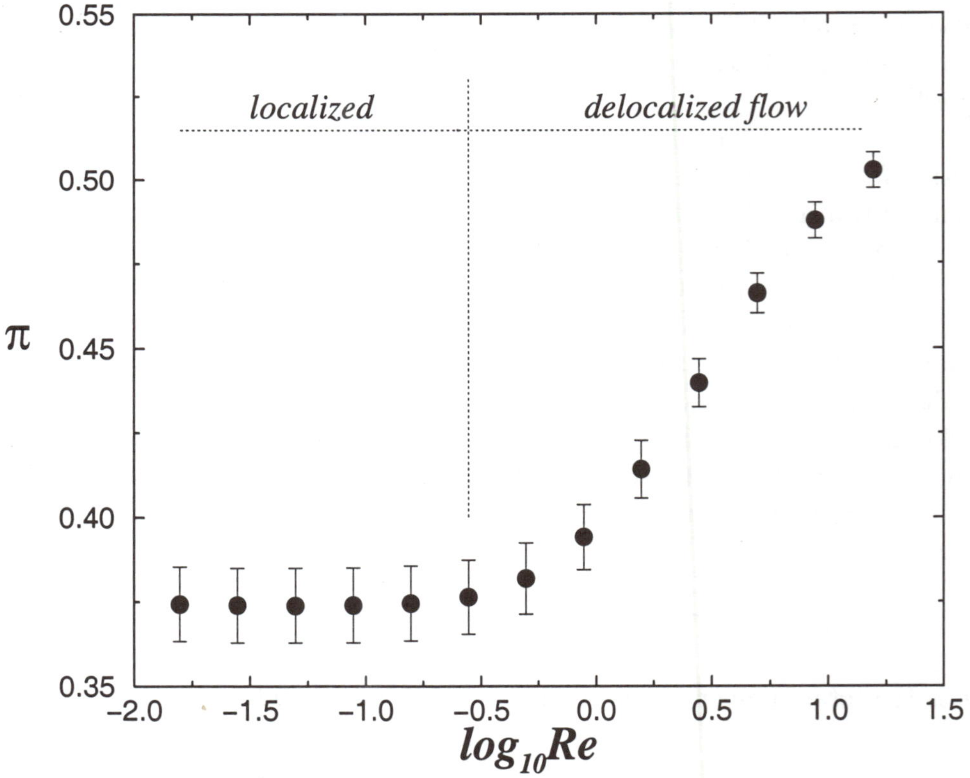


**Fig. 6** Dependence of the participation number $\pi$ on the Reynolds number $Re$ ($\varepsilon = 0.9$). These simulations have been performed with ten lattice realizations.

respectively. From the definition Eq. (8), $\pi = 1$ indicates a limiting state of equal partition of kinetic energy ($q_i = 1/n$, $\forall i$). On the other hand, a sufficiently large system ($n \to \infty$) should correspond to a "localized" flow field, $\pi \approx 0$. The function $\pi$ has been calculated for 10 pore space realizations generated with a porosity $\varepsilon = 0.9$. As shown in Fig. 6, the participation number remains constant, $\pi \approx 0.37$, for low $Re$ up to a transition point at about $Re \approx 0.3$. Above this point, the flow becomes gradually less "localized" ($\pi$ increases) as $Re$ increases. This transition reflects the onset of convective effects in the flow, and the significant changes in $\pi$ above the transition point indicate the sensitivity of the system to these nonlinearities.

## ACKNOWLEDGMENTS

We thank our co-authors whose work is reported here, particularly M. P. Almeida, S. V. Buldyrev, N. V. Dokholyan, S. Havlin, P. R. King, Y. Lee, H. A. Makse, G. Paul and B. Suki. We also thank BP, CNPq, FUNCAP and NSF for financial support.

## REFERENCES

1. D. Ben-Avraham and S. Havlin, *Diffusion and Reactions in Fractals and Disordered Systems* (Cambridge University Press, Cambridge, 2000).
2. A. Bunde and S. Havlin (eds.), *Fractals and Disordered Systems* (Springer-Verlag, New York, 1996).
3. M. Sahimi, *Rev. Mod. Phys.* **65**, 1393 (1993).
4. M. Sahimi, *Applications of Percolation Theory* (Taylor and Francis, London, 1994).
5. P. R. King, J. S. Andrade Jr., N. V. Dokholyan, S. V. Buldyrev, S. Havlin, Y. Lee and H. E. Stanley, *Physica* **A266**, 107 (1999).
6. P. R. King, S. V. Buldyrev, N. V. Dokholyan, S. Havlin, Y. Lee, G. Paul and H. E. Stanley, *Physica* **A274**, 60 (1999).
7. P. M. Adler, *Porous Media: Geometry and Transport* (Butterworth-Heinemann, Stoneham MA, 1992).
8. R. H. Ingram Jr. and T. J. Pedley, in *Handbook of Physiology. The Respiratory System. Mechanics of Breathing* (American Physiological Society, Bethesda, 1986).
9. D. MacDonald, *Blood Flow in Arteries* (Williams and Wilkins, Baltimore, 1974).
10. J. S. Andrade Jr., A. M. Alencar, M. P. Almeida, J. Mendes Filho, S. Buldyrev, S. Zapperi, H. E. Stanley and B. Suki, *Phys. Rev. Lett.* **81**, 926 (1998).
11. A. Coniglio, C. Nappi, L. Russo and F. Peruggi, *J. Phys.* **A10**, 205 (1977).
12. H. Makse, S. Havlin and H. E. Stanley, *Nature* **377**, 608 (1995).
13. H. A. Makse, J. S. Andrade Jr., M. Batty, S. Havlin and H. E. Stanley, *Phys. Rev.* **E58**, 7054 (1998).
14. S. Havlin, R. Selinger, M. Schwartz, H. E. Stanley and A. Bunde, *Phys. Rev. Lett.* **61**, 1438 (1988).
15. S. Havlin, M. Schwartz, R. Blumberg Selinger, A. Bunde and H. E. Stanley, *Phys. Rev.* **A40**, 1717 (1989).
16. D. Stauffer and A. Aharony, *Introduction to Percolation Theory* (Taylor and Francis, Philadelphia, 1994).
17. A. Geiger and H. E. Stanley, *Phys. Rev. Lett.* **49**, 1895 (1982).
18. E. T. Gawlinski and H. E. Stanley, *J. Phys.* **A14**, L291 (1981).
19. H. E. Stanley, *J. Phys.* **A10**, L211 (1977).
20. A. Coniglio, *Phys. Rev. Lett.* **46**, 250 (1981).
21. H. E. Stanley, J. S. Andrade Jr., S. Havlin, H. A. Makse and B. Suki, *Physica* **A266**, 5 (1999).
22. D. Avnir, O. Biham, D. Lidar and O. Malcai, *Science* **279**, 39 (1998).
23. O. Malcai, D. A. Lidar, O. Biham and D. Avnir, *Phys. Rev.* **E56**, 2817 (1997).
24. P. J. Reynolds, H. E. Stanley and W. Klein, *Phys. Rev.* **B21**, 1223 (1980).

25. H. E. Stanley, *Rev. Mod. Phys.* **71**, S358 (1999).
26. P. Grassberger, *Physica* **A262**, 251 (1999).
27. H. J. Herrmann and H. E. Stanley, *Phys. Rev. Lett.* **53**, 1121–1124 (1984).
28. R. Pike and H. E. Stanley, *J. Phys.* **A14**, L169 (1981).
29. H. J. Herrmann and H. E. Stanley, *J. Phys.* **A21**, L829 (1988).
30. N. V. Dokholyan, Y. Lee, S. V. Buldyrev, S. Havlin, P. R. King and H. E. Stanley, *J. Stat. Phys.* **93**, 603 (1998).
31. N. V. Dokholyan, S. V. Buldyrev, S. Havl in, P. R. King, Y. Lee, H. E. Stanley, *Physica* **A266**, 55 (1999).
32. F. A. L. Dullien, *Porous Media — Fluid Transport and Pore Structure* (Academic Press, New York, 1979).
33. A. Coniglio, in *Hydrodynamics of Dispersed Media*, eds. J. P. Hulin, A. M. Cazabat, E. Guyon and F. Carmona (Elsevier, North-Holland, 1990), p. 193.
34. M. Murat and A. Aharony, *Phys. Rev. Lett.* **57**, 1875 (1986).
35. Y. Lee, J. S. Andrade Jr., S. V. Buldyrev, N. Dokholyan, S. Havlin, P. R. King, G. Paul and H. E. Stanley, *Phys. Rev.* **E60**, 3425 (1999).
36. J. S. Andrade Jr., S. V. Buldyrev, N. V. Dokholyan, S. Havlin, P. R. King, Y. K. Lee, G. Paul and H. E. Stanley, *Phys. Rev.* **E62**, 8270 (2000).
37. J. S. Andrade Jr., A. D. Araújo, S. V. Buldyrev, S. Havlin and H. E. Stanley, *Phys. Rev.* **E63** (2001).
38. J. S. Andrade Jr., M. P. Almeida, J. Mendes Filho, S. Havlin, B. Suki and H. E. Stanley, *Phys. Rev. Lett.* **79**, 3901 (1997).
39. J. S. Andrade Jr, U. M. S. Costa, M. P. Almeida, H. A. Makse and H. E. Stanley, *Phys. Rev. Lett.* **82**, 5249 (1999).
40. H. Makse, J. S. Andrade Jr. and H. E. Stanley, *Phys. Rev.* **E61**, 583 (2000).
41. S. V. Patankar, *Numerical Heat Transfer and Fluid Flow* (Hemisphere, Washington DC, 1980).
42. S. Russ and B. Sapoval, *Phys. Rev. Lett.* **73**, 1570 (1994).

# SOCIOPHYSICS – A REVIEW OF RECENT MONTE CARLO SIMULATIONS

D. STAUFFER
*Instituto de Física, Universidade Federal Fluminense,*
*Av.Litorânea s/n, Boa Viagem, Niterói 24210-340, RJ, Brazil*
visiting from:
*Institute for Theoretical Physics, Cologne University,*
*D-50923 Köln, Euroland*

**Abstract**

Computational models for social phenomena are reviewed: Bonabeau et al. for the formation of social hierarchies, Donangelo and Sneppen for the replacement of barter by money, Solomon and Weisbuch for marketing percolation, and Sznajd for political persuasion. Finally we review how to destroy the internet.

## 1. HIERARCHIES OF BONABEAU ET AL.

How come some people like Don Antonio Coniglio are rich full professors, able to eat spaghetti napolitano all the time, while others are nearly starved to death as associate professors? Older ages justified nobility by the grace of god, and statistical physicists in our computer age explain about everything as being random. In this sense, Bonabeau et al.[1] explain the difference between the ruling classes and the proletariat as arising from fights between them, with initially random outcomes. People diffuse randomly on a square lattice, and if one moves to a place already occupied by another, they fight, with one of them winning randomly, and the other losing. Initially the probability to win is 50 percent; but later the probability is the higher, the larger was in the recent past the fraction of victories by this person over anybody. After some time, some people have won most fights and thus have a high probability to win again; some have lost most fights and have a high probability to lose again. In this way, a self-organized hierarchy has emerged, not by the grace of god but by the grace of random numbers.

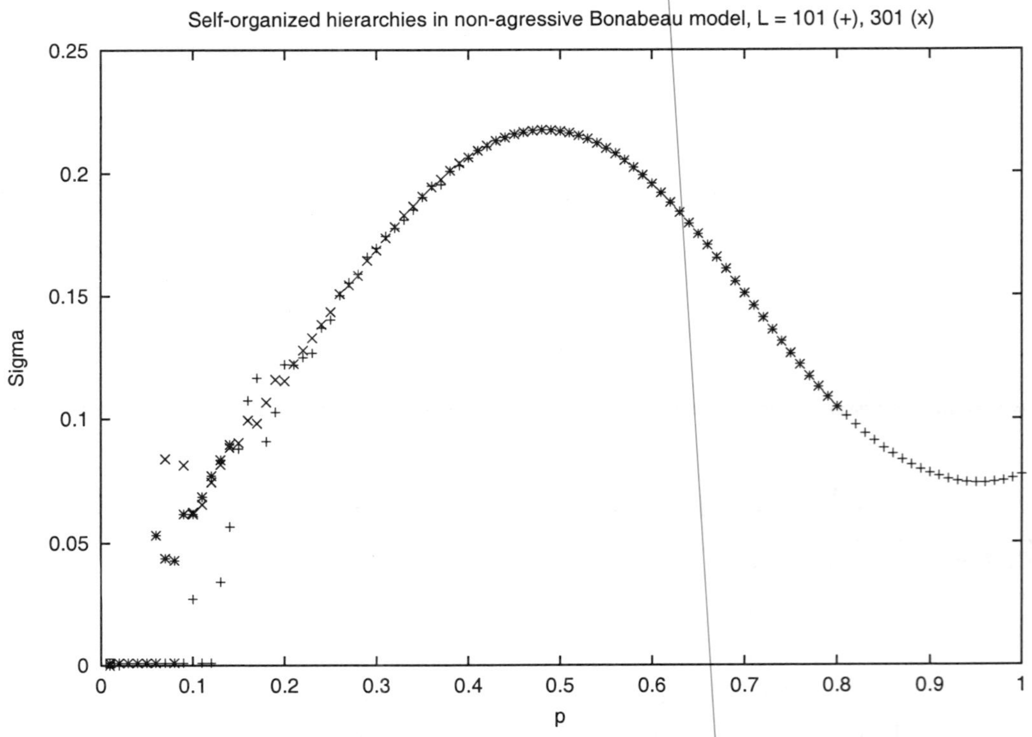


**Fig. 1** Bonabeau model: Variation of amount $\sigma$ of hierarchization as a function of population density for bible-following people (Lucas 14, 31).

Just as in an Ising ferromagnet below the Curie temperature the spontaneous magnetization will flip provided we wait long enough, also in this model the ruling class does not remain on top forever. Even the Camorra Napoletana changes and is more and more ruled by women. But taking snapshots of the social hierarchies at any moment, we see that some people are likely winners and some are likely losers, with an order parameter given by the variance $\sigma^2 = \langle (p - 1/2)^2 \rangle$ of the probability $p$ to win, with $\langle p \rangle = 1/2$. If we vary the concentration of people on the lattice, we gradually, without a sharp phase transition,[2] move from weak hierarchization at low densities (nomadic societies?) to strong hierarchization at high densities (city life?). A nonmonotonic behaviour is seen in Fig. 1, where people try to avoid losing fights by going preferrably to empty sites or to sites occupied by a weak person.

## 2. ORIGIN OF MONEY FOR DONANGELO AND SNEPPEN

Even before agriculture and city life, people traded goods which they needed against goods of which they had a surplus. This is called barter. What happens if my trading partner needs what I have, but I have already what this partner offers? Donangelo and Sneppen[3] assumed that I am intelligent, remember what goods I was asked for in the past, and now am willing to accept from my trading partner a good which I have already in my inventory but which was in the recent past the good which I was able to sell most often through the usual barter. This good then is a precursor of money: I accept it because I hope I can trade it soon for something I need, even though it has no direct value for me at present. This money is still of direct value for others who lack it, and thus differs from later forms of money which are of only symbolic value for both sides, like shells, or dollar bills.

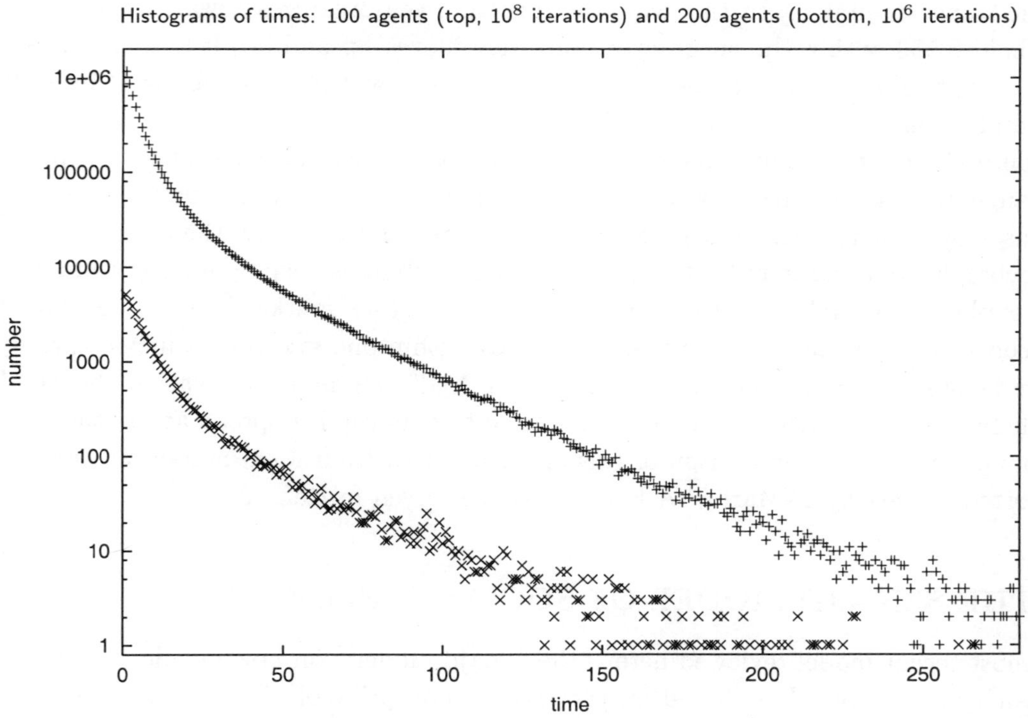


**Fig. 2** Donangelo-Sneppen model: Distribution of times over which one currency remains the leading one, for two different market sizes.

Again, the leading form of money changes with time, just as in Napoli the Roman coins of 2000 years ago were replaced by those of the Holy Roman Empire of German Nation under Federico II (from the Staufer dynasty) and then by the Italian Lire, to be replaced soon by the Euro. In a computationally simplified model[4] the traders are assumed to be less intelligent and in contrast to Ref. 3 do not keep lists of missing needed goods; instead they first randomly select something, and only then check if they need it and if the trading partner has it. Then the time during which one good stayed on as the most desired one, i.e. was the leading "currency", normally is quite short; its histogram decreases monotonically, perhaps with a (stretched) exponential (Fig. 2).

## 3. SOCIAL PERCOLATION OF SOLOMON AND WEISBUCH

Assume that Hollywood makes a movie which is a box office success; then it is likely to produce a sequel of lower quality $q$. On the other hand, after a commercial disaster, the film producers may try to increase the quality $q$ of their next movie to reach a large fraction of the movie-going population. Similar effects of self-organization can happen in other fields of marketing, politics, etc. Solomon and Weisbuch[5] set the customers onto a square lattice and give each site $i$ a quality expectation value $p_i$ randomly distributed between zero and unity. A site $i$ goes to the movie if informed about its quality $q$ by one of the four neighbours who went to this movie before, and if $q > p_i$. Similar to invasion percolation, the spread of the movie through the continent starts at the West Coast where one border of the lattice is informed about the movie. Then, if the movie was a success and reached the east coast, i.e. if the cluster of movie-goers percolated through the lattice, the quality $q$ of the next movie is decreased by $\varepsilon$, while after a flop (no percolating cluster) the quality increases by the same amount $\varepsilon$. In this way, the quality $q$ self-organizes towards the percolation threshold 0.593 within the accuracy $\varepsilon$, and then moves up and down by $\varepsilon$ if the lattice is large enough: Every flop is followed by a box-office hit, which is then followed by a another flop, and so on.

Monte Carlo simulations concentrated on the case[6] where also the quality expectations $p_i$ change in time: If person $i$ went to the movies, the corresponding quality expectation $p_i$ increases by $\varepsilon$ for the next movie; otherwise it decreases by $\varepsilon$. Now the dynamics becomes more complicated and may lead to blockades or instabilities keeping $q$ away from $p_c$. They were explained by Ref. 7. Later applications refer to stock markets[8] or to the competition between several movies.[9,10] The microdynamics within one sweep of a movie through the lattice shows the importance of Don Antonio's "red" bonds in percolation theory: If the people belonging to such a red bond refuse to go to a movie, it's spread across the continent is stopped there.[11] A comparison of this approach with traditional marketing theories was supported in part by K-Mart.[12] A longer review is given in Ref. 13.

## 4. THE SZNAJD MODEL OF SOCIAL INFLUENCE

The most recent model reviewed here is the Sznajd model[14] on how people convince others of their own opinion. It is based on the trade union principle "united we stand, divided we fall". If the two people forming a nearest–neighbour bond, or the four people forming a plaquette of nearest neighbours, all share the same of two possible opinions, then and only then they convince their neighbours of this opinion. In contrast to many voter models where a site accepts the majority opinion of its four neighbours, in the Sznajd model the flow of opinion is directed outward to the neighbours, not inward from the neighbours. Many variants were studied for the square lattice.[15] In both one and two dimensions, most variants let the system end up in a fixed point where all opinions are the same, or where the opinions are ordered antiferromagnetically. The times needed to reach this fixed point are sometimes but not always distributed log-normally. These results do not change much if the square lattice is diluted, using random or long-range correlated percolation.[16] The dynamics is similar[17] to spinodal decomposition of Ising ferromagnets at low temperatures. The model was applied to stock markets[18] and to the distribution of votes in Brazilian elections.[17,19] Without much success we combined the above model of Bonebeau et al. (for the initial fights within a party about who will be a candidate) with this Sznajd model (describing candidate-voter and voter-voter interactions).

Two of the rules discussed in Ref. 15 show a phase transition as a function of the initial distribution of opinions: If a fraction $p$ initially favours "up", then in an infinite lattice the fixed points have all spins up for $p > 1/2$ and all spins down for $p < 1/2$. For finite $L \times L$ square lattices and rule IIa of Ref. 15, Fig. 3 shows how the width of this transition diminishes with increasing lattice size.

## 5. HOW TO DESTROY THE INTERNET

How many computers have to crash before the internet is cut into at least two large parts? This question was discussed by several groups;[20–23] this section is roughly a translation of my review in Physikalische Blätter 2001. Are there "red" computers in the Coniglio sense of percolation, such that omission of one of these computers splits the internet? The answer seems to be: No.

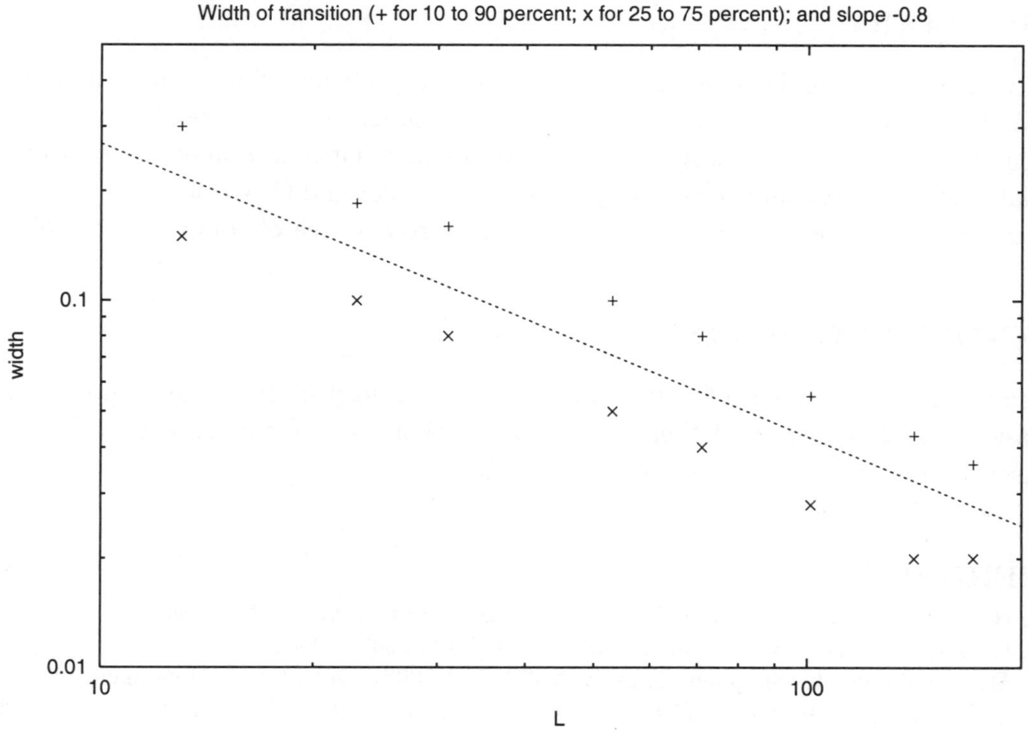


**Fig. 3** Sznajd model: The width of the phase transition is defined as the difference between the initial concentrations $p$ of "up" opinions which lead to 10 and 90 percent of all fixed points to be all "up", for the plus symbols. For the smaller width symbolized by x, the fractions are 25 and 75 percent. (Rule IIa; only horizontal bonds convince their neighbours.)

In principle, all computers connected to the internet should be able to reach all other such computers. Of course, if my workstation lost its connection to the institute server, the world is cut off from me. Similar accidents happen all the time with all computers. However, then there is still a large fraction of all computers which still can reach each other through the internet, in addition to many isolated clusters or workstations, just like in percolation theory in the presence of an infinite network. The internet would be destroyed only if this infinite network is split, such that e.g. the northern hemisphere cannot reach the southern hemisphere of Earth. A precise definition of an unsplit "infinite" internet is difficult,[24] but one may simply look at the fraction $P_\infty$ of computers in the largest computer cluster still connected through the internet. If this fraction is positive the internet is not yet destroyed.

All these papers[20–23] use different methods and models but agree in their final result: Failure of a randomly selected majority of computers does not destroy the internet, but destruction of a few percent suffices if they are selected according to their importance (number of connections). (After the destruction of the most important node in the selective case, one may then destroy what was before the second most important node,[21] or one may re-evaluate what is now the most connected node and then destroy it[20].) Either a structure similar to the (upper levels of the) real internet was used; or a percolation model was employed where each computer was connected to $k$ other computers, with the histogram of $k$ values decaying as $1/k^{2.5}$ and no geometric correlations at all. Both analytic theories similar to Bethe lattices, and Monte Carlo simulations were used. Less than ten percent need to be destroyed selectively, but more than 99 percent have to be destroyed randomly, to split the internet. Whether you like or dislike this result may depend on how many e-mails you get per day.

## 6. SUMMARY

Physics methods have been applied to sociological problems since some time, e.g. by W. Weidlich.[25] Other applications like the Galam models are reviewed in our book[26] or, particularly for Ising like models, in Ref. 27. As the International Conference on Computer Simulations and the Social Sciences, organized by G. Ballot and G. Weisbuch in Paris (18–20 September, 2000) showed, this field may become a growth area of computational physics.

## ACKNOWLEDGMENTS

Suzana and Paulo Murilo de Oliveira are responsible for hospitality to the author in Niterói, discussion, collaboration, and them as well as R. Donnagelo for a critical reading of this manuscript. Partial support came from FAPERJ.

## REFERENCES

1. E. Bonabeau, G. Theraulaz and J. L. Deneubourg, *Physica* **A217**, 373 (1995).
2. A. O. Sousa and D. Stauffer, *Int. J. Mod. Phys.* **C11**, 1063 (2000).
3. R. Donangelo and K. Sneppen, *Physica* **A276**, 572(2000) ; see also R. Donangelo, A. Hansen, K. Sneppen and S. R. Souza, *Physica* **A283**, 469 (2000); P. Bak, F. Norrelyke and M. Shubik, *e-print cond-mat 0009287.*
4. D. Stauffer and J. Radomski, *Physica* **A291**, 583 (2000).
5. S. Solomon and G. Weisbuch, *e-print adap-org/9909001.*
6. S. Solomon, G. Weisbuch, L. de Arcangelis, N. Jan and D. Stauffer, *Physica* **A277**, 239 (2000).
7. Z. F. Huang, *Int. J. Mod. Phys.* **C11**, 287 (2000).
8. Z. F. Huang, *Eur. Phys. J.* **B16**, 379 (2000).
9. E. Ahmed and H. A. Abdusalam, *Eur. Phys. J.* **B16**, 569 (2000).
10. G. Weisbuch and D. Stauffer, *Physica* **A287** 563 (2000).
11. G. Weisbuch and S. Solomon, *Int. J. Mod. Phys.* **C11**, 1263 (2000).
12. J. Goldenberg, B. Libai, S. Solomon, N. Jan and D. Stauffer, *Physica* **A284**, 335 (2000).
13. G. Weisbuch, S. Solomon and D. Stauffer, preprint.
14. K. Sznajd-Weron and J. Sznajd, *Int. J. Mod. Phys.* **C11**, 1157 (2000).
15. D. Stauffer, A. O. Sousa and S. Moss de Oliveira, *Int. J. Mod. Phys.* **C11**, 1239 (2000).
16. A. A. Moreira, J. S. Andrade Jr. and D. Stauffer, *Int. J. Mod. Phys.* **C12**, No. 1. (2001).
17. U. M. S. Costa, A. D. Araujo, A. T. Bernardes and D. Stauffer, *Int. J. Mod. Phys.* **C12**, preprint (2001).
18. K. Sznajd-Weron and R. Weron, *cond-mat/0101001* (2001).
19. R. N. Costa-Filho, M. P. Almeida, J. S. Andrade Jr. and J. E. Moreira, *Phys. Rev.* **E60**, 1067 (1999).
20. R. Albert, H. Jeong and A. L. Barabási, *Nature* **406**, 378 (2000).
21. A. Broder et al., *Comput. Netw.* **33**, (2000) 309.
22. R. Cohen, K. Erez, D. Ben-Avraham and S. Havlin, *Phys. Rev. Lett.* **85**, 4626 (2000) and preprint cond-mat/0010251.
23. D. S. Callaway, M. E. J. Newman, S. H. Strogatz and D. J. Watts, *Phys. Rev. Lett.* **85**, 5468 (2000).
24. D. Stauffer, *J. Irreprod. Results* **44**, 57 (1999).
25. W. Weidlich,*Sociodynamics: A Systematic Approach to Mathematical Modelling in the Social Sciences* (Harwood Academic Publishers, 2000).
26. S. Moss de Oliveira, P. M. C. de Oliveira and D. Stauffer, *Evolution, Money, War and Computers* (Teubner, Leipzig, 1999).
27. J. A. Hołyst, K. Kacperski and F. Schweitzer, in *Annual Reviews of Computational Physics* (World Scientific, Singapore 2001), Vol. 9, p. 275; K. Kacperski and J. A. Hołyst, *Physica* **A287**, 631 (2000).

# REMARKS ON THE NONUNIVERSALITY OF BOLTZMANN-GIBBS STATISTICAL MECHANICS

C. TSALLIS
*Centro Brasileiro de Pesquisas Fisicas,*
*Rua Xavier Sigaud 150, Rio de Janeiro 22290-180, RJ, Brazil*

*Lasciate ogne speranza, voi ch' intrate — Dante Alighieri (Inferno, III)*

**Abstract**

How general are Boltzmann-Gibbs statistical mechanics and standard Thermodynamics? What classes of systems have thermostatistical properties, in particular equilibrium properties, that are correctly described by these formalisms? The answer is far from trivial. It is, however, clear today that these relevant and popular formalisms are not universal, this is to say that they have a domain of validity that it would be important to define precisely. A few remarks on these questions are herein presented, in particular in relation to nonextensive statistical mechanics, introduced a decade ago with the aim to cover at least some (but most probably not all) of the natural systems that are out of this domain of validity.

## 1. INTRODUCTION

Once upon a time I heard Sam Edwards declaring in Brasilia (not to be confused with Brazil nor with Buenos Aires, and even less with Rio de Janeiro) that he thinks of theoretical physics as going on along three lines. The first of them is that in which the equations are known but not their answers; the second of them is when not even the equations are known; the third one concerns those cases for which both the equations and the answers are known [nobody is obliged to feel any identification with this line! But still, does this remind you of anything that you have ever seen or heard?]. Then, our distinguished colleague shared with the audience that he likes quite a lot the two first types, but not at all the third one. Having in my life already attempted to do, following masters like Onsager and il Signor Professor Dottor Antonio Coniglio, a few studies along the first line, and in undoubtedly too many occasions along the third line [dear Sam, please forgive my weaknesses!], I thought that, since I was already approaching the dangerous fifties and in order to make an attempt to be a "complete scientist" (one of my youth utopies!), I should try at least once the second line [I confess that, even in case of unsuccess, I was probably not going to lose my job!]. So, in Mexico City in 1985, I formulated in my mind, for my personal use, a path along which the usual formalism of statistical mechanics, this is to say our dear Boltzmann-Gibbs (BG) formalism, could be generalized. During three years, I was analyzing whether this was completely or only partially stupid ... or, perhaps, not even stupid after all! So, I decided to ask my friends! I discussed the question in Maceio with Hans J. Herrmann (present here, at this meeting) and Evaldo M. F. Curado. And also I discussed it later on, in Rio de Janeiro, with Silvio R. A Salinas (also present here, at this meeting) and some other distinguished Brazilian colleagues. They all stimulated me (quite irresponsibly, I suspect) to publish the idea in a serious international journal [very easy, of course, it was not their neck which was probably going to be cut by some of the erudite statistical physicists of the world!]. So I did in 1988, in the Journal of Statistical Physics.[1] Who knows whether Joel Lebowitz, editor of the journal, has not since then deeply regreted his friendly and wide generosity or momentary editorial inadvertence? In any case, it is now too late, the disaster has already happened, and since there are now hundreds of scientists foolish enough to follow along the same terrible path, I feel now obliged [it used to be "noblesse" which obliged, now it is "political correctness" ...] to give some minimal justification for my outrageous and irresponsible attitude! With the help of the Olympian gods, I will try in what follows to convince you to not yet kill all the so-called scientists who dared to give some credibility to the whole idea.

## 2. NONEXTENSIVE STATISTICAL MECHANICS

For simplicity, I will be calling Boltzmann-Gibbs (or, occasionally, Boltzmann-Gibbs-Shannon) entropy the expression

$$S = -k \sum_{i=1}^{W} p_i \ln p_i \equiv -k \langle \ln p_i \rangle \tag{1}$$

as well as its continuous version

$$S = -k \int d\mathbf{x}\, p(\mathbf{x}) \ln p(\mathbf{x}) \equiv -k \langle \ln p(\mathbf{x}) \rangle \tag{2}$$

and its quantum one

$$S = -k Tr \rho \ln \rho \equiv -k \langle \ln \rho \rangle \,. \tag{3}$$

For some discussions (see, for example, Ref. 2) it is important to make the distinction between the microcanonical form $S = k \ln W$, sometimes referred to as Boltzmann entropy (and corresponding to the particular case of equiprobability of expression (1), i.e. $p_i = 1/W (\forall i)$), and expression (2) ($\mathbf{x}$ being a point of the $N$-body system full phase space), sometimes referred to as Gibbs entropy. We shall not make such distinction here, since our central point concerns the use of a *logarithmic* or of a *nonlogarithmic* measure of *ignorance* (or of *lack of information*).

BG statistical mechanics can be founded in a variety of forms. Chronologically, it was developed as follows. (i) Boltzmann proposed it in terms of kinetic theory and the *molecular chaos* hypothesis, using Newtonian mechanics as the microscopic dynamical basis for Hamiltonian systems; (ii) Gibbs did so through the ensemble picture, i.e. essentially through a frame which came later to be called *information theory* (with Jaynes' interpretation), where $S$ appears as functional to be optimized in the presence of constraints if we wish to discuss equilibrium states; (iii) Fowler and Darwin did it through a steepest descent calculation; (iv) Khinchin founded it on the laws of large numbers; and, finally, (v) Balian and Balazs based it on a counting method for isolated systems. In all these manners the final outcome for thermal equilibrium (canonical ensemble) is one and the same, namely an *exponential* distribution, named Boltzmann factor, or Gibbs distribution, or, as we shall adopt here, the BG distribution, cornerstone of BG statistical mechanics. If we use the entropy given by Eq. (1), the BG distribution is given by

$$p_i = \frac{\exp(-\beta\varepsilon_i)}{Z} \; (\beta \equiv 1/kT) \tag{4}$$

with the partition function given by

$$Z \equiv \sum_{j=1}^{W} \exp(-\beta\varepsilon_j)\,, \tag{5}$$

where $\{\varepsilon_i\}$ are the eigenvalues associated with the Hamiltonian (and its associated boundary conditions). If we use Eq. (2) and $\varepsilon(\mathbf{x}) \propto x^2$ we obtain the celebrated Gaussian form.[3]

In the 1988 paper,[1] I followed Gibbs' path (manner (ii) mentioned above) in order to generalize BG statistical mechanics, but postulating that the entropy[4] to be optimized was a more general one, namely

$$S_q = k\,\frac{1-\sum_{i=1}^{W} p_i^q}{q-1} \; (q \in \mathcal{R})\,, \tag{6}$$

which reproduces the BG form in the $q = 1$ limit (i.e. $S_1 = S$). With the definition of *unnormalized q-expectation value* as $\langle \ldots \rangle_q \equiv \sum_{i=1}^{W} (\ldots) p_i^q$, Eq. (6) can be rewritten as follows

$$S_q = -k \langle \ln_q p_i \rangle_q\,, \tag{7}$$

where

$$\ln_q x \equiv \frac{x^{(1-q)}-1}{1-q} \; (\ln_1 x = \ln x)\,, \tag{8}$$

whose inverse function is given by

$$\exp_q(x) \equiv [1+(1-q)x]^{\frac{1}{1-q}} (\exp_1(x) = \exp(x))\,. \tag{9}$$

This (nonnegative) entropy becomes

$$S_q = k \ln_q W \tag{10}$$

for equiprobability. Also, it can be easily shown to satisfy, for two *independent* systems $A$ and $B$ (i.e., $p_{ij}^{A+B} = p_i^A \times p_j^B$),

$$\frac{S_q(A+B)}{k} = \frac{S_q(A)}{k} + \frac{S_q(B)}{k} + (1-q)\frac{S_q(A)}{k}\frac{S_q(B)}{k}\,, \tag{11}$$

which is *superextensive*, *extensive* and *subextensive* for $q < 1$, $q = 1$ and $q > 1$ respectively. It is from this property that the name *nonextensive statistical mechanics* stands. It is, as we shall mention again later on, very important to notice that this property does not forbidden that, for specific *nonindependent* systems (i.e., $p_{ij}^{A+B} \neq p_i^A \times p_j^B$), a special value $q^*$ for $q$ might exist such that $S_{q^*}(A+B) = S_{q^*}(A) + S_{q^*}(B)$. This is to say, that extensivity *might* still hold for special strongly-interacting (presumably very large) subsystems, but for an entropic form which might differ from the BG one! This important possibility must be kept in mind, since it would make possible to essentially maintain the Clausius idea of macroscopic thermodynamic entropy, but the connection with the microscopic world would have to be done not necessarily through the BG entropy, but through $S_{q^*}$ where $q^*$ needs not to be equal to unity. If this possibility turns out to be physically meaningful in some nontrivial cases, then it would be possible to "save the property" (thermodynamic extensivity for the entropy), just by "changing the concept"!

It is worthy mentioning that $S_q$ is simply related to the Renyi entropy[5] $S_q^R \equiv \frac{\ln \sum_{i=1}^W p_i^q}{1-q}$ as follows:

$$S_q^R = \frac{\ln\left[1 + (1-q)\dfrac{S_q}{k}\right]}{1-q}\,. \tag{12}$$

In the 1988 paper, the equilibrium distribution for the canonical ensemble (system in equilibrium with a thermostat) was derived by imposing, in addition to the constraint $\sum_{i=1}^W p_i = 1$, the usual constraint on the energy, i.e., $\langle \varepsilon_i \rangle \equiv \sum_{i=1}^W p_i \varepsilon_i = U$. A second possibility was there indicated, namely to impose $\langle \varepsilon_i \rangle_q \equiv \sum_{i=1}^W p_i^q \varepsilon_i = U$. This possibility was effectively explored in 1991 in a paper with Curado,[6] where the connection to thermodynamics was done as well. Later on, in 1998, with Mendes and Plastino,[7] the constraint was changed in such a way that several physically desirable features were incorporated. The constraint was written using *normalized q-expectation values*, namely

$$\langle\langle \varepsilon_i \rangle\rangle_q \equiv \sum_{i=1}^W P_i \varepsilon_i = U_q\,, \tag{13}$$

where the *escort* distribution $P_i$ is defined as follows

$$P_i \equiv \frac{p_i^q}{\sum_{j=1}^W p_j^q} \quad \left(\sum_{i=1}^W P_i = 1\right). \tag{14}$$

The equilibrium distribution thus obtained is given by

$$p_i = \frac{[1-(1-q)\beta_q(\varepsilon_i - U_q)]^{1/(1-q)}}{Z_q} = \frac{\exp_q[-\beta_q(\varepsilon_i - U_q)]}{Z_q} \quad \left(\beta_q \equiv \frac{1}{kT_q} \equiv \frac{\beta}{Z_q^{1-q}}\right) \tag{15}$$

with

$$Z_q \equiv \sum_{j=1}^W \exp_q[-\beta_q(\varepsilon_j - U_q)]\,, \tag{16}$$

$\beta$ being the Lagrange parameter. We easily verify that $q \to 1$ recovers the celebrated BG factor

$$p_i = \frac{e^{-\beta\varepsilon_i}}{\sum_j e^{-\beta\varepsilon_j}}. \tag{17}$$

For $q > 1$ a *power-law* tail emerges; for $q < 1$ the formalism imposes a high-energy *cutoff*, i.e., $p_i = 0$ whenever the argument of the power function becomes negative. One comment is worthy: this distribution is generically *not* an exponential law, i.e. it is generically *not* factorizable (under sum in the argument), and nevertheless is *invariant* under the choice of zero energy for the energy spectrum! (this is in fact one of the aside benefits of defining the constraints in terms of *normalized* distributions like the escort ones).

Equation (15) can be rewritten in the following convenient form:

$$p_i = \frac{\exp_q(-\beta' \varepsilon_i)}{Z'_q} \tag{18}$$

with $Z'_q \equiv \sum_{j=1}^{W}(-\beta'\varepsilon_j)$, where $\beta'$ is a simple monotonically increasing function of $\beta_q$.

If we assume $\varepsilon_{ij}^{A+B} = \varepsilon_i^A + \varepsilon_j^B$ and $p_{ij}^{A+B} = p_i^A p_j^B$ we obtain $U_q(A+B) = U_q(A) + U_q(B)$. However, is this compatible with the equilibrium distribution we are exhibiting here? Only if $q = 1$. But another question, more interesting, can be advanced, namely, is it possible to have $\varepsilon_{ij}^{A+B} \neq \varepsilon_i^A + \varepsilon_j^B$ and $p_{ij}^{A+B} = p_i^A p_j^B$ where the distributions are the present ones? Some preliminary indications do exist that suggest that such case can exist in the thermodynamic limit, this is to say for increasingly large subsystems. We have no mathematical argument proving this, but only some numerical indications, which are presently in progress. The whole problem involves, as one of its relevant ramifications, the validity of the zero$^{th}$ principle of thermodynamics out of BG statistical mechanics.

The paths followed by Darwin-Fowler, by Khinchin and by Balian-Balazs have been recently generalized to arbitrary $q$ by Abe and Rajagopal.[8] Analogously, the path followed by Boltzmann in terms of kinetic theory has also been recently generalized to arbitrary $q$ by Lima, Plastino and Silva.[9]

For more details about nonextensive statistical mechanics the reader is referred to Refs. 10–13. A sensible amount of applications and experimental verifications are now available. Just to see a few recent ones, the reader is referred to Ref. 14 for fully developed turbulence, to Ref. 15 for hadronic jets produced by electron-positron annihilation, to Ref. 16 for anomalous diffusion of cells of *Hydra viridissima*, and to Ref. 17 for quantum entanglement and information.

## 3. NONEXTENSIVITY AND MIXING

Boltzmann based on *ergodicity* his discussion on the foundations of the formalism he was revealing to us. More than half a century later, it started becoming clear, through the pioneering contributions of Krylov, that more fundamental than ergodicity was the *mixing* in the Riemann hypersurface (corresponding to fixed energy, and fixed values of other constants of motion if applicable) in the full phase space (with dimension of the order of $2dN$ for $N$ point particles) of an isolated system. It seems quite plausible today that it is the quick mixing (*exponentially* quick) in phase space, i.e., *strong chaos*, which is responsible for the validity of BG statistical mechanics. This happens, in particular, for short-range-interacting Hamiltonian large systems. If the system is classical, this corresponds to the existence of positive Lyapunov exponents $\{\lambda^{(n)}\}$, which ensure quick occupancy of the full accessible phase space. Through the Pesin theorem it is known that the Kolmogorov-Sinai entropy $K$, essentially given by $K \equiv \lim_{t\to\infty} S(t)/t$ (where, as time $t$ goes on, the occupancy of the full phase space increases if we started, at $t = 0$, at an infinitesimaly small region), satisfies

$$K = \sum_{\textit{all positive Lyapunov exponents}} \lambda^{(n)} . \tag{19}$$

If we have a one-dimensional case, there is only one Lyapunov exponent (noted $\lambda$) whose definition is given by $\xi = \exp(\lambda t)$, where $\xi \equiv \lim_{\Delta x(0)\to 0} \Delta x(t)/\Delta x(0)$, $\Delta x(t)$ being the difference between trajectories at time $t$. The sensitivity to the initial conditions $\xi$ satisfies the differential equation $\dot{\xi} = \lambda\xi$. If $\lambda$ vanishes, then we expect generically the following differential equation to be satisfied ($q^*$ being a special value determined by the specific dynamics):

$$\dot{\xi} = \lambda_{q^*}\xi^{q^*}, \tag{20}$$

hence,

$$\xi = [1 + (1 - q^*)\lambda_{q^*}t]^{\frac{1}{1-q^*}} = \exp_{q^*}(\lambda_{q^*}t)(\lambda_1 = \lambda)\,. \tag{21}$$

(This solution usually corresponds, in fact, to an upper bound of the sensitivity function $\xi$.) We see that the particular case $q^* < 1$ and $\lambda_{q^*} > 0$ yields slow, *power-law* mixing (i.e., *weak chaos*). It might well be that for large classes of systems the relevant full equation is

$$\dot{\xi} = \lambda_1\xi + (\lambda_{q^*} - \lambda_1)\xi^{q^*}, \tag{22}$$

hence[18]

$$\xi = \left\{1 - \frac{\lambda_{q^*}}{\lambda_1} + \frac{\lambda_{q^*}}{\lambda_1}\exp[(1 - q^*)\lambda_1 t]\right\}^{\frac{1}{1-q^*}} \tag{23}$$

which contains the previous cases as particular instances. If $q^* < 1$ and $\lambda_{q^*} \gg \lambda_1 > 0$, this solution makes a crossover from a power-law sensitivity at not too high values of $t$ to an exponential sensitivity at longer times. The crossover time $t_{cross} \equiv 1/(1 - q^*)\lambda_1$ diverges if $\lambda_1$ vanishes. It might well be that this is the structure which makes some systems to exhibit a crossover from an anomalous thermostatistical state to the BG one, when time is large enough. It is like $\lambda_1$ depending on the size $N$ of the system in such a way that $\lim_{N\to\infty} \lambda_1(N) = 0$, whereas $\lim_{N\to\infty} \lambda_{q^*}(N) > 0$. If the thermodynamical equilibrium limits are taken in the order $\lim_{N\to\infty} \lim_{t\to\infty}$ the system is driven to a BG equilibrium, but the order $\lim_{t\to\infty} \lim_{N\to\infty}$ drives the system to some different, anomalous equilibrium.

What can we say about the Kolmogorov-Sinai entropy if *all* Lyapunov exponents vanish? In such situation it is convenient to define a $q$-generalized Kolmogorov-Sinai entropy as follows:

$$K_q \equiv \lim_{t\to\infty} S_q(t)/t\,. \tag{24}$$

We then expect $K_{q^*}$ to be finite, whereas $K_q$ should vanish for all $q > q^*$, and diverge for all $q < q^*$. We similarly expect a nonextensive version of Pesin theorem, i.e., $K_{q^*} = \lambda_{q^*}$. Some aspects of this scenario have been numerically verified, others are consistent with our computational observations, others remain to be checked. Although we have no proofs, we have various evidences that large classes of physical systems hopefully follow the scenario we have presented here.

## 4. CONCLUSION

On the whole, the following picture emerges. Short-range interacting systems have positive Lyapunov exponents, hence they are strongly chaotic, quickly occupy the entire available phase space, and, if started out of thermal equilibrium, tend to relax exponentially quick to the equilibrium state (the relaxation times being of the order of the inverse maximum Lyapunov exponent). For such systems we have $q^* = 1$, $K_1$ is finite and the relevant thermodynamic entropy is the BG one, i.e., $S_1$. Consequently, the equilibrium distribution is an exponential of the energy. If the system is classical, everything commutes, in particular the kinetic and potential terms of the total energy, hence the distribution of velocities always is Maxwell's Gaussian. These are the basic features of BG statistical mechanics.

But systems like long-range-interacting ones and others evolve with a microscopic dynamics which yields vanishing Lyapunov exponents (at least in the thermodynamic limit), hence they are weakly chaotic, and only slowly and inefficiently occupy the *a priori* phase space. If started out of thermal equilibrium they might tend to it in a slow manner. For such systems we expect $q^* < 1$ (the exponent of the power-law sensitivity to the initial conditions being $1/(1-q^*)$), $K_{q^*}$ is finite and the relevant thermodynamic entropy is either $S_{q^*}$ or $S_{q^{**}}$ where $q^{**} = f(q^*)$ such that $f(x)$ is a monotonically decreasing function which satisfies $f(1) = 1$. If the system is classical, the velocity distribution hopefully is a power-law. These are the basic features expected for nonextensive statistical mechanics. Many things remain to be done, checked, proved. Not the least of them is to find the connection, at least for classical systems, between $q^*$ and $(\alpha, d)$ (probably $\alpha/d$).

## ACKNOWLEDGMENTS

I have benefitted from interesting conversations with F. Baldovin, E. P. Borges, A. Rapisarda, V. Latora and M. Baranger, which I am very happy to acknowledge. This work was partially supported by CNPq, PRONEX and FAPERJ (Brazilian agencies).

## REFERENCES

1. C. Tsallis, *J. Stat. Phys.* **52**, 479 (1988).
2. J. L. Lebowitz, *Phys. Today* **46**, 32 (1993); *Physica* **A263**, 516 (1999).
3. First published by De Moivre in 1733, then by Laplace in 1774, then by Adrain in 1808 and finally by Gauss in 1809, when he made the celebrated connection with the theory of nonsystematic errors. See S. M. Stigler, *Statistics on the Table — The History of Statistical Concepts and Methods* (Harvard University Press, Cambridge, 1999).
4. As I discovered along the years, this entropic form (sometimes with a different prefactor, adapted to binary representations), as well as close to twenty (!) other entropic forms, already existed in the literature of cybernetics and control theory. This type of functional was first published by Harvda and Charvat in 1967, then by Vajda in 1968, then by Daroczy in 1970, and finally by myself in 1988, when it was used to generalize the BG formalism.
5. According to Csiszar, first introduced by Schutzenberger in 1954, and then by Renyi in 1961. The Renyi entropy is always extensive but generically nonconcave. In contrast, the entropy $S_q$ always has a definite concavity but generically is nonextensive. Only the BG entropy $S_1 = S_1^R = S$ is both extensive and concave.
6. E. M. F. Curado and C. Tsallis, *J. Phys.* **A24**, L69 (1991) [Corrigenda: **24**, 3187 (1991) and **25**, 1019 (1992)].
7. C. Tsallis, R. S. Mendes and A. R. Plastino, *Physica* **A261**, 534 (1998).
8. S. Abe and A. K. Rajagopal, *Phys. Lett.* **A272**, 341 (2000); *J. Phys.* **A33**, 8733 (2000); *Europhys. Lett.* **52**, 610 (2000); *J. Phys.* **A33**, 8723 (2000); "Reexamination of Gibbs' Theorem and Nonuniqueness of Canonical Ensemble Theory," *Physica A*, in press (2001); *Proc. IUPAP Workshop on New Trends on Fractal Aspects of Complex Systems 16–20 October 2000, Maceio-AL, Brazil*, ed. M. L. Lyra (Elsevier, Amsterdam, 2001).
9. R. Silva, A. R. Plastino and J. A. S. Lima, *Phys. Lett.* **A249**, 401 (1998); J. A. S. Lima, R. Silva and A. R. Plastino, "On Tsallis' Nonextensive Thermostatistics and the H-Theorem," *Phys. Rev. Lett.*, in press [cond-mat/0101030] (26 March 2001).
10. S. R. A. Salinas and C. Tsallis, "Nonextensive Statistical Mechanics and Thermodynamics," *Braz. J. Phys.* **29**, (1999). [http://sbf.if.usp.br/WWW_pages/Journals/BJP/Vol29/Num1/index.htm].
11. S. Abe and Y. Okamoto, *Nonextensive Statistical Mechanics and Thermodynamics*, Series on Lecture Notes in Physics (Springer-Verlag, Heidelberg, 2001).

12. P. Grigolini, C. Tsallis and B. J. West, "Quantum and Classical Complexity and Nonextensive Statistical Mechanics," *Chaos, Solitons and Fractals*, in press (2001).
13. See *http://tsallis.cat.cbpf.br/biblio.htm* for regularly updated bibliography on the subject.
14. T. Arimitsu and N. Arimitsu, *Phys. Rev.* **E61**, 3237 (2000); *J. Phys.* **A33**, L235 (2000); C. Beck, *Physica* **A277**, 115 (2000); C. Beck, G. S. Lewis and H. L. Swinney, *Phys. Rev.* **E63**, 035303 (2001).
15. I. Bediaga, E. M. F. Curado and J. Miranda, *Physica* **A286**, 156 (2000); C. Beck, *Physica* **A286**, 164 (2000).
16. A. Upadhyaya, J.-P. Rieu, J. A. Glazier and Y. Sawada, "Anomalous Diffusion and Non-Gaussian Velocity Distribution of Hydra Cells in Cellular Aggregates," *Physica A*, in press (2001).
17. S. Abe and A. K. Rajagopal, *Physica* **A289**, 157 (2001); C. Tsallis, S. Lloyd and M. Baranger, "Peres Criterion for Separability Through Nonextensive Entropy," *Phys. Rev. A*, in press [quant-ph/0007112] (1 April 2001); C. Tsallis, P. W. Lamberti and D. Prato, "A Nonextensive Critical Phenomenon Scenario for Quantum Entanglement," *Physica A*, in press (2001), *Proceeding IUPAP Workshop on New Trends on Fractal Aspects of Complex Systems, 16–20 October 2000, Maceio-AL, Brazil*, ed. M. L. Lyra (Elsevier, Amsterdam, 2001) [cond-mat/0012502].
18. C. Tsallis, G. Bemski and R. S. Mendes, *Phys. Lett.* **A257**, 93 (1999).

# EFFECT OF DAMAGE ON THE ROUGHNESS OF PLANAR CRACKS: THE CASE OF THE RANDOM FUSE MODEL

STEFANO ZAPPERI
*INFM Sezione di Roma 1, Università "La Sapienza", P.le A. Moro 2, 00185 Roma, Italy*

HANS J. HERRMANN
*PMMH-ESPCI, 10 Rue Vauquelin, 75231 Paris Cedex 05, France*

STÉPHANE ROUX
*Laboratoire Surface du Verre et Interfaces Unité Mixte de Recherche CNRS/Saint-Gobain, 39 Quai Lucien Lefranc, BP. 135, F-93303 Aubervilliers Cedex, France*

**Abstract**

We investigate the effect on the roughness of microcrack nucleation ahead of a propagating planar crack and study the structure of the damage zone. To this end we consider a quasi-two dimensional random fuse model, confining the crack between two horizontal plates. The two-dimensional geometry introduces a characteristic length in the problem, limiting the crack roughness. The damage ahead of the crack does not appear to change the scaling properties of the model, which are well described by gradient percolation.

## 1. INTRODUCTION

Experiments have shown that in several materials under different loading conditions, the crack front tends to roughen and can often be described by self-affine scaling.[1,2] In particular, *in plane* roughness of a planar crack in PMMA was observed to scale with an exponent $\zeta = 0.63 \pm 0.03$.[3] The origin of this result is still not understood theoretically. It has been proposed to identify the crack front with a deformable line pushed by the external stress through a random toughness landscape. The deviations of the crack front from a flat line are opposed by the elastic stress field, through the stress intensity factor.[4] Thus the problem can be related to the theory of interface depinning in random media and the roughness exponent computed by numerical simulations and renormalization group (RG) calculations.[5,6] However, the agreement within this theoretical approach and experiments are quite poor: simulations predict $\zeta = 0.35$[8,9] and RG gives $\zeta = 1/3$,[9] both quite far from the experimental result.

A different approach to crack propagation in disordered media, considers the problem in the framework of lattice models.[10] The elastic medium is replaced by a network of bonds obeying discretized equations until the stress reaches a failure threshold. The disorder in the medium can be simulated by a distribution of thresholds or by bond dilution. Models of this kind have been widely used in the past to investigate several features of fracture of disordered media, such as the failure stress,[11–14] fractal properties[10,16] and avalanches.[15–19] The advantage of lattice models with respect of interface models is that the former allow for nucleation of microcracks ahead of the main crack.

In this paper, we present numerical simulations of a planar crack using the random fuse model.[11] We employ a quasi two-dimensional geometry, considering two horizontal plates separated by a network of vertical bonds.[20] A similar setup was used in a spring model,[21] and the roughness was studied only in the high velocity regime for crack motion. The experiments of Ref. 3 were instead performed at low velocity so that a quasistatic model seems more appropriate. A similar study has been undertaken for the beam model[22] and it was found that the damage ahead of the crack does not change the roughness exponent. Here we reach the same conclusion, although in our case the crack is not self-affine, but possibly self-similar. We study the damage zone close to the crack and find that several of its features can be described by gradient percolation.[23] In addition we find that the two-dimensional geometry introduces a characteristic length limiting the crack width.

## 2. THE MODEL

In the random fuse model, each bond of the lattice represents a fuse, that will burn when its current overcome a threshold.[11–14] The currents flowing in the lattice are obtained solving the Kirchhoff equation with appropriate boundary conditions. In this paper, we consider two horizontal tilted square lattices of resistors connected by vertical fuses (see Fig. 1). The conductivity of the horizontal resistors is chosen to be unity, while the vertical fuses have conductivity $\sigma$. A voltage drop $\Delta V$ is imposed between the first horizontal rows of the plates. To simulate the propagation of a planar crack, we allow for failures of vertical bonds

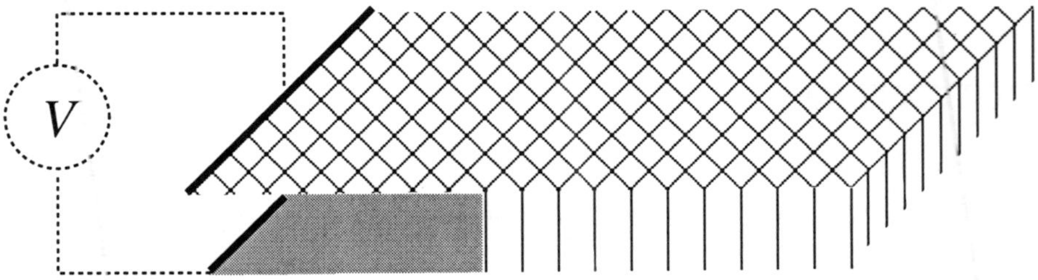


**Fig. 1** The geometry of the model. The horizontal bonds have unitary conductivity while the vertical bond have conductivity $\sigma$. A planar crack is present at the center of the system and a voltage drop is applied at the boundaries.

only and assign to each of them a random threshold $j_i^c$, uniformly distributed in the interval $[1:2]$. When the current in a bond $i$ overcomes the random threshold, the bond is removed from the lattice and the currents in the lattice are recomputed, until all the currents are below the threshold. The voltage drop is thus increased until the weakest bond reaches the threshold.

The quasistatic dynamics we are using should correspond to the small constant displacement rate at the boundary of the crack used in experiments.[3] In order to avoid spurious boundary effect, we start with a preexisting crack occupying the first half of the lattice (see Fig. 1) and employ periodic boundary conditions in the direction parallel to the crack surface. In addition, once an entire row of fuses has failed, we shift the lattice backwards one step in the direction perpendicular to the crack surface, to keep the crack always in the middle of the lattice.

A few analytical results can be directly be obtained from the model.[20] In particular, we can show that the problem has a characteristic length $\xi \simeq 1/\sqrt{2\sigma}$. This result derives directly from the structure of the Kirchhoff equations reading as

$$\sum_{nn}(V_{i+nn} - V_i) + \sigma(V_i' - V_i) = 0 \tag{1}$$

where the sum runs over the nearest neighbors of node $i$ and $V_i'$ is the voltage of the corresponding node in the opposite plate. Due to symmetry we can chose $V_i' = -V_i$ and solve the equations only for one of the plates. Equation (1) represents a discretization of a Laplace equation with a "mass term"

$$\nabla^2 V - \xi^{-2} V = 0\,, \tag{2}$$

where $\xi = 1/\sqrt{2\sigma}$. It is possible to derive this result exactly, without resorting to the continuum limit, by computing the current in a one-dimensional geometry.[20] The continuum limit, however, allow for a simple estimate of the current transfer function $G(x - x')$, giving the change of the current in a bond at $(x', 0)$ following the removal of a bond in $(x, 0)$ where $y = 0$ denotes the position of the planar crack. It is found that $G(x) \sim \log(x)$, for $x \ll \xi$ and $G(x) \sim \exp(-x/\xi)$ for $x \gg \xi$.[20]

## 3. SIMULATIONS

In order to analyze the effect of crack nucleation ahead of the main crack, we first simulate the model confining the ruptures to the crack surface. In this way, our model reduces to a connected interface moving in a random medium with an effective stiffness given by the solution of the Kirchhoff equations. The results are then compared with simulations of the unrestricted model, where ruptures can occur everywhere in the lattice. In both cases the crack width increases with time up to a crossover time at which it saturates. The height-height correlation function $C(x) \equiv \langle (h(x) - h(0))^2 \rangle$, where the average over different realizations of the disorder, is shown in Fig. 2. From this figure it is apparent that the structure of the crack is similar in the two cases. The only difference lies in the higher saturation width that is observed when microcracks are allowed to nucleate ahead of the main crack.

Next, we analyze the behavior of the crack as a function of $\sigma$ which should set the value of the characteristic length to $\xi \simeq 1/\sqrt{2\sigma}$. In this study we restrict our attention to the general model with crack nucleation. We compute the global width $W \equiv (\langle h^2 \rangle - \langle h \rangle^2)^{1/2}$, averaging over several realizations of the disorder (typically 10), as a function of time for

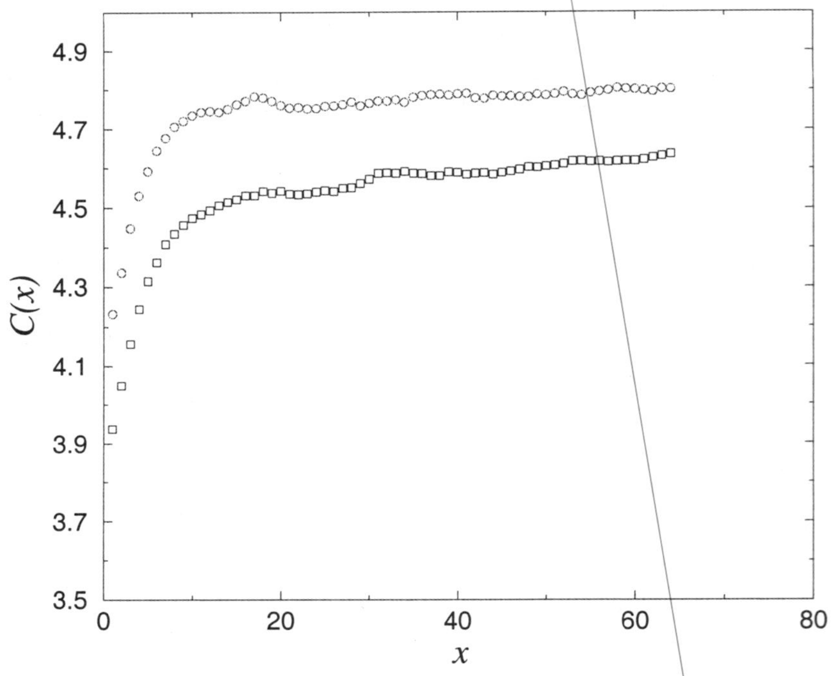


**Fig. 2** Comparison between the height correlation function of a connected crack (bottom curve) and that of a unrestricted crack (upper curve).

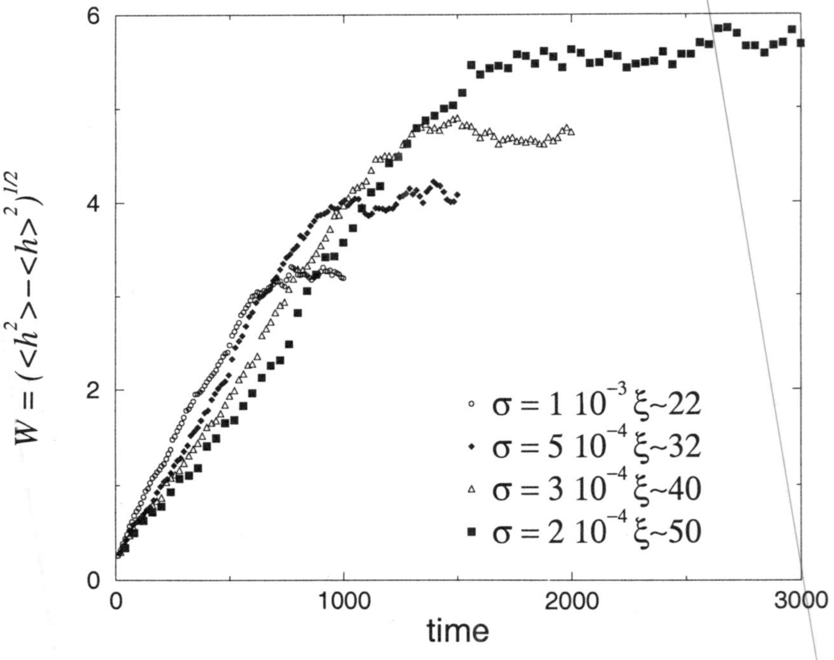


**Fig. 3** The global width as a function of time for different values of $\sigma$ and $\xi = 1/\sqrt{2\sigma}$.

different values of $\xi$. Figure 3 shows that $W$ increases linearly in time until saturation. However, we can not obtain a reliable estimate of the scaling with $\xi$.

Finally we study the damage $D(y)$ (i.e. the number of burnt fuses) as a function of the distance from the crack. The long-range nature of the Green function suggests that a mean-field approach could be suitable to derive the functional form of the damage profile. The

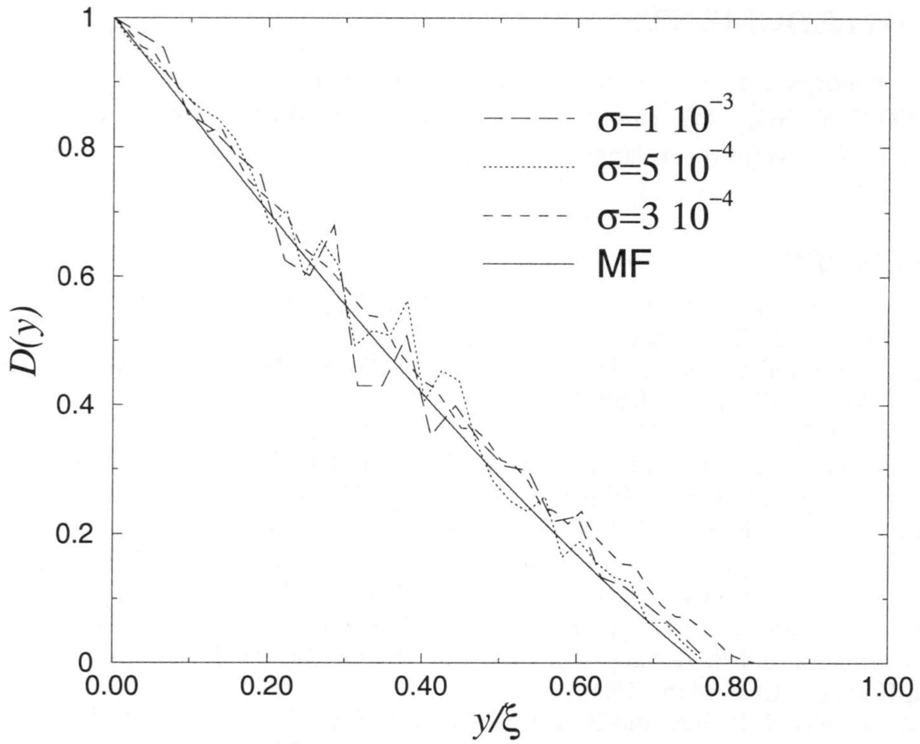


**Fig. 4** The average concentration $D(y)$ of burnt fuses as a function of the reduced distance $y/\xi$ from the crack for different values of $\sigma$. The numerical results are in excellent agreement with the mean-field solution.

result discussed in Ref. 20 is given by the solution of $d^2D/ds^2 = D(2-D)$, with appropriate boundary conditions, where $s \equiv y/\xi$. The mean-field result is compared with simulations in Fig. 4, which shows that the damage is indeed a function of $s = y/\xi$.

## 4. CONCLUSIONS

In this paper, we have studied the propagation of planar cracks in the random fuse model. This model allows to investigate the effect on the crack front roughness of the microcracks nucleating ahead of the main crack. The study was restricted to a quasi two-dimensional geometry and could apply to cases in which the material is very thin in the direction perpendicular to the crack plane.

In two dimensions, the geometry of the lattice induces a characteristic length $\xi$ limiting the roughness and microcrack nucleation does not appear to be relevant. In addition, for length scales smaller than $\xi$ the Green function decays very slowly, suggesting the validity of a mean-field approach. We study the problem numerically, computing the scaling of the crack width with time and $\xi$, and analyze the damage ahead of the crack. The results suggest an interpretation in terms of gradient percolation,[23] as it is also indicated by mean-field theory. The limited range of system sizes accessible to simulations does not allow for a definite confirmation of these results.

The present analysis does not resolve the issue of the origin of the value of the roughness exponent for planar cracks in heterogeneous media. While microcrack nucleation is irrelevant in the present context, three-dimensional simulations are needed to understand whether this is true in general. In principle, one could still expect that microcrack nucleation in three dimensions would change the exponent.

## ACKNOWLEDGMENTS

S. Z. acknowledges financial support from EC TMR Research Network under contract ERBFMRXCT960062. We thank A. Baldassarri, M. Barthelemy and J. R. Rice and A. Vespignani for useful discussions.

## REFERENCES

1. B. B. Mandelbrot, D. E. Passoja and A. J. Paullay, *Nature* (*London*) **308**, 721 (1984).
2. For a review, see E. Bouchaud, *J Phys.* **C9**, 4319 (1997).
3. J. Schmittbuhl and K. Måløy, *Phys. Rev. Lett.* **78**, 3888 (1997); A. Delaplace, J. Schmittbuhl and K. J. Måløy, *Phys. Rev.* **E60**, 1337 (1999).
4. H. Gao and J. R. Rice, *ASME J. Appl. Mech.* **56**, 828 (1989).
5. T. Nattermann, S. Stepanow, L. H. Tang and H. Leschhorn, *J. Phys. II* (*France*) **2**, 1483 (1992).
6. O. Narayan and D. S. Fisher, *Phys. Rev.* **B48**, 7030 (1993).
7. S. Ramanathan, D. Ertaş and D. S. Fisher, *Phys. Rev. Lett.* **79**, 873 (1997).
8. J. Schmittbuhl, S. Roux, J. P. Vilotte and K. J. Måløy, *Phys. Rev. Lett.* **74**, 1787 (1995); A. Tanguy, M. Gounelle and S. Roux, *Phys. Rev.* **E58**, 1577–1590, (1998).
9. S. Ramanathan and D. Fisher, *Phys. Rev. Lett.* **79**, 877 (1997); *Phys. Rev.* **B58**, 6026 (1998).
10. H. J. Herrmann and S. Roux (eds.), *Statistical Models for the Fracture of Disordered Media* (North-Holland, Amsterdam, 1990).
11. L. de Arcangelis, S. Redner and H. J. Herrmann, *J. Phys. Lett.* (*Paris*) **46**, L585 (1985).
12. P. Duxbury, P. D. Beale and P. L. Leath, *Phys. Rev. Lett.* **57**, 1052 (1986).
13. L. de Arcangelis and H. J. Herrmann, *Phys. Rev.* **B39**, 2678 (1989).
14. B. Kahng, G. G. Batrouni, S. Redner, L. de Arcangelis and H. J. Herrmann, *Phys. Rev.* **B37**, 7625 (1988).
15. A. Hansen and P. C. Hemmer, *Phys. Lett.* **A184**, 394 (1994).
16. F. Tzschichholz and H. J. Herrmann, *Phys. Rev.* **E51**, 1961 (1995).
17. S. Zapperi, P. Ray, H. E. Stanley and A. Vespignani, *Phys. Rev. Lett.* **78**, 1408 (1997).
18. G. Caldarelli, F. Di Tolla and A. Petri, *Phys. Rev. Lett.* **77**, 2503 (1996)
19. S. Zapperi, A. Vespignani and H. E. Stanley, *Nature* **388**, 658 (1997).
20. S. Zapperi, H. J. Herrmann and S. Roux, *E. Phys. J.* **B17**, 131 (2000).
21. T. Fukuhara and H. Nakanishi, *J. Phys. Soc. Japan* **67**, 4064 (1998).
22. J. Åström, M. J. Alava and J. Timonen, *Phys. Rev.* **E62**, 2878 (2000).
23. B. Sapoval, M. Rosso and J. F. Gouyet, *J. Phys. Lett.* (*Paris*) **46**, L149 (1985).

# Author Index